2023 8th South-East Europe Design Automation, Computer Engineering, Computer Networks and Social Media Conference (SEEDA-CECNSM 2023)

Piraeus, Greece
10 – 12 November 2023

IEEE Catalog Number: CFP23SHA-POD
ISBN: 979-8-3503-8368-3

**Copyright © 2023 by the Institute of Electrical and Electronics Engineers, Inc.
All Rights Reserved**

Copyright and Reprint Permissions: Abstracting is permitted with credit to the source. Libraries are permitted to photocopy beyond the limit of U.S. copyright law for private use of patrons those articles in this volume that carry a code at the bottom of the first page, provided the per-copy fee indicated in the code is paid through Copyright Clearance Center, 222 Rosewood Drive, Danvers, MA 01923.

For other copying, reprint or republication permission, write to IEEE Copyrights Manager, IEEE Service Center, 445 Hoes Lane, Piscataway, NJ 08854. All rights reserved.

*** This is a print representation of what appears in the IEEE Digital Library. Some format issues inherent in the e-media version may also appear in this print version.

IEEE Catalog Number: CFP23SHA-POD
ISBN (Print-On-Demand):   979-8-3503-8368-3
ISBN (Online):   979-8-3503-8367-6

Additional Copies of This Publication Are Available From:

Curran Associates, Inc
57 Morehouse Lane
Red Hook, NY 12571 USA
Phone: (845) 758-0400
Fax: (845) 758-2633
E-mail: curran@proceedings.com
Web: www.proceedings.com

Keynote 1: A multilayer approach in the AI cybersecurity

14:45 - 15:45, Day 1,
Chair: Sarandis Mitropoulos, Ionian University,
Conference Room: (Main Amphitheatre-MA)

Nineta Polemi, Professor, University of Piraeus, GR

Abstract: Various standards are being developed by numerous bodies – such as the European Telecommunications Standards Institute (ETSI), the European Committee for Standardization, the International Organization for Standardization (ISO), the National Institute for Standards and Technology (NIST), the Institute of Electrical and Electronics Engineers (IEEE) and the Open Web Application Security Project – along with recommendations and white papers by cybersecurity organisations – such as ENISA, the European Union Agency for Law Enforcement Cooperation, the European Defence Agency, the European Cyber Security Organisation and the Centre for European Policy Studies (CEPS) – and international bodies such as the OECD and the UN. These provide policy and technical measures (design principles, integration platforms, test cases), practices and further research needed to secure AI products (e.g., software, hardware, systems, services) and ensure their human-centric design.

Numerous existing traditional cybersecurity practices and solutions (methodologies, tools, recommendations) can be used to guide AI stakeholders in undertaking appropriate traditional controls. These good practices are presented in different documents which address various layers of the ICT environments (e.g., physical, network, informatics, data, services) that host AI products, making it difficult for the AI stakeholders to determine the ones appropriate for their environment. Furthermore, the dynamic nature of AI imposes some open issues and additional cybersecurity measures required in undertaking additional effective measures. Additional cybersecurity practices are also needed when AI products target a specific economic sector (e.g. health, energy, automotive) to meet sectoral security requirements. Finally, further research activities are needed to enforce the resilience and security of the AI-based products.

General Session: On-going projects and considerations I

16:15 - 17:55, Day 1,
Chair: Dimitrios Kallergis, University of West Attica,
Conference Room: (Main Amphitheatre-MA)

Enhancing Human Factors in Cybersecurity Training and Ensuring AI Trustworthiness
Dr Kitty Kioskli, CEO & Cofounder at trustilio BV & Research Fellow at the University of Essex...N/A

Par-ICT CENG: Enhancing ICT research infrastructure in Central Greece to enable processing of Big data from sensor stream, multimedia content, and complex mathematical modeling and simulations
Athanasios Kakarountas, Associate Professor, Dpt. of Computer Science and Biomedical Informatics, University of Thessaly...N/A

ERMIS – Smart digital applications and tools for the effective promotion and highlighting of the biodiversity of the Ionian Islands", results and challenges
George Voutos, MSC, PhD, Ionian University...N/A

MELTOPENLAB Project Panos Filippopoulos, University of Peloponnese, Greece...N/A

Session A1: Innovations in eHealth and Medical Devices

18:00 - 19:15, Day 1,
Chairs: Athanasios Kakarountas, University of Thessaly,
(Conference Room: MA)

Feasibility of an mHealth closed-loop system for the optimization of Parkinson's disease treatment
Dimitris Gatsios, Georgios Rigas, Georgios Bourazanis and Spyridon Konitsiotis...1

Real-Time Brain-Computer Interface Control of a Wheelchair Using Motor Imagery Commands
Kosmas Glavas, Georgios Prapas, Aimilia Ntetska, Pinelopi Adamakidou, Pantelis Angelidis and
Markos Tsipouras...5

A Comparative Testing of Yolov8-based algorithm for cancer biopsy images: Case study on Pannuke
pan-cancer dataset
Panagiotis N. Smyrlis, Odysseas Tsakai, Konstantinos Vogklis, Nikolaos Giannakeas, George Fragulis
and Markos Tsipouras...10

Brain-de-fer A brain-computer interface game for mind contest
Konstantinos Georgitsaros, Konstantinos Palegkas, Pantelis Angelidis and Markos Tsipouras...15

Session B1: Smart agricultural and environmental technologies

18:00 - 19:15, Day 1,
Chairs: Panagiotis Philippopoulos, University of Peloponnese,
(Conference Room: 001)

Supervisors for Gas Compressor Stations with Compression and Valve Faults
Dimitrios Fragkoulis, Fotis Koumboulis, Nikolaos Kouvakas and Antonios Menexis...22

TinyML-based Event Detection: An Edge-Cloud Approach for Smart Agriculture over LoRa WSNs
Aristeidis Karras, Christos Karras, Anastasios Giannaros, Konstantinos Giotopoulos, Dimitrios
Tsolis, Konstantinos Oikonomou and Spyros Sioutas...28

Ship Engine Data Analysis for the Application of Machine Learning Algorithms
Theodoros Dimitriou, Emmanouil Skondras, Christos Hitiris, Cleopatra Gkola, Ioannis S.
Papapanagiotou, Dimitrios J. Vergados, Constantinos Vergopoulos, Stratos Koumantakis,
Angelos Michalas and Dimitrios D. Vergados...38

Acceleration of GANs for Potato Crop Disease Identification via FPGA
Theodora Sanida, Argyrios Sideris, Maria Vasiliki Sanida, Michael Dossis and Minas Dasygenis
(remote presentation)...44

An Efficiency CNN Solution for Olive Disease Management Through FPGA
Theodora Sanida, Argyrios Sideris, Maria Vasiliki Sanida, Michael Dossis and Minas Dasygenis
(remote presentation)...48

Session A2: Artificial intelligent technologies

09:00 - 10:15, Day 2,
Chair: Georgios Drakopoulos, Ionian University,
(Conference Room: MA)

Q-Delegation: VNF Load Balancing through Reinforcement Learning-Based Packet Delegation
Alexandros Zervopoulos, Luís Miguel Campos and Konstantinos Oikonomou...52

Cognitive Email Analysis with Automated Decision Support for Business Email Compromise Prevention
Anastasios Papathanasiou, Georgios Germanos, Nicholas Kolokotronis and Euripidis Glavas...58

Higher Order Probabilistic Analysis Of Network Trajectories Of Intelligent Agents In Thespian
Georgios Drakopoulos and Phivos Mylonas...63

Session B2 (SS1): Technologies Essentials in Education

09:00 - 10:15, Day 2,
Chairs: Theodoros Karvounidis, University of Piraeus
Eleni Seralidou, University of Piraeus,
(Conference Room: 002)

Stimulating primary school students' interest in Data Structures through ESA's Astro Pi challenges
Athanasios Karakostas and Akrivi Vlachou...70

Teachers' Evaluation of 3D Design and 3D Printing Activities in Secondary Education
Eleni Seralidou, Theodoros Karvounidis and Christos Douligeris...76

Nanotechnology as a Tool for Computational Thinking Skills using Open Hardware, Embedded Systems and Repository Platform, in Industry 4.0 Era
Konstantinos Kalovrektis, Ioannis Dimos, Apostolos Xenakis and Athanasios Kakarountas...83

"Athens Museum Explorer": A pilot application of cultural proposals for a smart city
Anastasia Gasidou, Dimitrios Kotsifakos and Christos Douligeris...89

Precision education through the eyes of humanistic teaching in the L2 classroom
Ioanna Moustaka, Spyridon Doukakis and Marina Mattheoudakis...96

Keynote 2:
Networked Cyber-Physical-Social-Systems enabling Smart Internet of Things:
A Theoretical yet Pragmatic Approach

10:20 - 11:20, Day 2,
Chair: Christos Douligeris, University of Piraeus,
Conference Room: (Main Aphitheater-MA)

General Session: On-going projects and considerations II

11:50 - 13:30, Day 2,
Chair: Theodoros Karvounidis, University of Piraeus, GR,
Conference Room: (Main Amphitheatre MA)

VOXreality: Voice driven interaction in XR spaces, Olga Chatzifoti, Spyros Borotis (Maggioli)...N/A

AI4Gov: Trusted AI for Transparent Public Governance fostering Democratic Values , Sotiris Athanasopoulos and Spyros Borotis (Maggioli)...N/A

REWIRE – Cybersecurity Skills Alliance – a New Vision for Europe
Pinelopi Kyranoudi, PhDc, Technical University of Crete...N/A

CyberSecPro – a Collaborative, Multi-modal and Agile Professional Cybersecurity Training Program for A Skilled Workforce In the European Digital Single Market and Industries
Pinelopi Kyranoudi, PhDc, Technical University of Crete...N/A

CHAISE: Empowering   Europe   with   Blockchain   Skills
Dimitrios Kotsifakos, University of Piraeus and KEK Aegaleo, Greece...N/A

Industrial PhD in Computer and Telecommunications Sciences
Prof. Chrysostomos Stylios, Scientific Coordinator, Department of Informatics & Telecommunications, University of Ioannina...N/A

Session A3: Security and Reliability

14:30 - 15:45, Day 2,
Chairs: Daniele Rossi, University of Pisa, Italy,
Vasileios Tenentes: University of Ioannina,
(Conference Room: MA)

Cybersecurity and Democracy: A Review
Fotios Roumpies and Athanasios Kakarountas...102

Access to personal data is still tempting for mobile apps even after the GDPR implementation
Gerasimos Magoulas and Spyros Polykalas...110

Reliability Analysis of Fault Tolerant Memory Systems
Yagmur Yigit, Leandros Maglaras, Mohamed Amine Ferrag, Naghmeh Moradpoor and Giorgos Lambropoulos...116

Accelerate Processing of Image with the Keccak-512 Algorithm on Cryptoprocessor
Argyrios Sideris, Theodora Sanida, Maria Vasiliki Sanida, Michael Dossis and Minas Dasygenis (remote presentation)...122

Session B3 (SS2): Teaching and Learning in VET Laboratories

14:30 - 15:45, Day 2, Chairs: Kotsifakos Dimitrios, Department of Informatics, University of Piraeus Dimitrios Magetos , Department of Informatics, University of Piraeus, GR, (Conference Room: 002)

A Leveraging Matterport for Industry-Focused Mobile Applications: Augmented Reality Training for Vocational Education and Training
Dimitrios Kiriakos, Dimitrios Kotsifakos and Yannis Psaromiligkos...126

Optimizing Player Engagement in an Educational Virtual Game through Fuzzy Logic-based Challenge Adaptation
Akrivi Krouska, Christos Troussas, Yorghos Voutos, Phivos Mylonas and Cleo Sgouropoulou...133

"Technician of Refrigeration, Ventilation, and Air Conditioning Installations": A New Approach of the Modern Curricula in the Mechanical Sector of the 3rd Class of the Vocational School
Konstantinos Korakis and Michael Dossis...139

Educational Virtual Worlds for Vocational Education and Training Laboratories
Dimitrios Magetos, Dimitrios Kotsifakos and Christos Douligeris...145

Session A4: Networking and Communications

16:15 - 17:30, Day 2, Chairs: Dimitrios Kallergis, University of West Attica, (Conference Room: MA)

Area Allocation for Electric Vehicle Coverage Path Planning
Nikolaos Baras, Antonios Chatzisavvas, Dimitris Ziouzios, Ioannis Vanidis and Minas Dasygenis ...151

Modeling Network Traffic and Exploring Distribution Fitting: A Case Study on the Spotify
Odysseas Karadimas, Aikaterini Florou, Spiridoula V. Margariti, Eleftherios Stergiou and Chrysostomos Stylios...155

Vehicle Density in mmWave 5G V2X Highway Communication Systems: A Channel Coding Approach
Dimitrios Chatzoulis, Costas Chaikalis, Dimitrios Kosmanos, Apostolos Xenakis and Kostas Anagnostou...161

An Early Warning Opportunistic Interference Method in Tactical Voice and Data Communications
Nikiforos Kontopoulos, Dimitrios Kotsifakos and Christos Douligeris...166

Placing Multi-component Applications in the Multi-access Edge Computing
Asterios Papamichail, Athanasios Tsipis and Konstantinos Oikonomou...172

Session B4 (SS3): Securing Supply Chain Traceability Using Distributed and Embedded Security Mechanisms

16:15 - 17:30, Day 2, Chairs: Thomas Dasaklis, Hellenic Open UniversityVasileiosTenentes, University of Ioannina
Georgios Spathoulas, University of Thessaly, Greece - Norwegian University of Science & Technology, Norway,
(Conference Room: 002)

Design and Evaluation of a Peripheral for Integrity Checking to Improve RAS in RISC-V Architectures
Daniele Rossi, Nicasio Canino, Stefano Di Matteo, Sergio Saponara and Vasileios Tenentes...178

Compressing time series towards lightweight integrity commitments
Aggeliki Katsika, Konstantinos Papageorgiou, Alexandros Fakis, Athanasios Kakarountas,
Fotis Andritsopoulos, Vassilis Plagianakos and Georgios Spathoulas...184

Embedded Platforms for Trusted Edge Computing towards Quality Assurance along the Supply Chain
Vasileios Tenentes, Athanasios Xynos, Christos Zonios, Asimina Koutra, Christina Dilopoulou,
Yiorgos Tsiatouhas and Daniele Rossi...191

Uninterrupted Trust: Continuous Authentication in Blockchain-Enhanced Supply Chains
Vangelis Malamas, Dimitris Koutras and Panayiotis Kotzanikolaou...197

Session A5: Hardware and emerging technologies

17:35 - 18:50, Day 2,
Chair: Michalis Psarakis, University of Piraeus,
(Conference Room: MA)

Edge Artificial Intelligence in Large-Scale IoT Systems, Applications, and Big Data Infrastructures
Aristeidis Karras, Anastasios Giannaros, Christos Karras, Konstantinos C. Giotopoulos, Dimitrios
Tsolis and Spyros Sioutas...203

Synthesis-Embedded Verification
Michael Dossis...211

Study and Development of a High-Speed Fused Filament Fabrication 3D Printer
Ioannis Christodoulou, Vasiliki Alexopoulou and Angelos Markopoulos...218

Assessing Swapping Policies as a Detailed Placement Approach
Ioannis Arvanitakis, George Kranas, Michael Dossis, Antonios Dadaliaris and Athanasios
Kakarountas...224

ARM64 Architecture: A Review in Virtualization Technology and Cloud Computing Maturity, in the
context of Environmental Sustainability
Georgios Lambropoulos, Sarandis Mitropoulos and Christos Douligeris...230

Session B5: Social networks and Digital Health (remote presentations)

17:35 - 18:50, Day 2,
Chair: Roza Mavropodi, University of Piraeus,
(Conference Room: 002)

Efficient Categorization of Pneumonia Diagnosis Using Low-Power Embedded Devices
Theodora Sanida, Maria Vasiliki Sanida, Argyrios Sideris, Michael Dossis and Minas Dasygenis...236

The role of social media before and during a holiday travel: generational and gender differences
Ifigeneia Mylona, Dimitrios Amanatidis, Michael Dossis, Irene Kamenidou and Spyridon Mamalis...240

The role of hashtags in Social Networks: The case of social mobilization in Greece
Georgia Gioltzidou, Michael Dossis, Theodoros Chrysafis, Ifigeneia Mylona, Fotini Gioltzidou and Dimitrios Amanatidis...245

A Multi-class Classification Approach for Anemia Level Prediction with Machine Learning Models
Maria Trigka, Elias Dritsas and Phivos Mylonas...250

Recognition of Greek Orthodox Hymns Using Audio Fingerprint Techniques
Konstantinos Karasavvidis, Dimitris Kampelopoulos, Lazaros Moysis, Achilles Boursianis, Spyridon Nikolaidis, Panagiotis Sarigiannidis and Sotirios Goudos...256

The 8th South-East Europe Design Automation, Computer Engineering, Computer Networks and Social Media Conference (SEEDA-CECNSM 2023) took place in Piraeus, Greece, November 10th - 12th, 2023.

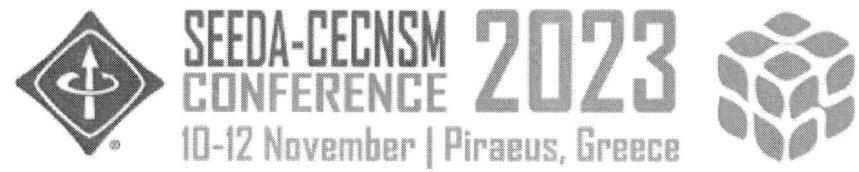

Preface

The 8th South-East Europe Design Automation, Computer Engineering, Computer Networks and Social Media Conference (SEEDA-CECNSM 2023), in its 8th year, provided an insight into the unique world stemming from the interaction between the fields of computer engineering, networks and design automation (DA). SEEDA-CECNSM 2023 provided a technical forum for experts from the engineering industry and academia to exchange ideas, innovations, and present results of on-going research in the most state-of-the-art areas.

This year's special focus was on the challenging issues related to all the opportunities of computer engineering, networks and design automation in the era of the integration of Internet of Things, Social Networks, Cloud Computing and Cyber-Physical Systems with an emphasis on security. The need for an efficient educational approach at all levels has also received special attention.

SEEDA-CECNSM was first held in Kastoria, Greece, at the premises of the Technological Educational Institute (TEI) of Western Macedonia (now University of Western Macedonia), under the leadership of Prof. Michael Dossis. Since then, it has been held, beyond Kastoria, in physical, online, or virtual formats in Piraeus, Preveza, and Ioannina.

This year, after a competitive selection process, 44 papers are included in the program divided into 10 sessions. Two of the sessions were special sessions – thanks to the special session chairs for an excellent job in attracting very high-quality papers. All the papers (regular and special session papers) were sent to at least three reviewers.

This year's program was complemented by two General Sessions, in which results of ongoing projects were presented. The projects covered the most modern problems in the conference topics, with a special focus on security issues in various environments, smart government, and digital culture.

Moreover, we were honored to have two prestigious scholars giving keynote speeches at the conference: Prof. Nineta Polemi (Department of Informatics, University of Piraeus, Greece) and Prof. Symeon Papassileiou (Department of Electrical and Computer Engineering, National Technical University of Athens, Greece)).

We would like to thank everyone who participated in the development of the SEEDA-CECSNM 2023 program. We want to give special thanks to the Technical Program Committee and the reviewers for their diligence and concern for the quality of the program, and also for their detailed feedback to the authors. I would personally like to thank all the volunteers of NetLab – University of Piraeus for their dedication and support of every single detail for the success of this conference.

The General co-chairs Conference Chairs Ioannis Voyiatzis, University of West Attica, Markos G. Tsipouras, University of Western Macedonia, Michael Dossis, University of Western

Macedonia, and Alexandros T. Tzallas, University of Ioannina, the TPC Co-Chairs Athanasios Kakarountas, University of Thessaly, Michalis Psarakis, University of Piraeus, Sarandis Mitropoulos, Ionian University and Theodoros Karvounidis University of Piraeus, the Publicity & Social Media Co-Chairs Leandros Maglaras, Edinburgh Napier University, Mohamed Amine Ferrag TII, UAE, Periklis Chatzimisios, International Hellenic University and Rosa Mavropodi, University of Piraeus, the Publication Chair Dimitrios Kallergis, University of West Attica, the Panels and Workshops Co-Chairs Cleo Sgouropoulou, University of West Attica, Helge Janicke, Australian Cyber Security Cooperative Research Centre, Dimitrios Kotsifakos, University of Piraeus and Dimitrios Kosmanos, University of Thessaly, the Special Sessions Co-Chairs Spyridon Doukakis, Ionian University and Konstantinos Kalovrektis, University of Thessaly, the Finance and Registration Co-Chair Theodoros Karvounidis, Univ. of Piraeus, and the Keynote Co-Chairs Alexandros T. Tzallas, University of Ioannina, Konstantinos Oikonomou, Ionian University and Costas Chaikalis, University of Thessaly as well as Eleftherios and Adriana Loukis, graphics and designs chairs, who have put many hours of work for the success of this conference.

Finally, and most importantly, we thank all the authors, who are the primary reason that this conference is so exciting. The interactions with them in the three days of the conference were thought provoking, forward looking and inspiring.

The Department of Informatics of the University of Piraeus, once more, has put this conference under its aegis and the University of Piraeus Research Center (UPRC) has provided financial support. The Greek Computer Society (ΕΠΥ) has helped in the financial organization of the conference and in the registration procedure. The University of Western Macedonia, the University of Thessaly and the University of West Attica have supported the conference in various ways. The support of the IEEE Greece Local Section needs to be acknowledged as well. Through their support we were able to publish the proceedings in IEEE Xplore.

We are confident that the friendly and warm environment of the University of Piraeus gave us the opportunity to exchange ideas, discuss in depth important scientific problems and create firm and long-lasting friendships. We are looking forward to the 2024 edition.

November 2023   **Christos Douligeris, General Co-Chair**

Conference Chairs

- Christos Douligeris, University of Piraeus
- Ioannis Voyiatzis, University of West Attica
- Markos G. Tsipouras, University of Western Macedonia
- Michael Dossis, University of Western Macedonia
- Alexandros T. Tzallas, University of Ioannina

TPC Co-Chairs

- Athanasios Kakarountas, University of Thessaly
- Michalis Psarakis, University of Piraeus
- Sarandis Mitropoulos, Ionian University
- Theodoros Karvounidis University of Piraeus

Panels and Workshops Co-Chairs

- Cleo Sgouropoulou, University of West Attica
- Helge Janicke – Research Director at the Australian Cyber Security Cooperative Research Centre
- Dimitrios Kotsifakos, University of Piraeus
- Dimitrios Kosmanos, University of Thessaly

Special Sessions Co-Chairs

- Spyridon Doukakis, Ionian University
- Konstantinos Kalovrektis, University of Thessaly

Finance and Registration Co-Chairs

- Theodoros Karvounidis, Univ. of Piraeus

Publication Chair

- Dimitrios Kallergis, University of West Attica

Keynote Co-Chairs

- Alexandros T. Tzallas, University of Ioannina
- Konstantinos Oikonomou, Ionian University
- Costas Chaikalis, University of Thessaly

Publicity & Social Media Co-Chairs

- Leandros Maglaras, Edinburgh Napier University
- Mohamed Amine Ferrag – Principal Researcher, TII, UAE
- Periklis Chatzimisios, International Hellenic University (GREECE) and Department of Electrical and Computer Engineering, University of New Mexico (USA)
- Rosa Mavropodi, University of Piraeus

Digital Co-Chairs

- Zacharenia Garofalaki, UNIWA, and UNIPI
- Eleni Seralidou, University of Piraeus

Awards Co-Chairs

- Jianmin Jiang – Distinguished Professor at Shenzhen University
- Christos Troussas, University of West Attica
- Apostolos Xenakis University of Thessaly

Local Organizing Committee

- Maria Eftychia Angelaki, University of Piraeus
- Panagiotis Gotsiopoulos, University of Piraeus
- Gerasimos Kaloumenos, University of Piraeus
- Dimitra Tzoumpa, University of Piraeus

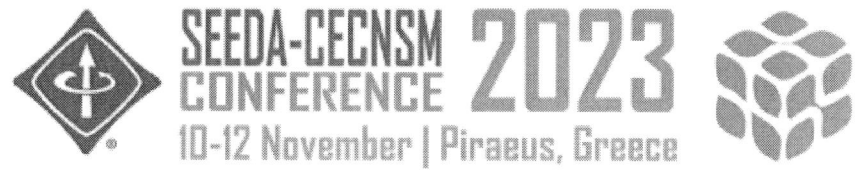

Technical Program Committee

- Al Husein Sami Abosaleh, Open Lab: Athens
- Dimitris Agiakatsikas, University of Piraeus Research Centre (UPRC)
- Panayiotis Alefragis, University of Peloponnese
- Dimitrios Amanatidis, Hellenic Open University
- Antonios Andreatos, Hellenic Air Force Academy
- Constantinos Angelis, University of Ioannina
- Vasileios Aspiotis, UNIVERSITY OF IOANNINA
- Edgar Batista, Universitat Rovira i Virgili
- Christos Bellos, LIME TECHNOLOGY IKE
- Fernando Bobillo, University of Zaragoza
- George Caridakis, Intelligent Interaction research group, ii.aegean.gr
- Thomas Dasaklis, Hellenic Open University
- Michael Dossis, University of Western Macedonia
- Spyridon Doukakis, Ionian University
- Christos Douligeris, University of Piraeus
- Gregory Doumenis, University of Ioannina
- Soultana Ellinidou, Université libre de Bruxelles
- Themis Exarchos, Ionian University
- Christos Georgiadis, University of Macedonia, Department of Applied Informatics
- Apostolos Gkamas, University of Ioannina
- Evripidis Glavas, University of Ioannina
- Christos Gogos, University of Ioannina
- Athanasios Kakarountas, University of Thessaly
- Dimitrios Kallergis, University of West Attica
- Petros Karvelis, Department of Informatics and Telecommunications, University of Ioannina
- Theodoros Karvounidis, University of Piraeus
- Katia Lida Kermanidis, Ionian University, Department of Informatics
- Vasileios Konstantakos, Aristotle University of Thessaloniki
- Charalampos Konstantopoulos, University of Piraeus
- Sotirios Kontogiannis, Dept. of Mathematics, University of Ioannina
- Dimitrios Kotsifakos, Uiversity of Piraeus
- Panayiotis Kotzanikolaou, University of Piraeus
- Vasiliki Liagkou, University of Ioannina
- Leandros Maglaras, Edinburgh Napier University
- Manolis Maragoudakis, University of the Aegean
- Angelos Markopoulos, National Technical University of Athens
- Antoni Martínez-Ballesté, Universitat Rovira i Virgili
- Angelos Michalas, University of Western Macedonia, Greece

- Sarandis Mitropoulos, Ionian University
- Phivos Mylonas, University of West Attica
- Petros Nicopolitidis, Aristotle University of Thessaloniki
- Spyridon Nikolaidis, Aristotle University of Thessaloniki
- Emmanouil Oikonomou, University of Ioannina
- Stergios Palamas, Ionian University
- Athanasios Papadimitriou, Dept. of Digital Systems, University of the Peloponnese, Greece
- Constantinos Patsakis, University of Piraeus
- Dimitrios Patsos, University of Piraeus
- Petros Potikas, National Technical University of Athens
- Michael Psarakis, University of Piraeus
- Athanasios Rizos, Airbus Defence and Space GmbH
- Konstantinos Sakkas, University of Ioannina
- Eleni Seralidou, University of Piraeus
- Panagiotis N. Smyrlis, University of Western Macedonia
- Anna Sotiropoulou, Dept. of Informatics, Ionian University
- Georgios Spathoulas, University of Thessaly
- Evaggelos Spyrou, University of Thessaly
- Yannis Stamatiou, University of Patras
- Chrysostomos Stylios, University of Ioannina
- Vasileios Tenentes, University of Ioannina
- Christos Troussas, University of Western Attica
- Chris Tselikis, HAI
- Markos Tsipouras, Department of Electrical and Computer Engineering, University of Western Macedonia, Kozani, Greece
- Fotios Vartziotis, Dept. of Informatics and Telecommunications, University of Ioannina
- Panagiotis Vlamos, Ionian University, Department of Informatics
- Gerasimos Vonitsanos, UNIVERSITY OF PATRAS
- Manolis Wallace, ΓΑΒ LAB – Knowledge and Uncertainty Research Laboratory

SEEDA-CECNSM CONFERENCE 2023
10-12 November | Piraeus, Greece

General Co-Chairs
- Christos Douligeris, University of Piraeus
- Ioannis Voyatzis, University of West Attica
- Markos G. Tsipouras, University of Western Macedonia
- Michael Dossis, University of Western Macedonia
- Alexandros T. Tzallas, University of Ioannina

Technical Program Committee Co-Chairs
- Athanasios Kakarountas, University of Thessaly
- Michalis Psarakis, University of Piraeus
- Sarandis Mitropoulos, Ionian University
- Theodoros Karvounidis University of Piraeus

Publicity & Social Media Co-Chairs
- Leandros Maglaras, Edinburgh Napier University
- Mohamed Amine Ferrag – Principal Researcher, TII, UAE
- Periklis Chatzimisios, International Hellenic University (GREECE) and Department of Electrical and Computer Engineering, University of New Mexico (USA)
- Rosa Mavropodi, University of Piraeus

Publication Chair
- Dimitrios Kallergis, University of West Attica

Panels and Workshops Co-Chairs
- Cleo Sgouropoulou, University of West Attica
- Helge Janicke – Research Director at the Australian Cyber Security Cooperative Research Centre
- Dimitrios Kotsifakos, University of Piraeus
- Dimitrios Kosmanos, University of Thessaly

Finance and Registration Committee Co-Chairs
- Theodoros Karvounidis, University of Piraeus

Digital Co-Chairs
- Zacharenia Garofalaki, University of West Attica and University of Piraeus
- Eleni Seralidou, University of Piraeus

Local Organization Committee
- Maria Eftychia Angelaki, University of Piraeus
- Panagiotis Gotsiopoulos, University of Piraeus
- Gerasimos Kaloumenos, University of Piraeus
- Dimitra Tzoumpa, University of Piraeus

SEEDA-CECNSM 2023
Programme

10-12 November, 2023

University of Piraeus (www.unipi.gr)

Address: 80, M. Karaoli & A. Dimitriou St., GR18534, Piraeus, Greece

Day 1	November 10, 2023	
14:00 - 18:00	**Registration** **(Lobby)**	
14:30 - 14:45	**Opening** **(Main Amphitheatre)**	
14:45 - 15:45	**Keynote 1** **A multilayer approach in the AI cybersecurity** **Nineta Polemi, Professor, University of Piraeus,GR** **Chair: Sarandis Mitropoulos, Ionian University** **(Main Amphitheatre-MA)**	
15:45 - 16:15	**Coffee Break I**	
16:15 - 17:55	**General Session** **On-going projects and considerations I** **Chair: Dimitrios Kallergis, University of West Attica** **(Main Amphitheatre-MA)**	
18:00 - 19:15	**Session A1:** **Innovations in eHealth and Medical Devices** **Chairs: Athanasios Kakarountas, University of Thessaly** **(Conference Room: MA)**	**Session B1:** **Smart agricultural and environmental technologies** **Chairs: Panagiotis Philippopoulos, University of Peloponnese** **(Conference Room: 001)**
20:00 - 23:00	**Gala dinner** **(Salty (N.A.S) Restaurant, Microlimano (https://www.salty.gr/))**	

Day 2	November 11, 2023
08:45 - 17:00	Registration (Lobby)

09:00 - 10:15	Session A2: Artificial intelligent technologies Chair: Georgios Drakopoulos, Ionian University (Conference Room: MA)	Session B2 (SS1): Technologies Essentials in Education Chairs: Theodoros Karvounidis, University of Piraeus Eleni Seralidou, University of Piraeus (Conference Room: 002)

10:20 - 11:20	Keynote 2 Networked Cyber-Physical-Social-Systems enabling Smart Internet of Things: A Theoretical yet Pragmatic Approach Symeon Papavassiliou, Professor, NTUA,GR Chair: Christos Douligeris, University of Piraeus (Main Aphitheater-MA)
11:20 - 11:50	Coffee Break II
11:50 - 13:30	General Session On-going projects and considerations II Chair: Theodoros Karvounidis, University of Piraeus, GR (Main Amphitheatre MA)
13:30 - 14:30	Lunch

14:30 - 15:45	Session A3: Security and Reliability Chairs: Daniele Rossi, University of Pisa, Italy, Vasileios Tenentes: University of Ioannina (Conference Room: MA)	Session B3 (SS2): Teaching and Learning in VET Laboratories Chairs: Kotsifakos Dimitrios, Department of Informatics, University of Piraeus Dimitrios Magetos Department of Informatics, University of Piraeus, GR (Conference Room: 002)

15:45 - 16:15	Coffee Break III

	Session A4: Networking and Communications Chairs: Dimitrios Kallergis, University of West Attica (Conference Room: MA)	Session B4 (SS3): Securing Supply Chain Traceability Using Distributed and Embedded Security Mechanisms Chairs: Thomas Dasaklis, Hellenic Open University Vasileios Tenentes, University of Ioannina Georgios Spathoulas, University of Thessaly, Greece - Norwegian University of Science & Technology, Norway (Conference Room: 002)
16:15 - 17:30		
17:35 - 18:50	Session A5: Hardware and emerging technologies Chair: Michalis Psarakis, University of Piraeus (Conference Room: MA)	Session B5: Social networks and Digital Health (remote presentations) Chair: Roza Mavropodi, University of Piraeus (Conference Room: 002)
18:50 - 19:00	closing session	

Day 3	November 12, 2023
09:15 - 12:00	Registration (Lobby)
09:30 - 10:45	Individual face-to-face communications
10:45 - 13:00	Individual face-to-face communications
13:00 - 13:30	Closing session

Feasibility of an mHealth closed-loop system for the optimization of Parkinson's disease treatment

Dimitrios Gatsios
Capemed
Ioannina, Greece
d.gatsios@capemed.gr

Georgios Rigas
Capemed
Ioannina, Greece
g.rigas@capemed.gr

Georgios Bourazanis
Capemed
Department of Informatics and Telecommunications
University of Ioannina
Ioannina, Greece
pint00084@uoi.gr

Spyridon Konitsiotis
Faculty of Medicine
University of Ioannina
Ioannina, Greece
skonitso@uoi.gr

Abstract—In this work we present the design of a treatment management system for Parkinson's disease (PD). The system is based on patient and device reported outcomes that combined with patient's health record data provide recommendations to treating physicians. The recommendations are based on a two-level system. In the first level patient data are incorporated in a clinical decision support system (CDSS) which provides a number of treatment alternatives based on expert rules and evidence-based guidelines. The second level is based on Pharmacokinetic and Pharmacodynamic (PK/PD) models enabling the evaluation of the expected benefit of each recommendation. The main components and requirements of the specific system are discussed and a real life example supporting its potential is presented.

Index Terms—Parkinson's disease, clinical decision support system, mHealth.

I. INTRODUCTION

Parkinson's disease (PD) is a progressive movement disorder. More than 6 million had PD in 2016 with the projections being that more than 12 million will suffer from it within the next 20 years [1]. Especially in low and middle income countries misdiagnosis is a very common issue. The management of Parkinson's is largely based on response to medication as reflected in improved control of symptoms. Patients' and their informal caregivers' feedback, along with the clinical examination during the visits are the main source of information for clinicians which rely on these patients reports to plan and adjust their personalised treatment.

Such personalised approaches are data driven and most effective when they take advantage of modern, widely used mhealth which includes wearable devices, mobile apps and patient portals directly and constantly updating the Electronic Medical Records (EMR) [2] with ecologically valid details about their symptoms. Health literacy and patient activation

are promoted by mhealth and are important tools for the self-management of the disease.

Clinical Decision Support Systems (CDSS) are on the other hand used to augment and support physicians in their complex and difficult decision-making processes [3]. Recommendations and suggestions in CDSS are based on clinical knowledge extracted from clinical practice Guidelines published by Movement Disorders Societies, patient-specific data from EMRs (including medication and comorbidity), patient symptoms (motor also with medical devices, non-motor as reported by patients and relatives), and any other relevant information, e.g regarding adherence to treatment, disease-relates beliefs etc.

The concept of incorporating automated symptoms monitoring and CDSS in PD management was promoted in the PD_Manager EU project [4]. PD_Manager proposed a system architecture for tailoring treatment suggestions to physicians. The provided suggestions were based on an expert-based decision support system that could provide either treatment modifications or dose/timing changes. However, the expected improvement of each intervention using a typical rule-based system cannot be evaluated in advance. To evaluate the expected response to potential treatment changes, information from the patient health record (i.e. medication prescription) objective symptom evaluation and Pharmacokinetic/Pharmacodynamic (PK/PD) models must be combined [5] to produce a model predicting the response in any change. Ursino et al. [6] have presented methods for the estimation of PK/PD models combining PK, Basial Gaglia models and measurements of finger taps as a measure of levodopa effect. Thomas et al. [7] have employed a similar approach for levodopa infusion optimization utilizing Treatment Response Scale [8] instead of finger taps. In this work the PK/PD models presented in [7] are utilized in order to evaluate the feasibility of evaluating oral medication treatment suggestions tailored by the proposed CDSS system. The goal of this manuscript is to: i) present the concept of a closed-loop treatment optimization solution,

---

PRIME is co-financed by the European Regional Development Fund of the European Union and Greek national funds through the Operational Program Competitiveness, Entrepreneurship and Innovation, under the call RESEARCH – CREATE – INNOVATE (project code: T2EΔK- 05199).

ii) present the main components and the main requirements and ii) evaluate the feasibility of the solution and its potential for providing reliable estimations of expected improvement in motor symptoms for each treatment suggestion.

II. MATERIALS AND METHODS

The proposed concept is presented in Figure 1. The requirements of such a system shall include:

- **Demographics and Medical History.** A FHIR (Fast Health Interoperability Resources) compatible EMR is required for the CDSS to evaluate current patient status and provide recommendations. The history should include also structured and standardized medication schedules.
- **Long term Continuous objective home monitoring.** The system shall receive longitudinal home based patient objective measurements of motor symptoms. The objective measurements should provide an estimation of symptom time profile in established scales such as Unified Parkinson's Disease Rating Scale (UPDRS) or Unified Dyskinesia Rating Scale.
- **Patient Reported Outcome (PROMS).** Any automated medication change system should consider medication adherence as well as patient's perspective on symptom tolerance. In order to evaluate the efficacy of current prescription data related to adherence are required. Medication adherence could be considered for providing confidence intervals in all estimations. Moreover, higher priority should be considered for management of symptoms the patient has reported as most problematic.
- **Integrated in clinical workflows.** In order systems to have success and acceptance by physicians and nurses they should be seamlessly integrated in their daily practice.

The proposed solution that meets the main requirements consists of the following main components:

Home Monitoring Device. Patient telemonitoring has emerged as a necessary and extremely useful tool especially in the era of COVID-19. A number of telemonitoring solutions have demonstrated their efficacy including mobile apps and symptom monitoring devices [9]. The proposed solution requires both objective (device reported) as well as subjective (patient reported) outcomes for the evaluation of the efficacy of the current treatment and the patient satisfaction with that. For the remote objective evaluation of PD motor symptoms a number of medical devices with the specific intended use are available nowadays. Such devices include the PKG [1], Sense4Care [2] and PDMonitor® [3]. In this paper we have adopted the PDMonitor® device since: i) it provides an estimation of UPDRS Part III score which is evaluated as a pharmacodynamic parameter and there is evidence regarding the minimum clinically important changes [UPDRS], ii) it is well suited to the proposed closed loop solution since

[1]www.pkgcare.com/
[2]www.sense4care.com
[3]www.pdneurotechnology.com

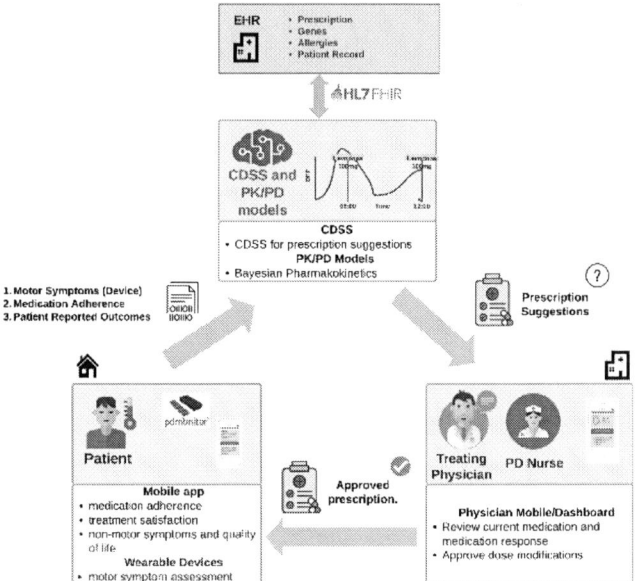

Fig. 1: Flowchart of PD Management suggestions generation.

the specific device is intended to be used for continuous monitoring as a patient-owned device compared to the rest of the devices which are used as Holter-like devices. The PDMonitor® is used once per month (or more frequently depending on the severity of the disease and the efficacy of the current treatment) and the data are automatically uploaded to the cloud. The feasibility of the use of PDMonitor® device has been evaluated by 133 patients (private sales) and 52 private physicians in Greece. Physicians instructed patients to wear a device for PD motor symptom telemonitoring for one week per month, during their awake hours. For the patients using the device for 12 months, the adherence was above 70% [10]. Therefore, considering also the current population the long-term use of similar devices is feasible.

Patient Reported Outcomes (PROMs). As discussed above, PROMs are required in order to evaluate non-motor symptoms [11] and also the satisfaction with specific medication doses. PDMonitor® provides an integrated way of collecting medication adherence and patient reported symptoms (OFF, Dyskinesia and Tremor). Therefore, the PROMs are also provided by the PDMonitor® Cloud platform. Interestingly, the PDMonitor® medication adherence module is used by more than 70% of the patients when the symptom status in more than 29% [10].

PD Management Platform. It is a SmartOnFhir [12] application which coordinates the data exchange between the different system components. The Management platform may also include a table of patient identifiers on the different subsystems in order to allow the correct data exchange.

CDSS Engine. The CDSS system is based on expert rules like the ones proposed in PD_manager [4]. This was, to the best of our knowledge, the only holistic mHealth CDSS for PD as the other systems target specific symptoms. The validity

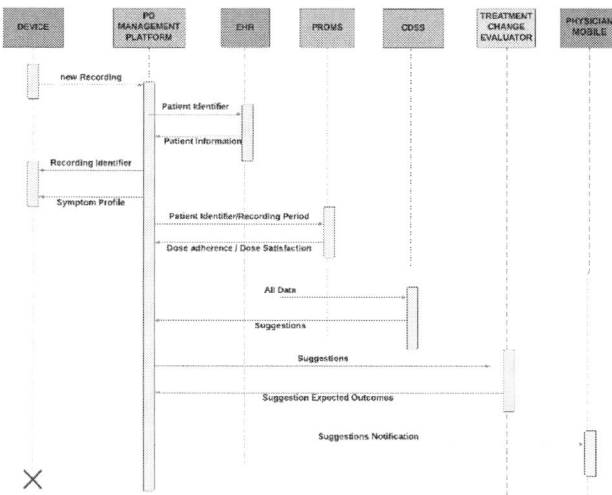

Fig. 2: Proposed closed-loop solution.

and feasibility of a manager CDSS system has been evaluated in previous works by Bohanec et al. [13] and Boshkoska et al. [14]. The suggested CDSS takes as input the current medication treatment of the patient (I), present motor (PDMonitor) and non-motor symptoms and epidemiologic factors (I). Then the models determine whether the current medical treatment is effective or not and suggest one or more alternative changes of medications.

Treatment Change Evaluator. This is the module where PK/PD models combined with the patient's treatment outcomes can provide an estimation of the expected response of the patient PK/PD on any new change. Changes that have clinically meaningful changes (OFF time, UPDRS Part III or IV) could be considered as candidate changes. The candidate changes are those recommended to treating physicians for consideration.

Physician App. The specific application can provide the required medical record history, a report of the current patient status and also notifications with treatment recommendations. Recommendations are prioritized based on their estimated benefit and patient reported dissatisfaction.

FHIR compatible EMR. An essential component of the specific system is an interoperable EMR that is able to provide patient record information including current prescription (FHIR MedicationRequest[4]).

In the proposed solution when a recording is performed a new patient treatment evaluation is triggered in the PDManagement Platform as depicted in Figure 2 which describes the data flow between the main system components. The PDManagement platform (SmartOnFhir) collects the required data from all connected systems in order to feed the CDSS. The CDSS system based on the expert rules and patient data provides a number of treatment suggestions. The Treatment Change Evaluator is used to evaluate the expected benefit of

[4]https://www.hl7.org/fhir/medicationrequest.html

TABLE I: PK/PD Parameters

V_1	27
V_2	27
Q	0.58
CL	0.9
γ	2.5
$E_{\max}$	$-\dfrac{(\max(\text{UPDRS})-\min(\text{UPDRS}))}{\max(\text{UPDRS})}$

those suggestions including a medication change with dose add/remove/increase/decrease. PDManagement is also used as the back-end for the Physician mobile app (or dashboard).

III. RESULTS

A. Evaluation of the treatment change evaluator

In this section the feasibility of using PK/PD models and sensor-based outputs for estimating the expected effect of a medication change was evaluated. R and RxODE were used for the estimation of PK/PD parameters. Data from a real-world case were used. In the specific case the physician performed a medication change and the motor symptoms response was measured with the PDMonitor before and after the specific change. Similarly to [7] a two-compartment pharmacokinetic model is considered. The UPDRS Part III outcome was also considered for the pharmacodynamics effect [15]. The differential equations for the PK/PD are the following:

$$\frac{dA_1}{dt} = -A_1 \cdot \text{KA} \tag{1}$$

$$\frac{dA_2}{dt} = \text{KA} \cdot A_1 - K_{11} \cdot A_2 + K_{12} \cdot A_3 \tag{2}$$

$$\frac{dA_3}{dt} = A_2 K_{21} - A_3 \cdot K_{22} \tag{3}$$

$$\frac{dA_4}{dt} = \text{KEO} \cdot \frac{A_2}{V_1} - \text{KEO} \cdot A_4 \tag{4}$$

$$EU = E_0 \cdot \left(1 + \frac{E_{max} \cdot A_4^\gamma}{A_4^\gamma + EC_{50}^\gamma} \right) \tag{5}$$

where $K_{11} = Q/V_1 + \text{CL}/V_1$, $K_{12} = K_{22} = Q/V_2$ and $K_{21} = Q/V_1$. The EU is the estimated UPDRS score. The parameters of the model used are presented in Table I. The medication prescription was also used for the estimation of the parameters. Similarly to [7] the KEO and EC50 parameters were optimized sequentially (Nelder-Mead method). Using the estimated parameters, the expected UPDRS score of the new prescription (a new dose added) was estimated and compared with the actual one measured by the device. The comparison is presented in Figure 2. The model predicted a change of **-5.9** in UPDRS score whereas the actual one was **-6.8**.

IV. DISCUSSION

A solution for a closed loop treatment optimization of PD patients is proposed. The system is expected to improve patients' Quality of Life and reduce the physician/PD-nurse burden. The main system components and requirements are

979-8-3503-8368-3/23 $31.00 © 2023 IEEE

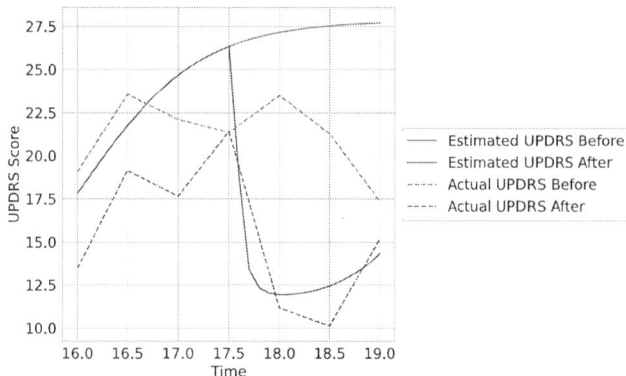

Fig. 3: Estimated and expected UPDRS Part III for the changed dose.

presented. Moreover, the feasibility of each component is discussed and the novel Treatment Change Evaluator is evaluated with a specific real-world case. The PK/PD model used in this work can be further improved incorporating information about dyskinesias as additional effect and also by using a Bayesian approach for the estimation of the parameters. The adoption of PDMonitor in real world settings can lead to the streamlined collection of a large dataset with hundreds of cases with medication changes and their actual effect. Future work will include the evaluation of the accuracy of this model, as well as whether this approach can be applied in complex medication prescriptions. Moreover, this model can be used to evaluate the percentage of patients that under current practice could have treatment changes with potential significant improvements. This will have a tremendous impact on the utility of tele-monitoring devices for close loop treatment solutions which are currently only encountered in limited clinical scenarios such as glucose meters. However, the success of the specific solution requires the efficient collection of patient medication adherence and satisfaction with current prescription (per dose). Proper strategies for patient empowerment and patient activation are therefore required and mhealth aims also at that. In the same context adoption from physicians and PD-nurses (depending on the health care system) is crucial. Probably the proposed mobile app and/or web-based application is the most straightforward method to integrate such a system in current clinical workflow. Feasibility, usability and user acceptance however needs to be further evaluated to ensure the success of the proposed solution. A longitudinal randomized controlled trial study is required in order to evaluate the whole solution compared to the standard of care and provide sound evidence.

V. CONCLUSIONS

A closed loop treatment optimization system relying on body worn sensors is proposed and initially seems feasible. The proposed solution is expected to improve patient's quality of life and also reduce physician and healthcare system burden. The essential components and requirements of the system are described and their respective feasibility is evaluated. Finally, the next steps towards the implementation, validation and adoption of the specific system have been discussed.

REFERENCES

[1] E. R. Dorsey *et al.*, "Global, regional, and national burden of Parkinson's disease, 1990–2016: a systematic analysis for the Global Burden of Disease Study 2016," *The Lancet Neurology*, vol. 17, no. 11, pp. 939–953, nov 2018. [Online]. Available: http://www.thelancet.com/article/S1474442218302953/fulltext

[2] C. Hansen, A. Sanchez-Ferro, and W. Maetzler, "How Mobile Health Technology and Electronic Health Records Will Change Care of Patients with Parkinson's Disease," *Journal of Parkinson's disease*, vol. 8, no. s1, pp. S41–S45, 2018. [Online]. Available: https://pubmed.ncbi.nlm.nih.gov/30584169/

[3] R. T. Sutton, D. Pincock, D. C. Baumgart, D. C. Sadowski, R. N. Fedorak, and K. I. Kroeker, "An overview of clinical decision support systems: benefits, risks, and strategies for success," *npj Digital Medicine 2020 3:1*, vol. 3, no. 1, pp. 1–10, feb 2020. [Online]. Available: https://www.nature.com/articles/s41746-020-0221-y

[4] K. M. Tsiouris, D. Gatsios, G. Rigas, D. Miljkovic, B. Koroušić Seljak, M. Bohanec, M. T. Arredondo, A. Antonini, S. Konitsiotis, D. D. Koutsouris *et al.*, "Pd_manager: an mhealth platform for parkinson's disease patient management," *Healthcare technology letters*, vol. 4, no. 3, pp. 102–108, 2017.

[5] G. Rigas, P. Bougia, D. Baga, M. Tsipouras, A. Tzallas, E. Tripoliti, S. Tsouli, M. Chondrogiorgi, S. Konitsiotis, and D. I. Fotiadis, "A decision support tool for optimal Levodopa administration in Parkinson's disease," in *Information Technology and Applications in Biomedicine (ITAB), 2010 10th IEEE International Conference on*. IEEE, 2010, pp. 1–6.

[6] M. Ursino, E. Magosso, G. Lopane, G. Calandra-Buonaura, P. Cortelli, and M. Contin, "Mathematical modeling and parameter estimation of levodopa motor response in patients with parkinson disease," *PLOS ONE*, vol. 15, no. 3, p. e0229729, 2020. [Online]. Available: https://journals.plos.org/plosone/article?id=10.1371/journal.pone.0229729

[7] I. Thomas, M. Alam, D. Nyholm, M. Senek, and J. Westin, "Individual dose-response models for levodopa infusion dose optimization," *International journal of medical informatics*, vol. 112, pp. 137–142, apr 2018. [Online]. Available: https://pubmed.ncbi.nlm.nih.gov/29500011/

[8] D. Nyholm *et al.*, "Duodenal levodopa infusion monotherapy vs oral polypharmacy in advanced Parkinson disease," *Neurology*, vol. 64, no. 2, pp. 216–223, jan 2005. [Online]. Available: https://pubmed.ncbi.nlm.nih.gov/15668416/

[9] Sánchez-Ferro *et al.*, "Minimal Clinically Important Difference for UPDRS-III in Daily Practice," pp. 448–450, jul 2018. [Online]. Available: https://www.ncbi.nlm.nih.gov/pmc/articles/PMC6336372/

[10] G. Rigas, N. Tachos, N. Kostikis, S. Kontogiannis, E. Kostoulas, A. Bousis, S. Konitsiotis, and D. I. Fotiadis, "Real-world evidence and feasibility of telemedicine in Parkinson's disease," in *Movement Disorders.*, 2022.

[11] D. J. van Wamelen, P. Martinez-Martin, D. Weintraub, A. Schrag, A. Antonini, C. Falup-Pecurariu, P. Odin, and K. Ray Chaudhuri, "The Non-Motor Symptoms Scale in Parkinson's disease: Validation and use," *Acta neurologica Scandinavica*, vol. 143, no. 1, pp. 3–12, jan 2021. [Online]. Available: https://pubmed.ncbi.nlm.nih.gov/32813911/

[12] M. L. Braunstein, "Health Informatics on FHIR: How HL7's API is Transforming Healthcare," 2022. [Online]. Available: https://link.springer.com/10.1007/978-3-030-91563-6

[13] M. Bohanec *et al.*, "A decision support system for Parkinson disease management: expert models for suggesting medication change," *Journal of Decision Systems*, vol. 27, no. sup1, pp. 164–172, may 2018. [Online]. Available: https://www.tandfonline.com/doi/full/10.1080/12460125.2018.1469320

[14] B. M. Boshkoska *et al.*, "Decision Support for Medication Change of Parkinson's Disease Patients," *Computer Methods and Programs in Biomedicine*, vol. 196, p. 105552, nov 2020.

[15] A. Marsot, R. Guilhaumou, J. P. Azulay, and O. Blin, "Levodopa in Parkinson's Disease: A Review of Population Pharmacokinetics/Pharmacodynamics Analysis," *Journal of pharmacy pharmaceutical sciences : a publication of the Canadian Society for Pharmaceutical Sciences, Societe canadienne des sciences pharmaceutiques*, vol. 20, no. 0, pp. 226–238, 2017. [Online]. Available: https://pubmed.ncbi.nlm.nih.gov/28719359/

979-8-3503-8368-3/23 $31.00 © 2023 IEEE

Real-Time Brain-Computer Interface Control of a Wheelchair Using Motor Imagery Commands

Kosmas Glavas
Department of Electrical and Computer Engineering
University of Western Macedonia
Kozani, Greece
k.glavas@uowm.gr

Georgios Prapas
Department of Electrical and Computer Engineering
University of Western Macedonia
Kozani, Greece
g.prapas@uowm.gr

Aimilia Ntetska
Department of Electrical and Computer Engineering
University of Western Macedonia
Kozani, Greece
antetska@uowm.gr

Pinelopi Adamakidou
Department of Electrical and Computer Engineering
University of Western Macedonia
Kozani, Greece
p.adamakidou@uowm.gr

Pantelis Angelidis
Department of Electrical and Computer Engineering
University of Western Macedonia
Kozani, Greece
paggelidis@uowm.gr

Markos G. Tsipouras
Department of Electrical and Computer Engineering
University of Western Macedonia
Kozani, Greece
mtsipouras@uowm.gr

Abstract—**The Brain-Computer Interface (BCI) field is rapidly growing and offers a wide range of uses in different domains, such as healthcare and daily life. This paper discusses the development of a wheelchair operated by BCI using Motor Imagery (MI) mental commands for turning left and right and blinking for stopping and starting. The Emotiv Insight is used to collect raw electroencephalography (EEG) data, while a Linear Discriminant Analysis (LDA) algorithm is used to classify the mental commands. OpenViBE software is employed to process and classify brain signals. The system is evaluated on 5 subjects in real-time, and 3 experiments are conducted to test starting, stopping, and turning. The results showed that the subjects can command the wheelchair operated by BCI with a high degree of precision.**

Index Terms—**Brain-Computer Interface, BCI, Motor Imagery (MI), Wheelchair commanded by BCI, Emotiv Insight, EEG.**

I. INTRODUCTION

Brain-Computer Interface (BCI) [1] [2] is a rapidly developing field that aims to translate human brain activity into computer commands. The basic idea behind BCI is to measure electrical signals generated by the brain, typically through non-invasive methods like electroencephalography (EEG), and convert them into specific actions or commands.

BCI systems can be categorized as invasive or non-invasive depending on the level of invasiveness of the measurement techniques used. Invasive BCIs [3] require implanting electrodes directly into the brain, while non-invasive BCIs [4]

[5] use sensors placed on the scalp or skin to detect brain activity. Invasive BCI systems offer higher spatial and temporal resolution and greater signal specificity but require invasive surgery and may raise ethical concerns. Non-invasive BCI systems are safer, easier to use, and more versatile, but have lower resolution and greater susceptibility to noise and interference.

Wheelchairs commanded by BCI are an emerging application of BCI technology that has the potential to greatly improve the quality of life of individuals with severe motor disabilities [6] [7]. These systems, offer increased mobility and independence, by using brain signals to control the movement of a wheelchair to navigate without relying on traditional input devices like joysticks. However, there are several challenges in developing these BCI systems such as high signal accuracy, robustness to environmental noise, and user training.

Motor imagery (MI) [8] [9] is a cognitive process where an individual mentally simulates a movement or action without physically executing it. This process can produce detectable changes in brain activity that can be recorded using EEG. Motor imagery-based BCIs have the potential to provide non-invasive control signals for external devices, but achieving high decoding accuracy remains a challenge due to individual differences in brain activity and susceptibility to noise.

In this study, a wheelchair operated by BCI that can move forward, turn left and right, and stop is implemented. MI is used for turning left and right and Electrooculogram (EOG) signals, blinking, are used for stopping and starting the wheelchair. Emotiv Insight headset is employed to acquire

---

This work has been financed by the European Union and Greek national funds through Operational Program Competitiveness, Entrepreneurship and Innovation under the call RESEARCH—CREATE—INNOVATE: "Intelli–WheelChair" (Project Code: T2EΔ K-02438).

979-8-3503-8368-3/23 $31.00 © 2023 IEEE

the EEG data recorded from 5 subjects who took part in 3 experiments that tested the ability of the proposed work to 1) start and stop, 2) start, stop, and turn right, and 3) start, stop, and turn left.

II. RELATED WORK

Jiang *et al.* [10] designed a hybrid wheelchair commanded by BCI with EEG and Electromyography (EMG) signals. An Emotiv Epoc headset was used, to record the raw EEG and EMG signals with 14 electrodes. A GUI was developed for the movement of the wheelchair and 3 different modes were employed; manual, brain, and multiple control. The degree of freedom of this system is 4; moving forward, backward, and turning left and right. The data are processed and epoched in time windows of 1.5 seconds. The system includes an EEG headset, a computer, a microcontroller, and a wheelchair. 5 subjects participated in the study testing the wheelchair in real time.

Xie and Li [11] developed a wheelchair operated by BCI with MI commands. A 24-electrode EEG cap was employed to acquire the raw EEG data for 2 mental states, left and right MI. The brain signals were processed with the Common Spatial Patterns (CSP) technique and classified with Support Vector Machines (SVM). To validate the system's offline accuracy dataset I from the 2004 BCI competition III was used. To further evaluate the BCI system 6 subjects were recorded. The average classification accuracy of dataset I is 87.7698% and for the subjects is 57.5% ranging from 81.6% to 30.0%. This work didn't evaluate the system in real-time.

Espiritu *et al.* [12] conducted an experiment in which mental commands were used to command a wheelchair instead of the standard joystick control. Emotiv Insight, a 5-channel EEG device was utilized to record the raw EEG data. Emotiv software was employed to process and classify the brain signals and to train the subject for mental commands. 4 MI mental commands were used for the movement of the wheelchair commanded by BCI; push, pull, left, and right corresponding to moving forward, backward, and turning left and right. To stop the wheelchair neutral state was used. To evaluate the experiment in real-time 1 subject trained and tested the system. To assess the success of the experiment, a command response delay for action metric was used for every movement.

III. MATERIALS AND METHODS

The goal of the proposed study is to develop a wheelchair operated by BCI to assist individuals with disabilities. The challenges for such a system are many; safety, connectivity, and high-performance. In this section, the design of the system, the hardware, and the software are presented.

1) System Design: The BCI system comprises Emotiv Insight, a 5-channel EEG device to record the raw EEG data, a laptop, and a wheelchair. OpenViBE is employed to process and classify the signals and a Python script is used to send the appropriate commands to the wheelchair depending on the outcome of the classifier. Figure 1 shows the electric wheelchair of the system.

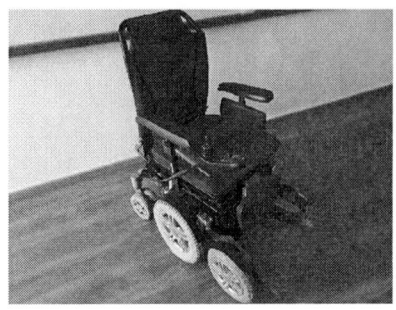

Fig. 1. The electric wheelchair used in this study. The traditional input device (joystick) is modified and it can receive commands from a laptop via USB and serial port communication.

2) Dataset: To assess the BCI system in real-time 5 subjects, 2 males and 3 females, participated in the experiment. The average age of the participants is 24.8 ranging from 23 to 30 years old. The volunteers signed a consent document in order to participate in the study. Every participant included in the study has both good mental and physical health, as well as good eyesight.

3) EEG Device: The EEG device utilized in this work is Emotiv Insight (Figure 2) which has 5 EEG semi-dry sensors (AF3, AF4, T7, T8, and Pz) and connects to the computer via Bluetooth. It is a wireless device with high accuracy and compact size. It is easy to use and can be placed comfortably on the user's head. Furthermore, it is an affordable and accessible solution for measuring brain activity non-invasively. Also, the device has motion sensors such as Accelerometer, Gyroscope, and Magnetometer. The sampling frequency of Emotiv Insight is 128Hz.

Fig. 2. The commercial 5-channel EEG device used in this work; Emotiv Insight. The channels of this headset are AF3, AF4, T7, T8, and Pz.

4) Signal Processing: The EEG device connects with the laptop via Bluetooth and then it's paired with EMOTIVPRO software. Through EMOTIVPRO, a Lab Streaming Layer (LSL) stream is initialized to transmit the raw EEG data to OpenViBE. LSL is an open-source tool that enables the streaming, receiving, synchronizing, and recording of time series data from many network acquisition devices. Then the raw EEG data are recorded and pre-processed by applying a 4th-order bandpass filter between 9Hz to 40Hz to remove noise. The recordings are 6 minutes for each mental command;

MI for left and right hand movement and blinking. After the recording phase has ended, spectral features are extracted using Fast Fourier Transform (FFT). Then, the EEG spectrum is divided into four frequency bands; Alpha between 9Hz to 13Hz, Low Beta between 13Hz to 20Hz, High Beta between 20Hz to 30Hz, and Gamma between 30Hz to 40Hz. These bands are selected since they are the main frequency bands of the brain and they are well established in the EEG analysis literature. Then the signals are segmented into 4-second epochs with a 1-second overlap, balancing the need for a fast system response to the user's mental commands and the need for sufficient information for accurate classification. Finally, the average spectral amplitude per band is computed and used as input to the classifier.

5) Classification: Linear Discriminant Analysis (LDA) is the classification algorithm utilized in this study. LDA [13] [14] is a widely used statistical technique for classification in machine learning. Its objective is to identify a linear combination of features that effectively separates two or more classes while minimizing the overlap between them. This makes LDA a highly effective classification method for BCI applications. In a three-class problem, it produces three linear discriminant functions, and the highest score determines the predicted class label.

The classification accuracy of the 3 classes is presented in Table I.

TABLE I
CLASSIFICATION ACCURACY FOR THE 3 CLASSES

Sub	Right MI	Left MI	Blinking
1	94.8%	84.5%	93.5%
2	92.4%	79.0%	95.0%
3	74.2%	94.8%	100%
4	89.7%	93.8%	95.0%
5	90.7%	82.5%	100%
Average	88.3%	86.9%	96.7%

Table I presents the classification accuracy for the 3 mental commands; right and left MI and blinking. The average accuracy for right MI is 88.3% ranging from 74.2% to 94.8%, for left MI is 86.9% varying from 82.5% to 94.8% and for blinking is 96.7% with accuracy from 93.5% to 100%.

6) Commanding the Wheelchair: The Degree of Freedom (DoF) of the system is 4; moving forward, stopping, turning left, and right. Blinking corresponds to the first 2 movements and MI mental commands correspond to the other 2 movements. To separate blinking from stopping and moving forward a counter is employed that rises every time a blink is detected. If the counter is an even number (exactly divisible by 2) then the Python script sends a forward command to the wheelchair. Otherwise, if the counter is an odd number (not completely divisible by 2) then a stop command is transmitted to the wheelchair. Turning is only available when the system is stopped for safety reasons. The speed of the system is steady and set at a quarter of the maximum speed of the wheelchair when the joystick is used.

The Python script acts as a mediator between the LSL stream and the wheelchair controller, receiving the outcome of the classifier via the LSL every 4 seconds and through threading simultaneously sending the appropriate commands to the wheelchair controller.

IV. RESULTS

To validate the proposed wheelchair operated by BCI in real-time, 5 volunteers with no previous experience in BCI applications participated in 3 different experiments. The experimental phase is approximately 1 hour for each participant.

Initially, the participants are familiarized with the equipment and tested the speed of the wheelchair by commanding it with the joystick to get comfortable with the system. Following this, they are given time to train in the mental commands. During the training, subjects attempt to execute the 3 mental commands without being seated in the wheelchair. When the training is completed the first experiment is conducted. Experiment 1 is starting and stopping the wheelchair using intentional blinks. Users have to stop and start the wheelchair 2 times in predefined positions (Figure 3). For this particular experiment, 3 evaluation metrics are utilized which included measuring the distance from the predetermined stop positions for both stop 1 and stop 2, as well as calculating the number of stops during each trial. Every subject has 5 trials to complete the experiment.

Fig. 3. In Experiment 1, participants are required to stop the wheelchair at two specific locations. The starting position is marked by the first orange cone, while the first stopping position is indicated by the initial set of orange cones. The second stopping position is designated by the final set of orange cones. The objective is to assess the precision of the wheelchair in stopping at these predetermined positions.

Table II presents the average results for the start/stop experiment. The participants are instructed to perform only 2 stops at the predefined positions. The table shows the average number of stops, the average distance from the first Stop position, and the average distance from the second Stop position for each subject. The distance is measured in centimeters (cm). The average number of Stops for all participants is 2.4 ranging from 3 to 2. The subject with the most stops is Sub 3 while

979-8-3503-8368-3/23 $31.00 © 2023 IEEE

TABLE II
EXPERIMENT 1 AVERAGE RESULTS. THE OPTIMAL NUMBER OF STOPS IS 2. THE AVERAGE DISTANCE IS MEASURED IN CENTIMETERS (CM)

Sub	Avg # of Stops	Avg Distance 1	Avg Distance 2
1	2.6	9.40 cm	7.74 cm
2	2.4	5.00 cm	5.68 cm
3	3.0	12.26 cm	1.76 cm
4	2.0	19.5 cm	17.26 cm
5	2.0	6.14 cm	23.6 cm
Average	2.4	10.46 cm	11.20 cm

TABLE III
AVERAGE RESULTS OF EXPERIMENT 2 FOR EVALUATING START-STOP-RIGHT TURNS

Sub	Avg Stops	Avg Commands	Avg Wrong Turns
1	3.8	9.8	1.0
2	3.0	9.4	1.2
3	3.8	10.6	0.8
4	3.4	14.4	4.0
5	3.8	12.2	1.4
Average	3.5	11.3	1.7

Subs 4 and 5 managed to stop the wheelchair 2 times in all 5 trials. The average distance from Stop 1 for all subjects is 10.46cm. The best-performing participant is Sub 2 with a 5 cm average distance while the worst is Sub 4 with 19.5 cm. Lastly, the average distance from the second Stop for all participants is 11.20 varying from 1.76 cm achieved by Sub 3 to 23.6 cm by Sub 5. The results indicate that the participants could command the wheelchair for starting and stopping with good precision.

Experiments 2 and 3 are performed to evaluate the system's ability to execute right and left turns. The participants are instructed to perform a total of 4 turns, specifically 4 right turns for Experiment 2 and 4 left turns for Experiment 3. To reach the end of the path, subjects have to execute 3 stop/start actions and 4 turns, thereby necessitating a minimum of 7 mental commands. Each participant has 5 trials for each of the experiments. The path that the participants are instructed to follow for Experiments 2 and 3 is presented in Figure 4.

Fig. 4. Experiments 2 and 3 predetermined route. To complete the task as per the experimental instructions, the subjects are expected to execute a sequence of turns, forward movements, and stops. The 4 orange cones represent the starting and the ending point of the path. The other two single orange cones mark the locations where the participants must stop the movement of the wheelchair and perform the turns; 2 turns at the first cone and 2 at the second.

Table III presents the average results of 5 tries for the second experiment. Participants must execute MI right turns to complete the predetermined route.

The metrics of this experiment are average stops, average number of commands, and average number of wrong turns.

The average score for the first metric among all participants is 3.5, with scores ranging from 3 to 3.8. For the second metric, the average score is 11.3 varying from 14.4 achieved by Sub 4 to 9.4 by Sub 2. The average wrong turns (left turns) for all participants is 1.7. Sub 4 has the most average wrong turns (4) while Sub 3 has the least with only 0.8. Overall, all participants finished the route on every try and manipulated the wheelchair operated by BCI with great accuracy. Subs 1 and 3 managed to finish 1 try with the minimum number of commands (7) and along with Sub 2 they are the best-performing participants of this experiment.

Experiment 3's average results are displayed in Table IV. The same evaluation metrics with Experiment 2 are employed since subjects perform the same number of commands and have to finish the same route with the only modification being that now they have to turn left 4 times.

TABLE IV
AVERAGE RESULTS OF EXPERIMENT 3 FOR EVALUATING START-STOP-LEFT TURNS

Sub	Avg Stops	Avg Commands	Avg Wrong Turns
1	3.2	9.2	1.0
2	3.8	10.6	1.2
3	4.1	13.4	2.6
4	3.2	10.4	0.8
5	3.6	7.6	0.0
Average	3.6	10.2	1.1

The average number of stops for all participants is 3.6; Sub 3 has the most stops with 4.1 and Subs 1 and 4 have the least with 3.2. The average number of commands is 10.2 ranging from 7.6 by Sub 5 to 13.4 by Sub 3. Lastly, the average number of wrong turns is 1.1 varying from 0 by Sub 5 to 2.6 by Sub 3. Sub 5 is the best-performing participant in this experiment and managed to finish the route with the minimum amount of commands 2 times. Subs 1 and 2 also managed to finish the path with 7 commands 1 time. The worst-performing participant is Sub 3 who is one of the best in Experiment 2. In contrast, Sub 5 had the second lowest performance in the second experiment but achieved the highest performance in the third. The average results of the left MI turns experiment are better than the one with the right MI turns.

V. DISCUSSION

In this study, a real-time BCI is implemented to command a wheelchair. To validate the proposed system, 5 subjects participated in 3 experiments. The raw EEG data are captured with Emotiv Insight that connects to a laptop via Bluetooth. All participants signed a consent form to be able to take part in the study. Significant emphasis is placed on safeguarding the privacy of every subject's data. All data is stored on a computer, and no user information is retained except for their age. When collecting real-time data, all information is kept exclusively on the local computer, guaranteeing subject anonymity and data security.

Table V presents a comparative analysis of this study with other published papers in the literature.

TABLE V
COMPARATIVE ANALYSIS OF THIS WORK WITH THE LITERATURE.

Authors	Device	Sub	Results
Jiang *et al.* [10]	Emotiv Epoc	5	Accuracy of left/right detection
Xie and Li [11]	24-electrode EEG cap	6	Offline Classification Online Classification
Espiritu *et al.* [12]	Emotiv Insight	1	Command Response Delay for Action
This work	Emotiv Insight	5	Classification results Experiment 1 results Experiment 2 results Experiment 3 results

Jiang *et al.* [10] designed a wheelchair operated by BCI and 5 subjects participated in the evaluation of this system. Emotiv Epoc was used to record the brain signals. Although real-time assessment was conducted, only the accuracy of detecting left/right turns was reported as an evaluation metric. Xie and Li [11] designed a wheelchair commanded by BCI that uses MI commands. The system used a 24-electrode EEG cap to acquire brain signals. To evaluate their work they used the classification results from dataset I from the 2004 BCI Competition III as an offline metric and 6 subjects classification accuracy as an online metric. This system was not tested in real-time. Espiritu *et al.* [12] designed a wheelchair operated by BCI that can execute five movements. The raw EEG signals were obtained using Emotiv Insight, and 1 subject participated in 30 trials to test the system. The Command Response Delay for Action was used as the metric to evaluate the system's performance. This study involved the development of a wheelchair operated by BCI capable of executing 4 movements, including going forward, stopping, turning right, and turning left. Emotiv Insight is utilized to capture brain signals from 5 subjects who participated in 3 experiments. Participants assessed the proposed system for 15 trials, 5 for each experiment. Classification accuracy, Experiment 1 results, Experiment 2 results, and Experiment 3 results are employed for the evaluation of the BCI system. This work utilizes more evaluation metrics to provide a more comprehensive assessment of the system for commanding the BCI-controlled wheelchair in real-time.

VI. CONCLUSION AND FUTURE WORK

This work presents the design of a wheelchair that can be controlled using BCI. EEG signals are acquired using the Emotiv Insight headset which is connected to a computer

via Bluetooth. The DoF of the system is 4; moving forward, stopping, turning right, and turning right. To turn, users must execute MI right or left commands, and to start or stop they must perform 1 intentional blink. Raw EEG data are processed using the OpenViBE software, and LDA algorithm is used for classifying the mental commands. Then a Python script sends the appropriate command to the wheelchair via USB.

The study involved 5 participants who completed 3 experiments. The first experiment involved stopping and starting the wheelchair, while the second experiment involved combining starting and stopping with turning right. The third experiment involved starting, stopping, and turning left. The study's results demonstrate that the participants are able to control the BCI system accurately, and all of them successfully commanded the wheelchair using their brain signals.

In the future, the aim is to enhance the wheelchair operated by BCI by adding a backward movement feature and integrating an obstacle detection system to ensure safety. Additionally, the study will involve more participants to obtain more accurate and reliable results.

REFERENCES

[1] L. F. Nicolas-Alonso and J. Gomez-Gil, "Brain computer interfaces, a review," *sensors*, vol. 12, no. 2, pp. 1211–1279, 2012.

[2] B. He, H. Yuan, J. Meng, and S. Gao, "Brain–computer interfaces," *Neural engineering*, pp. 131–183, 2020.

[3] N. Birbaumer, "Breaking the silence: brain–computer interfaces (bci) for communication and motor control," *Psychophysiology*, vol. 43, no. 6, pp. 517–532, 2006.

[4] L. Tonin and J. d. R. Millán, "Noninvasive brain–machine interfaces for robotic devices," *Annual Review of Control, Robotics, and Autonomous Systems*, vol. 4, pp. 191–214, 2021.

[5] F. Cincotti, D. Mattia, F. Aloise, S. Bufalari, G. Schalk, G. Oriolo, A. Cherubini, M. G. Marciani, and F. Babiloni, "Non-invasive brain–computer interface system: towards its application as assistive technology," *Brain research bulletin*, vol. 75, no. 6, pp. 796–803, 2008.

[6] Z. Al-Qaysi, B. Zaidan, A. Zaidan, and M. Suzani, "A review of disability eeg based wheelchair control system: Coherent taxonomy, open challenges and recommendations," *Computer methods and programs in biomedicine*, vol. 164, pp. 221–237, 2018.

[7] A. Palumbo, V. Gramigna, B. Calabrese, and N. Ielpo, "Motor-imagery eeg-based bcis in wheelchair movement and control: A systematic literature review," *Sensors*, vol. 21, no. 18, p. 6285, 2021.

[8] N. Padfield, J. Zabalza, H. Zhao, V. Masero, and J. Ren, "Eeg-based brain-computer interfaces using motor-imagery: Techniques and challenges," *Sensors*, vol. 19, no. 6, p. 1423, 2019.

[9] M. Lotze and U. Halsband, "Motor imagery," *Journal of Physiology-paris*, vol. 99, no. 4-6, pp. 386–395, 2006.

[10] L. Jiang, E. Tham, M. Yeo, Z. Wang, and B. Jiang, "Motor imagery controlled wheelchair system," in *2014 9th IEEE Conference on Industrial Electronics and Applications*. IEEE, 2014, pp. 532–535.

[11] Y. Xie and X. Li, "A brain controlled wheelchair based on common spatial pattern," in *2015 International symposium on bioelectronics and bioinformatics (ISBB)*. IEEE, 2015, pp. 19–22.

[12] N. M. D. Espiritu, S. A. C. Chen, T. A. C. Blasa, F. E. T. Munsayac, R. P. Arenos, R. G. Baldovino, N. T. Bugtai, and H. S. Co, "Bci-controlled smart wheelchair for amyotrophic lateral sclerosis patients," in *2019 7th International Conference on Robot Intelligence Technology and Applications (RiTA)*. IEEE, 2019, pp. 258–263.

[13] P. Xanthopoulos, P. M. Pardalos, T. B. Trafalis, P. Xanthopoulos, P. M. Pardalos, and T. B. Trafalis, "Linear discriminant analysis," *Robust data mining*, pp. 27–33, 2013.

[14] A. J. Izenman and A. J. Izenman, "Linear discriminant analysis," *Modern multivariate statistical techniques: regression, classification, and manifold learning*, pp. 237–280, 2008.

A Comparative Testing of Yolov8-based algorithm for cancer biopsy images: Case study on Pannuke pan-cancer dataset

Panagiotis N. Smyrlis
Department of Electrical &
Computer Engineering
University of Western Macedonia
GR50100 Kozani, Greece
pan.smyrlis@gmail.com

Odysseas Tsakai
Q Base R&D
Science & Technology Park of Epirus
University of Ioannina Campus
GR45110 Ioannina, Greece
o.tsakai@gmail.gr

Konstantinos Vogklis
Department of Tourism
Ionian University
GR49100 Corfu, Greece
voglis@ionio.gr

Nikolaos Giannakeas, Alexandros Tzallas*
Department of Informatics &
Telecommunications
University of Ioannina
GR47100 Arta, Greece
giannakeas@uoi.gr, tzallas@uoi.gr

George F. Fragulis
Department of Electrical &
Computer Engineering
University of Western Macedonia
GR50100 Kozani, Greece
gfragulis@uowm.gr

Markos G. Tsipouras
Department of Electrical &
Computer Engineering
University of Western Macedonia
GR50100 Kozani, Greece
mtsipouras@uowm.gr

Abstract—**Yolo architecture has presented promising results in various segmentation and detection tasks, including biomedical image analysis. This work studies the effectiveness and applicability of the method on cancer data, by comparative and experimental testing on the complex pan-cancer benchmark Pannuke for the recent Yolov8 algorithm variation. The experimental protocol encompasses a comprehensive exploration of various v8 approaches and subsequently provides an evaluative overview of the models employing a plug-and-play utilization paradigm. The results illustrate the performance characteristics of each separate model and explores the effect of the architecture in conjunction with the model development load and tradeoff.**

Index Terms—**Yolo, Pannuke, Segmentation, Biomedical Image, Biopsy image, Digital Pathology**

I. INTRODUCTION

Deep Learning evolution has trigerred large research attention and effort in cancer detection. Accurate Deep Learning models can facilitate and support the decision procedure, providing useful insight on pathology presence and quantification. Large Cancer datasets [16, 9], available nowadays, can support the development of robust algorithms for more generic solutions, working on multi-organ data collections and pan-cancer sets.

Image analysis is an open field of interest for the scientific community studying cancer. Visual diagnostic data, such as microscopy[8], dermoscopy [15], radiographic [1] images are under research for their property to become useful knowledge and optimize the diagnosis process. Sophisticated deep learning methods have been published in literature and presented promising results out of such imagery collections. The aim is to highlight the pathologies of interest, as also to quantify and measure the maligant cases, if any.

Cell segmentation methods have been studied to contribute the diagnosis decision support, while various deep learning architectures have been tested presenting descent performance capability. Mask-RCNN [5] followed a learnable object proposals' approach to support the detection procedure, composing a two-stage algorithm. Contrary, HoVerNet [4] adopts an one-stage model procedure along minor-overhead post-process actions to form the final solution. The same one-stage general schema is followed in STARDIST [12], and CPP-Net [2] and Micro-Net [11]. Also, the Yolo series [6] has published state-of-the-art results and applicability, while its recent versions attempt to minimize the training cost, providing a solution that relies on reduced trainable parameters and training time.

In this work, a comparison of different Yolov8 flavors was experimentally tested using Pannuke [3] pan-cancer dataset. We present the panoptic quality evaluation results of all 5 versions and we argue on the effectiveness and the tradeoff of this architecture. In addition, the experimental process

This research has been financed by the European Union and Greek national funds through the Operational Program Competitiveness, Entrepreneurship and Innovation, under the call RESEARCH – CREATE – INNOVATE (T2EDK-03660). Also, we acknowledge the support from the project "Immersive Virtual, Augmented, and Mixed Reality Center of Epirus" (MIS:5047221), from Action "Reinforcement of the Research and Innovation Infrastructure", funded by the Operational Programme "Competitiveness, Entrepreneurship, and Innovation" (NSRF 2014–2020) and co-financed by Greece and the European Union (European Regional Development Fund).

979-8-3503-8368-3/23 $31.00 © 2023 IEEE

included tests on transfer weight learning, by initializing the model using pretrained weights on common objects [10].

II. MATERIALS AND METHODS

A. PanNuke Dataset

The Pannuke dataset is one of well known and studied microscopy imagery collections. In this, 7901 individual 256x256 image pathces were extracted, out of The Cancer Genome Atlas [14] program Whole Slide Images. The images are of x20 and x40 magnification, while the distributed set is of x40 magnification by respective resampling. It is a multi-organ dataset, containing slides from 19 distinct cancerous tissue kinds, thus trying to feed pan-cancer detection and quantification applications.

The authors have additionally presented a cancer detection taxonomy to distinguish cells as a 5-class problem that may fit all distinct tissue kinds included. The respective annotation constitutes of 189,744 labelled nuclei instances in total, providing precise, pixel-level masks and tissue type information. The taxonomy categories are Neoplastic, Non-Neoplastic Epithelial, Inflammatory, Connective and Dead cells while the different tissue types are Adrenal Gland, Bile Duct, Bladder, Breast, Cervix, Colon, Esophagus, Head and Neck, Kidney, Liver, Lung, Ovarian, Pancreatic, Prostate, Skin, Thyroid, Uterus. Pannuke is considered as a pan-cancer dataset.

To annotate the WSIs, a fully convolutional neural network was set to produce cell detections, which are then evaluated by pathologists. The FCNN was trained using data from 4 well known public datasets. The detections were finally relabelled to Pannuke novel categorization taxonomy. Then, 2000 visual fields were randomly sampled from TCGA [14] and a local hospital to generate object detections. Subsequently, evaluation and retrain procedure was repeatedly performed to result in final and accurate detections. NuClick tool was employed to induct nuclei masks from the previous detection predictions.

Pannuke provides pluralism on tissue types, maligant scenarios and total number of instances. It is also highly unbalanced and a complex dataset, so being challenging to research robust model training solutions. A snapshot of labelled dataset patches is shown in Figure 1.

B. The Yolov8 Model

The trainable v8 model is the recent update of yolo series algorihtm [13]. In this version, 5 distinct variations are distributed, Nano, Small, Medium, Large, Extra Large. Starting from Nano, each version offers increased number of trainable parameters, and the accuracy-cost tradeoff may be studied. Yolov8 has presented state-of-the-art results to segmentation and detection tasks containing common objects. The authors argue for the increase of precision as a function of Yolo trainable parameters. Additionally, the smaller versions attempt to fit in minimal available resources circumstances. The number of parameters for all versions are presented in Table I.

Yolo remains an one stage detector / segmentor algorithm, and additionally this version supercedes the anchor-boxes philosophy for object localization and makes no use of predefined

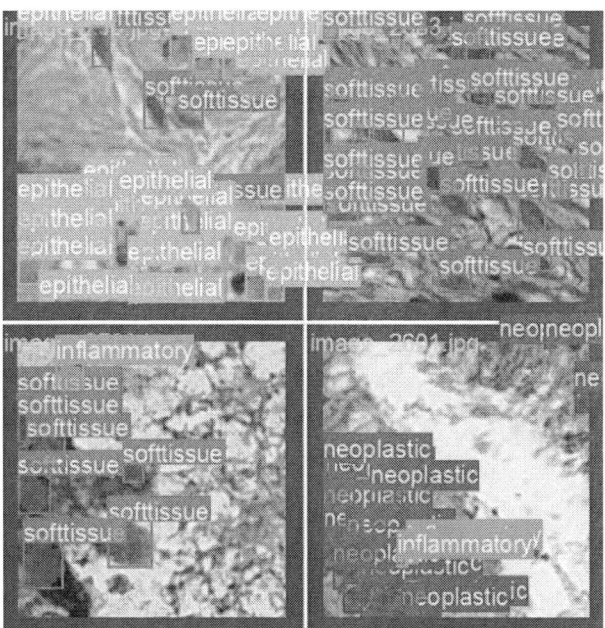

Fig. 1. Visualisation of four individual 256x256 patches in Pannuke, together with the respective class annotation highlight.

object proposals. For all v8 different versions, a respective semantic segmentation variation is distributed which embeds the mask prediction logic in the respective architecture.

A modified CSP-Darknet53 algorithm is used, which provides a larger feature map to the network. Subsequently, a Cross Stage Partial Bottleneck module of 2 convolutions is placed. Two segmentation heads follow, which learn to predict the segmentation masks on a given image. The algorithm makes use of CIoU, DFL and binary cross entropy loss function to the training process. The distinct losses respond to the measure of box, mask and classification accuracy.

The network is supplied by the Yolo framework data transformation methods which also facilitate data augmentation and on the processed data. These are HSV transformation, rotation, translation, perspective manipulation, scaling, shearing, flipping, mosaicing, image mixing (MixUp) and partial image mixing (CutMix) and CutOut.

For the experimental procedure, the publicly available Yolov8 implementation [7] was used [7]. The training process is done for 256x256 image size (Pannuke patches size), batch size of 16, used a learning rate of 0.001111 with a 0.9 momentum and 0.0005 weight decay and ran for 100 epochs.

C. Postprocessing

After obtaining the mask predictions, a filtering process is done using class thresholds for the instance detections. The

TABLE I
V8 VERSIONS TRAINABLE PARAMETERS

Model Name	Yolo-n	Yolo-s	Yolo-m	Yolo-l	Yolo-x
Parameters	3.2M	11.2M	25.9M	43.7M	68.2M

Fig. 2. Visualisation of four individual 256x256 patches in Pannuke with the respective class annotation highlight.

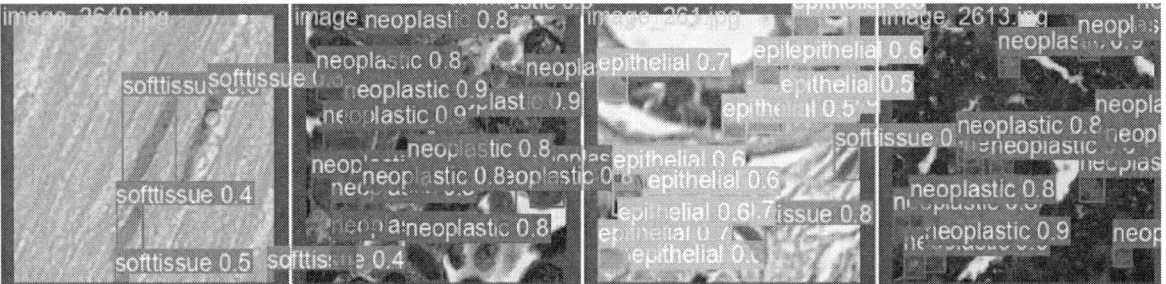

Fig. 3. Visualisation of the model predictions on Figure 2 together with the respective class annotation highlight.

predicted instances which were scored lower than this are not considered as detected and the algorithm ignores them. The threshold for all five classes was set to 0.15.

III. RESULTS

For the experimental procedure, the publicly available Yolov8 implementation [7] was used. The training process was done for 256x256 image size (Pannuke patches size), batch size of length 16, used a learning rate of 0.001111 with a 0.9 momentum and 0.0005 weight decay and ran for 100 epochs. The publicly available 3-fold Pannuke split fed the training phase.

After the model predictions and postprocessing, Panoptic Quality metrics were calculated. The obtained results are shown in the tables II, III and IV. For each Yolov8 variant, we report two scores, one after training using the pretrained weights and one after training from scratch. Table II shows the multi-class panoptic quality per tissue, as also the average of all tissues. In Table III we observe the Binary Panoptic quality for the same indices, while in Table IV we notice the Panoptic Quality per class across all tissues. As observed in all results, the trainable parameters' do affect the algorithm's accuracy given the hyperparameter set. The models with the lowest number of parameters (Nano, Small) were observed to score less. The overall panoptic quality ranges between 39% to 44% while Binary Panoptic Quality from 50.17% to 56.04%.

Regarding pretrained models the results in both Table II and III indicate that small sized models (nano, small) gain some advantage when used pretrained. However the bigger-sized models don't seem to obtain substantial benefit.

A sample visual snapshot of the algorithm's prediction performance is seen in Figures 2 and 3. In Figure 2 the actual annotation is presented, while in 3 the Yolov8 predictions are resided. As observed in figures, a large number of true instances were succesfully identified in the results figure.

IV. CONCLUSION

This work presents experiments on Pannuke dataset using all Yolov8 algorithm versions. The evaluation included transfer weight learning tests by pretrained weights initialization. The pretrain weights were inducted by common objects' imagery of the COCO dataset. The obtained results for the recent Yolo architecture on multi-organ imagery datasets, such as Pannuke gave us promising results of the applicability on biomedical image segmentation problems.

REFERENCES

[1] Hafiz Muhammd Ali Bhatti et al. "Multi-detection and segmentation of breast lesions based on mask rcnn-fpn". In: *2020 IEEE International Conference on Bioinformatics and Biomedicine (BIBM)*. IEEE. 2020, pp. 2698–2704.

[2] Shengcong Chen et al. "CPP-net: Context-aware polygon proposal network for nucleus segmentation". In: *IEEE Transactions on Image Processing* 32 (2023), pp. 980–994.

[3] Jevgenij Gamper et al. "Pannuke dataset extension, insights and baselines". In: *arXiv:2003.10778* (2020).

[4] Simon Graham et al. "Hover-net: Simultaneous segmentation and classification of nuclei in multi-tissue histology images". In: *Medical image analysis* 58 (2019), p. 101563.

TABLE II
MPQ PER TISSUE

Tissue #	Tissue Type	Yolo-n	Yolo-n pt	Yolo-s	Yolo-s pt	Yolo-m	Yolo-m pt	Yolo-l	Yolo-l pt	Yolo-x	Yolo-x pt
0	Adrenal_gland	44.25%	44.18%	47.33%	46.53%	48.33%	47.22%	48.64%	47.39%	48.82%	47.89%
1	Bile-duct	41.09%	41.89%	43.73%	44.75%	45.28%	45.15%	46.01%	45.05%	45.70%	45.38%
2	Bladder	50.37%	49.99%	52.63%	50.86%	53.78%	53.14%	55.15%	53.42%	54.91%	54.05%
3	Breast	43.66%	44.08%	45.96%	46.16%	46.99%	47.43%	47.46%	47.30%	47.81%	47.70%
4	Cervix	40.73%	40.44%	43.10%	43.08%	43.44%	43.37%	45.09%	42.99%	45.00%	45.83%
5	Colon	33.99%	35.10%	36.57%	37.59%	37.81%	38.21%	38.35%	39.02%	38.73%	39.00%
6	Esophagus	43.03%	43.07%	45.92%	45.56%	48.43%	48.25%	49.28%	48.02%	48.64%	48.30%
7	HeadNeck	39.75%	39.62%	43.04%	43.31%	43.72%	43.49%	43.73%	44.18%	44.31%	43.74%
8	Kidney	43.94%	42.79%	46.60%	46.89%	48.55%	49.16%	49.00%	48.11%	48.91%	48.83%
9	Liver	43.10%	42.80%	44.83%	46.08%	47.03%	45.58%	46.97%	45.38%	47.73%	47.75%
10	Lung	33.55%	33.07%	34.76%	36.62%	37.04%	36.37%	38.59%	36.60%	37.90%	36.24%
11	Ovarian	37.27%	37.44%	38.72%	42.39%	43.60%	40.04%	44.23%	42.82%	43.31%	43.17%
12	Pancreatic	37.17%	38.01%	39.77%	41.31%	42.03%	42.04%	43.95%	42.53%	43.75%	42.63%
13	Prostate	41.36%	43.36%	43.80%	45.53%	45.63%	45.42%	46.43%	45.03%	46.62%	45.91%
14	Skin	29.84%	28.85%	31.00%	32.95%	33.13%	32.21%	34.00%	34.08%	34.10%	34.58%
15	Stomach	35.45%	35.11%	37.19%	37.55%	39.64%	38.89%	40.77%	38.79%	41.23%	38.90%
16	Testis	38.66%	39.27%	41.09%	43.29%	43.29%	44.48%	45.80%	43.99%	44.34%	43.53%
17	Thyroid	36.34%	36.61%	38.13%	40.55%	39.41%	40.83%	39.85%	39.70%	40.37%	40.44%
18	Uterus	30.42%	32.07%	34.99%	36.33%	39.51%	35.03%	38.57%	38.89%	36.73%	39.63%
19	Mean	39.16%	39.36%	41.54%	42.49%	43.51%	42.96%	44.31%	43.33%	44.15%	43.87%

TABLE III
BPQ PER TISSUE

Tissue #	Tissue Name	Yolo-n	Yolo-n pt	Yolo-s	Yolo-s pt	Yolo-m	Yolo-m pt	Yolo-l	Yolo-l pt	Yolo-x	Yolo-x pt
0	Adrenal_gland	56.37%	56.47%	58.91%	59.48%	60.77%	60.73%	61.16%	60.58%	61.64%	61.42%
1	Bile-duct	51.43%	52.17%	54.01%	55.48%	56.43%	56.24%	57.20%	56.40%	57.70%	56.91%
2	Bladder	57.25%	56.31%	59.76%	59.45%	61.42%	60.86%	63.22%	61.55%	62.01%	61.33%
3	Breast	53.01%	52.74%	54.50%	55.37%	56.19%	56.18%	56.97%	57.38%	57.29%	57.42%
4	Cervix	52.27%	53.15%	55.48%	55.95%	57.32%	55.72%	58.37%	57.91%	58.21%	58.99%
5	Colon	40.56%	41.95%	43.48%	44.86%	45.47%	45.37%	46.13%	47.27%	46.53%	47.15%
6	Esophagus	50.13%	49.06%	52.70%	52.78%	54.67%	55.79%	56.26%	55.18%	56.27%	55.64%
7	HeadNeck	50.38%	50.33%	52.73%	53.02%	54.52%	54.22%	55.56%	54.17%	55.30%	54.30%
8	Kidney	54.53%	55.16%	57.12%	57.92%	57.90%	59.73%	59.96%	58.79%	59.86%	59.80%
9	Liver	57.12%	56.82%	58.99%	58.77%	61.11%	60.60%	62.04%	60.15%	62.40%	61.47%
10	Lung	47.24%	46.83%	49.02%	50.59%	52.31%	49.51%	53.36%	50.95%	53.35%	50.94%
11	Ovarian	44.61%	44.99%	46.01%	50.07%	51.47%	46.86%	52.08%	51.58%	52.05%	52.01%
12	Pancreatic	45.20%	47.61%	48.83%	51.79%	52.68%	52.59%	53.75%	52.73%	53.67%	53.38%
13	Prostate	51.59%	51.92%	53.24%	54.64%	55.46%	54.13%	56.04%	54.46%	56.16%	55.27%
14	Skin	43.78%	44.50%	45.68%	48.99%	49.69%	47.27%	48.71%	49.90%	49.65%	50.87%
15	Stomach	49.89%	49.87%	52.65%	52.27%	55.56%	54.34%	56.81%	55.36%	57.52%	55.00%
16	Testis	51.35%	50.76%	52.95%	54.49%	55.28%	56.46%	57.10%	55.37%	56.49%	57.38%
17	Thyroid	53.62%	53.13%	55.67%	55.75%	56.53%	56.44%	57.53%	55.83%	58.35%	57.02%
18	Uterus	42.86%	45.03%	44.99%	46.76%	49.58%	46.56%	51.48%	48.10%	50.30%	49.41%
19	Mean	50.17%	50.46%	52.46%	53.60%	54.97%	54.19%	55.99%	54.93%	56.04%	55.56%

TABLE IV
PER CLASS PANOPTIC QUALITY

Class #	Class Name	Yolo-n	Yolo-n pt	Yolo-s	Yolo-s pt	Yolo-m	Yolo-m pt	Yolo-l	Yolo-l pt	Yolo-x	Yolo-x pt
0	Neoplastic	45.69%	45.13%	48.09%	48.24%	49.65%	48.68%	50.44%	48.97%	50.87%	50.18%
1	Inflammatory	34.61%	35.57%	36.84%	37.57%	38.53%	38.31%	38.93%	38.76%	38.91%	38.74%
2	Softtissue	34.61%	35.57%	36.84%	37.57%	38.53%	38.31%	38.93%	38.76%	38.91%	38.74%
3	Dead	8.92%	10.25%	10.79%	11.96%	10.93%	12.98%	10.31%	13.34%	10.77%	11.95%
4	Epithelial	38.35%	40.81%	41.59%	44.41%	43.50%	44.82%	44.10%	46.57%	44.35%	46.16%

[5] Kaiming He et al. "Mask r-cnn". In: *Proceedings of the IEEE international conference on computer vision*. 2017, pp. 2961–2969.

[6] Muhammad Hussain. "YOLO-v1 to YOLO-v8, the Rise of YOLO and Its Complementary Nature toward Digital Manufacturing and Industrial Defect Detection". In: *Machines* 11.7 (2023), p. 677.

[7] Glenn Jocher, Ayush Chaurasia, and Jing Qiu. "YOLO by Ultralytics". In: *URL: https://github.com/ultralytics/ultralytics* (2023).

[8] Athanasios Kanavos et al. "Breast Cancer Classification of Histopathological Images using Deep Convolutional Neural Networks". In: *2022 7th South-East Europe Design Automation, Computer Engineering, Computer Networks and Social Media Conference (SEEDA-CECNSM)*. IEEE. 2022, pp. 1–6.

[9] Neeraj Kumar et al. "A multi-organ nucleus segmentation challenge". In: *IEEE transactions on medical imaging* 39.5 (2019), pp. 1380–1391.

[10] Tsung-Yi Lin et al. "Microsoft coco: Common objects in context". In: *Computer Vision–ECCV 2014: 13th European Conference, Zurich, Switzerland, September 6-12, 2014, Proceedings, Part V 13*. Springer. 2014, pp. 740–755.

[11] Shan E A Raza et al. "Micro-Net: A unified model for segmentation of various objects in microscopy images". In: *Medical image analysis* 52 (2019), pp. 160–173.

[12] Uwe Schmidt et al. "Cell detection with star-convex polygons". In: *Medical Image Computing and Computer Assisted Intervention–MICCAI 2018: 21st International Conference, Granada, Spain, September 16-20, 2018, Proceedings, Part II 11*. Springer. 2018, pp. 265–273.

[13] Juan Terven and Diana Cordova-Esparza. "A comprehensive review of YOLO: From YOLOv1 to YOLOv8 and beyond". In: *arXiv preprint arXiv:2304.00501* (2023).

[14] Katarzyna Tomczak, Patrycja Czerwińska, and Maciej Wiznerowicz. "Review The Cancer Genome Atlas (TCGA): an immeasurable source of knowledge". In: *Contemporary Oncology/Współczesna Onkologia* 2015.1 (2015), pp. 68–77.

[15] Halil Murat Ünver and Enes Ayan. "Skin lesion segmentation in dermoscopic images with combination of YOLO and grabcut algorithm". In: *Diagnostics* 9.3 (2019), p. 72.

[16] Quoc Dang Vu et al. "Methods for segmentation and classification of digital microscopy tissue images". In: *Frontiers in bioengineering and biotechnology* (2019), p. 53.

Brain-de-fer
A brain-computer interface game for mind contest

Konstantinos D. Georgitsaros, Konstantinos B. Palegkas, Pantelis Angelidis and Markos G. Tsipouras

Electrical and Computer Engineering
University of Western Macedonia
Campus ZEP Kozani, Greece
ece01451@uowm.gr; ece01530@uowm.gr; pangelidis@uowm.gr; mtsipouras@uowm.gr

Abstract—In this paper "Brain-de-fer" is presented, which is a game where two players compete against each other using their attention in a "bras-de-fer" match. A brain-computer interface is implemented to allow the players to play based on their attention. Two different EEG devices were tested for this application, the Muse S headband and the Mindwave mobile. To quantify the attention value from the raw EEG values, five classifiers were trained and tested.

Index Terms—electroencephalography (EEG), brain-computer interface (BCI), BCI game, human-computer interface (HCI), Muse S headband, MindWave mobile, Unity

I. INTRODUCTION

Brain-Computer Interfaces (BCI) is a continuously evolving area, that already demonstrates profound impact on various aspects of society and human experience. Using a BCI a user can control mechanical or electrical devices without using their nerve system or generally any movement. In the field of Biomedicine, BCI enables the study of the brain activity. Some capabilities of this technology include prediction of epilepsy episodes [1], enabling paralyzed people to control the movement of their wheelchair [2] or (like in this case) create games that are both fun to play and help cope with diseases like Adolescent Hyperactivity Disorder (ADHD) or dementia. For such an interface to work correctly, amongst other systems, there must be signal processing and filtering systems that can correctly distinguish between voluntary actions and noise that seems like an action (involuntary). For that reason, there has been developed a pipeline that should be followed when designing such an interface. The main components of the pipeline are [3]:

- Signal acquisition
- Feature extraction
- Feature translation
- Device output

BCI is divided in two categories: 1) invasive and 2) non-invasive [4]. The difference between the two is that in the invasive category the electrodes are placed inside the skull compared to non-invasive where the electrodes are placed on top of the skull. The Invasive category provides steady and noise-free signals, but it is extremely expensive and neuro-surgery needs to be performed which is extremely difficult both for the doctor and the patient. The advantages of non-invasive are that it is cost effective, portable and easy to use, although hair, bones and brain tissues can reduce the accuracy of the signal. The most common non-invasive technique is Electroencephalography (EEG) which is also used in this paper. Many companies sell commercial devices that besides being easily obtainable, can also provide an easy way for students to experiment with EEG.

In the current paper the "Brain-de-fer" is presented, which is a game where the players compete using their focusing power. Players are enabled to use their focusing powers by means of a BCI. The game can be played by two players (against each other) or by one player (against a bot player). The players are represented by two in-game characters (referred as athletes) who fight in a bras-de-fer match. The game was developed in the Unity game engine with assets taken from Mixamo. In this work, two EEG devices were used and compared, the devices are: 1) Muse [5], 2) Mindwave [6].

II. RELATED WORK

Cheng et al. [7], presented a game which would help drivers reduce the chances of getting distracted. They did this by creating a game in which a player has to avoid obstacles while getting distracted by visual and audio distractions. For the game to work a racing wheel is needed and while playing the game the player wears Mindwave Neurosky Device which captures all the beta waves. The point of the game is to make the driver understand when he is distracted and prevent that from happening .

Another use case of EEG signals was in the study of Coulton et al. [8], where they created a game on cellphones. They used a relatively affordable and easy to use headset from Mindwave and their point was to create a game accessible everywhere by anyone. The game consisted of a maze with doors opening and closing according to the different values of attention and meditation the player accumulated. The point of the game is to measure the emotional states of each player while playing and create "mind maps" where attention and meditation levels are shown according to the position of the player .

Wang et al. [9] designed a parkour game where the player needs to jump to avoid objectives, collect coins and run. The game consists of four levels with the last one being an infinite loop level. The running speed of the player's avatar is determined by the attention level (the higher the faster) and whenever the player blinks the avatar will jump. The creators also claim that the game can help people who suffer from ADHD by helping them exercise concentration and also help people suffering from cerebral palsy to recover mentally.

III. DEVICES AND METHODS

A. Muse

The first Device that was used is Muse S headband which is a four-channel electroencephalographer, the electrodes are placed in the positions AF7, AF8, TP9 and TP10. The AF7 and AF8 are found in the anterior frontal region of the scalp which is responsible for decision-making, and emotional regulation, while the TP9 and TP10 are positioned in the temporal-parietal region of the scalp which is useful for studying processes related to auditory perception, memory, and language. The positions of electrodes on the scalp can be seen in figure 1. It has a sampling rate of 256 Hz and comes at a price of 400€. This specific devices was chosen due to availability, the cost and the ease of use.

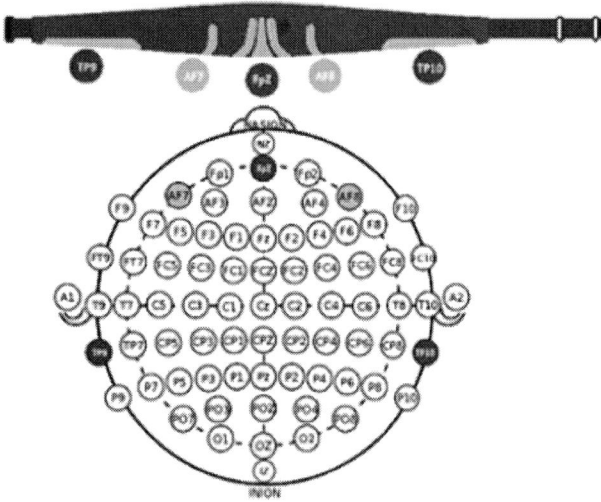

Fig. 1. Muse S and the placement of electrodes [24]

1) BlueMuse: To collect the data, the software BlueMuse is used. It provides a fast and easy access to the device and the ability to connect two or more devices at the same time which is critical for our application. After connecting the device to the computer via Bluetooth, it collects data from the headset through the Lab Streaming Layer (LSL) protocol [10].

2) Lab Streaming Layer: LSL [11] is an open source ecosystem that enables streaming and capturing data from certain devices. The streaming is asynchronous thus enabling the streaming of multiple sources. The data is retrieved either in packets or in a single measurement.

B. Mindwave

The second device that was used is Mindwave mobile. This device has one electrode placed on the FP1 position (Figure

2) where this area of the scalp is responsible for functions, including executive functions, decision-making, personality, and social behavior. The sampling rate is 512 hz and the connection is made via Bluetooth. To request the data, the serial port of the device needs to be specified, and then collect and decode the packets. A Python script was written using a library provided by BarkleyUS [12] to connect to the device and record data.

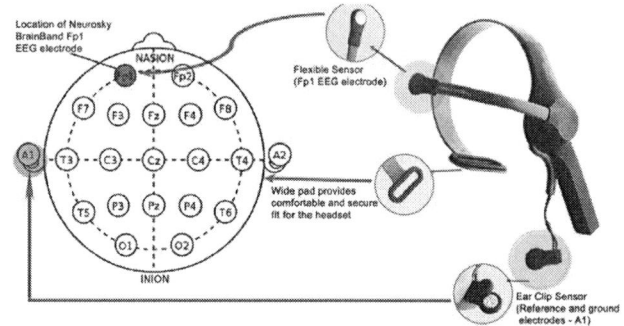

Fig. 2. Mindwave Mobile and the placement of electrode [25]

C. EEG Values

Each device records values of mV/s. Muse returns raw values, while Mindwave returns 11 values that refresh every second, and consist of brain waves, Delta (0.5 – 2.75Hz), Theta (3.5-6.75Hz), Low-alpha (7.5-9.25Hz), High-alpha (10-11.75Hz), Low-beta (13-16,75Hz), High-beta (18-29.75Hz), Low-gamma (31-39.75Hz), Mid-Gamma (41-49.75Hz). The three remaining are:

- RawValue, the value itself has no meaning but we can convert the value to mV/s by using the equation [13]:

$$\text{TrueValue} = \frac{\text{RawValue} \cdot \left(\frac{\text{InputVoltage}}{\text{ValueRange}} \right)}{\text{Gain}} \quad (1)$$

where TrueValue is measured in mV/s, InputVoltage equals to 1,8V, valueRange is 4096 and Gain is typically 2000+-5%. The TrueValue, depending on the headset, typically refreshes 128 times per second.
- Meditation, represents the calmness of mind, has range 0-100 and it refreshes once every second [14]
- Attention, represents the concentration of a player, has range 0-100 and it refreshes once every second [14]

D. Power Spectral Density

Power spectral density (PSD) calculation is employed in the processing of the signals. PSD is the distribution of power in the range of signal frequency. Fast Fourier Transform [15] was used to transform the signal to the frequency domain, then to calculate the length of the signal the equation $T = fs * recordingTime$ is utilised, with fs being the sampling frequency of the corresponding device [16].

$$S_{xx}(f) = \lim_{T \to \infty} \frac{1}{T} |\hat{x}_T(f)|^2 \quad (2)$$

E. Dataset

Two datasets of EEGs were developed, one for each device. The datasets contained measurements from ten healthy people. Each candidate's brainwaves were recorded two times, one when the person was at a relaxed state and one when the person was at a focused state. The duration of each recording is 60 seconds. An additional binary column was added to each dataset to represent the class (relaxed/focused).

F. Classification

Five classifier models are trained and tested for this application. Each model is trained and evaluated with the method 10-fold-cross-validation.

1) Random Forest: Random Forest which produces a set of trees, with random data pre-processing. To classify a prototype, the model employs a voting procedure [17].

2) Gaussian Naive Bayes: The next classifier that is used is Naive Bayes. It works by trying to maximize the $P(Y|X1, X2..., XD)$ possibility. The Gauss distribution was chosen [18].

$$P(X_i \mid Y_j) = \frac{1}{\sqrt{2\pi\sigma_{ij}^2}} e^{-\frac{(X_i - \mu_{ij})^2}{2\sigma_{ij}^2}} \qquad (3)$$

Where:

- x_i is the sample.
- σ_{ij}^2 is the variance of the sample.
- m_{ij} the mean of the sample.

3) K Nearest Neighbors: K nearest neighbors (KNN) classifies values according to the K–closer categories (neighbors) [19]. It is a simple and fast algorithm [20]. The K value needs to be chosen carefully because a high K value increases the chances of overfitting and a low K value increases the chances of underfitting.

4) Support Vector Machines: Support Vector Machines (SVM) work by calculating a hyper-plane that maximizes the distance between the training data, that way the model improves upon generalization [20]. SVM can be altered to use nonlinear decision limit by changing the kernel, but in our case, a linear decision limit was enough due to the linear dataset.

5) MultiLayer Perceptron: Multilayer Perceptron (MLP) consists of two or more layers with each layer including multiple neurons [20]. The tested model included one hidden layer with 64 neurons and ReLu as the activation function.

IV. DESIGNING THE GAME

The game is mostly composed of one scene (Figure 3). This scene refers to the main frame that the user sees in the game. In this section, the scene will be described, and an explanation of how it was created will be provided. The scene is composed of three main items: 1) The room that houses the contest, 2) The audience in the background and lastly, 3) The stage where the two athletes compete in the foreground. The development process will be described in this order. Also the logic of the game and how the game computes the score will be mentioned at the end of this section.

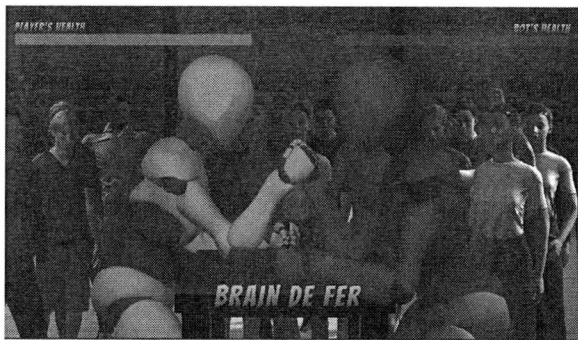

Fig. 3. Main view of the game

A. Room

The room (shown in figure 4) is designed as a half of a large rectangle with walkways in the back corners. It is developed using the "Pro Builder" extension in Unity. This extension brings some features from Blender into unity, allowing the users to make 3D objects easier. Only one half was made because the stage is positioned in the middle of the room and the camera can see only one half of the room behind the athletes and audience. There was no particular reason for selecting the textures, they just needed to fit the theme of the game and not be in contrast with the colors of the models.

Fig. 4. Design of the room

B. Audience and animations

The audience is composed of a group of pre-selected characters. At the start of the game a routine randomly selects a character and places it in a arc around the stage (figure 5). When a character is created (spawns), it chooses if it will be active or not. Being active means to randomly play an animation while the two players are fighting, also an active character needs to choose a player to be a fan of. Being a fan of the left player means that when that player is losing the character will play a losing (sad) animation, contrary, when that player is winning, it will play a winning (happy) animation, and vice versa when a character is a fan of the right player. The inactive characters only play an idle animation. Lastly for the animations, there is a pool of around 20 animations, where, while importing the characters each one was assigned 6 total animations: 3 random, an idle, a loosing and a winning.

Fig. 5. Spawning arcs for the characters

C. Athletes and stage

As mentioned before, the main focus point of the game is the two athletes fighting on a "bras-de-fer" match in the middle of the screen. The pose of the athletes and their hand movement was made using Unity's constraints and inverse kinematics system. Unity provides a way of applying certain constraint types to a set of bones (or to a single one), so that while the game is running the system will efficiently use inverse kinematics to compute how the bones and joints should be positioned and rotated. For the main function of the arms, an "Aim constraint" was applied, the constraint states that a face of the constrained bone will be pointing at a defined target. The constraint was applied to the forearm of each model and while the game is running, the target moves according to the score of the game. The target's path can be seen in figure 6.

Fig. 6. Path of the target for the constraint

D. Game logic and scoring

The score of every round depends on the attention level of each player. On each frame the game reads the attention levels from both devices, compares them and adds or subtracts a point from the score. The technique used to obtain the attention values from the devices will be described later in this section.

The game score is represented by the athlete's hands position and starts with a value of 5, if at the current frame the left player wins, a point is added to the score (increasing to 10) and if the right player wins, a points is subtracted from the score (decreasing to 0). The hand positions that represent the score of the game is shown in figure 6. Lastly the exact value of points that are being added or subtracted are variable and depend on the hardware of the system running the game. If the game is running at a high frame rate, fewer points will be added, contrary, if the game is running at a low frame rate more points will be added. This is done so the round takes the same time on every computer system.

As mentioned in section III-B the devices are connected using Bluetooth and the attention values are read using a driver written in python. The driver also acts as a bridge allowing data to be transferred from the devices (python side) to C# (Unity side).

The main working principle of the bridge is that the python side writes the attention values in a temporary file and saves it. After that the unity side can open the file whenever is needed, read those values and save them to local variables.

At the python side there is an infinite loop running, iterating every second. The main function of this loop is to request the values of attention, meditation and signal quality from both EEG devices and write them to the file. On the other side, unity opens the file on every frame, reads the values and saves them at local variables.

This working principle allow asynchronous communication between both sides but the main problem that arises is that both sides will try to open the file in the same time causing an IO exception error due to mutual exclusion of system resources. To address this problem we implemented an exception handling statement (try – catch block) to each side, so if a side tries to access the file while the other side is using it, will skip the current iteration and try again in the next one. A overview of this technique can be seen in Fig 7

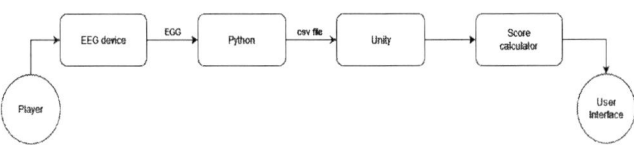

Fig. 7. Data flow from player to game

E. Final notes on game design

Lastly, the source of the assets (models, animations and textures) must be noted. The humanoid models and their animations were obtained from "Mixamo" [21] and the textures for the room were provided by "Poly haven" [22]. Blender was also utilized to create game objects. Mixamo is a platform made by Adobe, the platform provides humanoid models and a plethora of animations for them. Moreover the platform provides a tool allowing users to upload a custom humanoid character and automatically add bones and joints to it (this process is referred as "rigging") [23]. Poly Haven is a curated

979-8-3503-8368-3/23 $31.00 © 2023 IEEE

public asset library where visual effects artists and game designers can upload and download high quality assets. All assets are under CC0 license, meaning they are free for any use.

V. RESULTS

Testing started from early stages of development. Initially, only one player was able to play and the opponent was represented with a "Bot" function. The Bot returned a random value, using a normal distribution in a predefined range:

- Easy (30:70)
- Medium (40:80)
- Hard (70:100)

Initial testing focused on gathering player feedback and opinions about the game and the way it is played. After the development of the athletes model and their movement, players showed high levels of immersion on the game. The game was also showcased at "Thessaloniki International Fair" (TIF) at the period 9 to 17 of September 2023. At this fair, more than 50 people played the game and positive feedback was returned along with some suggestions on improvements. A Picture of two players playing the game is shown in figure 8.

Fig. 8. Showcase of the game at the fair

VI. RESULTS AND FEATURE IMPROVEMENTS

The results are split in two parts the first part being the results of the Muse device both on the training of the models and on the Signal processing, while the second part presents the comparison between the Mindwave's device value and the model's value.

A. Muse results

1) Machine learning: Initially, the game is tested employing the Muse device:

The results from table I show that all classification schemes resulted in moderate values, mainly due to the noise in the recordings. During the creation of the datasets it was noticed that a slight movement of the person when being measured, produced a lot of noise.

TABLE I
TRAINING SCORES OF THE MODELS USING THE MUSE DATASET

Models	Scores
Random Forest	67.355
Naive Bayes	58.55
KNN	66.08
SVM	67.007
MLP	66.638

2) Signal Processing: Four filters were applied for the signal processing of the EEG data to capture each brain wave independently. A bandpass filter in the range of 8-50 HZ, a lowpass filter with a top of 12 Hz to capture the Alpha waves, a bandpass filter in the range of 12-30 Hz to capture Beta waves and lastly a bandpass filter in the range of 30-50 Hz to capture the Gamma waves.

After capturing the rhythms, the mean of each one is calculated on the four channels. It was noticed that certain frequencies had extremely high power compared to the rest and thus a filter was used to remove outlier values. A comparison of the data before and after applying the filter is show below (Fig 9).

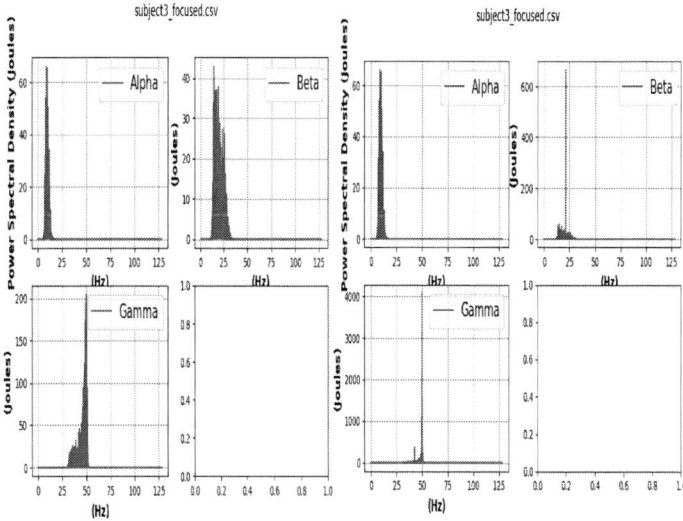

Fig. 9. The three rhythms before and after the application of RemoveOddValues filter

The obtained results are presented in fig 10.

B. Mindwave

1) Mindwave Concentration value: The Mindwave Device provides the percentage of concentration; to evaluate this, ten candidates have been measured. To compare the results each measurement has been plotted and the mean for both the focused and relaxed states have been calculated.

In table II 7/10 candidates have higher average focused value than relaxed, although in cases like 3 and 7 the candidate couldn't focus at all but still maintained higher focused value. In Fig 11 the difference between the focused and relaxed state can be clearly seen.

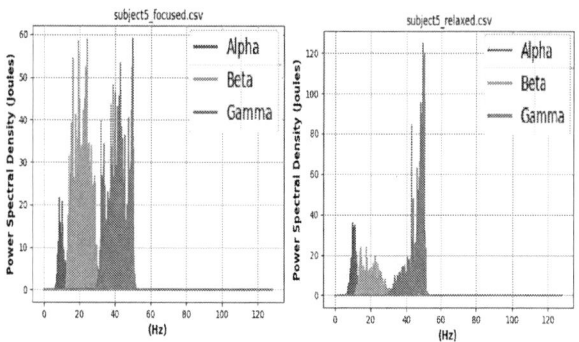

Fig. 10. Noisy measurements

TABLE II
THE MEAN VALUES OF FOCUSED AND RELAXED STATES

Candidates	Focused mean (Higher is better)	Relaxed (Lower is better)
1	66.999	29.327
2	53.249	65.383
3	29.607	27.591
4	57.021	37.691
5	36.921	55.862
6	62.540	56.754
7	38.240	32.894
8	74.609	33.159
9	59.173	47.904
10	39.431	57.988

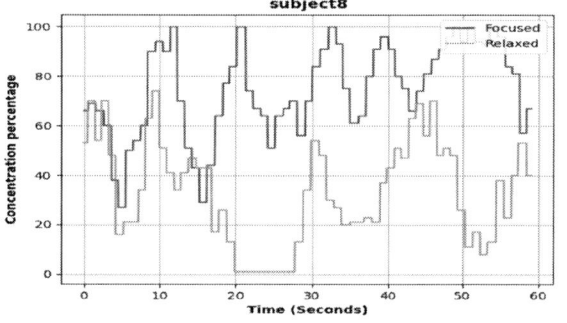

Fig. 11. Comparison between the focused and relaxed states

```
attention   :  0.004497128412653082
meditation  :  -0.00011742832049805224
low-alpha   :  -8.604461532120601e-07
high alpha  :  -4.903967056581117e-06
low-beta    :  -1.4429461826128476e-06
high-beta   :  -1.4850356770738874e-06
low gamma   :  3.035463354581242e-06
mid-gamma   :  6.29582728853754e-07
raw_value   :  -3.0686080833779785e-05
delta       :  6.558598542703181e-08
theta       :  9.274179103196852e-07
```

Fig. 12. Coefficients of the Mindwave dataset

TABLE III
TRAINING SCORES OF THE MODELS USING THE MINDWAVE DATASET

Models	Scores
Random Forest	89.899
Naive Bayes	59.539
KNN	72.614
SVM	66.643
MLP	66.239

TABLE IV
COMPARISON OF MEAN VALUES BETWEEN MINDWAVE AND OUR METHOD

Method	Mean Concentration percentage	Mean Relaxation
Headset	58.747	24.095
RandomForest	63.726	12.081

2) Model Value: To train the models, a linear regression model is used to determine the coefficients between each attribute Fig 12.

The Random Forest model presents the better performance as seen in table III. To evaluate the proposed method, using the RF classifier, measurements have been taken from candidates being concentrated and relaxed. Each measurement lasted two minutes, and the results are presented in Fig 13 vs the Mindwave output value.

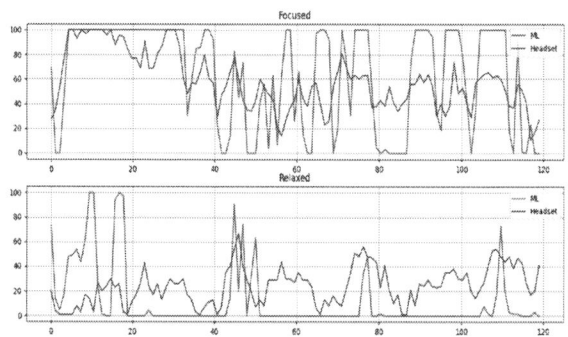

Fig. 13. Comparison between the proposed method and Mindwave output value

VII. DISCUSSION

As seen in Figure 13 our model produces more spikes than the smoother value of Mindwave . The comparison between the mean values proves a clear difference between the relaxation and concentration states. Thus, the proposed method can discriminate between concentration and relaxation, in a simple and robust way.

REFERENCES

[1] A. T. Tzallas, M. G. Tsipouras and D. I. Fotiadis, "Epileptic Seizure Detection in EEGs Using Time–Frequency Analysis," in IEEE Transactions on Information Technology in Biomedicine, vol. 13, no. 5, pp. 703-710, Sept. 2009, doi: 10.1109/TITB.2009.2017939.

[2] A. Palumbo, V. Gramigna, B. Calabrese, and N. Ielpo, "Motor-Imagery EEG-Based BCIs in Wheelchair Movement and Control: A Systematic Literature Review," Sensors, vol. 21, no. 18, p. 6285, Sep. 2021, doi: 10.3390/s21186285.

[3] Shih JJ, Krusienski DJ, Wolpaw JR. Brain-computer interfaces in medicine. Mayo Clin Proc. 2012 Mar;87(3):268-79. doi: 10.1016/j.mayocp.2011.12.008. Epub 2012 Feb 10. PMID: 22325364; PMCID: PMC3497935.

[4] K. Glavas, G. Prapas, K. D. Tzimourta, N. Giannakeas, and M. G. Tsipouras, "Evaluation of the User Adaptation in a BCI Game Environment," Applied Sciences, vol. 12, no. 24, p. 12722, Dec. 2022, doi: 10.3390/app122412722.

[5] "Muse S (Gen 2) — MuseTM EEG-Powered Meditation & Sleep Headband." Accessed: Oct. 07, 2023. [Online]. Available: https://choosemuse.com/products/muse-s-gen-2

[6] "MindWave." Accessed: Oct. 07, 2023. [Online]. Available: https://store.neurosky.com/pages/mindwave

[7] J. Cheng, G. Mabasa and C. Oppus, "Prolonged distraction testing game implemented with ImpactJS HTML5, Gamepad and Neurosky," 2014 International Conference on Humanoid, Nanotechnology, Information Technology, Communication and Control, Environment and Management (HNICEM), Palawan, Philippines, 2014, pp. 1-6, doi: 10.1109/HNICEM.2014.7016184.

[8] Paul Coulton, Carlos Garcia Wylie, and Will Bamford. 2011. Brain interaction for mobile games. In Proceedings of the 15th International Academic MindTrek Conference: Envisioning Future Media Environments (MindTrek '11). Association for Computing Machinery, New York, NY, USA, 37–44. https://doi.org/10.1145/2181037.2181045

[9] P. Wang, Y. Yang and J. Li, "Development of Parkour Game System Using EEG Control," 2018 International Symposium on Computer, Consumer and Control (IS3C), Taichung, Taiwan, 2018, pp. 258-261, doi: 10.1109/IS3C.2018.00072.

[10] J. Kowaleski, "BlueMuse." Jun. 26, 2023. Accessed: Jun. 27, 2023. [Online]. Available: https://github.com/kowalej/BlueMuse

[11] "Lab Streaming Layer." Accessed: Jun. 27, 2023. [Online]. Available: https://labstreaminglayer.org/

[12] "mindwave-python." Barkley, May 17, 2023. Accessed: Jun. 29, 2023. [Online]. Available: https://github.com/BarkleyUS/mindwave-python

[13] "How to convert raw values to voltage? / Science / Knowledge Base - NeuroSky - Home Page Support." Accessed: Jun. 30, 2023. [Online]. Available: http://support.neurosky.com/kb/science/how-to-convert-raw-values-to-voltage

[14] "thinkgear_communications_protocol [NeuroSky Developer - Docs]." Accessed: Jun. 29, 2023. [Online]. Available: https://developer.neurosky.com/docs/doku.php?id=thinkgear_communications_protocol

[15] Heckbert, Paul. Fourier transforms and the fast Fourier transform (FFT) algorithm. Computer Graphics, 1995, 2.1995: 15-463.

[16] Youngworth, Richard N.; GALLAGHER, Benjamin B.; STAMPER, Brian L. An overview of power spectral density (PSD) calculations. Optical manufacturing and testing VI, 2005, 5869: 206-216.

[17] D. R. Edla, K. Mangalorekar, G. Dhavalikar, and S. Dodia, "Classification of EEG data for human mental state analysis using Random Forest Classifier," Procedia Comput. Sci., vol. 132, pp. 1523–1532, 2018, doi: 10.1016/j.procs.2018.05.116.

[18] H. Kamel, D. Abdulah and J. M. Al-Tuwaijari, "Cancer Classification Using Gaussian Naive Bayes Algorithm," 2019 International Engineering Conference (IEC), Erbil, Iraq, 2019, pp. 165-170, doi: 10.1109/IEC47844.2019.8950650.

[19] F. Lotte, M. Congedo, A. Lécuyer, F. Lamarche, and B. Arnaldi, "A review of classification algorithms for EEG-based brain–computer interfaces," J. Neural Eng., vol. 4, no. 2, pp. R1–R13, Jun. 2007, doi: 10.1088/1741-2560/4/2/R01.

[20] J. H. Friedman, "On Bias, Variance, 0/1—Loss, and the Curse-of-Dimensionality," Data Min. Knowl. Discov., vol. 1, no. 1, pp. 55–77, Mar. 1997, doi: 10.1023/A:1009778005914.

[21] Adobe TM, Mixamo, Accessed: Oct. 7, 2023. [Online]. Available: https://www.mixamo.com/#/ .

[22] Poly Haven TM, Poly Haven, Accessed: Oct. 7, 2023. [Online]. Available: https://polyhaven.com/

[23] Blackman, S. (2014). Rigging with Mixamo. In: Unity for Absolute Beginners. Apress, Berkeley, CA. https://doi.org/10.1007/978-1-4302-6778-2_12

[24] "Fig. 2. Comparison of electrode placement between MUSE and the...," ResearchGate. Accessed: Jul. 04, 2023. [Online]. Available: https://www.researchgate.net/figure/Comparison-of-electrode-placement-between-MUSE-and-the-international-10-20-system-a_fig2_352019337

[25] "Neurosky home use devices for neurofeedback training/therapy," BIOFEEDBACK NEUROFEEDBACK THERAPY. Accessed: Sep. 27, 2023. [Online]. Available: https://biofeedback-neurofeedback-therapy.com/neurosky/

Supervisors for Gas Compressor Stations with Compression and Valve Faults

Dimitrios G. Fragkoulis
Robotics, Automatic Control and
Cyber-Physical Systems
Laboratory,
Department of Digital Industry
Technologies,
School of Science,
National and Kapodistrian
University of Athens,
Euripus Campus, 34400, Greece
dfragkoulis@dind.uoa.gr

Fotis N. Koumboulis
Robotics, Automatic Control and
Cyber-Physical Systems
Laboratory,
Department of Digital Industry
Technologies,
School of Science,
National and Kapodistrian
University of Athens,
Euripus Campus, 34400, Greece
fkoumboulis@dind.uoa.gr

Nikolas D. Kouvakas
Robotics, Automatic Control and
Cyber-Physical Systems
Laboratory,
Department of Digital Industry
Technologies,
School of Science,
National and Kapodistrian
University of Athens,
Euripus Campus, 34400, Greece
nkouvak@dind.uoa.gr

Antonis N. Menexis
Robotics, Automatic Control and
Cyber-Physical Systems
Laboratory,
Department of Digital Industry
Technologies,
School of Science,
National and Kapodistrian
University of Athens,
Euripus Campus, 34400, Greece
amenexis@dind.uoa.gr

Abstract—**A Discrete Event System (DES) based approach is developed for the derivation of the model and safe performance of a gas compressor station (GCS), in the presence of all possible actuator faults. The safe performance goal is surge avoidance, despite the presence of actuator faults, and is modelled as four rules, being translated to four regular languages. The respective supervisor automata are developed to realize the four regular languages. The physical realizability and nonblocking of the synchronous product of the supervisor automata and the automaton of the system is proved. Implementation of the supervisors in the Structured Text language for PLCs is provided.**

Keywords— Discrete Event Systems, Gas compression, Safe performance, Supervisory Control

I. INTRODUCTION

GCSs are used to protect gas networks from pressure drops due to consumption ([1,2]). Clearly, the reliability of the total gas network depends on the performance of these stations [3]. The safety of the station depends, among other factors, by the avoidance of the phenomenon of surge. Therefore, the protection of the system is guaranteed via a recycle valve that maintains the required gas flow [1]. Many methods have been used to address the surge phenomenon alongside with the recycle valve. Model based methods are met quite often in the literature ([4]-[6]). Discrete Event Systems (DES) and Supervisor Control Theory (SCT) are used in gas storage ([7]) and in gas networks ([1], [7] and [8]) for system modelling and supervisor design. The study of gas networks in the presence of actuator faults is essential for the efficient control of the system. In [8], the problem of avoiding the surge phenomenon, in the presence of compressor faults has been studied. The issue of supervisor design in the presence of compression and valve faults has not yet been studied.

In the present paper, the surge avoidance problem for a GCS, in the presence of compression and valve faults, will be studied. The DES model of the station has been presented in [1], without considering faults, and in [8] for the case where compression faults are considered. In [1], a monolithic approach for the surge avoidance has been studied. In [8], the surge avoidance problem

is studied using a modular supervisor architecture, taking into account faults of one device. In comparison to [8], the present paper provides a new faulty model, where the possibility of actuator faults is considered for both actuators of the system. Regarding the rule describing the desired performance of the system, it is important to mention that, in comparison to the monolithic rule of [1] and the modular set of rules of [8], the present paper introduces two extra rules to guarantee the desired performance in the presence of all actuator faults. Hence, it contributes towards the actuator fault handling in GCSs.

Here, the model of the system will be developed by giving the analytic DES models of the station subsystems (compressor, recycle valve, surge sensor) in the presence of all actuator faults. This is the 1st contribution of the paper. The desired behavior of the system is expressed by four rules. The first two rules come from the monolithic specification proposed in [1]. The last two rules cover the case of all actuator faults. This is the 2nd contribution. It is important to mention that in [8] only the compressor fault is considered to take place in specific states. On the other hand, in the present paper, the compressor faults and the recycle valve faults are considered to take place in every state of the system. The desired rules are formulated by regular languages in in analytic forms. The four languages are realized by several supervisor automata. The supervisors are implemented using Structured Text (ST) language for PLCs, being widely used in controlling of recycle valves [2]. Finally, the nonblocking property of the controlled system (regarding nonblocking see [9,10]) and the controllability property regarding the total system is proved, via the physical realizability (PR) of the synchronous product of the supervisors and the system. Regarding controllability and PR, see [11]. This the last contribution of the paper.

In Section II, taking into account possible faults in the two actuators of the GCS, a DES model is developed. In Section III, the desired behavior is formed in the form of rules and then translated to regular languages. Next, in Section IV, the supervisor realizing each language is developed. In Section V, the resulting controlled automaton, the PR property, and the nonblocking property are proved. Finally, in Section VI, the issue of the implementation of the supervisors is investigated, using ST language.

II. THE DES MODEL WITH ACTUATOR FAULTS

A. Description of the GCS

GCS consists of three distinct parts, the compressor, the recycle valve, and the surge sensor. Surge is an "undesired state" provoked by low gas supply and providing reverse gas flow. As already mentioned, the control objective of the controlled system is surge avoidance. DES models of the subsystems were given in [9,10]. Furthermore, the models in the presence of faults are presented, considering that Fault Detection mechanisms are installed to provide the necessary information regarding the faults.

B. The automaton of the compressor

Following [1] and [8], Figure 1 depicts the compressor state diagram.

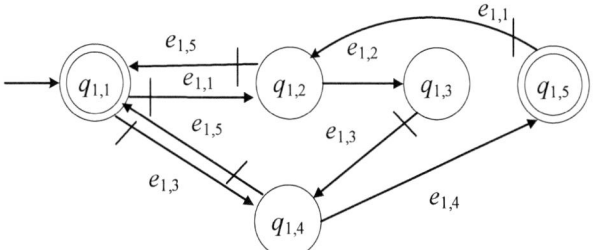

Fig. 1. Compressor state diagram

The model of the compressor is the 6-tuple (more about 6-tuple automata, see [11]-[14])
$$\mathbf{G}_1 = (\mathbf{Q}_1, \mathbf{E}_1, f_1, \mathbf{H}_1, x_{1,0}, \mathbf{Q}_{1,m}).$$
$\mathbf{Q}_1 = \{q_{1,1}, q_{1,2}, q_{1,3}, q_{1,4}, q_{1,5}\}$ is the state set. The description of the states is in Table I.

TABLE I.   COMPRESSOR STATES

Symbol	Description
$q_{1,1}$	Normal mode.
$q_{1,2}$	Increasing speed.
$q_{1,3}$	Maximum value of speed
$q_{1,4}$	Decreasing speed.
$q_{1,5}$	Minimum value of speed

$\mathbf{E}_1 = \{e_{1,5}, e_{1,4}, e_{1,3}, e_{1,2}, e_{1,1}\}$ is the alphabet. $\mathbf{E}_{1,c} = \{e_{1,5}, e_{1,3}, e_{1,1}\}$ is the controllable event set. $\mathbf{E}_{1,uc} = \{e_{1,4}, e_{1,2}\}$ is the uncontrollable events set. The description of the events is in Table II.

TABLE II.   COMPRESSOR EVENTS

Symbol	Description
$e_{1,1}$	Increase speed command.
$e_{1,2}$	Speed caught upper limit
$e_{1,3}$	Decrease speed command.
$e_{1,4}$	Speed caught lower limit.
$e_{1,5}$	Transition to normal mode command.

$x_{1,0} = q_{1,1}$ is the initial state and $\mathbf{Q}_{1,m} = \{q_{1,1}, q_{1,5}\}$ is the marked states set ([9,10]). $\mathbf{L}(\mathbf{G}_1)$ is the closed behavior of the automaton and $\mathbf{L}_m(\mathbf{G}_1)$ is the marked behavior. Regarding the closed and the marked behavior, see [9,10]). It holds that $\overline{\mathbf{L}_m(\mathbf{G}_1)} = \mathbf{L}(\mathbf{G}_1)$, thus $\mathbf{G}_1$ is a nonblocking automaton.

C. The automaton of the recycle valve

Following [1] and [8], the valve state diagram is presented in Figure 2.

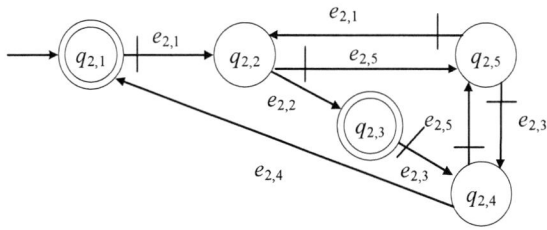

Fig. 2. Recycle valve state diagram

The model of the valve is
$$\mathbf{G}_2 = (\mathbf{Q}_2, \mathbf{E}_2, f_2, \mathbf{H}_2, x_{2,0}, \mathbf{Q}_{2,m}).$$
$\mathbf{Q}_2 = \{q_{2,1}, q_{2,2}, q_{2,3}, q_{2,4}, q_{2,5}\}$ is the state set. The description of the states is in Table III.

TABLE III.   RECYCLE VALVE STATES

Symbol	Description
$q_{2,1}$	Fully closed mode.
$q_{2,2}$	Opening mode
$q_{2,3}$	Fully opened mode
$q_{2,4}$	Closing mode.
$q_{2,5}$	Stopped mode (stuck between the fully opened and closed mode)

$\mathbf{E}_2 = \{e_{2,5}, e_{2,4}, e_{2,3}, e_{2,2}, e_{2,1}\}$ is the alphabet. $\mathbf{E}_{2,c} = \{e_{2,5}, e_{2,3}, e_{2,1}\}$ is the controllable event set. $\mathbf{E}_{2,uc} = \{e_{2,4}, e_{2,2}\}$ is the uncontrollable event set. The description of the events is in Table IV.

979-8-3503-8368-3/23 $31.00 © 2023 IEEE

TABLE IV. RECYCLE VALVE EVENTS

Symbol	Description
$e_{2,1}$	Open command
$e_{2,2}$	Indication of fully opened valve.
$e_{2,3}$	Close command.
$e_{2,4}$	Indication of fully closed valve.
$e_{2,5}$	Stop command.

$x_{2,0} = q_{2,1}$ and $\mathbf{Q}_{2,m} = \{q_{2,1}, q_{2,3}\}$. It holds that $\overline{\mathbf{L}_m(\mathbf{G}_2)} = \mathbf{L}(\mathbf{G}_2)$. Thus, $\mathbf{G}_2$ is a nonblocking automaton.

D. The automaton of the surge sensor

Following [1] and [8], sensors state diagram is depicted in Figure 3.

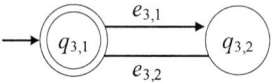

Fig. 3. Surge sensor state diagram

The surge sensor model is
$$\mathbf{G}_3 = (\mathbf{Q}_3, \mathbf{E}_3, f_3, \mathbf{H}_3, x_{3,0}, \mathbf{Q}_{3,m}).$$
$\mathbf{Q}_3 = \{q_{3,1}, q_{3,2}\}$ is the state set. The description of the states is in Table V.

TABLE V. SURGE SENSOR STATES

Symbol	Description
$q_{3,1}$	No surge.
$q_{3,2}$	Surge

$\mathbf{E}_3 = \{e_{3,1}, e_{3,2}\}$ is the alphabet. $\mathbf{E}_{3,c} = \varnothing$ is the controllable event set. $\mathbf{E}_{3,uc} = \{e_{3,1}, e_{3,2}\}$ is the uncontrollable event set. The description of the events is in Table VI.

TABLE VI. SURGE SENSOR EVENTS

Symbol	Description
$e_{2,1}$	Surge detection.
$e_{2,2}$	Surge expiration.

$x_{3,0} = q_{3,1}$ and $\mathbf{Q}_{3,m} = \{q_{3,1}\}$. It holds that $\mathbf{L}(\mathbf{G}_3) = \overline{(e_{3,1}e_{3,2})^*}$. Thus, $\mathbf{G}_3$ is a nonblocking automaton as.

E. The faulty models

In [1] and [7], the recycle valve have been given, with faults in the opening and closing process, is modelled. Next, a faulty model for both actuators of the system (compressor and recycle valve) will be considering that a fault may occur in any state of the system, even in the idle mode of the actuator. A Fault Detection (FD) mechanism is considered to the system's, using a sensor, indicating the presence of a possible fault (e.g., sensors providing info regarding the behavior of the mechanical systems). Based on the faulty model developed in [11], the model of the fault is the 6-tuple form is

$$^{n}\mathbf{G}_F = (^{n}\mathbf{Q}_F, {}^{n}\mathbf{E}_F, {}^{n}f_F, {}^{n}\mathbf{H}_F, {}^{n}x_{F,0}, {}^{n}\mathbf{Q}_{F,m}); n = \{1, 2\}.$$

where for $n = 1$ the origin of the fault is the compressor and for $n = 2$ the origin of the fault is the valve. The set of states is $^{n}\mathbf{Q}_F = \{^{n}q_{F,1}, {}^{n}q_{F,2}\}$. In Table VII, the description of the states of the automaton of the fault is given.

TABLE VII. STATES OF THE AUTOMATON OF THE FAULT

Symbol	Description
$^{n}q_{F,1}$	No fault has been detected.
$^{n}q_{F,2}$	A fault has been detected.

$^{n}\mathbf{E}_F = \{^{n}e_{F,1}, {}^{n}e_{F,2}\}$ is the alphabet. $^{n}\mathbf{E}_{F,c} = \varnothing$ is the controllable event set and $^{n}\mathbf{E}_F = \{^{n}e_{F,1}, {}^{n}e_{F,2}\}$ is the uncontrollable event set. The description of the events is in Table VIII.

TABLE VIII. EVENTS OF THE AUTOMATON OF THE FAULT

Symbol	Description
$^{n}e_{F,1}$	A fault takes place.
$^{n}e_{F,2}$	The repair of the fault took place.

The marked behavior is $\mathbf{L}_m(^{n}\mathbf{G}_F) = \left((^{n}e_1)(^{n}e_2)\right)^*$. The closed behavior is $\mathbf{L}(^{n}\mathbf{G}_F) = \overline{\mathbf{L}_m(^{n}\mathbf{G}_F)}$.

In the presence of faults, the model of each subsystem is described by the synchronous product of the automaton of the model of faults and the automaton of the actuator. Regarding the synchronous product of two or more automata, see [9]. The automaton of the compressor with faults is
$$\mathbf{G}_{F,1} = \mathbf{G}_1 \parallel {}^{1}\mathbf{G}_F .$$
The set of the states is $\mathbf{Q}_{F,1} = \mathbf{Q}_1 \times {}^{1}\mathbf{Q}_F$ and the alphabet is $\mathbf{E}_{F,1} = \mathbf{E}_1 \cup {}^{1}\mathbf{E}_F$. Considering the faults, the recycle valve is
$$\mathbf{G}_{F,2} = \mathbf{G}_2 \parallel {}^{2}\mathbf{G}_F .$$
$\mathbf{Q}_{F,2} = \mathbf{Q}_2 \times {}^{2}\mathbf{Q}_F$ is the state set. The alphabet is $\mathbf{E}_{F,2} = \mathbf{E}_2 \cup {}^{2}\mathbf{E}_F$.

F. The automaton with actuators faults

The two actuators and the sensor have disjoint alphabets, i.e., there is no common event among the alphabets. $\mathbf{G}$ is the total automaton, being the following shuffle (see [9])
$$\mathbf{G} = \mathbf{G}_{F,1} \parallel \mathbf{G}_{F,2} \parallel \mathbf{G}_3 .$$
The state set of $\mathbf{G}$ is
$$\mathbf{Q} = \mathbf{Q}_{F,1} \times \mathbf{Q}_{F,2} \times \mathbf{Q}_3 = \mathbf{Q}_1 \times {}^{1}\mathbf{Q}_F \times \mathbf{Q}_2 \times {}^{2}\mathbf{Q}_F \times \mathbf{Q}_3 = .$$
$$= \bigcup_{i=1}^{5} \left(\bigcup_{j=1}^{7} \left(\bigcup_{k=1}^{2} \{(q_{1,i}, q_{F,2,j}, q_{3,k})\} \right) \right).$$
$|\mathbf{Q}| = 200$ and $x_0 = (q_{1,1}, {}^{1}q_{F,1}, q_{2,1}, {}^{2}q_{F,1}, q_{3,1})$ is the initial state. $\mathbf{Q}_{F,m} = \{(q_{1,1}, {}^{1}q_{F,1}, q_{2,1}, {}^{2}q_{F,1}, q_{3,1})\}$ is the marked state set. The alphabet of the automaton $\mathbf{G}$ is $\mathbf{E} = \mathbf{E}_1 \cup {}^{1}\mathbf{E}_F \cup \mathbf{E}_2 \cup {}^{2}\mathbf{E}_F \cup \mathbf{E}$. The controllable and the

uncontrollable alphabets are $\mathbf{E}_{F,c} = \mathbf{E}_{1,c} \cup \mathbf{E}_{2,c}$ and $\mathbf{E}_{F,uc} = \mathbf{E}_{1,uc} \cup \mathbf{E}_{2,uc} \cup \mathbf{E}_{F,1} \cup \mathbf{E}_{F,2} \cup \mathbf{E}_3$, respectively.

$$\mathbf{L}(\mathbf{G}) = P_{F,1}^{-1}\big(\mathbf{L}(\mathbf{G}_{F,1})\big) \cap P_{F,2}^{-1}\big(\mathbf{L}(\mathbf{G}_{F,2})\big) \cap P_3^{-1}\big(\mathbf{L}(\mathbf{G}_3)\big),$$

where $P_{F,1}$, P_{F2}, and P_3 are the projections of $\mathbf{E}^*$ to $\mathbf{E}_{F,1}^*$, $\mathbf{E}_{F,2}^*$, and $\mathbf{E}_3^*$, respectively. The definition of the projection can be found in [7] and [8]. Finally, it holds that

$$\mathbf{L}_m(\mathbf{G}) = P_{F,1}^{-1}\big(\mathbf{L}_m(\mathbf{G}_{F,1})\big) \cap P_{F,2}^{-1}\big(\mathbf{L}_m(\mathbf{G}_{F,2})\big) \cap P_3^{-1}\big(\mathbf{L}_m(\mathbf{G}_3)\big).$$

III. SAFE PERFORMANCE

The safe performance specs of the system are four rules:
1. a) In surge mode, the speed of the compressor must not be increased.
    b) In no surge mode, the recycle valve may close.
2. a) In surge mode, the valve must not close.
    b) The valve may open only in the low-speed compressor mode.
3. In case of a valve fault, the speed of the compressor must not change (decrease or increase)
4. In case of a fault in the compressor, the mode of the recycle valve must not change (close or open).

Rules 1 and 2, that guarantee surge avoidance, have already been presented in [1] as a monolithic rule and in [8] in the current form. Rules 3 and 4 are introduced here to handle actuator faults. Next, the four rules will be translated to a set of four regular languages.

The desired language for the 1st rule is:

$$\mathbf{K}_{D,1} = P_{S,1}^{-1}\big(\overline{\mathbf{K}_1}\big) \cap \mathbf{L}_m(\mathbf{G}) = P_{S,1}^{-1}\big(\mathbf{K}_1\big) \cap \mathbf{L}_m(\mathbf{G}), \quad (1)$$

where

$$\mathbf{K}_1 = \overline{\big((e_{3,2} + e_{2,3} + e_{1,1})^* e_{3,1} e_{3,1}^* e_{3,2}\big)^*},$$

$\mathbf{E}_{S,1} = \{e_{3,2}, e_{3,1}, e_{2,3}, e_{1,1},\}$ and the projection of $\mathbf{E}^*$ to $\mathbf{E}_{S,1}^*$ is denoted as $P_{S,1}$.

The desired regular language for the 2nd rule is:

$$\mathbf{K}_{D,2} = P_{S,2}^{-1}\big(\overline{\mathbf{K}_2}\big) \cap \mathbf{L}_m(\mathbf{G}) = P_{S,2}^{-1}\big(\mathbf{K}_2\big) \cap \mathbf{L}_m(\mathbf{G}), \quad (2)$$

where

$$\mathbf{K}_2 = \overline{\big((e_{3,2} + e_{2,3} + e_{1,4})^* e_{3,1} e_{3,1}^* \big(\varepsilon + e_{1,4}(e_{1,4} + e_{3,1})^*\big) e_{3,2}\big)^*},$$

$\mathbf{E}_{S,2} = \{e_{3,2}, e_{3,1}, e_{1,4}, e_{2,3}\}$ and the projection of $\mathbf{E}^*$ to $\mathbf{E}_{S,2}^*$ is denoted as $P_{S,2}$.

The desired regular language for the 3rd rule is:

$$\mathbf{K}_{D,3} = P_{S,3}^{-1}\big(\overline{\mathbf{K}_3}\big) \cap \mathbf{L}_m(\mathbf{G}) = P_{S,3}^{-1}\big(\mathbf{K}_3\big) \cap \mathbf{L}_m(\mathbf{G}), \quad (3)$$

where

$$\mathbf{K}_3 = \overline{\big((e_{1,1} + e_{1,3} + {}^2e_{F,2})^* {}^2e_{F,1} {}^2e_{F,1}^* {}^2e_{F,2}\big)^*},$$

$\mathbf{E}_{S,3} = \{e_{1,1}, e_{1,3}, {}^2e_{F,1}, {}^2e_{F,2}\}$ and the projection of $\mathbf{E}^*$ to $\mathbf{E}_{S,3}^*$ is denoted as $P_{S,3}$.

The desired regular language for the 4th rule is:

$$\mathbf{K}_{D,4} = P_{S,4}^{-1}\big(\overline{\mathbf{K}_4}\big) \cap \mathbf{L}_m(\mathbf{G}) = P_{S,4}^{-1}\big(\mathbf{K}_4\big) \cap \mathbf{L}_m(\mathbf{G}), \quad (4)$$

where

$$\mathbf{K}_4 = \overline{\big((e_{2,1} + e_{2,3} + {}^1e_{F,2})^* {}^1e_{F,1} {}^1e_{F,1}^* {}^1e_{F,2}\big)^*},$$

$\mathbf{E}_{S,4} = \{e_{2,1}, e_{2,3}, {}^1e_{F,1}, {}^1e_{F,2}\}$ the projection of $\mathbf{E}^*$ to $\mathbf{E}_{S,4}^*$ is denoted as $P_{S,4}$.

According to [16,17], the desired behavior described by the rules 1-4 will be imposed to $\mathbf{G}$ by a supervisor-based control scheme. The design of the supervisors will be developed using $\mathbf{K}_1$, $\mathbf{K}_2$, $\mathbf{K}_3$ and $\mathbf{K}_4$, being simpler than the respective languages $\mathbf{K}_{D,1}$, $\mathbf{K}_{D,2}$, $\mathbf{K}_{D,3}$ and $\mathbf{K}_{D,4}$.

IV. SUPERVISORS

A. 1st supervisor

Let

$$\mathbf{S}_1 = (\mathbf{Q}_{S,1}, \mathbf{E}_{S,1}, f_{S,1}, \mathbf{H}_{S,1}, x_{S,1,0}, \mathbf{Q}_{S,1}).$$

$\mathbf{Q}_{S,1} = \{q_{S,1,1}, q_{S,1,2}\}$ is set the states of $\mathbf{S}_1$. $\mathbf{E}_{S,1}$ is the alphabet of $\mathbf{S}_1$. $x_{S,1,0} = q_{S,1,1}$ is the initial state, and $\mathbf{L}_m(\mathbf{S}_1) = \mathbf{K}_1$. The complexity of $\mathbf{S}_1$ is $(2,4,6)$ (regarding the complexity of an automaton see [15]). Figure 4 depicts the state diagram of $\mathbf{S}_1$.

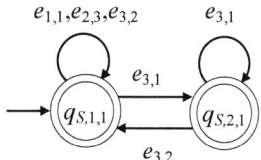

Fig. 4. $\mathbf{S}_1$

B. 2nd supervisor

Let

$$\mathbf{S}_2 = (\mathbf{Q}_{S,2}, \mathbf{E}_{S,2}, f_{S,2}, \mathbf{H}_{S,2}, x_{S,2,0}, \mathbf{Q}_{S,2}).$$

$\mathbf{Q}_{S,2} = \{q_{S,2,1}, q_{S,2,2}, q_{S,2,3}\}$ is set the states of $\mathbf{S}_2$. $\mathbf{E}_{S,2}$ is the alphabet of $\mathbf{S}_2$. $x_{S,2,0} = q_{S,2,1}$ is the initial state, and $\mathbf{L}_m(\mathbf{S}_2) = \mathbf{K}_2$. The complexity of $\mathbf{S}_2$ is $(3,4,10)$. Figure 5 depicts its state diagram.

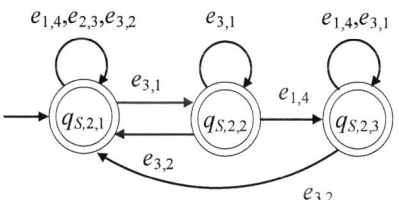

Fig. 5. $\mathbf{S}_2$

C. 3rd supervisor

Let

$$\mathbf{S}_3 = (\mathbf{Q}_{S,3}, \mathbf{E}_{S,3}, f_{S,3}, \mathbf{H}_{S,3}, x_{S,3,0}, \mathbf{Q}_{S,3}).$$

$\mathbf{Q}_{S,3} = \{q_{S,3,1}, q_{S,3,2}\}$ is the state set. $\mathbf{E}_{S,3}$ is the event set. $x_{S,3,0} = q_{S,3,1}$ is the corresponding initial state, and $\mathbf{L}_m(\mathbf{S}_3) = \mathbf{K}_3$. The complexity of $\mathbf{S}_3$ is $(2,4,6)$. Figure 6 depicts its state diagram.

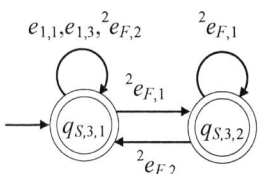

Fig. 6. $\mathbf{S}_3$

D. 4th supervisor

Let

$$\mathbf{S}_4 = (\mathbf{Q}_{S,4}, \mathbf{E}_{S,4}, f_{S,4}, \mathbf{H}_{S,4}, x_{S,4,0}, \mathbf{Q}_{S,4}).$$

$\mathbf{Q}_{S,4} = \{q_{S,4,1}, q_{S,4,2}\}$ is set the states of $\mathbf{S}_4$. $\mathbf{E}_{S,4}$ is the alphabet of $\mathbf{S}_4$. $x_{S,4,0} = q_{S,4,1}$ is the corresponding initial state, and $\mathbf{L}_m(\mathbf{S}_4) = \mathbf{K}_4$. The complexity triad of $\mathbf{S}_4$ is $(2,4,6)$ and Figure 7 depicts its state diagram.

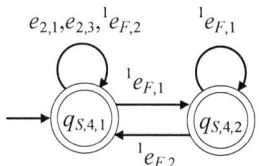

Fig. 7. $\mathbf{S}_4$

V. THE CONTROLLED GCS

The controlled automaton of the gas station is of the form

$$\mathbf{G}_c = \mathbf{S}_1 \parallel \mathbf{S}_2 \parallel \mathbf{S}_3 \parallel \mathbf{S}_4 \parallel \mathbf{G} \quad (5)$$

The total complexity of the proposed supervisory scheme is $(9,16,28)$ which is quite near to the complexity of the system being $(12,12,18)$.

The closed behavior of $\mathbf{G}_c$ is

$$\mathbf{L}(\mathbf{G}_c) = \mathbf{L}(\mathbf{G}) \cap P_{S,1}^{-1}(\overline{\mathbf{K}_1}) \cap P_{S,2}^{-1}(\overline{\mathbf{K}_2}) \cap P_{S,3}^{-1}(\overline{\mathbf{K}_3}) \cap P_{S,4}^{-1}(\overline{\mathbf{K}_4}) \quad (6)$$

The marked behavior of $\mathbf{G}_c$ taking into consideration the relations (1)-(4) is

$$\mathbf{L}_m(\mathbf{G}_c) = \mathbf{L}_m(\mathbf{G}) \cap P_{S,1}^{-1}(\mathbf{K}_1) \cap P_{S,2}^{-1}(\mathbf{K}_2) \cap P_{S,3}^{-1}(\mathbf{K}_3) \cap$$
$$\cap P_{S,4}^{-1}(\mathbf{K}_4) = \mathbf{K}_{D,1} \cap \mathbf{K}_{D,2} \cap \mathbf{K}_{D,3} \cap \mathbf{K}_{D,4} \quad (7)$$

From the last equation (6) it is obvious that the performance of the controlled automaton $\mathbf{G}_c$ is satisfactory.

It remains to prove a) the controllability of $\mathbf{K}_1$, $\mathbf{K}_2$, $\mathbf{K}_3$ and $\mathbf{K}_4$, via the physical realizability of the synchronous product of the respective supervisors and $\mathbf{G}$, and b) $\mathbf{G}_c$ is a nonblocking automaton.

Proposition 1: The synchronous product of the designed supervisor scheme is PR, with respect to $\mathbf{G}$, through (5).

Proof: From the alphabet of $\mathbf{E}_{S,1}$ and $\mathbf{E}_{S,2}$ it is concluded that the uncontrollable events in both supervisors are the events $e_{3,1}$ and $e_{3,2}$. Also, from $\mathbf{S}_1$ and $\mathbf{S}_2$ we get $e_{3,1}, e_{3,2} \in \mathbf{H}_{S,1}(q_{S,1,2}) \cap \mathbf{H}_{S,1}(q_{S,1,1})$ and $e_{3,1}, e_{3,2} \in \mathbf{H}_{S,2}(q_{S,2,2})$

$\cap \mathbf{H}_{S,2}(q_{S,2,3}) \cap \mathbf{H}_{S,2}(q_{S,2,1})$. Thus, according to [11] the supervisors $\mathbf{S}_1$ and $\mathbf{S}_2$ are PR. From the alphabet of the supervisor $\mathbf{E}_{S,3}$ it is concluded that the uncontrollable events are the events $^2e_{F,1}$ and $^2e_{F,2}$. Also, from the active events sets of $\mathbf{S}_3$ it is observed that $^2e_{F,1}, {}^2e_{F,2} \in \mathbf{H}_{S,3}(q_{S,3,2}) \cap \mathbf{H}_{S,3}(q_{S,3,1})$. Thus, according to [11] the supervisor $\mathbf{S}_3$ is PR. From the alphabet of the supervisor $\mathbf{E}_{S,4}$ it is concluded that the uncontrollable events are the events $^1e_{F,1}$ and $^1e_{F,2}$. Also, from the active events sets of $\mathbf{S}_4$ it is observed that $^1e_{F,1}, {}^1e_{F,2} \in \mathbf{H}_{S,4}(q_{S,4,1}) \cap \mathbf{H}_{S,4}(q_{S,4,2})$. Thus, according to [11] the supervisor $\mathbf{S}_4$ is PR. ∎

Proposition 2: The nonblocking property of $\mathbf{G}_c$ is true.

Proof: The nonblocking property of all the subsystems of the gas station will be proved, guaranteeing the nonblocking property of $\mathbf{G}_c$ as the subsystems has disjoint sets.

Regarding $\mathbf{G}_1$ it holds that in the non-marked states $q_{1,2}$ and $q_{1,4}$ the transition to the marked state $q_{1,1}$ is always feasible through $e_{1,5} \notin (\mathbf{E}_{S,4} \cup \mathbf{E}_{S,3} \cup \mathbf{E}_{S,2} \cup \mathbf{E}_{S,1})$. The transition of the non-marked state $q_{1,3}$ to $q_{1,4}$, through the event $e_{1,3}$ is also always feasible as $e_{1,3} \notin (\mathbf{E}_{S,3} \cup \mathbf{E}_{S,2} \cup \mathbf{E}_{S,1})$ and also in $\mathbf{S}_4$ the automaton can always return to state $q_{S,4,1}$ (it is guaranteed by the PR property). Hence, there is always an allowable transition from a non-marked to a marked state of $\mathbf{G}_1$. Finally, in the faulty model $^1\mathbf{G}_F$, the return to the marked non faulty state is guaranteed through the PR property.

Regarding $\mathbf{G}_2$ it holds that in the non-marked states $q_{2,2}$ the transition to the marked states $q_{2,3}$ is always feasible through $e_{2,2} \notin (\mathbf{E}_{S,4} \cup \mathbf{E}_{S,2} \cup \mathbf{E}_{S,1} \cup \mathbf{E}_{S,3})$. Similarly, the transition from $q_{2,4}$ to $q_{2,1}$ through $e_{2,4} \notin (\mathbf{E}_{S,4} \cup \mathbf{E}_{S,3} \cup \mathbf{E}_{S,1} \cup \mathbf{E}_{S,2})$ is always feasible. The transition from $q_{2,5}$ to $q_{2,2}$ through $e_{2,1} \notin (\mathbf{E}_{S,2} \cup \mathbf{E}_{S,3} \cup \mathbf{E}_{S,1})$ is also always feasible (it is guaranteed by the PR property). Finally, in the faulty model $^2\mathbf{G}_F$ the return to the marked non faulty state is unobstructed through the PR property.

The proof of the PR property covers the nonblocking property of $\mathbf{G}_3$. The proof has been completed. ∎

VI. ST IMPLEMENTATION FOR PLCs

The proposed supervisor scheme is implemented in ST language being suitable for industrial computer platforms (ICPs), operating as control units, like industrial PCs, PLCs and SCADA. The software package, used for the development of the ST code, is CODESYS V35 ([19]-[21]). ST is one of the most popular programming languages in industrial environments. The main tasks of ICPs are i) read the inputs (signals from the system), ii) determine the signal that changed in comparison to their previous value (event), iii) execute the main program constructed by appropriate "if-else" commands,

979-8-3503-8368-3/23 $31.00 © 2023 IEEE

iv) output appropriate signals resulting from the confirmation of the –"if-else" commands.

It is important to mention that before the first execution of the main program, all variables corresponding to states and events of the supervisor must be initialized. Thus, the initial values of all variables must be set. For instance, the value of the variable representing the initial state must be True, while the values of all other variables must be False. In Figure 8, the ST code of the main program of supervisor S_1 is presented. The main programs of supervisors S_2, S_3 and S_4 are of similar structure.

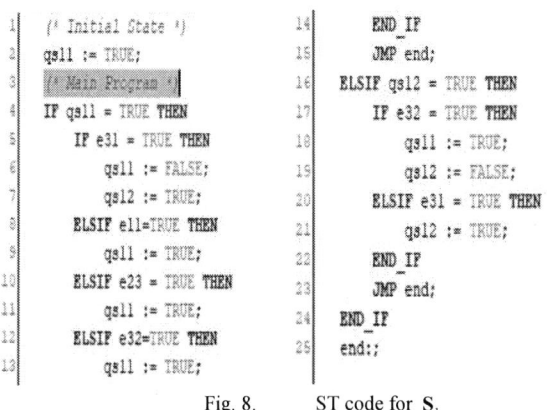

Fig. 8. ST code for S_1

CONCLUSIONS

GCSs with all possible actuator faults have been modelled. The desired behavior for the surge phenomenon avoidance has been imposed in the form of four rules and four regular languages. The desired languages have been realized in the form of the supervisor automata. The PR property of the synchronous product of the supervisor automata and the total automaton of the system, as well as nonblocking, have been proved. The implementation of the supervisors, using ST for PLCs, has been presented.

ACKNOWLEDGMENT

MSc Program "Advanced Control Systems and Robotics", NKUA, Greece.

REFERENCES

[1] V. Saeidi, A. A. Afzalian and D. Gharavian, "Modular Hierarchical Supervisory Control for Surge Avoidance in a Gas Compressor Station," *Smart Grid Conference* (SGC 2015), 23-24 Dec, Tehran, Iran, 2015.

[2] M. Shahidehpour, Y. Fu and T. Wiedman, "Impact of Natural Gas Infrastructure on Electric Power Systems," *Proceeding of the IEEE*, Vol.93, No.5, May 2005.

[3] D. Marques and M. Morari, "On-line Optimization of Gas Pipeline Networks," *Automatica*, Vol. 24, No. 4, pp. 455-469, 1988.

[4] S. Y. Yoon, Z. Lin and P. E. Allaire, "Control of Surge in Centrifugal Compressors by Active Magnetic Bearings-Theory and Implementation," *Springer-Verlag*, London 2013.

[5] J. T. Gravdahl and O. Egeland, "Compressor Surge and Rotating Stall-Modeling and Control," *Springer-Verlag*, London 1999.

[6] V. D. Phan, C. P. Vo, H. V. Dao and K. K. Ahn, "Robust Fault-Tolerant Control of an Electro-Hydraulic Actuator With a Novel Nonlinear Unknown Input Observer," *IEEE Access*, vol. 9, pp. 30750-30760, 2021.

[7] F. N. Koumboulis, D. G. Fragkoulis, I. Kalkanas, and G. F. Fragulis, "Supervisor Design for a Pressurized Reactor Unit in the Presence of Sensor and Actuator Faults," *Electronics*, vol. 11, no. 16, p. 2534, 2022.

[8] N. D. Kouvakas, F. N. Koumboulis, D. G. Fragkoulis and E. Zoto, "Supervisory control for surge prevention in a gas compressor station with actuator faults," *2022 International Conference on Engineering and Emerging Technologies (ICEET)*, Kuala Lumpur, Malaysia, 2022

[9] W. M. Wonham and C. Kai, Supervisory control of discrete-event systems, *Springer International Publishing*, 2019.

[10] C. G. Casandras and S. Lafortune, Introduction to Discrete Event Systems, *Kluwer Academic Publishers*, 1999.

[11] F. N. Koumboulis, D. G. Fragkoulis and S. Arapakis, "Supervisor design for an assembly line in the presence of faults," *2022 IEEE 27th International Conference on Emerging Technologies and Factory Automation (ETFA)*, Stuttgart, Germany, 2022.

[12] F. N. Koumboulis, D. G. Fragkoulis and A. A. Michos, "Modular Supervisory Control for multi-floor manufacturing processes," *Control Theory and Technology*, Springer, 2023.

[13] F. N. Koumboulis, D. G. Fragkoulis and P. Georgakopoulos, "A Distributed Supervisor architecture for a General Wafer Production System," *Sensors*, vol. 23, no. 9, p. 4545, 2023.

[14] F. N. Koumboulis, D. G. Fragkoulis and B. Siake, "Safe operation and coordination of a modular production system via supervisor automata," *31st Mediterranean Conference on Control and Automation (MED2023)*, June 26 – 29, Limassol, Cyprus, 2023.

[15] L. Guo, A. S. Vincentelli and A. Pinto, "A complexity metric for concurrent finite state machine based embedded software," *8th IEEE International Symposium on Industrial Embedded Systems (SIES)*, June 19-21, Porto, Portugal, 2013.

[16] M.H. de Queiroz and J.E.R. Cury, "Modular control of composed systems," *Proc. of the ACC*, Chicago, USA, 2000.

[17] M.H. de Queiroz and J.E.R. Cury, "Modular supervisory control of large scale discrete-event systems," *Discrete Event Systems: Analysis and Control*, Kluwer Academic Publishers, pp. 103-110. (Proc. WODES 2000, Ghent, Belgium), 2000.

[18] X. Lu, L. Piétrac and E. Niel, "A new approach of modeling supervisory control for manufacturing systems based on SysML," *2017 22nd IEEE International Conference on Emerging Technologies and Factory Automation (ETFA)*, Limassol, 2017.

[19] https://www.helpme-codesys.com/ (last access: 31/10/2023)

[20] A. Velazquez, F. Martell, I. Y. Sanchez, and C. A. Paredes, "Cyberphysical System Modeled with Complex Networks and Hybrid Automata to Diagnose Multiple and Concurrent Faults in Manufacturing Systems," *Applied Sciences*, vol. 13, no. 19, p. 10603, Sep. 2023.

[21] L. Prenzel, and J. Provost, "PLC implementation of symbolic, modular supervisory controllers," *IFAC-PapersOnLine*, vo. 51(7), pp. 304-309, 2018.

TinyML-based Event Detection: An Edge-Cloud Approach for Smart Agriculture over LoRa WSNs

Aristeidis Karras*, Christos Karras*, Anastasios Giannaros *, Konstantinos C. Giotopoulos[†],
Dimitrios Tsolis[‡], Konstantinos Oikonomou[§], Spyros Sioutas*
*Computer Engineering and Informatics Department, University of Patras, Patras, Greece
{akarras, c.karras, giannaros, sioutas}@ceid.upatras.gr
[†]Department of Management Science and Technology, University of Patras, Patras, Greece
kgiotop@upatras.gr
[‡]Department of History and Archaeology, University of Patras, Patras, Greece
dtsolis@upatras.gr
[§]Department of Informatics, Ionian University, Corfu, Greece
okon@ionio.gr

Abstract—**In modern agriculture, the capability to promptly detect and respond to specific events is crucial. This study centres on the transformative potential of TinyML for enhancing event detection in Smart Agriculture, particularly when integrated with LoRa-based Wireless Sensor Networks (WSNs). In this work, we underscore the unique advantages of utilizing TinyML at the edge—bypassing the latency and overhead associated with cloud-centric models and ensuring immediate, on-site analytical insights. Employing LoRa WSNs as the backbone provides a seamless, low-power, and expansive data communication framework. Detailed experiments demonstrate TinyML's efficacy in accurately predicting agricultural events while reducing computational and energy consumption. In conclusion, the synergy between TinyML and LoRa WSNs offers a promising approach for fine-tuned, real-time event detection, aiming for sustainable and high-yield agricultural practices.**

Index Terms—**Edge AI, TinyML, IoT, Event Detection, LoRa, LPWAN, WSN, Edge Computing, Cloud Computing, Smart Agriculture**

I. INTRODUCTION

In recent years, the difficulties of agricultural operations, particularly in dairy production, have increased. Faced with these complexities, many dairy production organizations and entities have adopted automation, not only in farming activities but also in the production processes. The ultimate success of such automation—whether in terms of productivity or product quality—is interconnected with the health of livestock. Monitoring every aspect of the production lifecycle is imperative, not only to elevate productivity but to ensure product safety. Therefore, the necessity for an environment-friendly, non-invasive animal health monitoring system is evident.

At the core of this progressive automation lies the Internet of Things (IoT), which facilitates the collection, transmission, and storage of vital data parameters. In this landscape, LoRa (Long Range) has emerged as a pivotal technology, offering power-efficient, long-range wireless communication tailored for the unique demands of IoT deployments, especially in the vast expanses of agricultural setups. Traditional short-range radio technologies such as ZigBee and Bluetooth falter when

long-range transmissions are needed. Meanwhile, cellular solutions, while extending coverage, are often prohibitive in terms of energy consumption. This has been conducted in the era of Low Power Wide Area Networks (LPWAN) [1], with LoRa standing out as a promising option.

Cloud computing, having dominated the previous decade as a central paradigm, finds its prowess challenged by the explosive growth of IoT. Addressing the several challenges posed by IoT often falls outside the traditional sector of cloud computing. This gap has created interest in Edge computing, which brings computation closer to the data sources—at the very edge of the network. Transitioning from Cloud to Edge, Fog computing has emerged as a bridging paradigm, seamlessly connecting Cloud infrastructure and IoT.

In this technological landscape, we introduce the SAF system—a comprehensive LoRa-based solution tailored for dairy farm monitoring [2]. SAF, an acronym for "safe for food and animals," underscores our dedication to ensuring the safety and quality of dairy products from the farm to the consumer's table. Our system harnesses the synergy between Edge and Cloud computing. By employing TinyML for event detection, we analyze data at the Edge, reducing the data relayed to the Cloud, ensuring both energy and bandwidth efficiency [3]. This is augmented by an algorithm that identifies events within a dataset and another that assigns computational tasks to the most power-rich peer in a peer-to-peer network. This dynamic allocation ensures that the peer responsible for event detection is also the one uploading data to the cloud, optimizing energy consumption across the network.

In our interconnected world, food supply chains have gained remarkable complexity. Food Production is not just controlled by merely one entity or nation; it is overseen by a worldwide alliance. International standardization of food production, safety protocols, and comprehensive process oversight is essential. The SAF system [2], with its advanced technology and infrastructure, serves as a model in this domain. It monitors livestock health, farm microclimates, production, storage, and transportation logistics. Each step is documented, and every

979-8-3503-8368-3/23 $31.00 © 2023 IEEE

data point is identifiable, ensuring transparency and traceability.

Our primary research contribution revolves around the SAF system—a holistic solution to track and monitor the entire dairy production process. This solution's significance can't be overstated; it promises to revolutionize dairy production quality, ensuring health, safety, and a formidable competitive stance in both national and global markets.

The remainder of this study is organized as follows: In Section II introduces the SAF System, highlighting its transformative role in dairy production. Section III delves into the technological pillars of smart agriculture, with special emphasis on LoRa Wireless Sensor Networks and the potential of Edge Computing and TinyML. In Section IV, we explore the analytics and event detection, including applications of TinyML and the efficiency of LoRa. Experimental results are presented in Section V. The study concludes with Section VI, summarizing key insights and suggesting directions for future research.

Acronyms are defined upon their initial appearance in the text. Additionally, a Glossary of Symbols summarizing this work is provided in the table I.

TABLE I: Glossary of Symbols

Symbol	Description
A	Anomaly Score
$T_p(t)$	Predictive Trajectory
S	Stress Index
T_{sheep}	Sheep's Temperature
T_{pred}	Predicted Normal Temperature
ΔT	Temperature Difference
T_{fridge}	Fridge Temperature
H_{fridge}	Fridge Humidity
MSI	Milk Safety Index
D	Decision Function
P_{sleep}	Power consumed in sleep mode
t_{awake}	Time device spends awake
t_{sleep}	Time device spends in sleep mode
E_T	Total energy consumed
α	Network activity factor
$t_{\text{sleep, adjusted}}$	Adjusted sleep time
E_{event}	Energy threshold for event detection
T_s	Sensed temperature of sheep
T_{thresh}	Static fever threshold temperature
T_a	Ambient temperature
$T_{s,h}$	Historical average temperature of sheep
β	Weight factor for influence of historical temperature
$T_{\text{dyn-thresh}}$	Dynamic fever threshold
δ	Difference value for suspected threshold
$T_{\text{susp-thresh}}$	Suspected fever threshold
T_f	Fridge's internal temperature
$T_{\min}$	Minimum acceptable temperature for fridge
$T_{\max}$	Maximum acceptable temperature for fridge

II. System Overview

A. The SAF System: A Paradigm Shift in Dairy Production

Within the dairy industry's complex challenges, which span livestock health to product quality assurance, the SAF system stands as a pivotal advancement. By integrating modern technology with established agricultural methods, it offers significant potential to enhance dairy production for local and international markets.

Core Features & Technological Integration: At the core, SAF is a testament to the future of dairy farming. It leverages:

- Cutting-edge technologies to boost production and cement the unique quality of dairy yields.
- Collaborative practices that optimize the livestock food chain, from feed to final product.
- Advanced monitoring systems that ensure optimal livestock environments and health.

With the integration of TinyML, SAF excels in real-time monitoring of livestock health, prioritizing early detection of anomalies. This is further complemented by the Edge-Cloud computing paradigm, which assures prompt data-driven responses. What sets SAF apart is its adoption of LoRa technology. This long-range, low-power wireless platform is tailored for the expansive area of dairy farms, enabling seamless real-time data management.

Impact & Benefits:

- SAF's introduction promises to solve many of the dairy industry's longstanding challenges. By emphasizing preventive care, the likelihood of livestock ailments is drastically reduced, as is the risk of procuring subpar raw materials. The resultant products are thus top-notch, distinctive, and reliable. Beyond just product quality,
- SAF leads in a new era of cost-effectiveness, addressing both livestock health and product optimization. For consumers, it sets the stage for unwavering trust in their dairy choices, assuring both safety and value for money.

Architectural Overview: Underpinning SAF's prowess is its multifaceted architecture:

1) Dedicated repositories for both quality production and product traceability.
2) Intelligent decision-support mechanisms tailored for dairy production.
3) A suite of user-friendly mobile applications for stakeholders.

Ultimately, the SAF system stands as a testament to the synergy of diverse technologies, encompassing data repositories, product traceability mechanisms, advanced sensors, and the robustness of LoRa Wireless Sensor Networks. By integrating innovations such as TinyML, Edge-Cloud frameworks, and LoRa networks, SAF envisions a dairy industry setting new benchmarks in efficiency, safety, and product hygiene, creating a future prioritizing both livestock welfare and consumer trust.

B. Technological Foundations and Innovations in the SAF

The SAF system, while innovative, is founded on existing technologies, methodologies, and frameworks that have individually catalyzed advances in various sectors. Before diving into SAF's specifics, it's crucial to appreciate the technological foundations underpinning its success.

1) LoRa Technology and Wireless Sensor Networks: LoRa, or Long Range, technology has emerged as a leading candidate

979-8-3503-8368-3/23 $31.00 © 2023 IEEE   29

in the Internet of Things (IoT) landscape, particularly in applications demanding low-power and long-range communications. SAF's design capitalizes on LoRa Wireless Sensor Networks, particularly advantageous for vast agricultural settings where traditional Wi-Fi or GSM connections might be unstable or inefficient. With devices such as the Lilygo T-Beam LoRa transceiver, which combines both LoRa communication and GPS tracking, SAF has instituted a decentralized monitoring system, uniquely identifying every farm entity.

2) Event Detection using TinyML: SAF's efficacy in early problem identification is embedded in TinyML. This subset of machine learning optimizes models for microcontrollers, enabling real-time data processing without extensive computational power. When applied to SAF, TinyML algorithms analyze data from sensors, detecting anomalies or patterns, such as potential health issues in livestock, and implementing countermeasures.

3) Edge-Cloud Architectures in Smart Agriculture: Modern agricultural systems, including SAF, are moving towards Edge-Cloud frameworks. While cloud platforms like Google Firebase in SAF ensure worldwide data accessibility and reliable storage, edge computing processes a portion of the data closer to its source—be it a cow, a milk tank, or a farm. Such an approach reduces latency, guarantees faster decision-making, and cuts down data transfer costs.

4) SAF's Hierarchical Design: The SAF's architecture can be analogously visualized as a pyramid, where every level represents a specialized function. At the foundational level, the architecture predominantly concentrates on data acquisition and communication. At the top of the structure, the focus gravitates towards data management, followed by advanced analytics and event detection. This hierarchical design underscores the integral components of the architecture:

- **Sensors and Microcontrollers:** This is the base layer of SAF. Its primary role is real-time data collection, followed by the transmission of this data to the LoRa gateway and then through the SAF API.
- **TinyML for Event Detection:** Positioned with the sensors and microcontrollers, TinyML is employed for on-device event detection. Through this, anomalies or significant events in the data can be identified promptly.
- **SAF API:** Serving as the intermediary, the SAF API manages communication by converting sensor data into HTTP post requests which are directed to the database.
- **Database and Cloud:** Data relayed from the SAF API is stored as key-value pairs in a NoSQL database. Cloud solutions, like Google Firebase, provide data redundancy and availability.
- **Application Interface:** This uppermost layer provides an interface for stakeholders, presenting real-time data insights. This assists in identifying and addressing issues in the supply chain efficiently.

This layered approach, supplemented by the capabilities of TinyML, ensures that SAF effectively captures, processes, and presents data.

5) Monitoring Crucial Parameters: Central to SAF's operations is the consistent surveillance of key variables detailed in table II. From tracking farm animal temperatures—a vital health metric—to monitoring farm conditions like humidity, temperature, and methane concentration, SAF offers comprehensive insights. This real-time data collection aids in early problem detection and mitigation, invariably saving costs and ensuring product quality.

The system flow of data for our smart agriculture architecture is represented in Figure 1.

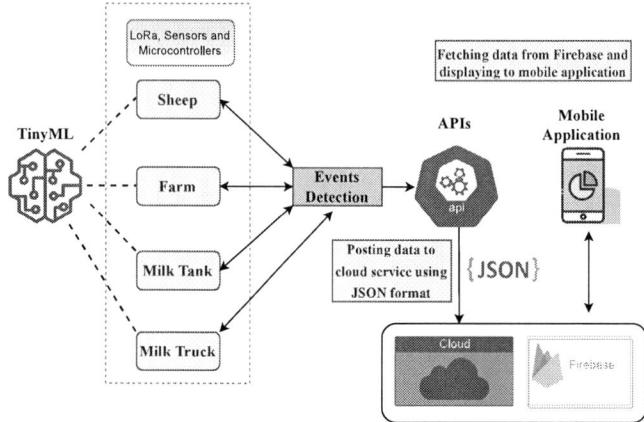

Fig. 1: Smart agriculture system data flow.

6) LoRa WSN Deployment in SAF: Technical Overview: The SAF's use of LoRa-based Wireless Sensor Network (WSN) demonstrates how modern communication technologies enhance agricultural operations. Below are the key technical details:

- **Node Configuration:**
  - SAF uses around 100 LoRa T-beam TTGO nodes per hectare for optimal coverage.
  - Three gateways, equipped with LoRa antennas, ensure strong data traffic and connectivity.
- **Power Management:**
  - Nodes have 200 mAh batteries and 40 cm^2 solar panels for continuous operation.
  - LoRa's Adaptive Data Rate (ADR) adjusts power based on network conditions.
- **Monitoring Precision:**
  - Livestock nodes use DS18B20 and MPU-6050 sensors for accurate temperature and movement data.
  - Environmental nodes use SHT31 and MQ-4 sensors to measure humidity, temperature, and methane accurately.
- **Data Protocols:**
  - SAF uses the LoRaWAN 1.0.3 protocol stack for broad compatibility.
  - Gateways connect securely to cloud databases using MQTT with TLS.
- **Data Security:**

TABLE II: SAF events of interest with TinyML and LoRa enhancements.

Entity	Event	Description
Animals	Deviation from the herd	Anomaly detected in animal movement patterns
	Unusual temperatures	Deviation from expected animal body temperatures
	Large percentage of slow animals	Many animals moving slower than usual
	Incoherent herd move	Erratic or uncoordinated herd movements
	Predictive trajectory anomaly	Detected from equation (2)
	Elevated stress index	Detected from equation (3)
	Potential sheep fever	Detected from equations (4) and (11)
Farm	Milk production drop	Sudden decrease in milk production
	High methane concentration	Above normal methane levels detected
	Low air humidity	Ambient humidity levels dropping below threshold
Tank	Unusual temperatures	Temperature deviating from expected tank conditions
	Unusual milk level	Unexpected drop or increase in milk level
	Unusual pH values	pH levels outside the acceptable range
Truck	Unusual pH values	Detected anomaly in milk pH during transit
	Large route deviation	Truck deviating significantly from expected route
Milk Fridge	Temperature anomaly	Detected from equation (12)
	Humidity anomaly	Unusual humidity levels affecting milk safety
LoRa Network	Predictive packet failure	Detected from equation (8)
	Excessive energy consumption	Based on equation (13)

- Data is encrypted with 128-bit AES, and each device uses unique session keys.
- A Message Integrity Code (MIC) verifies message authenticity.

III. BACKGROUND AND RELATED WORK

In the *Background and Related Work* section, we begin by discussing the advancements in agricultural practices through III-A. Next, III-B examines the use of wireless sensor networks in agriculture and their suitability for large areas. The importance of local data processing is highlighted in III-C, showing the advantages of not always depending on main servers. III-D reviews the literature on the combination of edge devices and cloud platforms, and III-E discusses methods for detecting important agricultural events promptly.

A. Smart Agriculture and Technological Evolution

The confluence of technology and modern agriculture is witnessing a pronounced emphasis on real-time data analytics. Distinct innovations have emerged as pivotal in this domain:

- **Internet of Things (IoT):** An interconnected web of devices facilitated by sensors, IoT has revolutionized communication protocols within agriculture. By minimizing the intermediaries between producers and consumers, it assures enhanced product integrity and equitable remuneration structures for farmers [4].
- **Machine Learning:** Through comprehensive analysis of data amassed from Wireless Sensor Networks (WSNs), machine learning algorithms project precise irrigation timings, superseding conventional schedules and curtailing undue water consumption [5].
- **Sensor Networks:** These networks, instrumental in accruing real-time agricultural data, leverage energy harvesting, obviating recurrent battery replacements. Furthermore, Low Power Wide Area (LPWA) technologies

ensure efficient long-range communication. This infrastructure is critical for monitoring potential agricultural disruptions, offering instantaneous alerts [6], [7].

Research studies by Shanmugasundaram et al. [8] and Telagam et al. [9] underscore the revolutionary potential represented by IoT, smart sensors, and data analytics in agriculture. A proposition in the literature proposes the development of a spatiotemporal semantic IoT data management framework [10]. This initiative acknowledges the imperative of adeptly managing voluminous datasets, particularly within critical sectors such as agriculture, thus aligning with the overarching vision of technology-infused precision in farming activities.

Within the evolving landscape of agricultural innovation, data mining techniques have become paramount for efficient processing and analyzing copious amounts of real-time data [11]. Ayaz et al. delve into the transformative impact of wireless sensors and IoT within the agricultural sector, elucidating their roles in areas like soil optimization, crop monitoring, and precise irrigation [12]. Concurrently, Elijah et al. articulate both the advantages and impediments of IoT and data analytics in bolstering agricultural efficiency and yield [13]. Further emphasizing the modern agricultural paradigm, Tikas et al. underscore the criticality of advanced farming methods in navigating the dual challenges of a burgeoning global populace and the constraints of limited resources [14].

B. LoRa Wireless Sensor Networks (WSNs) in Smart Agriculture

The integration of LoRa (Long Range) wireless technology in smart agriculture systems has garnered significant attention due to its distinctive capability to transmit data over extensive distances at a minimal cost. Such LoRa-empowered wireless sensor networks (WSNs) find their placements in proximity to crops and strategic agricultural locales, facilitating the real-time acquisition of pertinent crop and environmental infor-

979-8-3503-8368-3/23 $31.00 © 2023 IEEE

mation [15]. This concurrent data proves instrumental, both for immediate, latency-sensitive decisions and for extended agricultural planning [15].

A LoRa WSN-based agriculture system consists of perception and network layers. The perception layer captures environmental data and manages agricultural machinery, utilizing LoRa for extended transmission. Meanwhile, the network layer, dominated by LoRa gateways, interfaces with servers, transmitting both data and control instructions.

The integration of LoRa-based WSNs holds the promise of elevating agricultural practices, offering farmers avenues to optimize profitability while simultaneously minimizing hands-on land supervision [15]. Additionally, these networks act as conduits for channelling scientific insights to boost crop yields and infuse technological innovations into contemporary agricultural activities. The spectrum of agricultural and ambient parameters they can monitor is broad, encompassing metrics like soil moisture, ambient temperature, relative humidity, and nitrate concentrations [16], [17]. A noteworthy feature is the cloud-based accessibility of this collected data, ensuring it is available for scrutiny by both academicians and agricultural practitioners [16].

The integration of LoRa wireless technology into smart agriculture has shown a plethora of advantages, essential for modern-day farming. Some of these notable benefits are:

1) **Extended Range:** LoRa's capacity to transmit data over expansive distances, even up to several kilometers, without an internet connection makes it particularly suitable for vast agricultural terrains where traditional wireless modalities might be inadequate [18].

2) **Minimal Power Usage:** Owing to its design tailored for low power consumption, LoRa-based WSNs operate over extended periods, diminishing the necessity for frequent battery replacements [19], [20].

3) **Cost-Efficiency:** Relative to other wireless communication modalities, LoRa emerges as a more cost-effective option, aligning with the budgetary constraints of many farmers [18].

4) **Real-Time Surveillance:** LoRa-enabled WSNs' proficiency in instantaneously collecting and transmitting data offers farmers real-time insights into crop and environmental conditions, facilitating timely and crucial decision-making processes [21]–[23].

5) **Operational Efficacy:** Such networks empower farmers to boost profitability, reduce hands-on field monitoring, and receive invaluable scientific guidance, all contributing to enhanced crop yields and the seamless integration of the latest technological advancements into farming [24].

6) **Adaptable Scalability:** LoRa WSNs' architecture is inherently scalable, accommodating expansions or contractions based on the agricultural area and the number of sensors required [24].

7) **Edge Machine Learning Synergy:** The integration of LoRa with edge machine learning offers a confluence of real-time data transmission and on-site data processing.

This combination promotes faster response times and efficient bandwidth utilization, crucial for decision-making processes in smart agriculture [25], [26].

8) **Path Loss Advantages:** LoRa technology exhibits significant resilience to path loss, a common issue in long-range communication. This inherent resistance ensures reliable data transmission over vast distances, making it a robust choice for agricultural terrains with varying topographies [27].

In essence, the incorporation of LoRa-centric WSNs in smart agriculture epitomizes the transformative potential of technology in modern farming, optimizing efficiencies, refining decision-making, and achieving considerable cost savings.

C. Edge Computing and TinyML in Agriculture

In the evolving landscape of agricultural technology, edge computing, and TinyML stand out as transformative elements that can redefine traditional farming methodologies. This subsection delves into the foundational concepts of edge computing, elucidates the rising significance of TinyML, and underscores their combined potential in revolutionizing event detection within agriculture.

Edge Computing in Modern Agriculture Edge computing represents a paradigm shift in data processing by emphasizing on-site data processing, typically at the network's periphery. This method reduces the need for continuous data transmission to central data centers or clouds, thereby offering reduced latency, optimized bandwidth consumption, and enhanced data security. Such a model is indispensable in scenarios demanding prompt data analysis and action, a case in point being agriculture. Here, expeditious decisions based on real-time data can profoundly influence both the quantity and quality of crop output.

TinyML's Role in Agricultural Enhancement TinyML describes the use of machine learning algorithms on resource-limited devices like microcontrollers. These devices are low-power, cost-effective, and operate with limited memory and computational capabilities. Yet, TinyML is transforming agricultural practices by enabling smart sensors for real-time responses, such as tracking soil or atmospheric changes. Within agriculture, TinyML allows real-time monitoring of threats, from pests to crop diseases. Using TinyML on-edge devices like drones or soil sensors provides immediate data assessment, helping farmers to act swiftly, thus improving crop health and yield.

Combining edge computing with TinyML revolutionizes agriculture. Processing data on-site with compact devices provides farmers instant insights, guiding decisions for optimal crop health and yield.

D. Related Studies on Edge-Cloud Synergies

Edge computing in smart agriculture involves utilizing computing resources located near data sources, like sensors, to facilitate real-time data analysis. This methodology not only decreases latency but also bolsters data security and relieves demands on centralized cloud data centers [28]. Below are

979-8-3503-8368-3/23 $31.00 © 2023 IEEE

some pivotal developments exploiting edge computing in smart agriculture:

- Embedded AI for Air Quality Monitoring: Inexpensive smart probes equipped with sensors for gas, humidity, pressure, and temperature have been employed to scrutinize air quality, environmental conditions, and crop emissions. Through the incorporation of edge computing for data pre-processing, these sensors' capabilities are enhanced. Notably, embedded AI algorithms have been demonstrated to discern gases like NH3, CH4, and N20, relying on data from gas sensors [29].

- Adoption of UAVs: UAVs, armed with high-definition cameras and assorted sensors, have been utilized to collate real-time data, which encompasses detailed crop imagery. Such data is instrumental for expedited decision-making, trimming operational costs, and boosting crop yields. Moreover, precision agriculture leverages UAVs to refine agricultural protocols. Yet, incorporating UAVs into smart agriculture faces hurdles pertaining to technology adoption and implementation, especially concerning data procurement and image processing [30], [31].

- Security through Physical Unclonable Function (PUF): Although IoT devices are valuable tools for monitoring field conditions, they also introduce security concerns. Proposing a PUF-based hardware security primitive can enhance the authentication of Internet of Agro-Things (IoAT) devices. Such a security paradigm is notable for its lightweight nature, scalability, and robustness, with its efficacy hinging on intrinsic manufacturing discrepancies that obviate the replication of IoT devices [32].

- Integrating Blockchain with Edge Computing: The development of an astute climate and irrigation system, maneuverable via an Android application, has been advanced for judicious water consumption in small to medium-sized agricultural fields. Given the escalating concerns about data privacy and the increasing susceptibility of IoT devices to security threats, a combination of intelligent fuzzy logic and blockchain technology can offer real-time analysis while fortifying the network. Through blockchain's deployment, only authenticated devices gain access, ensuring the network's integrity [33].

E. Event Detection Mechanisms in Agriculture

In the field of agricultural research, event detection signifies the methodological approaches employed to discern and categorize distinct events within the broader agricultural ecosystem. While a significant portion of the existing literature emphasizes the role of IoT systems in event detection, there exists potential applicability of the detailed techniques within the agricultural ambit. Detailed below are some of the prominent mechanisms extracted from pertinent research:

- Complex Event Processing (CEP): CEP is an advanced methodology employed to identify complex patterns of events within streams of real-time data. The essence of CEP is the formulation of specific rules and patterns that describe agricultural events of interest. Utilizing these predefined patterns, CEP facilitates timely detection and appropriate response to emergent events [34], [35].

- Applications of Machine Learning (ML): Machine Learning techniques can be used to enhance fault tolerance in IoT frameworks. Through analysis of historical data pertaining to reactive fault responses, ML algorithms can anticipate and mitigate known error occurrences [35].

- Hierarchical Tagging Networks with Enhanced Attention Mechanisms: This innovative paradigm, designed for the collective detection of events, employs a hierarchical tagging system. Such a system can concurrently discern multiple events within individual textual entities. A supporting attention mechanism is integrated to synthesize information at both micro (sentence) and macro (document) levels [36].

- Attention Mechanisms for Structured Event Detection: This approach capitalizes on structured argument data to facilitate event detection. Its efficacy is underscored by its superior performance metrics on standard datasets, notably ACE 2005 [37].

- Graph-centric and Cross-Lingual Event Detection Techniques: This methodology transforms social messages into graph structures characterized by their heterogeneity. By utilizing a reinforced multi-relational graph neural framework, this approach captures the inherent semantics of social messages. Given the rise of social platforms in disseminating agricultural trends and updates, this method offers potential applicability in identifying agriculturally relevant events from social media.

In summary, though the literature on event detection in agriculture is limited, the discussed methodologies show potential for agricultural adaptation.

IV. METHODOLOGY: ANALYTICS AND EVENTS WITH TINYML

Event detection in IoT-based agriculture is paramount. With the infusion of TinyML into SAF's architecture, an unmatched ability emerges to predict, detect, and act upon events right at the data source.

A. TinyML-enhanced Event Detection

a) Anomaly Detection with TinyML:: Employing TinyML for anomaly detection helps discern any divergence from standard patterns. The equation for anomaly score A can be presented as:

$$A = D_{\mathrm{KL}}(p||r) \times \sum_{k=1}^{N} p_k \qquad (1)$$

A larger value of A signals an anomaly, prompting investigations.

b) Predictive Trajectory Analysis:: Historical data and machine learning combined can predict an animal's trajectory, aiding in preemptive interventions. The predictive trajectory T_p is:

$$T_p(t) = a_i \times t + b_i \qquad (2)$$

979-8-3503-8368-3/23 $31.00 © 2023 IEEE

c) Sensor Fusion for Holistic Analysis:: By fusing data from multiple sensors, a more comprehensive understanding of livestock health is achievable. The Stress Index S can be delineated as:

$$S = \text{THI} \times (v + r_i) \quad (3)$$

d) Sheep Fever Detection:: Temperature is a clear indicator of fever. By analyzing historical data, a predictive model can be generated. If the temperature T_{sheep} surpasses the predicted temperature T_{pred} by a certain margin, an alert for potential fever can be raised:

$$\Delta T = T_{\text{sheep}} - T_{\text{pred}} \quad (4)$$

If ΔT exceeds a pre-defined threshold, it signals potential fever.

e) Milk Fridge Monitoring:: For milk safety, it's essential to monitor the fridge temperature T_{fridge} and humidity H_{fridge}. The Milk Safety Index MSI can be calculated as:

$$MSI = \omega_4 \times T_{\text{fridge}} + \omega_5 \times H_{\text{fridge}} \quad (5)$$

An anomalous MSI prompts immediate attention, ensuring milk quality.

B. Decision Making with TinyML

The Decision Function D for the above metrics can be presented as:

$$D = \omega_1 \times A + \omega_2 \times S + \omega_3 \times |\vartheta_i| + \omega_6 \times \Delta T + \omega_7 \times MSI \quad (6)$$

Surpassing a predefined threshold for D activates the necessary interventions.

In summary, the above equations, fueled by TinyML in SAF's architecture, deliver a potent system for comprehensive livestock and produce monitoring, ensuring optimal health and product quality.

C. Predictive Mechanism for Unsuccessful Packet Transmission/Receive

In addition to simply monitoring current packet success rates, predicting future transmission failures can add a proactive dimension to event detection in LoRa WSNs. We propose a predictive mechanism that uses historical data and observed trends to predict potential disruptions or downtrends in packet transmission success.

Let ΔP_t represent the rate of change in the number of packets transmitted, and ΔP_r represent the rate of change in the number of packets received over time. A significant difference between these two rates may indicate a brewing issue in the network. The predictive failure metric, denoted as Φ, is given by:

$$\Phi = \frac{\Delta P_r}{\Delta P_t} \quad (7)$$

In a robust network, Φ should ideally be close to 1. Deviations from this value can be indicative of emerging problems. To this end, we can set a threshold Φ_{thresh}, e.g., 0.95, beyond which a predictive event is flagged:

$$\text{Predictive\_Packet\_Event} = \begin{cases} 1, & \text{if } \Phi < \Phi_{\text{thresh}} \\ 0, & \text{otherwise} \end{cases} \quad (8)$$

When the Predictive_Packet_Event is 1, proactive measures such as rerouting traffic, adjusting LoRa transmission power, or even tangible measures (like adjusting sensor node placement) can be initiated to proactively handle the predicted transmission/reception issues, ensuring a more resilient network.

D. Advanced Event Detection for Sheep Fever

For accurate detection of sheep fever, merely relying on a static temperature threshold might not suffice. Environmental factors, such as surrounding temperature, and historical data should be considered to adjust the threshold dynamically.

Let:

- T_a represent the surrounding temperature.
- $T_{s,h}$ be the historical average temperature of the sheep over the past week.
- β be a weight factor (between 0 and 1) representing the influence of the historical temperature on the threshold.

The dynamic fever threshold, $T_{\text{dyn-thresh}}$, can be computed as:

$$T_{\text{dyn-thresh}} = T_{\text{thresh}} + \beta(T_s - T_{s,h}) + (1 - \beta)(T_s - T_a) \quad (9)$$

This equation considers both the deviation from the sheep's historical average temperature and the difference between the sheep's current temperature and the ambient temperature.

Additionally, to allow for an early warning system, a suspected fever threshold $T_{\text{susp-thresh}}$ can be introduced, which is slightly below $T_{\text{dyn-thresh}}$:

$$T_{\text{susp-thresh}} = T_{\text{dyn-thresh}} - \delta \quad (10)$$

Where δ is a small value (for example, 0.5°C).

The advanced event detection mechanism for sheep fever then becomes:

$$\text{Fever\_Event} = \begin{cases} 2, & \text{if } T_s > T_{\text{dyn-thresh}} \\ 1, & \text{if } T_s > T_{\text{susp-thresh}} \text{ and } T_s \leq T_{\text{dyn-thresh}} \\ 0, & \text{otherwise} \end{cases} \quad (11)$$

Here, '2' indicates a confirmed fever event, '1' indicates a suspected fever, and '0' indicates no fever.

E. Event Detection for Milk Fridge

For a milk fridge, maintaining a specific temperature range is crucial. Let T_f represent the fridge's internal temperature, T_{min} be the minimum acceptable temperature, and T_{max} be the maximum acceptable temperature. A temperature-based event detection mechanism for the milk fridge can be:

$$\text{Fridge\_Event} = \begin{cases} 1, & \text{if } T_f < T_{\text{min}} \text{ or } T_f > T_{\text{max}} \\ 0, & \text{otherwise} \end{cases} \quad (12)$$

These mechanisms ensure the timely detection of events, making the best use of LoRa's capabilities while ensuring energy efficiency and prompt responses.

F. LoRa Event Detection and Energy Efficiency

LoRa (Long Range) technology has gained prominence in IoT solutions, primarily for its ability to achieve long-range communication with minimal power consumption. Efficient event detection mechanisms that utilize LoRa's low power consumption can significantly extend the lifetime of battery-operated devices. Here, we explore some state-of-the-art approaches to utilize LoRa's sleep modes and energy-saving features for better event detection.

G. LoRa Sleep Modes and Power Consumption

LoRa modules generally have different modes of operation, including transmit mode, receive mode, standby, and sleep mode. When not transmitting or receiving data, LoRa devices should ideally be in sleep mode to minimize power consumption. Let's denote the power consumed in sleep mode as P_{sleep}.

Given that t_{awake} is the time a LoRa device spends awake (either transmitting, receiving, or in standby) and t_{sleep} is the time it spends in sleep mode within a time frame T, the total energy consumed E_T can be expressed as:

$$E_T = P_{\text{awake}} \times t_{\text{awake}} + P_{\text{sleep}} \times t_{\text{sleep}} \quad (13)$$

1) LoRa Smart Energy Saving: To enhance the energy-saving potential of LoRa, we can introduce a Smart Energy Saving (SES) mechanism that dynamically adjusts the device's awake and sleep times based on network traffic and event detection needs.

Let α be a factor that represents the network's activity (ranging between 0 and 1, with 1 being the most active). Using α, we can dynamically adjust the device's sleep time as:

$$t_{\text{sleep, adjusted}} = t_{\text{sleep}} \times (1 - \alpha) \quad (14)$$

By integrating event detection needs, if E_{event} represents the energy threshold for detecting a specific event, a device should switch to awake mode when:

$$E_T < E_{\text{event}} \quad (15)$$

V. EXPERIMENTAL RESULTS

In this section, we evaluate our system using four parameters. We assess the performance based on i) actual and predicted anomaly scores, ii) actual and predicted trajectories, iii) actual and predicted animal temperatures, and iv) actual and predicted temperatures and humidities of the milk fridge.

Figure 2 illustrates the effectiveness of the TinyML model in predicting anomaly scores. It is clear from the tight alignment of the actual and forecasted values that the model is proficient in identifying discrepancies, guaranteeing prompt countermeasures.

Figure 3 showcases the model's capability to predict an animal's movement. By utilizing past data, the model provides reliable estimates of future trajectories, aiding in proactive interventions.

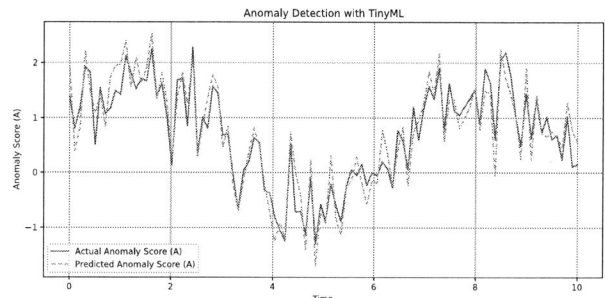

Fig. 2: Anomaly Detection with TinyML showcasing actual and predicted anomaly scores over time.

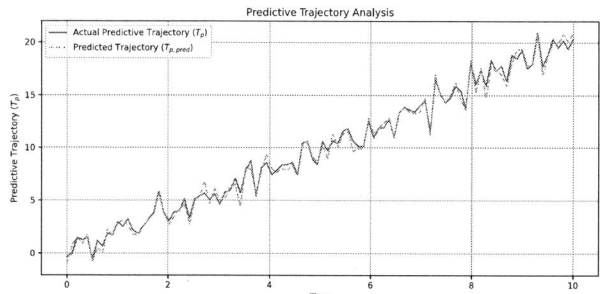

Fig. 3: Predictive Trajectory Analysis illustrating both the actual and estimated trajectories.

Figure 4 indicates the model's proficiency in predicting sheep's body temperature, an essential metric for fever detection. The alignment between actual and predicted temperatures underscores the model's accuracy, facilitating early fever detection.

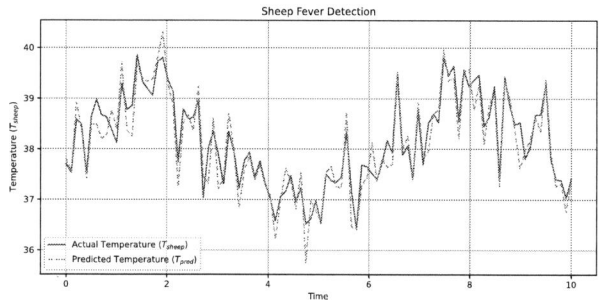

Fig. 4: Sheep Fever Detection plot comparing actual temperature against predicted values.

Moreover, The milk fridge monitoring plot, Figure 5, illustrates the importance of consistent temperature and humidity levels for milk quality. The near-matching between the predicted and actual metrics highlights the model's adeptness, ensuring milk remains in optimal conditions.

Lastly, in Figure 6, we showcase the dynamics of packet transmission over a specified duration. The transmitted packets, represented in blue, exhibit a steady linear progression,

979-8-3503-8368-3/23 $31.00 © 2023 IEEE

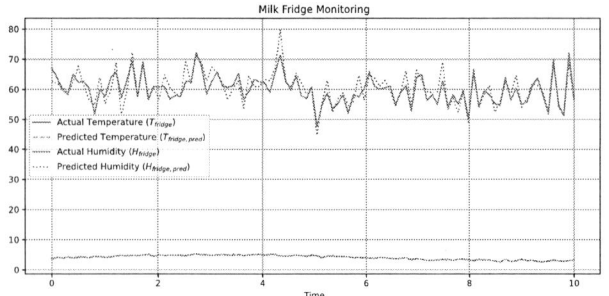

Fig. 5: Milk Fridge Monitoring detailing the variation in both temperature and humidity, with their respective predictions.

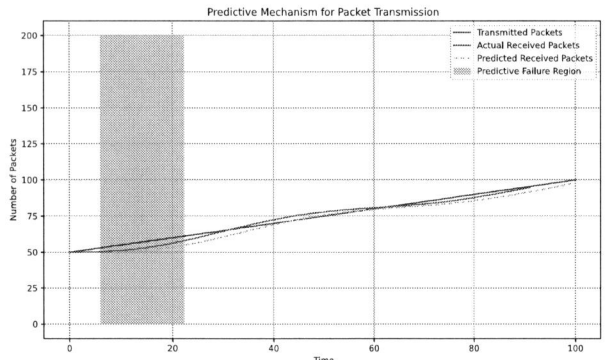

Fig. 6: Visualization of Predictive Mechanism for Packet Transmission over Time.

which shows a consistent data transmission strategy. In contrast, the actual received packets, visualized in green, display specific oscillations—variations that can be attributed to a variety of external factors such as interference, environmental conditions, or physical barriers. The orange dashed line captures the predicted number of received packets, formulated through an analysis of historical data. Notably, the red-shaded region is of paramount interest; it demarcates areas where the received packets significantly deviate from the anticipated trend, flagging potential predictive failures. Such visual insights underscore the importance of proactively addressing network disruptions, ensuring optimal packet transmission efficiency in LoRa WSNs enhanced with TinyML.

Ultimately, the use of TinyML-based solutions in smart agriculture has drastically changed the dynamics of event detection. To encapsulate the efficacy of this approach, we have measured the performance of various event detection methodologies deployed on LoRa WSNs using standard evaluation metrics. These metrics include accuracy, precision, recall, and the F1-score, providing a holistic view of the system's robustness and reliability. As shown in Table III, it is evident that TinyML-based solutions show promising performance across different event detection scenarios in smart agriculture. Noteworthy is the sheep fever detection method, which has the highest accuracy, showcasing the precision with which

TinyML can operate when provided with domain-specific data.

TABLE III: Performance Metrics for TinyML Event Detection.

Event Detection	Accuracy	Precision	Recall	F1-Score
Anomaly Detection	92.3	91.7	90.9	91.3
Predictive Trajectory Analysis	89.8	88.4	90.1	89.2
Integrated Sensor Analysis	93.5	94.1	92.9	93.5
Sheep Fever Detection	95.1	94.8	95.5	95.1
Milk Fridge Monitoring	91.7	92.2	91.3	91.8
Predictive Packet Transmission	88.4	87.9	88.9	88.4

VI. CONCLUSIONS AND FUTURE WORK

The culmination of TinyML-based event detection in an edge-cloud setting, especially when deployed over LoRa WSNs, offers a progressive plan for revolutionizing smart agriculture. The five pivotal experiments, ranging from anomaly detection to the pivotal task of milk fridge monitoring, are testimonies to the system's proficiency and the profound potential of TinyML in handling intricate event detection tasks. By optimizing data collection at the edge and harnessing the cloud's computational prowess, this model ensures rapid, data-driven decision-making, a prerequisite for the evolving demands of current agriculture. As challenges in agriculture intensify, systems such as the one proposed will be indispensable, utilizing technology to assure food security and sustainability.

The research outlined in this study on TinyML-based event detection using an edge-cloud approach in smart agriculture over LoRa WSNs sets the stage for several potential extensions. First, there is a clear opportunity to investigate the integration of advanced machine learning models tailored for limited-resource environments. These models could enhance the precision of event detection across various agricultural contexts. Moreover, the development of algorithms that can adapt to evolving environmental factors can ensure consistent system performance, even in changing agricultural conditions.

Furthermore, as the agricultural IoT landscape grows, ensuring compatibility between varied sensor platforms and communication systems becomes crucial. Building upon the established LoRa WSNs foundation, subsequent studies could focus on crafting a unified system that combines data from multiple networks and sensors. Such an integrated approach would allow for more thorough real-time data analysis, providing broader insights for effective agricultural decision-making. Ensuring energy efficiency within this framework remains a priority to guarantee both the longevity and eco-friendliness of the proposed innovations.

ACKNOWLEDGMENT

This study was completed in the framework of the project: "SAF: Safe for Animal and Food: Integrated System for Interactive Monitoring, Recording and Optimization of Animal Health and for the Safety and Quality of Animal Food", Case Study: Feta Cheese of Kalavryta (Designation of Origin). Contract No M16ΣYN − 00452, Agricultural Development Programme, Measure 16, Sub_Measure 16.1, Action 1.

REFERENCES

[1] K. Mekki, E. Bajic, F. Chaxel, and F. Meyer, "A comparative study of lpwan technologies for large-scale iot deployment," *ICT express*, vol. 5, no. 1, pp. 1–7, 2019.

[2] A. Karras, C. Karras, G. Drakopoulos, D. Tsolis, P. Mylonas, and S. Sioutas, "Saf: a peer to peer iot lora system for smart supply chain in agriculture," in *IFIP International Conference on Artificial Intelligence Applications and Innovations*, pp. 41–50, Springer, 2022.

[3] N. Schizas, A. Karras, C. Karras, and S. Sioutas, "Tinyml for ultra-low power ai and large scale iot deployments: A systematic review," *Future Internet*, vol. 14, no. 12, p. 363, 2022.

[4] S. Parween, R. S. Hameed, and K. Sinha, "Iot and its real-time application in agriculture," in *Handbook of Research on Knowledge and Organization Systems in Library and Information Science*, pp. 103–123, IGI Global, 2021.

[5] J. Cardoso, A. Glória, and P. Sebastião, "Improve irrigation timing decision for agriculture using real time data and machine learning," in *2020 International Conference on Data Analytics for Business and Industry: Way Towards a Sustainable Economy (ICDABI)*, pp. 1–5, 2020.

[6] J. Dai and M. Sugano, "Low-cost sensor network for collecting real-time data for agriculture by combining energy harvesting and lpwa technology," in *2019 IEEE Global Humanitarian Technology Conference (GHTC)*, pp. 1–4, 2019.

[7] C.-P. Balatsouras, A. Karras, C. Karras, D. Tsolis, and S. Sioutas, "Wichord: A chord protocol application on p2p lora wireless sensor networks," in *2022 13th International Conference on Information, Intelligence, Systems & Applications (IISA)*, pp. 1–8, 2022.

[8] N. Shanmugasundaram, G. Santhip Kumar, S. Sankaralingam, S. Vishal, and N. Kamaleswaran, "Smart agriculture using modern technologies," in *2023 9th International Conference on Advanced Computing and Communication Systems (ICACCS)*, vol. 1, pp. 2025–2030, 2023.

[9] N. Telagam, N. Kandasamy, and M. Arun Kumar, "Review on smart farming and smart agriculture for society: Post-pandemic era," *Green Technological Innovation for Sustainable Smart Societies: Post Pandemic Era*, pp. 233–256, 2021.

[10] M. San Emeterio de la Parte, J.-F. Martínez-Ortega, V. Hernández Díaz, and N. L. Martínez, "Big data and precision agriculture: a novel spatio-temporal semantic iot data management framework for improved interoperability," *Journal of Big Data*, vol. 10, no. 1, pp. 1–32, 2023.

[11] H. A. Issad, R. Aoudjit, and J. J. Rodrigues, "A comprehensive review of data mining techniques in smart agriculture," *Engineering in Agriculture, Environment and Food*, vol. 12, no. 4, pp. 511–525, 2019.

[12] M. Ayaz, M. Ammad-Uddin, Z. Sharif, A. Mansour, and E.-H. M. Aggoune, "Internet-of-things (iot)-based smart agriculture: Toward making the fields talk," *IEEE Access*, vol. 7, pp. 129551–129583, 2019.

[13] O. Elijah, T. A. Rahman, I. Orikumhi, C. Y. Leow, and M. N. Hindia, "An overview of internet of things (iot) and data analytics in agriculture: Benefits and challenges," *IEEE Internet of Things Journal*, vol. 5, no. 5, pp. 3758–3773, 2018.

[14] G. D. Tikas and K. Akhilesh, "Smart agriculture: A tango between modern iot-based technologies and traditional agriculture techniques," *Smart Technologies: Scope and Applications*, pp. 387–394, 2020.

[15] S. Mishra, S. Nayak, and R. Yadav, "An energy efficient lora-based multi-sensor iot network for smart sensor agriculture system," in *2023 IEEE Topical Conference on Wireless Sensors and Sensor Networks*, pp. 28–31, 2023.

[16] X. Jiang, J. F. Waimin, H. Jiang, C. Mousoulis, N. Raghunathan, R. Rahimi, and D. Peroulis, "Wireless sensor network utilizing flexible nitrate sensors for smart farming," in *2019 IEEE SENSORS*, pp. 1–4, 2019.

[17] V. V. Das, A. Sathyan, and D. D S, "Establishing lora based local agri-sensor network through sensor plugin modules and lorawan data concentrator for extensive agriculture automation," in *2022 IEEE 19th India Council International Conference (INDICON)*, pp. 1–6, 2022.

[18] S. Gore, S. Patil, and V. Khalane, "Intelligent farm monitoring system using lora enabled iot," in *2022 IEEE Bombay Section Signature Conference (IBSSC)*, pp. 1–6, 2022.

[19] S. Bagwari, A. Roy, A. Gehlot, R. Singh, N. Priyadarshi, and B. Khan, "Lora based metrics evaluation for real-time landslide monitoring on iot platform," *IEEE Access*, vol. 10, pp. 46392–46407, 2022.

[20] D. Wu and J. Liebeherr, "A low-cost low-power lora mesh network for large-scale environmental sensing," *IEEE Internet of Things Journal*, 2023.

[21] P. S. Raju, S. L. Priya, S. Ksheeraja, B. Menaga, and V. Ragul, "Green iot framework for deep forest surveillance," in *2023 International Conference on Signal Processing, Computation, Electronics, Power and Telecommunication (IConSCEPT)*, pp. 1–5, IEEE, 2023.

[22] P. S. Raju, S. Lakshmi Priya., S. Ksheeraja., B. Menaga., and V. Ragul., "Green iot framework for deep forest surveillance," in *2023 International Conference on Signal Processing, Computation, Electronics, Power and Telecommunication (IConSCEPT)*, pp. 1–5, 2023.

[23] M. Padmaja, V. S. Priya, C. H. N. Dontu, B. S. Devi, and P. P. Kumar, "Surveillance system for improved energy efficiency using lora technology," in *2023 4th International Conference on Signal Processing and Communication (ICSPC)*, pp. 138–142, IEEE, 2023.

[24] C. Zhang and J. Yang, "Lora-based smart greenhouse control system," in *2023 IEEE 2nd International Conference on Electrical Engineering, Big Data and Algorithms (EEBDA)*, pp. 948–952, 2023.

[25] T. N. Gia, L. Qingqing, J. P. Queralta, Z. Zou, H. Tenhunen, and T. Westerlund, "Edge ai in smart farming iot: Cnns at the edge and fog computing with lora," in *2019 IEEE AFRICON*, pp. 1–6, 2019.

[26] G. P. Lakshmi, P. Asha, G. Sandhya, S. V. Sharma, S. Shilpashree, and S. Subramanya, "An intelligent iot sensor coupled precision irrigation model for agriculture," *Measurement: Sensors*, vol. 25, p. 100608, 2023.

[27] R. Anzum, M. H. Habaebi, M. R. Islam, G. P. N. Hakim, M. U. Khandaker, H. Osman, S. Alamri, and E. AbdElrahim, "A multiwall path-loss prediction model using 433 mhz lora-wan frequency to characterize foliage’s influence in a malaysian palm oil plantation environment," *Sensors*, vol. 22, no. 14, 2022.

[28] S. Dhifaoui, C. Houaidia, and L. A. Saidane, "Cloud-fog-edge computing in smart agriculture in the era of drones: a systematic survey," in *2022 IEEE 11th IFIP International Conference on Performance Evaluation and Modeling in Wireless and Wired Networks (PEMWN)*, pp. 1–6, 2022.

[29] C. Bruno, A. Licciardello, G. A. M. Nastasi, F. Passaniti, C. Brigante, F. Sudano, A. Faulisi, and E. Alessi, "Embedded artificial intelligence approach for gas recognition in smart agriculture applications using low cost mox gas sensors," in *2021 Smart Systems Integration (SSI)*, pp. 1–5, 2021.

[30] W. Chen, B. Liu, H. Huang, S. Guo, and Z. Zheng, "When uav swarm meets edge-cloud computing: The qos perspective," *IEEE Network*, vol. 33, no. 2, pp. 36–43, 2019.

[31] H. Mei, K. Yang, Q. Liu, and K. Wang, "Joint trajectory-resource optimization in uav-enabled edge-cloud system with virtualized mobile clone," *IEEE Internet of Things Journal*, vol. 7, no. 7, pp. 5906–5921, 2019.

[32] V. K. V. V. Bathalapalli, S. P. Mohanty, E. Kougianos, V. P. Yanambaka, B. K. Baniya, and B. Rout, "A puf-based approach for sustainable cybersecurity in smart agriculture," in *2021 19th OITS International Conference on Information Technology (OCIT)*, pp. 375–380, 2021.

[33] L. Ting, M. Khan, A. Sharma, and M. D. Ansari, "A secure framework for iot-based smart climate agriculture system: Toward blockchain and edge computing," *Journal of Intelligent Systems*, vol. 31, no. 1, pp. 221–236, 2022.

[34] A. Power and G. Kotonya, "Complex patterns of failure: Fault tolerance via complex event processing for iot systems," in *2019 International Conference on Internet of Things (iThings) and IEEE Green Computing and Communications (GreenCom) and IEEE Cyber, Physical and Social Computing (CPSCom) and IEEE Smart Data (SmartData)*, pp. 986–993, 2019.

[35] A. Power and G. Kotonya, "Providing fault tolerance via complex event processing and machine learning for iot systems," in *Proceedings of the 9th International Conference on the Internet of Things*, IoT '19, (New York, NY, USA), Association for Computing Machinery, 2019.

[36] Y. Chen, H. Yang, K. Liu, J. Zhao, and Y. Jia, "Collective event detection via a hierarchical and bias tagging networks with gated multi-level attention mechanisms," in *Proceedings of the 2018 Conference on Empirical Methods in Natural Language Processing*, (Brussels, Belgium), pp. 1267–1276, Association for Computational Linguistics, Oct.-Nov. 2018.

[37] S. Liu, Y. Chen, K. Liu, and J. Zhao, "Exploiting argument information to improve event detection via supervised attention mechanisms," in *Proceedings of the 55th Annual Meeting of the Association for Computational Linguistics (Volume 1: Long Papers)*, pp. 1789–1798, 2017.

979-8-3503-8368-3/23 $31.00 © 2023 IEEE

Ship Engine Data Analysis for the Application of Machine Learning Algorithms

Theodoros Dimitriou[1], Emmanouil Skondras[1], Christos Hitiris[2], Cleopatra Gkola[3], Ioannis S. Papapanagiotou[1], Dimitrios J. Vergados[3], Constantinos Vergopoulos[4], Stratos Koumantakis[5], Angelos Michalas[2], Dimitrios D. Vergados[1]

[1]Department of Informatics, University of Piraeus, Piraeus, Greece, Email: {theodim, skondras, jpapapanagiotou, vergados}@unipi.gr
[2]Department of Electrical and Computer Engineering, University of Western Macedonia, Kozani, Greece,
Email: {c.hitiris, amichalas}@uowm.gr
[3]Department of Informatics, University of Western Macedonia, Kastoria, Greece, Email: {c.gkola, dvergados}@uowm.gr
[4]Internet Business Hellas, Athens, Greece, Email: vergopoulos@ibhellas.gr
[5]MAS S.A., Athens, Greece, Email: skoumantakis@maseurope.com

Abstract— **In Machine Learning (ML) the analysis and the preparation of data before their use is considered an important task, in order to improve the performance of the ML algorithms. Techniques like clustering and dimensionality reduction are applied for the preparation of the data offering several advantages for the ML algorithms to which the data will be used. Some indicative advantages include the improved performance and the faster training of the ML algorithms, the handling of missing or corrupted data, the detection of data overfitting and the visualization of the data. In this paper, a dataset that contains plenty of ship engine's data is analyzed. Specifically, methodologies for density estimation, clustering and dimensionality reduction are studied. Subsequently, such methodologies are applied to the aforementioned dataset, providing useful results about the structure of the dataset, about the correlation of its data, as well as about the importance of each feature included in the dataset.**

Keywords— *Machine Learning (ML), Ship Engine Data Analysis, Density Estimation, Data Clustering, Dimensionality Reduction*

I. INTRODUCTION

Machine learning (ML) [1] is a subfield of computer science, developed from the study of computational learning theory in artificial intelligence. In particular, machine learning explores the study and development of algorithms that are capable of learning from data and predicting or making decisions based on it. These algorithms create models based on experimental data, allowing them to make predictions of outcomes or make decisions expressed as the result.

ML algorithms require the use of training data, which allows computer systems to learn through trial and error procedures. In general, they rely on pattern recognition and inference which allow the decision making without the need of writing special programming code.

The preparation of the data for their subsequent use, is considered as one of the most important parts of ML. The collection, processing and evaluation of training data prepares them in order to be used by the ML algorithms in an efficient way. In particular, the removal of unnecessary data or outlier values, or the correction of incorrect data snapshots, improves the performance of the ML algorithms. More specifically, techniques like clustering and dimensionality reduction offer several advantages in the context of data preparation, including the following:

- Improved ML algorithms performance: By reducing noise and irrelevant information in the data, data preparation techniques can enhance the overall performance of ML models. This is particularly important when working with high-dimensional and noisy datasets.
- Faster training of ML algorithms: Dimensionality reduction techniques reduce the number of features that exist to a dataset, leading to faster training times for ML models. This is especially helpful for large datasets.
- Enhanced interpretability: Data preparation techniques complex datasets, making it easier to understand the relationships within the data. This can be crucial for model debugging and explaining model predictions to stakeholders.
- Handling missing data: Clustering techniques can be used to impute missing data by assigning data points to clusters with similar characteristics. This can improve the quality of the dataset and ensure that the model is trained on as much available information as possible.
- Detection and reduction of data overfitting: Dimensionality reduction can help reduce overfitting, a common problem in machine learning, by simplifying the model's representation of the data. This can lead to better generalization on unseen data.
- Data visualization: Clustering and dimensionality reduction techniques are valuable for data visualization. They allow you to project high-dimensional data into lower-dimensional spaces or visualize data clusters, making it easier to explore and gain insights from the data.

In summary, data preparation techniques like clustering and dimensionality reduction play a crucial role in improving the quality of data, simplifying complex datasets, and enhancing the performance of machine learning models. These techniques enable data scientists and machine learning practitioners to work with cleaner, more informative data, leading to more accurate and interpretable models.

It has to be noted that for accomplishing data analysis and preparations, statistical models and unsupervised ML algorithms are used. Unsupervised machine learning algorithms learn exclusively from unlabeled data, which is much more frequently available than labeled data [2]. These algorithms are suitable for clustering techniques by indicating associations between the data, as well as for creating labels on the data, which are then used to implement supervised learning tasks. More specifically, unsupervised clustering algorithms identify inherent groupings within the unlabeled data and then assign a label to each data. On the other hand, unsupervised correlation mining algorithms identify rules that accurately represent the relationships between data features [3]. The absence of labels that represent the desired behaviour of a model also means the absence of a stable reference point for evaluating the quality of the model.

In this paper, the Condition Based Maintenance of Naval Propulsion Plants (CBM) dataset [4] is used as a template for data analysis. In particular, the CBM dataset contains 11934 records of data about ship engines and thus it provides plenty of data that can be used for the training and the evaluation of ML algorithms. The analysis of the dataset is deemed to be useful for several research issues in the field of ML including the estimation of the density of the data, the clustering of the data and the dimensionality reduction. The analysis of data if performed using the Orange toolkit [5]. Orange provides a set of widgets that allows the application of machine learning algorithms, as well as the processing and the evaluation of the data that these algorithms use as input or produce as output. The paper processes the CBM dataset in terms of feature correlations, scatterplots, feature statistics, as well as performs clustering and dimensionality reduction using the k-Means and the PCA unsupervised ML algorithms.

The remainder of the paper is organized as follows: Section II presents the state of the art, including the description of density estimation, data clustering, dimensionality reduction and autoencoders. Subsequently, section III presents the simulation setup and the data analysis, while section IV concludes our paper.

II.   STATE OF THE ART

In the following subsections, the main categories of unsupervised ML techniques used for data analysis are described, including density estimation, data clustering, dimensionality reduction and autoencoders.

II.1. Density Estimation

In density estimation [6][7], the machine learning algorithm learns a function p: $\mathbb{R}^n \rightarrow \mathbb{R}$, where p(x) is regarded as a probability density function when x is continuous, or as a probability mass function when x is discrete, in the space from which the data were taken from. To accomplish this task, the algorithm must learn the structure of the observed data. The distribution of the data can be explicitly captured by density estimation. The distribution obtained can then be use to solve other problems. In particular, if we have performed density estimation to obtain a probability distribution p(x), we can use this distribution to solve the missing value imputation problem. If a value x_i is missing and all other values, denoted x_{-i}, are given, then we know that the distribution on it is given by $p(x_i \mid x_{-i})$. In practice, density estimation does not always enable us to solve all these related problems, because in many cases the required operations on p(x) are computationally intractable.

II.2. Clustering

Clustering [8] is a learning problem of assigning a label to input data by utilizing an unlabeled training dataset. The goal is to organize similar or related data into clusters. Just like in classification ML algorithms, each input is assigned to

a cluster. Unlike classification however, clustering is an unsupervised task.

There is no single definition of what a cluster is since it depends on the scope, while different algorithms could create different clusters. Certain algorithms look for continuous areas of dense instances, producing clusters that could take any shape. Other algorithms look for instances that are centered around a specific point, called the center of the cluster. Additionally, some algorithms produce hierarchical groups of clusters.

Clustering can be applied to a wide range of situations such as [9]:

- Data segmentation: Indicatively, clustering can be used for customer segmentation based on aspects such as their purchasing habits and behavior, in order to get insights on their needs, so that products and marketing campaigns can be tailored accordingly. Segmentation is also applicable to recommendation systems to suggest content based on same group users' choices.
- Dimensionality reduction: As a reduction technique, after the application of clustering to a dataset, each instance's association to each cluster can often be measured. Thus, the feature vector x of each datum can then be replaced by the vector of its cluster associations. When k clusters are produced, then the produced vector will have k dimensions. Often this vector has much less dimensions than the original feature vector x, but the information retained is enough to process the data further.
- Outlier detection: Any data value that is not close to any cluster is probably an outlier. A case that this technique can be used is to uncover uncommon behavior of website users by detecting excessive rate of requests. Outlier detection is especially useful in spotting fraudulent behavior and manufacturing defects.
- Semi-supervised learning: For datasets where only a few instances are labeled, clustering could be used to propagate these labels to the rest of the instances belonging the same cluster. This way the laborious task of manually assigning labels can be assisted, resulting in a greater number of labeled data, enabling improved performance on subsequently applied supervised learning algorithms.
- Image segmentation: After grouping image data based on their color, image segmentation can be performed by replacing each pixel's color with the average color of the cluster it belongs to. In this way the number of different colors in the image can be notably reduced. Image segmentation is often used in object detection and tracking systems, as it facilitates the detection of the outline of each object.

II.2.1. The K-Means algorithm

K-Means [10] is a popular unsupervised learning clustering algorithm. The algorithm creates a set of clusters and categorizes, in each cluster, the data it receives as input, based on the similarity between them. The steps of the K-Means algorithm are:

1. Definition of the number of groups (clusters) to be created: The number (k) of clusters to be produced is determined by the user.
2. Initialization: k points are randomly selected from the data, as initial centers of the groups (centroids).

979-8-3503-8368-3/23 $31.00 © 2023 IEEE

3. Placement-categorization of data into clusters: For each of the input data, its distance from the center of each cluster is calculated and it is placed in the cluster where the smallest distance is observed.
4. Update centers: Cluster centers are recalculated as the average of the data values placed in each cluster.
5. Iteration: Steps 3 and 4 are repeated until the centers of the clusters no longer change and convergence of the algorithm has been achieved.

II.2.2. The Density Based Spatial Clustering of Applications with Noise (DBSCAN) algorithm

The Density Based Spatial Clustering of Applications with Noise (DBSCAN) algorithm [11] identifies and clusters data located in high density areas, while ignoring data located in low density areas, as it considers them as noise. The DBSCAN algorithm starts by randomly selecting a value from the dataset. Afterwards, the distance of neighboring values from the selected value is examined. If the density of the neighboring values is high enough (higher than the minimum acceptable density set by the user), then a cluster is created, in which the selected value as well as the neighboring values are placed. Similarly, if the density of the neighboring values is not high enough, then the algorithm proceeds to examine the next value present in the dataset. The DBSCAN algorithm continues this process by gradually examining all the values present in the dataset. Finally, values that do not belong to a cluster are labeled as noise.

DBSCAN has the advantage that it can create clusters of arbitrary shape, in contrast to other centroid-based algorithms such as k-means that create hypersphere shaped clusters. But DBSCAN has the disadvantage that it has two hyperparameters that need to be set and choosing appropriate values for them can be a challenge. Moreover, because the minimum acceptable density is constant, the clustering algorithm cannot effectively deal with groups of different densities. Overall DBSCAN works well for cluster that are dense enough and well separated by regions of low-density.

II.3. Dimensionality Reduction

Several ML problems involve thousands or even millions of features for each training instance. This number of features makes training extremely slow, while at the same time finding a good solution becomes a difficult task. This problem is known as the curse of dimensionality. Usually, it is possible to significantly reduce the number of features, turning an unmanageable problem into a solvable one. Indicatively, the pixels at the borders of some images, such as the ones from the MNIST database [12], are most often white, so it is possible to completely drop them from the training dataset without losing much information. Also, two neighboring pixels often have a high correlation, so merging them into one pixel and assigning the average intensity of the two does not result in any significant loss of information.

However, dimensionality reduction [13] incurs a loss of information, so although training speed will increase, the system performance may become somewhat worse. Also, this increases the complexity of the process pipelines and therefore it becomes more difficult to maintain. Though it is not common, in some situations, dimensionality reduction can reduce some noise and filter out some unnecessary details and thus lead to improved performance for the ML algorithm.

In addition to speeding up the training process, reducing the number of the dimensions is also useful for visualizing data. Visualizing a high-dimensional training dataset down to two or three dimensions makes possible to plot a condensed version of the data on a graph. This enables important information to be detected by visually analyzing patterns of the studied dataset.

II.3.1. The Principal Component Analysis (PCA) algorithm

The Principal Component Analysis (PCA) algorithm [14] is used to decompose a multivariate dataset into a sequence of - mutually orthogonal - components that explain, in the maximum degree, the variance of the dataset. PCA performs linear dimensionality reduction using Singular Value Decomposition (SVD) [15] of the data, projecting them into a space of less dimensions than the original one. Before the SVD is applied to the input data, they are centered but not scaled.

II.3.2. The Recursive Feature Elimination (RFE) algorithm

The Recursive Feature Elimination (RFE) algorithm [16] is a recursive process that ranks features based on some measure of their importance. In each iteration, the importance of every feature is estimated. The feature with the lowest importance is then dismissed. Alternatively, it is possible to remove an entire group of features in each iteration (instead of a single feature) in order to make the process faster. However, removing only one feature at each iteration of the process is sometimes necessary because the relative importance of each feature may change when it is evaluated against a different subset of features (in the progressive elimination process), especially when features with a high association between them are evaluated. The final ranking is constructed by ordering the features by the order they were eliminated and then keeping only the features in the first positions.

II.4. Autoencoders

Autoencoders are unsupervised Artificial Neural Network (ANN) learning algorithms [17], used to learn compressed and coded data representation. They are mainly used for dimensionality reduction and unsupervised pretraining of feedforward neural networks. In general, autoencoders are applied to unlabeled data. They use approximation functions and are trained with backpropagation [18] and Stochastic Gradient Descent techniques [19]. Also, their purpose is to learn a compact representation of the input function using the same number of input and output units, but with typically fewer hidden units. The input function is learned by trying to recreate the input to the output, a process called encoding/decoding, where a simple autoencoder learns a low-dimensional representation of the input data by exploiting similar repeating patterns.

III. DATASET ANALYSIS RESULTS

In this section, we will refer to the processing of the CBM Naval Propulsion Plants dataset obtained from the Kaggle online community [4]. The propulsion system modelled in the simulation consists of metrics for the Propeller, Hull, Gas Turbine (GT), Gearbox and Controller parts which have been developed and accurately tuned with the operation of real propulsion plants during one year presenting similarities with real vessels. Furthermore, the performance decay of the GT components over time is considered.

The input data to the propulsion system simulator is defined by the parameters:

- Lever position (lp)
- Compressor decay state coefficient (kMc)
- Turbine decay state coefficient (kMt)

where each possible state of ship engine can be described by a combination of the above parameters (lp, kMc, kMt). The compressor and turbine decay ranges have been tested over a range of values with a step change of 0.001, in order to obtain the highest degree of fidelity of representation. In particular, for the analysis of the compressor decay condition, the kMc coefficient has 51 values in the range [1, 0.95], while the turbine coefficient has 26 values in the range [1, 0.975]. Therefore, the dataset contains 11934 (51x26x9) records.

The propulsion system simulator output has been acquired and stored in the dataset which includes a set of the following 16 features that affect the state of the system subject to decay:

1. Lever position (lp)
2. Ship speed (v) [knots]
3. Gas Turbine (GT) shaft torque (GTT)) [kN m]
4. Revolutions per minute gas turbine shaft (GT rate of revolutions (GTn)) [rpm]
5. Gas Generator rate of revolutions (GGn) [rpm]
6. Starboard Propeller Torque (Ts) [kN]
7. Port Propeller Torque (Tp) [kN]
8. High Pressure (HP) Turbine exit temperature (T48) [C]
9. Air temperature at the gas turbine compressor inlet valve (GT Compressor inlet air temperature (T1)) [C]
10. Air temperature at the outlet valve of the gas turbine compressor (GT Compressor outlet air temperature (T2)) [C]
11. HP Turbine exit pressure (P48) [bar]
12. Air pressure at the gas turbine compressor inlet valve (GT Compressor inlet air pressure (P1)) [bar]
13. Air pressure at the gas turbine compressor outlet valve (GT Compressor outlet air pressure (P2)) [bar]
14. Gas turbine exhaust gas pressure (GT exhaust gas pressure (Pexh)) [bar]
15. Turbine Injection Control (TIC) [%]
16. Fuel flow (mf) [kg/s]
17. Gas turbine compressor decay state coefficient (GT Compressor decay state coefficient) (values [0-1] with 0 specifying the highest decay level and 1 that there is no decay)
18. Gas turbine decay state coefficient (GT Turbine decay state coefficient) (similar to 17)

III.1. Dataset study

In this section, the CBM dataset is analyzed using the Orange toolkit [5]. These tools offer the ability to handle the data volume of the dataset and run a set of algorithms, including feature statistics, the k-Means and the PCA.

Regarding the CBM dataset we have two types of correlations: a) those that concern measurements of the same characteristic from different engine parts (eg torque) and b) measurements having a linear (or polynomial) relationship between them (eg gear lever position and ship speed). In the first case, the change in the degree of correlation could be due

to a failure of one of the two sensors, while in the second case it could be due to a malfunction of some part of the engine. Indicatively, for the gear level position and ship speed characteristics, a low correlation degree specifies that ship speed does not respond correctly to the speed lever position.

Figures 1 and 2 show excerpts of the table produced by the correlation algorithm. In each row of the table the correlation degree of characteristics pairs is given. We notice that some features pairs show a degree of correlation between them that is close to or equal to 1.0, which means that there is a dependence between them. Thus, only one feature from each pair can be used as input to the learning algorithms, since the presence of both does not affect the algorithms' output. Accordingly, the reduction of the input characteristics would mean faster learning. However, in real-life scenarios, a change in the correlation degree of features can be an indication of damage or malfunction. As a result, features' elimination based on correlations should be carefully considered.

1	+1.000	6 - Starboard Propeller Torque (Ts) [kN]	7 - Port Propeller Torque (Tp) [kN]
2	+1.000	1 - Lever position (lp) [ ]	2 - Ship speed (v) [knots]
3	+0.999	11 - HP Turbine exit pressure (P48) [bar]	13 - GT Compressor outlet air pressure (P2) [bar]
4	+0.999	3 - Gas Turbine shaft torque (GTT) [kN m]	7 - Port Propeller Torque (Tp) [kN]
5	+0.999	3 - Gas Turbine shaft torque (GTT) [kN m]	6 - Starboard Propeller Torque (Ts) [kN]
6	+0.999	11 - HP Turbine exit pressure (P48) [bar]	3 - Gas Turbine shaft torque (GTT) [kN m]
7	+0.999	11 - HP Turbine exit pressure (P48) [bar]	7 - Port Propeller Torque (Tp) [kN]
8	+0.908	11 - HP Turbine exit pressure (P48) [bar]	6 - Starboard Propeller Torque (Ts) [kN]
9	+0.908	11 - HP Turbine exit pressure (P48) [bar]	14 - Gas Turbine exhaust gas pressure (Pexh) [bar]

Figure 1. Beginning of correlation table.

87	+0.924	14 - Gas Turbine exhaust gas pressure (Pexh) [bar]	5 - Gas Generator rate of revolutions (GGn) [rpm]
88	+0.914	1 - Lever position (lp) [ ]	15 - Turbine Injection Control (TIC) [%]
89	+0.919	15 - Turbine Injection Control (TIC) [%]	2 - Ship speed (v) [knots]
90	+0.897	16 - Fuel flow (mf) [kg/s]	5 - Gas Generator rate of revolutions (GGn) [rpm]
91	+0.879	15 - Turbine Injection Control (TIC) [%]	5 - Gas Generator rate of revolutions (GGn) [rpm]
92	-0.047	10 - GT Compressor outlet air temperature (T2) [C]	17 - GT Compressor decay state coefficient.
93	-0.040	17 - GT Compressor decay state coefficient.	8 - HP Turbine exit temperature (T48) [C]
94	-0.038	18 - GT Turbine decay state coefficient.	8 - HP Turbine exit temperature (T48) [C]
95	+0.035	14 - Gas Turbine exhaust gas pressure (Pexh) [bar]	17 - GT Compressor decay state coefficient.
96	-0.032	15 - Turbine Injection Control (TIC) [%]	17 - GT Compressor decay state coefficient.
97	-0.018	15 - Turbine Injection Control (TIC) [%]	18 - GT Turbine decay state coefficient.

Figure 2. End of correlation table.

Figures 3 and 4 depict the lever position and the ship speed features in a scatterplot for a random sample of the data as it is and the same data with noise added respectively. As it is shown these features are not continuous but take discrete values.

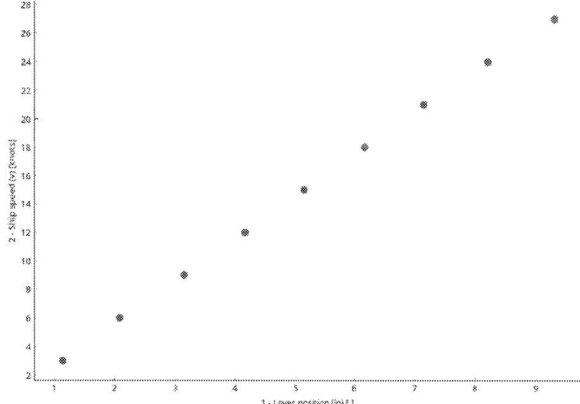

Figure 3. Scatterplot of lever position and ship speed.

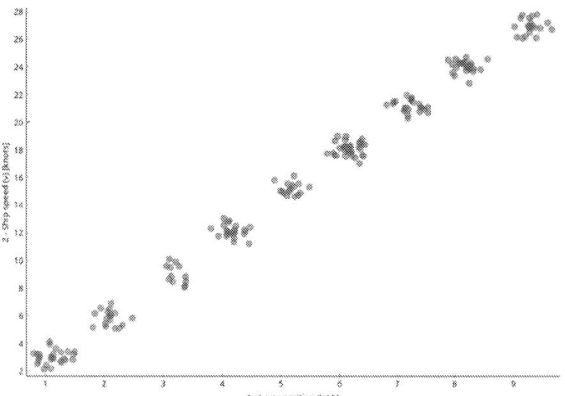

Figure 4. Scatterplot of lever position and ship speed with added noise.

Figure 5 shows in a scatterplot the features gas turbine rate of revolutions and gas generator rate of revolutions. As it is shown the values of both features include plenty different samples. Additionally, the scatterplot denotes that the measurements are accurate and do not include any noise.

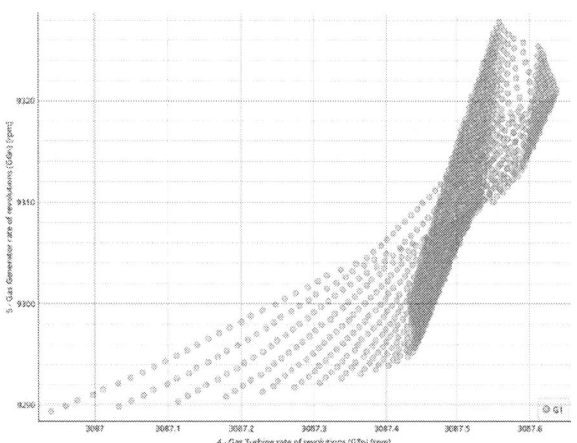

Figure 5. Scatterplot representation of a subset of data for gas turbine rate of revolutions and gas generator rate of revolutions features.

Figure 6 presents summary statistics of the dataset features. The distribution, mean, median, dispersion, minimum, maximum and percentage of missing values for each feature is given. Useful insights can be obtained for each feature from the analysis performed. Indicatively GT Compressor inlet air temperature and GT compressor inlet pressure contain fixed values. Also "HP Turbine exit temperature" and "Turbine Injection Control" features approach the normal distribution, while "Ship speed" approaches the uniform distribution.

For the application of the k-Means algorithm, the hyperparameter of the number of clusters should be set. However, the Orange toolkit has the ability to provide results for a set of cluster numbers within a given range. For each number of clusters, a silhouette score is returned, which indicates how well the number of clusters fits the data of the dataset on average (Figure 7). The silhouette score has values in the range [−1, +1], with higher scores indicating better clustering. As it is observed, the best performance is obtained using two clusters while acceptable silhouette scores are obtained using up to 12 clusters.

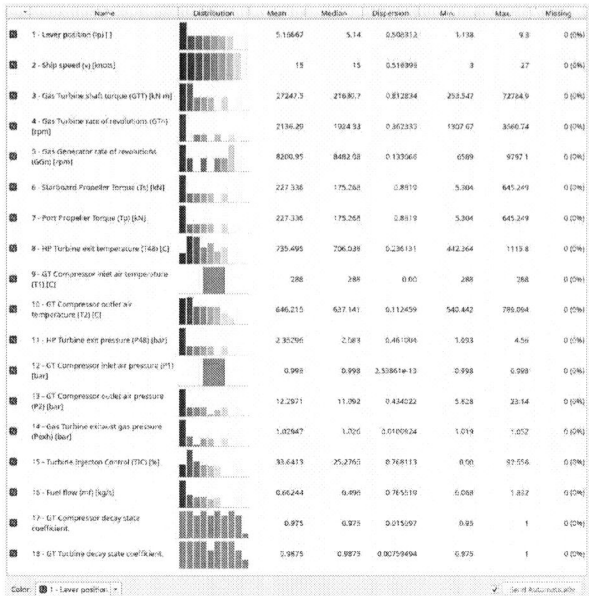

Figure 6. Statistics about the features of the dataset.

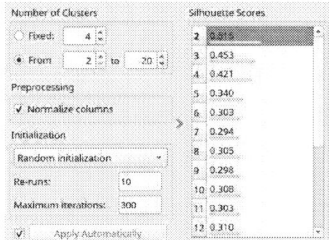

Figure 7. Silhouette index table for each number of groups.

By analyzing the results of clustering, we plot the dataset together with clustering labels into a scatterplot. The characteristics of the position of the lever position and the ship speed are used as coordinates, since they obtain distinct values and thus they are likely to be factors of attraction for the creation of the clusters. Each data cluster is depicted using a different color. By selecting seven clusters from k-Means and adding noise to the plot points to reveal the data at each point, it is obtained that the data corresponding to each plot point are dominated by only one color, as shown in Figures 8 and 9. Thus it becomes visually obvious that there is a strong influence to the data from the lever position and ship speed features.

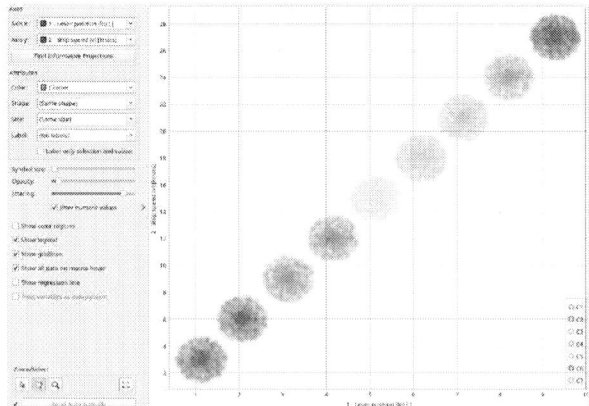

Figure 8. Scatterplot of lever position and ship speed colored according to the cluster to which they belong.

979-8-3503-8368-3/23 $31.00 © 2023 IEEE

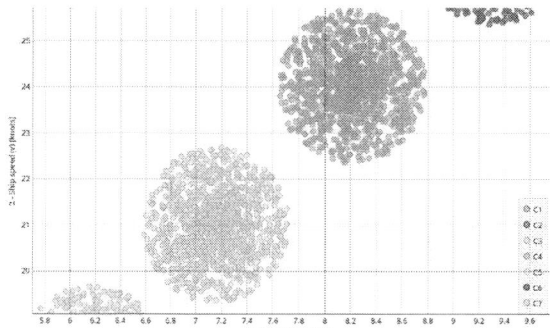

Figure 9. Zoom in the scatterplot of lever position and ship speed colored according to the cluster to which they belong.

Finally, by executing the PCA algorithm in the Orange toolkit, we observe that the cumulative variance of the dataset explained by the 4 first components, is equal to 99.5% (Figure 10). Also, the results for each component's variance confirm that the first four components are the most significant. This information is useful for reducing the number of dimensions of the dataset, in order to reduce the complexity of processing the dataset using ML algorithms. This could apply in cases where reduction of processing time is more important than a small loss of information, which will cause little alteration to the results of the ML algorithm.

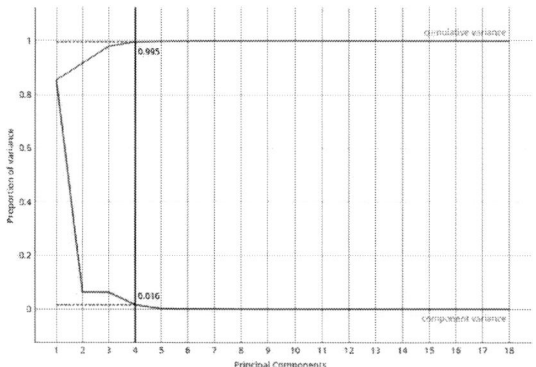

Figure 10. Cumulative variance of the data set as a function of the number of first components of the PCA algorithm.

IV. Conclusion

In this paper, the CBM Naval Propulsion Plants dataset, that contains ship engine's data, is analysed. In particular, the visual analysis of the dataset is performed using scatterplots, while ML algorithms for clustering and dimensionality reduction are applied through the Orange toolkit. Clustering of the data is achieved using the K-Means and the DBSCAN ML algorithms. Also, for the dimensionality reduction the PCA algorithm is used and its effectiveness is demonstrated. Useful results concerning the structure of the dataset are obtained, indicating the correlation among the data and the importance of each feature of the dataset. Future work includes the application of the aforementioned methodologies for density estimation, clustering and dimensionality reduction in data produced from ship engines' operation.

ACKNOWLEDGEMENTS

This research has been co-financed by the European Regional Development Fund of the European Union and Greek national funds through the Operational Program Competitiveness, Entrepreneurship and Innovation, under the call RESEARCH-CREATE-INNOVATE (project code: T2EDK-C.873).

References

[1] Sarker, I. H., "Machine learning: Algorithms, real-world applications and research directions", Computer Science, Springer, vol. 2, issue 3, no. 160, pp. 1-21, 2021.

[2] Russell, S., Norvig, P., "Artificial Intelligence: A Modern Approach", 2002.

[3] Alloghani, M., Al-Jumeily, D., Mustafina, J., Hussain, A., Aljaaf, A. J., "A systematic review on supervised and unsupervised machine learning algorithms for data science", Supervised and Unsupervised Learning for Data Science Journal, Springer, pp. 3-21, 2020.

[4] The Condition Based Maintenance of Naval Propulsion Plants (CBM) dataset: https://www.kaggle.com/datasets/elikplim/maintenance-of-naval-propulsion-plants-data-set , Accessed: October 2023.

[5] Orange toolkit: https://orangedatamining.com , Accessed: October 2023.

[6] Goodfellow, I., Bengio, Y., Courville, A., "Deep learning", MIT press, 2016.

[7] Dalmasso N., Pospisil T., Lee A. B., Izbicki R., Freeman P. E., Malz A. I., "Conditional density estimation tools in python and R with applications to photometric redshifts and likelihood-free cosmological inference", Astronomy and Computing, Elsevier, vol. 30, pp.1-14, 2020.

[8] Ezugwu A. E., Ikotun A. M., Oyelade O. O., Abualigah L., Agushaka J. O., Eke C. I., Akinyelu,A. A., "A comprehensive survey of clustering algorithms: State-of-the-art machine learning applications, taxonomy, challenges, and future research prospects", Engineering Applications of Artificial Intelligence, Elsevier, vol. 110, pp. 1-43, 2022.

[9] Géron, A., "Hands-on machine learning with Scikit-Learn, Keras, and TensorFlow: Concepts, tools, and techniques to build intelligent systems", O'Reilly Media, Inc., 2019.

[10] Pham T. A., Dang X. K., Vo,N. S., "Optimising Maritime Big Data by K-means Clustering with Mapreduce Model", International Conference on Industrial Networks and Intelligent Systems, Springer, Da Nang, Vietnam, pp. 136-151, April 2022.

[11] Han X., Armenakis C., Jadidi M., "DBSCAN optimization for improving marine trajectory clustering and anomaly detection", The International Archives of the Photogrammetry, Remote Sensing and Spatial Information Sciences, ISPRS Congress, vol. 43, pp. 455-461, 2020.

[12] The MNIST database: http://yann.lecun.com/exdb/mnist , Accessed: October 2023.

[13] Ashraf M., Anowar F., Setu J. H., Chowdhury A. I., Ahmed E., Islam A., Al-Mamun A., "A Survey on Dimensionality Reduction Techniques for Time-series Data", IEEE Access, vol.11, pp. 42909 - 42923, 2023.

[14] Park J., Oh J., "Analysis of Collected Data and Establishment of an Abnormal Data Detection Algorithm Using Principal Component Analysis and K-Nearest Neighbors for Predictive Maintenance of Ship Propulsion Engine", Processes Journal, MDPI, vol. 10, issue 11, pp. 1-13, 2022.

[15] Epps, B. P., Krivitzky, E. M., "Singular value decomposition of noisy data: noise filtering", Experiments in Fluids, Springer, vol. 60, pp. 1-23, 2019.

[16] Habibi A., Delavar M. R., Sadeghian M. S., Nazari B., Pirasteh, S., "A hybrid of ensemble machine learning models with RFE and Boruta wrapper-based algorithms for flash flood susceptibility assessment", International Journal of Applied Earth Observation and Geoinformation, Elsevier, vol. 122, pp. 1-18, 2023.

[17] Usama, M., Qadir, J., Raza, A., Arif, H., Yau, K. L. A., Elkhatib, Y., Al-Fuqaha, A., "Unsupervised machine learning for networking: Techniques, applications and research challenges", IEEE Access, vol. 7, pp. 65579-65615, 2019.

[18] Vlachas P. R., Pathak J., Hunt B. R., Sapsis T. P., Girvan M., Ott E., Koumoutsakos P., "Backpropagation algorithms and reservoir computing in recurrent neural networks for the forecasting of complex spatiotemporal dynamics", Neural Networks, Elsevier, vol 126, pp. 191-217, 2020.

[19] Netrapalli, P., "Stochastic gradient descent and its variants in machine learning", Journal of the Indian Institute of Science, Springer, vol. 99, issue 2, pp. 201-213, 2019.

Acceleration of GANs for Potato Crop Disease Identification via FPGA

Theodora Sanida[1], Argyrios Sideris[1], Maria Vasiliki Sanida[2], Michael Dossis[1] and Minas Dasygenis[1]

[1]*Department of Electrical & Computer Engineering, University of Western Macedonia, 50131, Kozani, Greece*
[2]*Department of Digital Systems, University of Piraeus 18534, Piraeus, Greece*
thsanida@uowm.gr, asideris@uowm.gr, sanidasilia@gmail.com, mdossis@uowm.gr, mdasyg@ieee.org

Abstract—**Effective management of plant diseases is essential for ensuring food protection, as a wide range of diseases can significantly reduce crop yields. Potatoes, a staple crop worldwide, are particularly susceptible to various diseases. The timely identification and early warning systems are pivotal in curbing the proliferation of potato diseases and ultimately increasing crop productivity. In recent times, Generative Adversarial Networks (GANs) have emerged as revolutionary technology within the realm of artificial intelligence and machine learning. These networks provide a diverse dataset of synthetic images, thereby facilitating the development of robust computer vision algorithms capable of accurately discerning and categorizing various potato crop diseases. In the scope of our work, we have implemented and accelerated GAN on a Xilinx FPGA SoC. This platform has exceptional efficiency in handling such challenges in terms of power consumption and performance. Our method has also achieved superior power and performance efficiency compared to conventional GPU and CPU setups, boasting an impressive 0.015 ms average time per image in the FPGA's configuration.**

Index Terms—**Potato crop diseases; Identification; GANs; Hardware; FPGA; Generative Adversarial Networks.**

I. INTRODUCTION

Ensuring food security remains a paramount global challenge, and the potato stands as a vital staple crop, poised to contribute significantly to food supplies due to its high yield and starch content. The potato ranks as the third most essential food crop worldwide, following wheat and rice, and serves as the primary staple for over a billion people globally. Nonetheless, the potato crop faces a constant threat from various diseases, which can severely impact production and result in substantial yield losses. Timely detection and warnings play a pivotal role in preventing large disease outbreaks, traditionally relying on the expertise of plant disease specialists or experienced farmers. However, this approach is inherently time-consuming, inefficient, labour-intensive, and tedious, making it challenging to implement widely. Hence, there is a pressing need for a simple, rapid, and reliable tool capable of automatically identifying various types of crop diseases [1]–[3].

Today, Generative Adversarial Networks (GANs) have emerged as a groundbreaking technology, particularly in the context of generating highly realistic and top-tier quality data. GANs have assumed a pivotal role in agriculture, offering innovative solutions to address critical challenges, notably the detection and prevention of crop diseases, with a special emphasis on those afflicting potato crops. The remarkable utility of GANs in agriculture is evident in their ability to produce synthetic images that serve a dual purpose. Firstly, these synthetic images provide an invaluable resource for training specialized machine learning models focused on disease detection. Leveraging GAN-generated images, these models can be fine-tuned to recognize even subtle disease symptoms, enhancing the accuracy and reliability of early disease identification. Secondly, GANs empower the development of robust computer vision algorithms that excel in precisely identifying and categorising diverse diseases affecting potato crops. Furthermore, GANs contribute significantly to the advancement of precision agriculture techniques. So, by analyzing extensive datasets comprising synthetic images, researchers can glean valuable insights into disease progression, the influence of environmental factors, and the consequences of varying agricultural management practices on disease development [4]–[6].

The newness of this investigation in the potato crop disease identification field is:

- We create and train a GAN model specifically tailored to generate novel and previously unseen images derived from potato crop disease data. This approach can complement real-world datasets, enhancing the accuracy and robustness of disease detection models.
- We implement a dedicated hardware architecture designed for a Xilinx FPGA to accelerate the GAN design. This hardware acceleration is marked by meticulous optimizations in various aspects, including host interaction, memory management, and kernel operations, which significantly enhance the overall efficiency of the GAN model, allowing for faster and more resource-efficient disease image generation.
- We comprehensively evaluate the newly generated images regarding both performance and image quality. This evaluation extends across different metrics in the hardware architecture, enabling us to fine-tune and optimize the GAN model's output. Moreover, we extend our scrutiny to other computing devices, such as GPU and CPU, in order to benchmark and compare the performance and effectiveness of our approach against conventional computing setups.

979-8-3503-8368-3/23 $31.00 © 2023 IEEE

This study's remaining parts are structured as follows: We give matching research investigations in Part II, and Part III analyses the materials and methods used in our methodology. Part IV describes the investigation's outcomes, and Part V summarises our research outcomes and future study.

II. RELATED WORK

The field of GANs, particularly their integration with FPGAs, has relatively limited prior research. This is primarily due to the novelty of this intersection within the research community. However, there have been previous studies focused on image reconstruction algorithms that have utilized hardware acceleration. These earlier works share a common problem domain with our research, providing an opportunity to draw parallels and assess our contributions in terms of result quality and acceleration speed in comparison to prior endeavors [7].

In [8], the authors introduced a hardware architecture tailored for the operation of GANs. They validated it on several GAN models, such as Wasserstein GAN (WGAN), deep convolutional GAN (DCGAN), and energy-based GAN (EBGAN). The experimental outcomes from their study showcased that their design can reach 2211 GOPS while operating at a frequency of 185 MHz. These experiments were conducted on an Intel Stratix 10SX FPGA device with sufficient visual outcomes, affirming the practicality and effectiveness of their hardware architecture. Their presented procedure can reach more than twofold hardware efficiency improvement over prior implementations. Additionally, their design contributed to a reduction in storage requirements in resource-constrained environments.

On [9], the authors proposed a GAN explicitly designed to synthesise person images. This GAN model generates high-quality person images while offering precise control over attributes and poses. The researchers conducted their experiments on the Xilinx ZCU102 platform, showcasing the adaptability of their solution to FPGA technology. Their presented network performs better quantitative and qualitative results compared with CPU and GPU. Moreover, the hardware performance based on the FPGA device can reach the most increased energy efficiency of 73.67 GOPS / W.

In contrast to the previous approaches, our work introduces a GAN model tailored for generating images, utilizing potato crop disease data to complement existing real-world collections. Our research incorporates a specialized hardware architecture meticulously designed for a Xilinx FPGA, which serves as a dedicated accelerator for our GAN model. A focus of our study lies in the pursuit of optimizations across various facets of the GAN model. These optimizations significantly boost its efficiency, facilitating expedited generation of disease images of potato crops. Furthermore, we evaluate by subjecting our approach to alternative platforms, including CPU and GPU setups. This evaluation enables us to gauge the efficiency and efficacy of our solution in comparison to conventional computing configurations.

III. MATERIALS AND METHODS

A. Collection of potato

The potato collection [10] utilized in our study constitutes a subset of the broader PlantVillage collection, comprising a total of 2,152 images. This collection encompasses three distinct categories, each portraying various facets of potato health and disease. Specifically, the collection includes 1,000 images dedicated to the early blight category, another 1,000 images representing the late blight category, and a subset of 152 images showcasing the pristine state of healthy potato plants. In Figure 1, a selection of images from the potato collection is visually depicted.

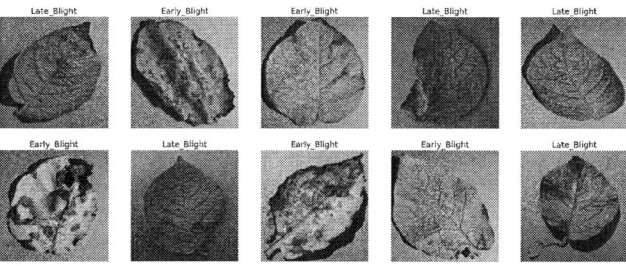

Fig. 1. The potato colection [10] samples.

B. GAN model

The discriminator in our model is designed with a well-structured 3-layer architecture that should fit within the FPGA's on-chip memory for attaining the highest data bandwidth, each layer featuring three key components to enhance its functionality. The layers consist of a Dense layer, allowing the discriminator to capture and process complex patterns and information while evaluating input data. Alongside the Dense layers, each layer has a ReLU activation function, which can be efficiently decoded into hardware. Furthermore, a dropout layer is incorporated after each ReLU activation, with a dropout rate set at 0.3. This dropout layer prevents overfitting and improves the model's generalisation ability.

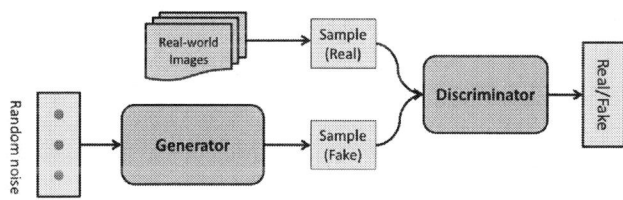

Fig. 2. The GAN model for potato crop diseases.

At the final layer of the discriminator, a Sigmoid activation function is employed for output processing. This function squashes the output values to a range between 0 and 1, effectively providing a probability-like score. Through this final layer, the discriminator makes its ultimate judgment regarding the authenticity of the input data, deciding whether it is real or generated by the generator.

The generator, tasked with producing synthetic images, follows a similar 3-layer layout to the discriminator. However, there is a pivotal distinction in that the generator's final layer incorporates a Tanh activation function with a pre-stored value table in the FPGA. This Tanh activation function is essential as it allows the generator to produce image data within the range of -1 to 1, often the desired output format for images. This characteristic makes the generator capable of generating high-quality synthetic images that closely resemble real data. The Gan model for potato crop diseases is illustrated in Figure 2, and in Figure 3, we can notice the loss attained for the discriminator and generator design during training.

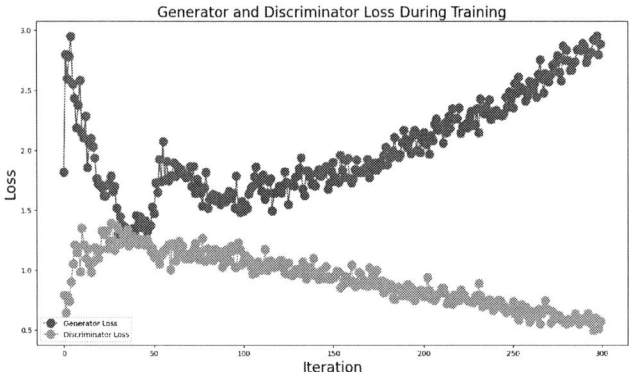

Fig. 3. Generator and discriminator loss during training.

C. Efficient FPGA acceleration for high-performance GAN model

The objective behind acceleration to FPGA was to enhance the GAN model's speed. We developed a memory-efficient neural network, as described in the section above, and then used a pipelined method to synthesize all of its layers in hardware in order to produce a fast reconfigurable approach. Our efforts included careful host code optimizations, extensive pipelining techniques within the FPGA, proficient buffer management for efficient data transfer between the kernels and host, and the establishment of accurate synchronization mechanisms between the kernels and host. These endeavours were guided by FPGA design principles aimed at achieving peak performance, and the resulting system is depicted in Figure 4.

In the initial phase of our FPGA accelerator design, we focused on aligning the host-side functions with the specific requirements of our target applications. Our accelerator primarily handles 64 × 64 pixel images, which are processed through a series of optimized operations. To ensure efficient data flow, we allocated the input image within C++ vectors, organizing it in contiguous memory to facilitate the accelerator's most efficient data transfer mechanisms. The final output of the accelerator is a processed 64 × 64 image.

In the subsequent stage of development, we prioritized optimizing data movement and memory format, recognizing their critical roles in achieving high-performance results. By strategically partitioning the Block RAMs (BRAMs) within

the FPGA fabric, we maximized data bandwidth while enabling the instantiation of additional DSPs for parallel processing. Simultaneous access to model weights was facilitated, enhancing overall efficiency. Furthermore, we adopted fixed-point arithmetic for Multiply-Accumulate (MAC) operations, significantly enhancing the hardware synthesis efficiency compared to floating-point arithmetic.

In our final stage of design, we sought to attain substantial throughput by introducing an increased boost to the fine-grained parallelism inside the Programmable Logic (PL) fabric in order to achieve a significant throughput. This was achieved by meticulously eliminating data dependencies and using pragma directives to build the generator model's design as a completely parallel and pipelined system, hence reducing latency. In a dataflow manner, each layer's operation overlapped with that of the next layer, enabling the smooth transfer of intermediate results down the processing stream without the need for extra memory resources.

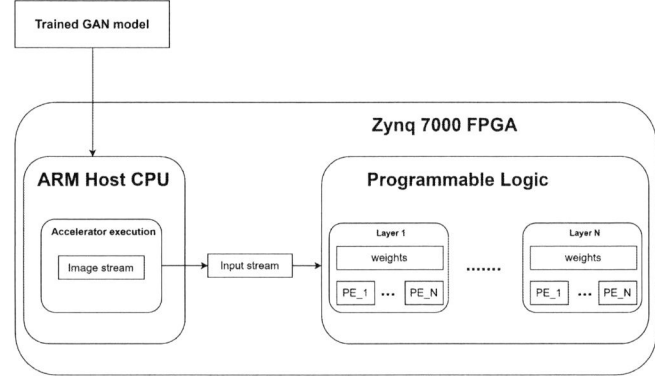

Fig. 4. Overall system design.

IV. EXPERIMENTAL OUTCOMES

A. System configuration

We employed a Xilinx ZC702 as our hardware platform of choice for the system configuration. This board boasts a Zynq-7000 SoC, featuring a Dual-core ARM Cortex-A9 processor as its central processing unit, complemented by 512 MB of DDR3 memory.

Table I depicts how our specialized hardware accelerator, which is installed on the FPGA board, uses its resources within the Xilinx ZC702 and shows the allocation of FPGA resources, which includes critical metrics such as FFs (Flip-Flops), LUTs (Look-Up Tables), BRAMs (Block RAMs), and DSPs (Digital Signal Processors). These metrics are essential in assessing the efficiency and scalability of our hardware accelerator. In addition to resource utilization, Table I presents crucial performance metrics, such as latency timing, indicating the time our hardware accelerator takes to process specific tasks or computations. Furthermore, we calculated and included the Frames Per Second (FPS) achieved by our system. FPS is a vital parameter in applications reflecting the system's ability to handle data at a certain rate.

TABLE I
RESOURCE UTILIZATION OF FPGA BOARD.

LUT	9972
FF	18954
BRAM	56
DSP	114
Latency	0.015 (ms)
FPS	75K

TABLE II
SYSTEM EVALUATION WITH OTHER COMPUTING DEVICES.

Computing device	Speed-up	Time per image (ms)	PPW	Power (W)
CPU (ARM A9)	1x	2.14	1x	3.4
GPU (NVIDIA Tesla P40)	68x	0.037	2.9x	82
FPGA (ZC702)	152x	0.015	142x	3.8

B. Generated images

In Figure 5, we observe the generated images, each corresponding to one of the three distinct categories of potato plants: late blight, early blight, and healthy.

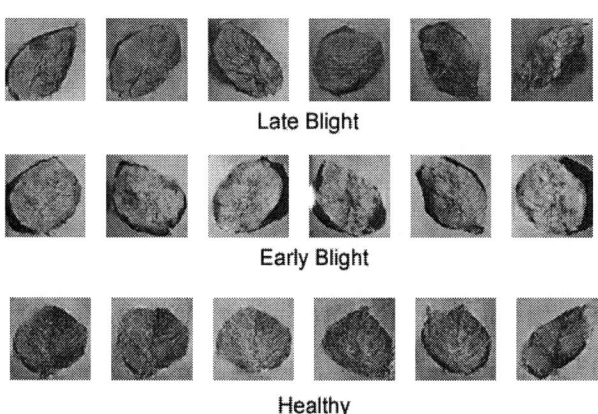

Fig. 5. The outcomes of the gan model for the each category.

C. System performance with other computing devices

Table II offers a comprehensive overview of the performance metrics for different computing devices within the experimental setup. The computing devices under consideration encompass the CPU (ARM Cortex-A9), GPU (NVIDIA Tesla P40), and FPGA (ZC702), representing a diverse range of computing architectures. The speed-up quantifies the performance improvement achieved by each device compared to the baseline CPU (ARM A9). This metric showcases the remarkable acceleration capabilities of the GPU and FPGA, with the FPGA exhibiting a notably high speed-up factor of 152x. The time per image provides insight into the processing speed of each device, measured in milliseconds. The FPGA (ZC702) outperforms the CPU and GPU by processing images in just 0.015 milliseconds, showcasing its suitability for time-critical applications. The PPW (Performance per Watt) evaluates the energy efficiency of each device by comparing its computational performance to its power consumption. The FPGA (ZC702) impressively achieves a PPW ratio of 142x, underlining its energy-efficient design. Lastly, the power consumption of each device is measured in watts. The FPGA (ZC702) consumes significantly less power (3.8 W) compared to the GPU (82 W) while delivering superior performance.

V. CONCLUSIONS AND FUTURE WORK

Our work has focused on implementing and accelerating GANs on a Xilinx FPGA SoC, a platform known for its exceptional efficiency in handling computational challenges in terms of both performance and power consumption. Notably, our FPGA-based design has outperformed traditional CPU and GPU setups, achieving an impressive average image processing time of just 0.015 milliseconds per image. This accomplishment underscores the potential of FPGA technology in the field of agricultural disease management, offering a highly efficient and effective solution for addressing the critical issue of potato crop diseases and contributing to global food security. In future work, we will explore techniques for optimizing machine learning models deployed on FPGAs, aiming to strike a balance between accuracy and resource efficiency. This will enable faster and more accurate disease detection while conserving FPGA resources.

ACKNOWLEDGMENTS

This work was partially supported by the University of Western Macedonia.

REFERENCES

[1] I. U. Haq, Z. Mukhtar, M. Anwar-ul Haq, and S. Liaqat, "Deciphering host–pathogen interaction during streptomyces spp. infestation of potato," *Archives of Microbiology*, vol. 205, no. 6, p. 222, 2023.

[2] U. Ahmad and L. Sharma, "A review of best management practices for potato crop using precision agricultural technologies," *Smart Agricultural Technology*, p. 100220, 2023.

[3] M. Riaz, N. Akhtar, L. A. Msimbira, M. Antar, S. Ashraf, S. N. Khan, and D. L. Smith, "Neocosmospora rubicola, a stem rot disease in potato: Characterization, distribution and management," *Frontiers in Microbiology*, vol. 13, p. 953097, 2022.

[4] X. Wang and W. Cao, "Gacn: Generative adversarial classified network for balancing plant disease dataset and plant disease recognition," *Sensors*, vol. 23, no. 15, p. 6844, 2023.

[5] T. Sanida, D. Tsiktsiris, A. Sideris, and M. Dasygenis, "A heterogeneous implementation for plant disease identification using deep learning," *Multimedia Tools and Applications*, vol. 81, no. 11, pp. 15 041–15 059, 2022.

[6] S. De, I. Bhakta, S. Phadikar, and K. Majumder, "Agricultural image augmentation with generative adversarial networks gans," in *International Conference on Computational Intelligence in Pattern Recognition*. Springer, 2022, pp. 335–344.

[7] C. Wang and Z. Luo, "A review of the optimal design of neural networks based on fpga," *Applied Sciences*, vol. 12, no. 21, p. 10771, 2022.

[8] W. Mao, P. Yang, and Z. Wang, "Fta-gan: A computation-efficient accelerator for gans with fast transformation algorithm," *IEEE Transactions on Neural Networks and Learning Systems*, 2021.

[9] S. Lin and Y. Zhang, "Controllable person image synthesis gan and its reconfigurable energy-efficient hardware implementation," in *2022 the 6th International Conference on Innovation in Artificial Intelligence (ICIAI)*, 2022, pp. 154–160.

[10] D. Hughes, M. Salathé *et al.*, "An open access repository of images on plant health to enable the development of mobile disease diagnostics," *arXiv preprint arXiv:1511.08060*, 2015.

979-8-3503-8368-3/23 $31.00 © 2023 IEEE

An Efficiency CNN Solution for Olive Disease Management Through FPGA

Theodora Sanida[1], Argyrios Sideris[1], Maria Vasiliki Sanida[2], Michael Dossis[1] and Minas Dasygenis[1]

[1]*Department of Electrical & Computer Engineering, University of Western Macedonia, 50131, Kozani, Greece*
[2]*Department of Digital Systems, University of Piraeus 18534, Piraeus, Greece*
thsanida@uowm.gr, asideris@uowm.gr, sanidasilia@gmail.com, mdossis@uowm.gr, mdasyg@ieee.org

Abstract—Olives, one of the world's oldest and most significant cultivated crops, face persistent threats from various diseases, devastatingly affecting olive production and quality. In this era of technological advancement, the integration of Convolutional Neural Networks (CNNs) has emerged as a promising tool for the early and accurate identification of diseases in agricultural crops. This work presents a novel solution of CNNs for the automated identification of diseases in olive plants in 3 categories (aculus olearius healthy and olive peacock spot). To further enhance the practicality and real-time applicability of the proposed system, an implementation of Field-Programmable Gate Arrays (FPGAs) is explored. This study delves into the FPGA-based deployment of the CNN solution, optimizing its computational efficiency for on-site disease diagnosis with a 98.82% accuracy rate. The results of this study reveal the potential of CNNs in revolutionizing olive disease management, enabling early intervention and precise treatment. Additionally, the FPGA implementation demonstrates the feasibility of deploying such advanced models where immediate decision-making is crucial for crop protection. This analysis paves the way for sustainable olive cultivation practices, ensuring the longevity and resilience of this invaluable crop in the face of evolving disease challenges.

Index Terms—Low-power; Hardware; Embedded device; FPGA; Olive crop diseases; CNN; Identification.

I. INTRODUCTION

Olive crop diseases constitute a critical challenge in global agriculture, as olive trees are a vital source of olive oil, one of the world's most cherished commodities. Efforts to combat diseases often involve a combination of preventive measures, including removing infected trees, using disease-resistant cultivars, and the application of chemical treatments, which can have environmental and economic implications. Moreover, as climate change introduces new environmental conditions and exacerbates existing challenges, the distribution of olive diseases is expected to evolve, making early detection and accurate diagnosis paramount. In this context, advanced technologies like artificial intelligence are being harnessed to develop innovative disease monitoring and management systems. These systems enable the timely identification of disease symptoms and the implementation of targeted control strategies, minimizing the impact on olive production and ensuring the sustainability of this precious crop [1]–[3].

Today, CNNs have demonstrated remarkable capabilities in image identification and categorization tasks. In the context of olive crop diseases, CNNs are being leveraged to automate the identification of disease symptoms in olive plants.

These advanced neural networks are trained on extensive datasets of olive leaf images, learning to identify subtle visual cues that indicate the presence of diseases. This automation enhances the speed and accuracy of disease diagnosis and reduces reliance on manual inspection, which can be time-consuming and subject to human error. To further enhance the practicality and real-time applicability of CNN-based disease detection, researchers are exploring the integration of Field-Programmable Gate Arrays (FPGAs). So, by deploying CNN models on FPGAs, the technology becomes more accessible where immediate decision-making is critical for mitigating disease outbreaks. The synergy between CNNs and FPGA technology represents a significant step forward in the battle against olive crop diseases. This innovative approach not only streamlines disease detection but also empowers farmers with timely and accurate information, allowing for targeted interventions and the preservation of olive orchards [4], [5].

In outline, the significant contributions of the work are as follows:

- We create a robust and highly accurate solution disease identification system that categorizes olive leaves into one of three pivotal types: aculus olearius, healthy, and olive peacock spot leaves. The CNN solution design reached a maximum rate of 98.82% in terms of accuracy.
- In order to accelerate the CNN solution layout, we implemented a Xilinx FPGA-specific hardware configuration technique. Optimizations distinguish this hardware acceleration in a variety of areas to improve the CNN solution design's overall performance.
- In order to optimize the CNN solution design's output, we conduct an exhaustive evaluation of the hardware configuration's various metrics. In addition, we examine other computing devices to compare our solution's performance and efficacy to that of conventional computing configurations.

The remaining components of this work have the following organizational structure: We provide corresponding investigation studies in Part II, and Part III analyses our methodology's constituent elements and techniques. Part IV explains the study's outcomes, and Part V summarises our investigation results and future investigation.

979-8-3503-8368-3/23 $31.00 © 2023 IEEE

II. RELATED WORK

The categorization of olive tree diseases through image analysis has garnered significant attention in recent research endeavours, with a substantial portion of this work harnessing the capabilities of artificial intelligence in various design paradigms. Specifically, deep learning techniques, particularly CNNs, have emerged as a cornerstone in the quest for accurate disease identification and classification within olive crop health. So, we will concentrate on research efforts that employ diverse methods to enhance and extend the capabilities of CNN-based processes dedicated to detecting and characterizing diseases that impact olive crops.

In [6], the authors developed a CNN system tailored to manage the categorization of olive tree diseases based on image analysis. The collection used for this study comprised 4138 images of olive leaves, categorized into 4 distinct types: 3 of these categories represented various diseases affecting the olive trees, while the fourth type denoted the category of healthy leaves. To ensure the robustness of their CNN system, the authors compared its performance against six well-established and widely identified pre-trained layouts, DenseNet201, ResNet50, MobileNetV2, VGG19, InceptionV3, and EfficientNetB0. So, the custom prediction model based on EfficientNetB0 emerged as the standout performer among these models, gaining an accuracy rate of 96.14%.

In [7], the authors designed a CNN method to categorize illnesses that might affect olive crops—the collection that was utilized for this study comprised 5571 images of olive leaves. These images include six disease categories and one healthy leaf category. In order to validate the efficacy of their method, the authors did an approximation of their implementation using seven different pre-trained layouts. These layouts were DenseNet, InceptionV3, MobileNet, AlexNet, VGG, GoogleNet, and ResNet. The MobileNet layout achieved the best results, with a 92.59

In disparity with prior methodologies, our study introduces a novel and highly customized CNN solution to address the real-world challenges of olive crop diseases. A distinctive hallmark of our work is the development of a specialized hardware architecture explicitly optimized for Xilinx FPGA, strategically serving as a dedicated accelerator to improve the efficiency of our CNN solution model. Moreover, our comprehensive evaluation methodology extends beyond the FPGA realm, as we rigorously assess the versatility and adaptability of our solution by subjecting it to alternative computing platforms [8].

III. MATERIALS AND METHODS

A. Olive leaf collection

In our study, we utilized an olive leaf collection comprising a whole of 3,400 images [9]. This collection encompasses three distinct categories crucial for disease identification: aculus olearius, healthy, and olive peacock spot leaves. Specifically, the collection of 890 images dedicated to the aculus

olearius category, 1,050 images representing the healthy category, and 1,460 images showcasing the characteristic symptoms of the olive peacock spot. The olive leaf collection was separated into 2 distinct groups, with 80% of the images assigned for training the model and the remaining 20% reserved for rigorous testing and validation. This partitioning ensures that our CNN-based model is trained on a diverse and representative dataset while providing an independent evaluation set to assess its generalization performance effectively. For a comprehensive overview of the distribution of leaf images based on their intended usage, please refer to Table I, which outlines the precise allocation of images for training and testing. Figure 1 illustrates a sample of images from the olive leaf collection.

TABLE I
THE TOTAL NUMBER OF OLIVE LEAF IN EACH CATEGORY FOR TRAINING, VALIDATION, AND TESTING.

Category	Total	Train	Test
Aculus Olearius	890	690	200
Healthy	1050	830	220
Olive Peacock Spot	1460	1200	260

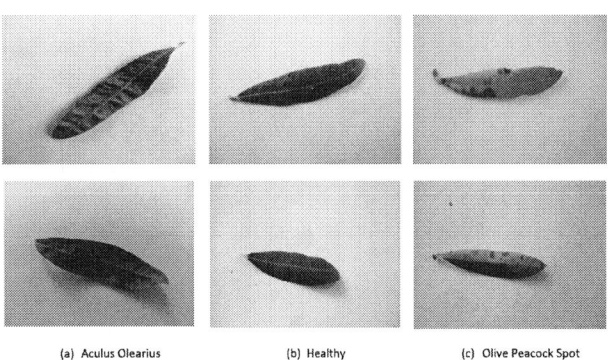

(a) Aculus Olearius   (b) Healthy   (c) Olive Peacock Spot

Fig. 1. The olive leaf collection [9] samples.

B. CNN solution

The first layer consists of 32 kernels with a size of 3x3. The ReLU activation function is used to introduce non-linearity to the model. Max-pooling with a 2x2 pool size reduces the spatial dimensions. The second convolutional layer employs 64 kernels of size 3x3. It also uses the ReLU activation function and includes dropout regularization with a rate of 0.25 to reduce overfitting. Max-pooling is applied again to downsample the feature maps. The third convolutional layer uses 64 kernels of size 3x3 and applies ReLU activation. Max-pooling further reduces the spatial dimensions, enhancing the model's ability to capture high-level features. The fourth layer introduces 128 kernels of size 3x3 with ReLU activation. Dropout regularization with a rate of 0.20 is included to mitigate overfitting. Max-pooling continues to downsample the feature maps. The fifth and final convolutional layer employs 256 kernels of size 3x3 with ReLU activation. Dropout regularization with a rate of 0.20 is again applied to prevent overfitting. Max-pooling is used for feature reduction.

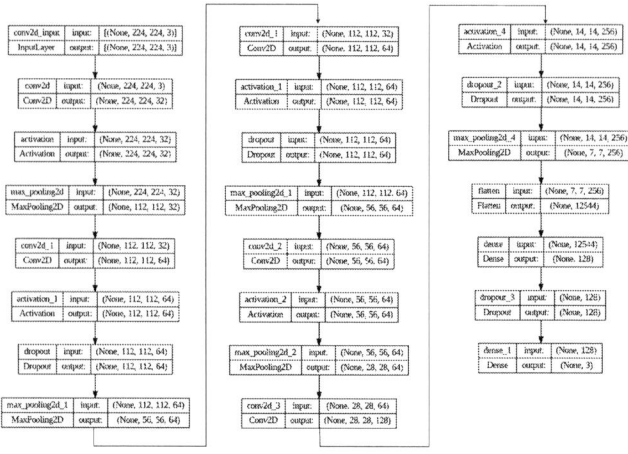

Fig. 2. CNN solution's layout for olive crop diseases.

Following the convolutional layers, the flatten layer transforms the multidimensional feature maps into a one-dimensional vector, organising the data for input into the fully connected layers. The first fully connected layer is a densely connected neural network layer with 128 kernels and uses the ReLU activation function. Dropout with a rate of 0.45 is applied to reduce overfitting. For identification, the final fully connected layer consists of three units corresponding to the three categories (aculus olearius, healthy, and olive peacock spot). It uses the softmax activation function to produce category probabilities, determining the predicted category of the input image.

The CNN model encompasses a total of 2,031,491 parameters, enabling its capacity to learn intricate patterns within the data. The proposed CNN solution was trained and tested using the NVIDIA Tesla P40 GPU. Figure 2 depicts our CNN solution's layout for olive crop diseases.

C. CNN solution acceleration on FPGA

The CNN solution leverages an embedded platform centred on the Xilinx PYNQ-Z1. This remarkably versatile device seamlessly amalgamates a dual-core ARM Cortex-A9 processor with a highly adaptable FPGA fabric, ingeniously encapsulated within a unified and compact unit. This amalgamation of processing power and reconfigurable hardware brings forth a potent combination for tackling computationally intensive tasks, as it melds the flexibility of a general-purpose processor with the parallel processing capabilities and customizable nature of the FPGA fabric. This synergy empowers CNN to harness the benefits of hardware acceleration, allowing it to swiftly execute complex computations, such as convolutions and neural network layers, with remarkable efficiency and speed. The Xilinx PYNQ-Z1 is the linchpin of this embedded platform, fostering a harmonious fusion of traditional CPU capabilities and FPGA's parallelism, rendering it exceptionally

adept at a wide spectrum of applications, including image processing, machine learning, and real-time data analysis [10].

The utilization of Vivado High-Level Synthesis (HLS) represents a pivotal component of our strategy to accelerate the performance of our proposed. Vivado HLS transforms high-level C, C++, or SystemC source code into a register-transfer level (RTL) implementation. This pivotal transformation bridges abstract algorithmic descriptions and hardware-specific RTL code, unlocking the potential for efficient hardware acceleration. Following the HLS process, a crucial step entails the generation of output files, which encapsulate the design files necessary to articulate the desired RTL language. These files encompass the meticulously crafted hardware description, configuration settings, and other essential components essential for synthesizing and deploying the accelerated CNN model. This process not only streamlines the integration of our CNN into FPGA hardware but also facilitates optimization and customization, enabling us to harness the total computational power of the FPGA platform.

IV. EXPERIMENTAL OUTCOMES

The proposed CNN solution demonstrates a remarkable 98.82% accuracy rate, and the confusion matrix in Figure 3 shows the identification effects for each category of olive crop disease on the test collection. This high rank of accuracy emphasizes the significance of the CNN model in specifically categorizing and identifying different types of olive crop diseases.

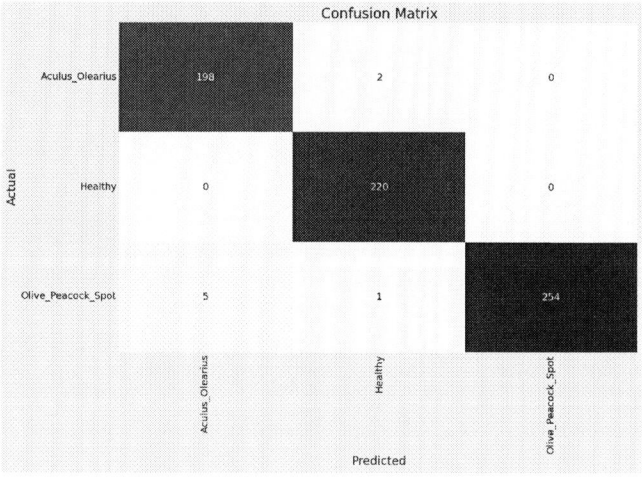

Fig. 3. Confusion matrix for olive crop diseases.

The ROC curve effectively encapsulates a model's categorization prowess by plotting its ability to correctly identify true positives (examples accurately categorized as positive) against its tendency to produce false positives (examples wrongly categorized as positive). The ROC curve for the CNN solution for olive crop diseases is illustrated in Figure 4.

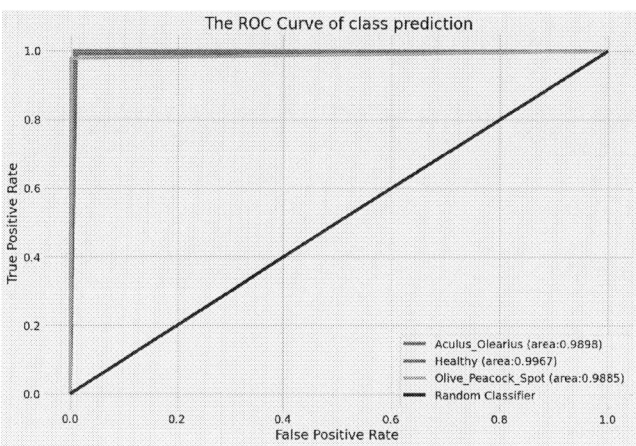

Fig. 4. ROC Curve for olive crop diseases.

Table II shows a comprehensive snapshot of the hardware cost, device utilization percentage and performance metrics associated with the acceleration of our CNN solution on the FPGA board. These metrics are pivotal in assessing the feasibility and efficiency of our hardware implementation.

TABLE II
HARDWARE COST OF THE CNN SOLUTION ACCELERATION ON FPGA BOARD.

BRAM	62	22.14 (%)
FF	3634	3.41 (%)
LUT	4282	8.04 (%)
DSP	98	44.54 (%)
FPS	78K	-
Latency	0.016 (ms)	-

Table III provides a comparison of the efficiency metrics of our proposed solution across different devices. These metrics offer insights into the solution's computational performance and power consumption characteristics when executed on distinct hardware platforms. When running on the NVIDIA Tesla P40 GPU, the solution exhibits a rapid processing speed, with a mere 0.016 milliseconds (ms) required per image for analysis. However, this efficiency comes at a relatively higher power consumption of 64.35 watts (W), reflecting the trade-off between speed and energy usage that is typical of GPU-based computations. On the CPU, specifically the Intel i7-6850K clocked at 3.6GHz, the solution processes images slightly slower, requiring 0.17 ms per image. However, it operates at a notably lower power consumption of 15.72W, showcasing a more energy-efficient performance compared to the GPU. In contrast, when executed on the FPGA platform, specifically the ZC702, the solution demonstrates an intermediate processing time of 0.078 ms per image. Importantly, it achieves this level of efficiency with remarkably low power consumption, consuming only 2.48W. This signifies the FPGA's ability to strike a balance between computational speed and energy efficiency, making it a compelling choice for applications demanding real-time processing with minimal power consumption.

TABLE III
EFFICIENCY WITH OTHER DEVICES.

Platform	Time per image (ms)	Power (W)
GPU - NVIDIA Tesla P40	0.016	64.35
CPU - i7-6850K 3.6GHz	0.17	15.72
FPGA - ZC702	0.078	2.48

V. CONCLUSIONS AND FUTURE WORK

Our study is centred on developing and accelerating an innovative CNN-based solution for the automated detection of diseases in olive crops using a Xilinx FPGA SoC with high performance and low power consumption. Our FPGA-based solution has an average latency of 0.078 milliseconds per image. This achievement highlights the potential of FPGA technology in providing a very effective solution to the serious problem of olive crop diseases, which contributes to global food security. This study lays the way for more sustainable olive production practices, maintaining the life and resilience of this priceless commodity in the face of growing disease threats. In future study, we will extend the capabilities of the CNN solution to categorize a broader range of diseases and pests that affect olive crops, and we will further optimize power consumption while maintaining performance, ensuring that our solution remains environmentally cost-effective.

ACKNOWLEDGMENTS

This work was partially supported by the University of Western Macedonia.

REFERENCES

[1] E. Anastasiou, A. T. Balafoutis, and S. Fountas, "Trends in remote sensing technologies in olive cultivation," *Smart Agricultural Technology*, vol. 3, p. 100103, 2023.

[2] R. Buonaurio, L. Almadi, F. Famiani, C. Moretti, G. E. Agosteo, and L. Schena, "Olive leaf spot caused by venturia oleaginea: An updated review," *Frontiers in Plant Science*, vol. 13, p. 1061136, 2023.

[3] V. Skiada, P. Katsaris, M. E. Kambouris, V. Gkisakis, and Y. Manoussopoulos, "Classification of olive cultivars by machine learning based on olive oil chemical composition," *Food Chemistry*, vol. 429, p. 136793, 2023.

[4] P. Bocca, A. Orellana, C. Soria, and R. Carelli, "On field disease detection in olive tree with vision systems," *Array*, vol. 18, p. 100286, 2023.

[5] M. Mamalis, E. Kalampokis, I. Kalfas, and K. Tarabanis, "Deep learning for detecting verticillium fungus in olive trees: Using yolo in uav imagery," *Algorithms*, vol. 16, no. 7, p. 343, 2023.

[6] H. El Akhal, A. B. Yahya, N. Moussa, and A. E. B. El Alaoui, "A novel approach for image-based olive leaf diseases classification using a deep hybrid model," *Ecological Informatics*, p. 102276, 2023.

[7] M. Lachgar, H. Hrimech, A. Kartit *et al.*, "Optimization techniques in deep convolutional neuronal networks applied to olive diseases classification," *Artificial Intelligence in Agriculture*, vol. 6, pp. 77–89, 2022.

[8] A. Sideris, T. Sanida, and M. Dasygenis, "Hardware acceleration design of the sha-3 for high throughput and low area on fpga," *Journal of Cryptographic Engineering*, pp. 1–13, 2023.

[9] S. Uğuz and N. Uysal, "Classification of olive leaf diseases using deep convolutional neural networks," *Neural computing and applications*, vol. 33, no. 9, pp. 4133–4149, 2021.

[10] A. Sideris, T. Sanida, and M. Dasygenis, "A novel hardware architecture for enhancing the keccak hash function in fpga devices," *Information*, vol. 14, no. 9, p. 475, 2023.

979-8-3503-8368-3/23 $31.00 © 2023 IEEE

Q-Delegation: VNF Load Balancing through Reinforcement Learning-Based Packet Delegation

Alexandros Zervopoulos
Department of Informatics
Ionian University
Corfu, Greece
azervop@ionio.gr

Luís Miguel Campos
PDMFC
Lisbon, Portugal
luis.campos@pdmfc.com

Konstantinos Oikonomou
Department of Informatics
Ionian University
Corfu, Greece
okon@ionio.gr

Abstract—While Network Function Virtualization (NFV) and Service Function Chaining (SFC) offer unprecedented capabilities to network operators, there are still performance concerns that need to be addressed, especially with the advent of network slicing requiring automated and efficient resource allocation. In this context, sharing the available networking and computational resources is an appealing option to minimize under-utilization, but this can lead to over-subscription of the available resources under heavy traffic loads. To overcome this, distributed load balancing mechanisms, such as *packet delegation*, can be utilized to move the decision-making further from centralized orchestrators and closer to the Virtual Network Functions (VNF) themselves, allowing them to quickly redirect incoming traffic to other servers when they are overloaded. However, the effectiveness of packet delegation can vary considerably depending on the selection of the target VNF that receives the delegated packets. As such, in this paper, the Q-Delegation algorithm is proposed, which treats this as a non-stationary multi-agent Reinforcement Learning (RL) problem, providing an adaptive way of selecting appropriate delegation targets based on the VNF's current queueing state through Q-Learning. Two variations of the Q-Delegation algorithm are proposed, either allowing an RL agent to act once a VNF's queue has exceeded a static threshold or enabling it to freely learn how to reroute packets to other VNFs regardless of their queue size. Both variations are experimentally evaluated and compared to other baseline algorithms under diverse topologies and traffic rates in a simulation setting, indicating that in most cases both variations of Q-Delegation can be effective at delegating packets and outperforming the baselines, though their effectiveness can vary based on how evenly the traffic is initially distributed.

Index Terms—Load balancing, load sharing, reinforcement learning, network function virtualization, service function chaining

I. INTRODUCTION

Network Function Virtualization (NFV) is one of the main cornerstones, alongside Software-Defined Networking (SDN), in modern networks [1] by decoupling network functions from dedicated hardware and migrating them to virtualized counterparts, i.e., Virtual Network Functions (VNFs), to provide unprecedented flexibility, scalability, and cost-efficiency. Nonetheless, challenges related to the performance of NFV still persist, requiring complex orchestration to monitor and optimize the network dynamically [2]. Particularly with the emergence of Service Function Chains (SFCs) and network slicing, where multiple virtual networks share the same phys-

ical infrastructure, ensuring consistent performance for each slice adds another layer of complexity [3].

Efficiently allocating resources and balancing the demands of various network functions is essential; hence, placement and load balancing of VNFs are some of the most widely investigated problems in NFV [4]. Due to their complexity, most of the proposed approaches address these problems using either (i) centralized orchestration techniques, such as [5], which introduce additional communication overhead, slowing down response times to anomalies, or (ii) dedicated load balancing VNFs, which add significant resource consumption as they need to be deployed across the network [2]. Especially with regards to load balancing, rapid response times are necessary, as overloaded VNFs will result in longer end-to-end times across SFCs, potentially violating service-level agreements. Furthermore, many orchestration approaches often require fine-tuning, which results in increased management complexity for network operators; ideally, networks should autonomously adapt to the current network conditions with minimal human interference [6].

Distributed load balancing methods can potentially address these issues [7], moving some of the decision-making closer to the VNFs to reduce communication overhead and speed up response times when they are overloaded. This distributed approach also promotes resource sharing, which can be useful for network slicing applications, sacrificing isolation for resource efficiency. One way to accomplish this is through the packet delegation mechanism introduced in previous work [8], which is a distributed method that can be used for load balancing by itself or alongside other more resource-demanding countermeasures. For this mechanism, a VNF Manager at each server can dynamically redirect incoming packets to different servers depending on its VNFs' queues. However, the problem of selecting the optimal VNF to receive a delegated packet has not been explored in previous work [8]. It turns out that this decision can affect the performance of packet delegation, which can become a complex non-stationary optimization problem. Furthermore, parameterization of the packet delegation mechanism can significantly affect performance, requiring different parameter values depending on the current network conditions.

In this light, this paper proposes *Q-Delegation*: the integration of the Q-Learning algorithm into packet delegation. By incorporating Reinforcement Learning (RL)-based techniques, packet delegation can become more flexible by autonomously deciding *where* to delegate a packet to. An additional variation is considered that allows agents to also decide *when* to delegate a packet, completely eliminating the need for parameterization of the original packet delegation mechanism. Q-Delegation utilizes multi-agent Reinforcement Learning (RL), treating packet delegation as a non-stationary problem where each agent acts using purely local information. Essentially, each VNF is treated as an RL agent that decides whether to process an incoming packet or forward it to a different VNF based on its current queue length. The proposed algorithm is experimentally evaluated in a simulation setting, yielding rather promising results with respect to end-to-end delay and server utilization, compared to baseline methods.

The structure of the remaining paper is laid out in the sequel. Section II formalizes the considered networking environment. The Q-Delegation algorithm is described in Section III. The results of the experimental evaluation are shown in Section IV. Section V contrasts past related work, while conclusions and future work are discussed in Section VI.

II. System Model

We consider an NFV network consisting of a set of physical servers, each hosting one or more VNFs; the set of all VNFs is denoted as F. Servers are interconnected through forwarding elements, such as switches or routers. VNFs can provide different kinds of functions based on their class $c \in C$, e.g., Firewalls, Intrusion Detection Systems (IDS), Network Address Translation (NAT), etc. Traffic packets are generated by clients, which have to traverse through a predefined SFC, i.e., an ordered sequence of VNF classes, before reaching their destination. For instance, a typical SFC in a cloud setting could be IDS → Firewall → NAT → Router.

Additionally, each server hosts a single VNF Manager, which is responsible for the lifecycle management of VNFs as well as forwarding incoming packets to the appropriate VNF. The set of all VNF Managers is denoted as M. Thus, the considered network can be modeled as an undirected graph $G = (V, E)$, where $V = F \cup M$ and E corresponds to the physical links between VNF Managers as well as the (virtual) links between VNFs and their Managers. To facilitate SFCs, traffic steering is required to route packets through the appropriate VNFs. This is typically accomplished using an SDN controller, which installs forwarding rules to the switches and VNF Managers according to the accepted SFC requests.

In this case, the VNF Managers and VNFs are modeled as M/M/1 queues, with interarrival and service times assumed to be exponentially distributed. The processing rate of node $v \in V$ is denoted by μ_v and it is assumed that $\mu_m > \mu_f$, for all $m \in M$, $f \in F$. Moreover, the VNF Manager utilizes the packet delegation mechanism, proposed in previous work [8]. In essence, the VNF Manager can autonomously decide whether to forward a packet to one of its VNFs or to delegate a packet to a VNF hosted on a different server with

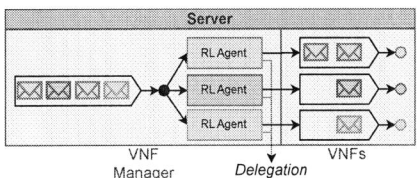

Fig. 1. The architecture of a server: the VNF Manager with an RL agent for each VNF. RL Agents can either forward an incoming packet to a local VNF or delegate packets to a different server.

the goal of redirecting load from overburdened VNFs. This decision is made by monitoring the VNFs' queue and, when the length of a VNF's queue exceeds a certain threshold θ, incoming packets are kept with probability k and delegated to a different server with a probability $1 - k$.

In addition, delegated packets are served with higher priority than the kept packets to counteract the additional propagation delay of delegated packets. A packet may only be delegated once, if a VNF Manager receives a delegated packet it has to keep it so as to avoid closed loops. In the original paper [8], the server to receive the delegated packet was randomly chosen from the alternative VNFs; in load sharing terminology, this represents the location policy [9]. The current work differs from [8], however, as it further expands on the problem of choosing an appropriate target to receive a delegated packet.

III. Packet Delegation through Reinforcement Learning

In this paper, an RL agent is instantiated for each VNF, which decides when to delegate a packet and which server should receive it, as shown in Fig. 1. This essentially forms an overlay network for a given SFC, as packets can be exchanged between VNF Managers regardless of the underlying topology, as depicted in Fig. 2. To achieve this, decision-making is modeled through a Markov Decision Process (MDP), which consists of a set of states and a state-dependent set of actions. Depending on the current state, the agent can take one of the available actions to transition to a new state; the agent is unaware of the transition dynamics. Instead, to inform the agent's decision-making policy, it receives a reward R according to some reward function when an action is taken. The agent's goal is to learn the optimal policy, which maximizes the accumulated reward over time, i.e., the return.

In the context of packet delegation, an MDP is designed so that the RL agent of each VNF Manager can be trained. The states of the MDP correspond to the queue length of the VNF, which is assumed to have a maximum capacity of κ, i.e. $S \subseteq \{0, 1, ..., \kappa\}$. The set of actions A is identical across all states and corresponds to the available delegation options, i.e., for a packet that requires processing by a VNF of class c, $A \subseteq F_c$. The agents receive a reward after some time, once the delegated packet arrives at the Manager of the selected VNF; the value of the reward corresponds to the length of the VNF's queue when the packet arrives at the Manager multiplied by -1. In this way, the agent is penalized

979-8-3503-8368-3/23 $31.00 © 2023 IEEE

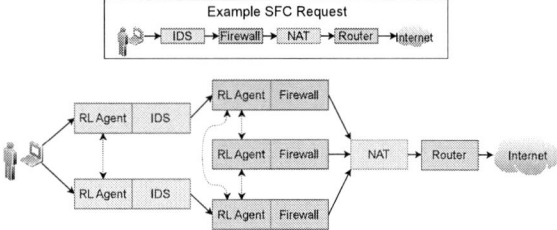

Fig. 2. An overlay network for an SFC of four VNF classes with multiple replicas potentially present for some VNF classes.

for delegating a packet to VNFs with a higher load in terms of queue length, which serves as an estimation of the delay. Upon receiving a reward, the agent transitions to a new state corresponding to the VNF's current queue length. It is worth noting that rewarding the agent with the VNF's queue length as soon as the packet arrives at its destination, rather than the time elapsed until the packet is processed, is important here; it reduces the delay between taking an action and receiving the corresponding reward, which can otherwise cause issues during training [10].

Q-Learning is utilized to train the agents, which approximates the optimal policy by mapping each state-action pair to an estimation of the expected return through an evaluation function $Q : S \times A \to \mathbb{R}$. The evaluation function can be initialized arbitrarily for all state-action pairs and then updated during runtime according to the received rewards. Upon a transition from (s, a) to s' with reward R, Q-Learning updates its evaluation function Q through the formula

$$Q(s,a) \leftarrow Q(s,a) + \eta \left[R + \gamma \max_{a'} Q(s', a') - Q(s,a) \right], \quad (1)$$

where $\eta \in [0, 1)$ denotes the learning rate, affecting the impact of each update on the previous estimation, and $\gamma \in [0, 1)$ the discount factor of future rewards, with higher values making the agent value long term returns over short term rewards. A constant value of η is used here, as it is more suitable for non-stationary problems [11]. To ensure that the agent keeps exploring state-action pairs, the ϵ-greedy policy is used, whereby each agent has a small probability $\epsilon \in [0, 1)$ of selecting a random action $a \in A$ and a probability $1 - \epsilon$ to select the best action according to Q, i.e., $\max_a Q(s, a)$.

This forms the basis of the Q-Delegation algorithm, with two different variations being considered here:

a) Static Q-Delegation (SQ-Delegation): when an incoming packet exceeds the static threshold θ of a VNF, the RL agent chooses one of the remaining VNFs to delegate the packet. This variation is closer to the original packet delegation mechanism, except there is no longer a need for a probability k to keep the packet; the decision is entirely up to the RL agent, given that the static threshold θ has been exceeded. This effectively reduces the state space to $S = \{\theta, \theta+1, ..., \kappa\}$ and the action space of VNF f to $A_f = F_c \setminus \{f\}$.

b) Dynamic Q-Delegation (DQ-Delegation): the RL agent is allowed to delegate an incoming packet even when the queue length is smaller than θ. This approach effectively

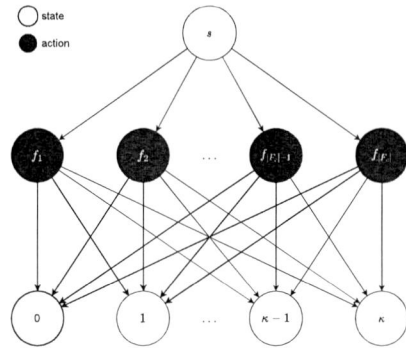

Fig. 3. The backup diagram for a DQ-Delegation agent: there are $|F_c|$ possible actions from any state $s \in S$ and it can transition to any state $s' \in S$ depending on the queueing and traffic dynamics.

results in the VNF Managers autonomously routing packets, regardless of the installed forwarding rules, eliminating the need for a static threshold θ. Accordingly, this expands the state space to $S = \{0, 1, ..., \kappa\}$ and the action space to $A = F_c$. In this case, if the VNF Manager selects one of its managed VNFs, it immediately receives a reward, based on the VNF's current queue length, and the packet is served with low priority.

DQ-Delegation can be considered more complex, as it utilizes a slightly larger state-action space. The DQ-Delegation MDP for VNF f is visually represented through the backup diagram depicted in Fig. 3; the backup diagram for SQ-Delegation is similar. Regardless, for both variations of Q-Delegation, the space complexity for the estimation function Q of each agent is $O(\kappa|F_c|)$. Since multiple VNFs may be instantiated on a single server and a VNF Manager has to instantiate an RL agent for each one, the space complexity for a Manager becomes $O(n\kappa|F_c|)$, where n denotes the VNF capacity of a server. Therefore, the memory requirement for each VNF Manager remains reasonable for moderate to large networks.

One drawback of the adopted multi-agent approach is that each RL agent cannot evaluate the long-term consequences of its decisions when considering routing across an SFC. Due to the MDP states representing only the VNF's own queue length, the agent cannot associate its own routing decisions with the queue sizes encountered at a VNF of the next class. As such, the RL agents can only be effective at load balancing within VNFs of the same class; load balancing decisions cannot be adjusted to account for a bottleneck formed at a later processing stage. This issue could be alleviated by designing a much more sophisticated MDP to represent the emergent routing overlay, as shown in Fig. 2. This would be more aptly suited for centralized RL decision-making, which could be easily integrated into an SDN-based approach. Furthermore, since the agents' states rely only on local information, they could be trained offline and deployed in an online setting, although tabular RL methods like Q-Learning can struggle to generalize to different operating conditions, especially if rewards are not available in the online setting; deep learning

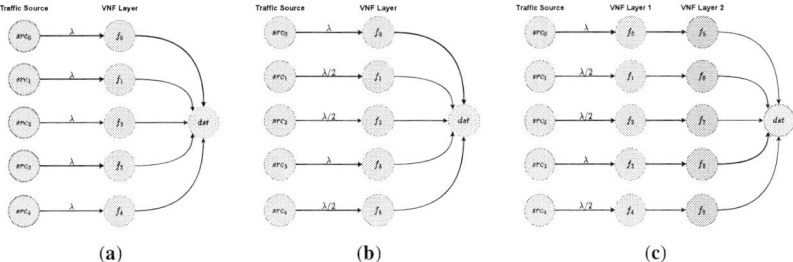

Fig. 4. The three different evaluation scenarios considered: (a) five VNF replicas with even traffic; (b) five VNF replicas with uneven traffic; and (c) two layers of five VNF replicas with initially uneven traffic.

methods would likely be more effective in this case.

IV. EVALUATION

A. Setup

The proposed SQ-Delegation and DQ-Delegation algorithms are experimentally evaluated in a simulation setting using the OMNeT++ simulator [12]. Three different scenarios are considered, capturing different traffic and topology characteristics, as shown in Fig. 4. Scenario (a) assesses the performance of a single-VNF deployment with five replicas, where traffic is evenly distributed across all replicas with rate λ. Scenario (b) is similar but traffic is unevenly distributed, with two replicas receiving traffic of rate λ and the rest receiving traffic of rate $\lambda/2$. Scenario (c) is investigated to highlight the effects of packet delegation when VNF chaining is considered as it combines the two previous scenarios into a simple SFC of two VNF classes. The traffic of the first layer is unevenly distributed, whereas the traffic of the second layer should be more evenly distributed due to delegations among the VNFs of the first layer.

The proposed variations of Q-Delegation are compared against three simple baseline algorithms that utilize the original packet delegation mechanism, i.e., a VNF Manager delegates an incoming packet with probability $1 - k$, once the queue length of a VNF exceeds threshold θ:

a) No Delegation: packets are not delegated, so VNF Managers act as standard M/M/1 queues.

b) Closest Distance: the packet is delegated to the closest replica in terms of link delay.

c) Random: the packet is delegated to a random replica.

d) Shortest Queue: the packet is delegated to the replica with the fewest packets enqueued at that moment; the remote queue state is acquired instantly in the simulation, disregarding queueing and other delays, which is somewhat unrealistic. A more practical approach would involve periodically advertising the average queue length [9], however, the instantaneous acquisition of the current queue state more closely resembles the instant reward that RL agents receive during the simulation, so it can be considered a fairer comparison.

The processing rate for all VNFs f and their Managers m is set to $\mu_f = \mu_m/2 = 1 \, \text{packet/s}$. For the Q-Delegation variations, the VNF queues have a capacity of $\kappa = 4000 \, \text{packets}$. Regarding the underlying topology, each link between VNFs

has a random delay between $0.1\,\text{s}$ and $0.6\,\text{s}$. Across all scenarios, traffic is generated for $30.000\,\text{s}$, and the simulation ends when all packets reach their destination. The SQ-Delegation and baseline algorithms are executed for $\theta \in \{10, 20\}$; for the baseline algorithms, the probability of delegating a packet is set to $1 - k = 0.4$. For both variations of Q-Delegation, the ϵ-greedy policy is used with $\epsilon = 0.15$, whereas the parameters used in update rule (1) are set to $\eta = 0.3$ and $\gamma = 0.5$, which were selected experimentally so as to yield the lowest end-to-end delay in high-traffic scenarios, although performance was not sensitive to most (η, γ) permutations. When using DQ-Delegation, the function Q of agent f is initialized across all states $s \in S$ according to $Q_f(s, a) = -s/2$, if $a = f$, and $Q_f(s, a) = 0$, otherwise, in order to promote the exploration of delegation options and speed up training.

B. Results

Initially, the mean end-to-end delay is investigated for the considered algorithms, which is measured as the mean time taken for a packet to reach its destination from the time it is generated at a source node. The end-to-end delay for all three scenarios is depicted in Fig. 5 for different traffic arrival rates λ.

Under even traffic conditions, most of the algorithms perform rather well compared to the No Delegation baseline, with SQ-Delegation and Shortest Queue providing the best performance across all traffic rates, regardless of θ; DQ-Delegation performs reasonably well for lower traffic rates, while for $\lambda > 1.2$, it results in a performance decrease as it struggles to identify good delegation targets. Under uneven traffic conditions, however, DQ-Delegation outperforms even SQ-Delegation and Shortest Queue for $\lambda <= 1.2$, as it can delegate packets more often and the uneven traffic conditions allow it to easily identify better delegation targets. In the chaining scenario, the performance of all algorithms closely resembles the uneven traffic scenario, except delay values are slightly increased due to the additional processing stage, with DQ-Delegation performing the best for most λ values, and SQ-Delegation along with Shortest Queue having comparable performance.

Overall, SQ-Delegation seems to consistently perform very well across all scenarios and traffic conditions, converging to the performance of Shortest Queue for high traffic rates. Furthermore, a hybrid Q-Delegation approach could be considered

979-8-3503-8368-3/23 $31.00 © 2023 IEEE

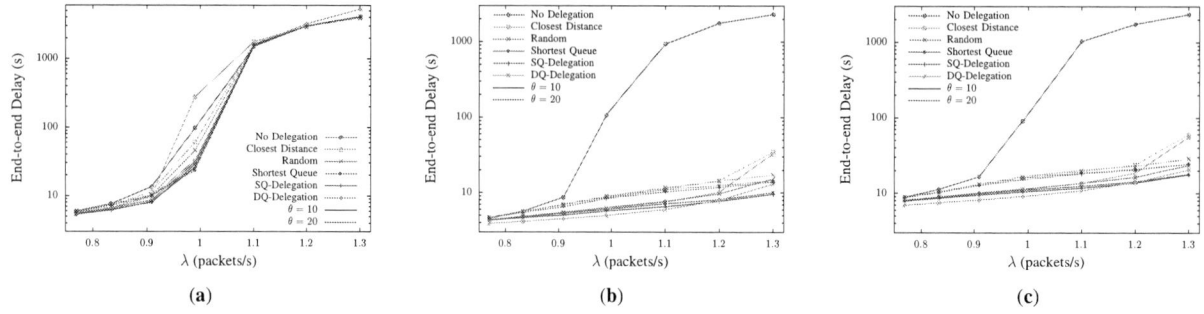

Fig. 5. The mean end-to-end delay acquired by each algorithm for different traffic rates λ across the scenarios (a), (b) and (c).

advantageous, where DQ-Delegation is used for lower traffic rates and SQ-Delegation is used for higher traffic rates. This would be feasible as both approaches share a similar MDP, so training could be performed for both using the same data, but this would double the memory requirements for storing both Q evaluation functions, making it impractical for larger networks.

Regarding the other baselines, a few interesting comments can be made. The Closest Distance baseline performs rather poorly across all scenarios, in some cases providing worse performance than the No Delegation baseline, as it is rather simplistic and in cases where two overloaded nodes are close, they start delegating packets to each other, introducing additional delays at no performance gain. The Random baseline offers decent performance across all scenarios and traffic conditions, which is noteworthy, given its simplistic approach. The results indicate that in most cases, selecting an appropriate value of θ relies on the traffic pattern, rather than the particular algorithm. For instance, under uneven traffic, just about all algorithms can identify a good delegation target, so lower values of θ are better; on the other hand, under even traffic, delegation targets need to be selected more carefully, as there is greater risk in overloading the receiving server, so higher values of θ seem better for higher traffic rates.

As a load balancing metric, the mean delay is often not sufficient due to outliers, i.e., a small number of requests having extremely high delay. As such, the Empirical Cumulative Distribution Function (ECDF) of the end-to-end delay for each scenario is depicted in Fig. 6 for the case of $\lambda = 0.99$. The end-to-end delay of the No Delegation baseline is omitted for visualization purposes due to its significantly worse performance. In scenario (a), it can be observed that DQ-Delegation results in a tighter upper bound, despite having a lower mean delay, while SQ-Delegation yields consistently low delay. Again, the other two scenarios do not exhibit significant performance differences, with DQ-Delegation providing consistently lower delay than the other algorithms, while SQ-Delegation also results in consistently better performance than Shortest Queue, even for the first scenario.

Finally, to better understand the decision-making process of the DQ-Delegation RL agents, the policy of VNF RL agents f_0 and f_5 that resulted from training in each scenario is visualized in Fig. 7. The decision-making process of f_0

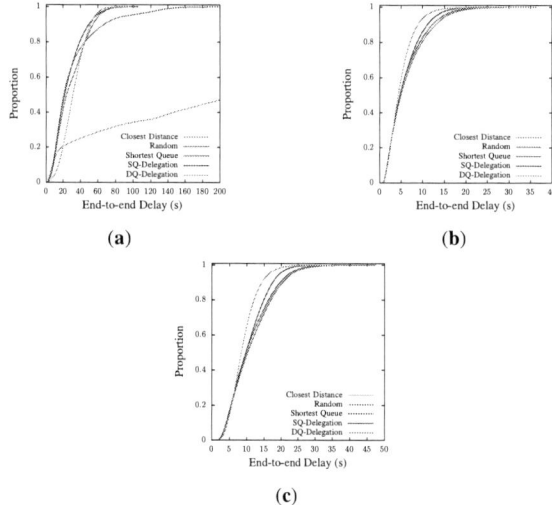

Fig. 6. ECDF of the end-to-end delay for $\lambda = 0.99$ and $\theta = 10$ across scenarios (a), (b) and (c).

and f_5 is visualized from the third scenario, indicating the VNF selected to delegate packets for different queue sizes and traffic rates; only the states explored during simulation are visualized. This set of figures suggests that the RL agents can learn to delegate packets at different thresholds θ dynamically according to the system's incoming traffic rate, preferring less overloaded VNFs as targets (e.g., f_0 tends to delegate packets to f_1), which can result in better performance, as observed in some of the scenarios. However, it does seem that the agent is too eager to delegate packets, which is not very effective at high traffic rates, as already showcased in Fig. 5.

V. RELATED WORK

One of the most well-established distributed load balancing techniques is adaptive load sharing, which was extensively studied by Eager et al. [9] in homogeneous systems. Their paper shows that homogeneous distributed systems with load sharing policies exhibit significantly improved performance than those without. Even though the assumption of homogeneous traffic may seem unrealistic, in modern networks multiple forms of load balancing are typically encountered, such as Equal-Cost Multi-Path Routing [13]; thus, by the time traffic actually reaches servers, the system is often close to

979-8-3503-8368-3/23 $31.00 © 2023 IEEE

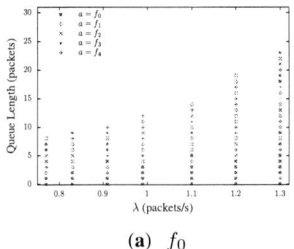

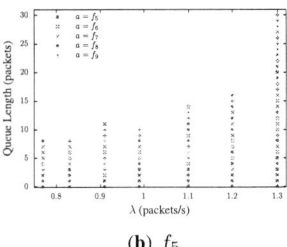

(a) f_0 **(b)** f_5

Fig. 7. Visualization of the policies derived from scenario (c) by the RL agents of VNFs f_0 and f_5. The color and shape indicate the action a taken to maximize Q based on the current queue length.

homogeneous. Packet delegation is effectively an application of load sharing in the context of NFV, which is a generalization of the originally considered setting, as factors related to the underlying topology and SFCs need to be considered. Additionally, the proposed Q-Delegation algorithm is similar to the *threshold* probing policy proposed in [9], but rather than probing for a server's state, Q-Delegation uses that state to inform future decisions through RL.

RL methods have been considered for load sharing policies in the past, even when purely local information is available. One of the most notable contributions is the paper by Schaerf et al. [14], which proposes a multi-agent RL algorithm for load sharing policies utilizing local information. Their work shares similarities to the algorithm proposed in this paper, although their agents are rewarded upon job completion, which can introduce delay between an action being taken and the corresponding reward being granted, which can hinder training [10]. Additionally, the NFV setting in the current paper considers aspects related to VNF chaining that can potentially be exploited to improve performance.

VI. Conclusion and Future Work

In this paper, the Q-Delegation algorithm was proposed to provide an appropriate location policy for packet delegation in the context of NFV. Two variations of Q-Delegation were experimentally evaluated across diverse network topologies and traffic conditions. SQ-Delegation turned out to outperform DQ-Delegation as well as the rest of the considered baselines in most cases. DQ-Delegation has significant performance gains when the network is operating under unevenly distributed traffic, likely because learning is easier in this context.

While the proposed variations of Q-Delegation perform well, the current simulation setting does not entirely consider the communication overhead that is required to award and train the RL agents. This is a typical issue when applying RL to a real environment [15], which can be circumvented to some extent by training the RL agents offline and then deploying them.

The use of more sophisticated RL methodologies is likely to address most of these issues. In particular, a more centralized SDN-based approach would allow for a single-agent setting that is generally easier to train, while also allowing for the development of an MDP that incorporates the routing across

SFCs. This has multiple advantages, as it would make deep learning-based methods easier to apply allowing for more complex state-action spaces that can generalize to unseen topologies and deployment conditions.

Acknowledgment

This work was supported by the EU ECSEL project DAIS (Distributed Artificial Intelligent Systems). DAIS (https://dais-project.eu/) has received funding from the ECSEL Joint Undertaking (JU) under grant agreement No 101007273. The JU receives support from the European Union's Horizon 2020 research and innovation programme and Sweden, Spain, Portugal, Belgium, Germany, Slovenia, Czech Republic, Netherlands, Denmark, Norway and Turkey.

The document reflects only the authors' view, and the Commission is not responsible for any use that may be made of the information it contains.

References

[1] A. M. Medhat, T. Taleb, A. Elmangoush, G. A. Carella, S. Covaci, and T. Magedanz, "Service function chaining in next generation networks: State of the art and research challenges," *IEEE Communications Magazine*, vol. 55, no. 2, pp. 216–223, 2016.

[2] H. U. Adoga and D. P. Pezaros, "Network Function Virtualization and Service Function Chaining Frameworks: A Comprehensive Review of Requirements, Objectives, Implementations, and Open Research Challenges," *Future Internet*, vol. 14, no. 2, 2022.

[3] J. Ordonez-Lucena, P. Ameigeiras, D. Lopez, J. J. Ramos-Munoz, J. Lorca, and J. Folgueira, "Network slicing for 5g with sdn/nfv: Concepts, architectures, and challenges," *IEEE Communications Magazine*, vol. 55, no. 5, pp. 80–87, 2017.

[4] J. Sun, Y. Zhang, F. Liu, H. Wang, X. Xu, and Y. Li, "A survey on the placement of virtual network functions," *Journal of Network and Computer Applications*, vol. 202, p. 103361, 2022.

[5] M. Hamdan, E. Hassan, A. Abdelaziz, A. Elhigazi, B. Mohammed, S. Khan, A. V. Vasilakos, and M. N. Marsono, "A comprehensive survey of load balancing techniques in software-defined network," *Journal of Network and Computer Applications*, vol. 174, p. 102856, 2021.

[6] C. Benzaid and T. Taleb, "AI-Driven Zero Touch Network and Service Management in 5G and Beyond: Challenges and Research Directions," *IEEE Network*, vol. 34, no. 2, pp. 186–194, 2020.

[7] M. Randles, D. Lamb, and A. Taleb-Bendiab, "A comparative study into distributed load balancing algorithms for cloud computing," in *2010 IEEE 24th International Conference on Advanced Information Networking and Applications Workshops*. IEEE, 2010, pp. 551–556.

[8] A. Zervopoulos and K. Oikonomou, "Load balancing of virtual network functions through packet delegation," in *2023 International Balkan Conference on Communications and Networking (BalkanCom)*, 2023, pp. 1–6.

[9] D. L. Eager, E. D. Lazowska, and J. Zahorjan, "Adaptive load sharing in homogeneous distributed systems," *IEEE transactions on software engineering*, no. 5, pp. 662–675, 1986.

[10] J. A. Arjona-Medina, M. Gillhofer, M. Widrich, T. Unterthiner, J. Brandstetter, and S. Hochreiter, "Rudder: Return decomposition for delayed rewards," *Advances in Neural Information Processing Systems*, vol. 32, 2019.

[11] R. S. Sutton and A. G. Barto, *Reinforcement learning: An introduction*. MIT press, 2018.

[12] A. Varga, *OMNeT++*. Berlin, Heidelberg: Springer Berlin Heidelberg, 2010, pp. 35–59.

[13] C. Hopps, "Analysis of an equal-cost multi-path algorithm," Tech. Rep., 2000.

[14] A. Schaerf, Y. Shoham, and M. Tennenholtz, "Adaptive load balancing: A study in multi-agent learning," *Journal of artificial intelligence research*, vol. 2, pp. 475–500, 1994.

[15] G. Dulac-Arnold, N. Levine, D. J. Mankowitz, J. Li, C. Paduraru, S. Gowal, and T. Hester, "Challenges of real-world reinforcement learning: definitions, benchmarks and analysis," *Machine Learning*, vol. 110, no. 9, pp. 2419–2468, 2021.

Cognitive Email Analysis with Automated Decision Support for Business Email Compromise Prevention

1st Anastasios Papathanasiou
Department of Informatics and Telecommunications
University of Ioannina
Ioannina, Greece
anastasios.papathanasiou@gmail.com

2nd Georgios Germanos
Department of Informatics and Telecommunications
University of Peloponnese
Tripolis, Greece
germanos@uop.gr

3rd Nicholas Kolokotronis
Department of Informatics and Telecommunications
University of Peloponnese
Tripolis, Greece
nkolok@uop.gr

4th Euripidis Glavas
Department of Informatics and Telecommunications
University of Ioannina
Ioannina, Greece
eglavas@uoi.gr

Abstract—In the realm of cybersecurity, Business Email Compromise attacks have emerged as a significant threat to organizations. This paper introduces a forward-looking approach, the Cognitive Email Analysis with Automated Decision Support mechanism, tailored to confront the challenges of Business Email Compromise prevention. Cognitive Email Analysis with Automated Decision Support mechanism leverages cognitive analysis, artificial intelligence, and automated decision support to discern linguistic nuances and contextual cues within email communications. By doing so, it aims to detect Business Email Compromise threats and enhance users' decision-making capabilities, fostering a proactive cybersecurity stance. This paper presents a comprehensive conceptual framework of the Cognitive Email Analysis with Automated Decision Support mechanism, highlighting its potential to revolutionize the landscape of Business Email Compromise prevention.

Index Terms—Business Email Compromise, Cognitive Analysis, Threat Detection, Sentiment Analysis, AI-driven Defense, Email Security, Contextual Understanding

I. INTRODUCTION

Business Email Compromise (BEC) is a type of cybercrime that involves fraudulently convincing victims to wire large amounts of funds or send valuable data to criminally controlled accounts. The attack can be identified and categorized into five broad categories which are, Chief Executive Officer (CEO) Fraud, Bogus Invoice Scheme, Account Compromise, Lawyer/Attorney Impersonation, and Data Theft [1]. The major techniques used by attackers and criminals for performing a BEC attack are usually Credential-grabbing and Email-only Method. BEC attacks pose a high degree of risk to companies and organizations that rely on financial transactions in their work [2]. The psychology behind this scam has been studied, and personality traits play an important role in Social Engineering-based attacks [3]. The best defense for countermeasure is a well-informed workforce, and countermeasures used include training and awareness programs, Phishing attack trainings, using of Sender Policy Framework (SPF), DomainKeys Identified Mail (DKIM), Domain-based Message

Authentication, Reporting and Conformance (DMARC) anti-spoofing and email authentication techniques [4].

The financial and reputational repercussions of BEC attacks underscore the critical importance of robust prevention mechanisms. Traditional BEC prevention strategies often rely on rule-based email filtering, behavioral analysis, and machine learning algorithms. While these methods have made strides in detecting certain forms of attacks, their effectiveness diminishes when confronted with intricately woven linguistic patterns and contextual cues characteristic of BEC attacks.

Addressing the shortcomings of existing approaches, we propose the Cognitive Email Analysis with Automated Decision Support (CEA-ADS) mechanism as an innovative solution. CEA-ADS amalgamates the power of cognitive analysis, Artificial Intelligence (AI)-driven linguistic pattern recognition, and automated decision support to proactively counter BEC attacks. By interpreting linguistic subtleties and contextual insights, CEA-ADS transcends traditional methods, offering a holistic and context-aware defense against BEC threats.

This paper unfolds by delving into the theoretical underpinnings of BEC attacks, contextualizing the challenges posed by linguistic sophistication, and substantiating the need for a multifaceted approach to BEC prevention. In section II, we introduce the conceptual framework of CEA-ADS, outlining its key components and the synergy between cognitive analysis and automated decision support. Furthermore, in section III, we explore the potential implications of CEA-ADS's integration within the broader landscape of cybersecurity.

While the empirical validation of CEA-ADS remains a crucial aspect that we plan to undertake in future work, the focus of this paper is to articulate the theoretical foundations and conceptual constructs of the proposed solution. By bridging the gap between linguistic comprehension and technical analysis, CEA-ADS aspires to fortify organizations against the ever-evolving threat of BEC attacks.

979-8-3503-8368-3/23 $31.00 © 2023 IEEE

II. METHODS

The proposed CEA-ADS mechanism for BEC prevention is founded on a multidimensional approach that synergizes cognitive analysis, AI-driven decision support, and contextual understanding. While the empirical validation of CEA-ADS is planned for future research, this section outlines the envisioned methodologies that constitute the backbone of the proposed solution.

A. Research Question

Central to the development of CEA-ADS is a fundamental research question that guides its conceptualization and implementation:

"Can cognitive analysis and AI-driven decision support be harnessed to enhance the accuracy of BEC threat detection and provide contextually relevant recommendations to users?"

This research question encapsulates the core objective of CEA-ADS: to explore the integration of cognitive analysis techniques with advanced AI-driven decision support systems in order to mitigate the risks posed by BEC attacks. By addressing this research question, CEA-ADS aims to bridge the gap between linguistic subtleties, technical analysis, and human intuition in the domain of cybersecurity.

B. Cognitive Analysis of Linguistic Nuances

It has been suggested that cognitive analysis can be harnessed to enhance the accuracy of BEC threat detection and provide contextually relevant recommendations to users. A systematic review of BEC phishing detection techniques, including the use of machine learning algorithms and features for detection has been presented [5]. Then, an email summarization approach that leverages transformer-based machine learning to analyze psychological triggers, detect malicious intent, and create representative summaries of emails to assist users in identifying phishing emails has been proposed [6]. Moreover, researchers proposed an automatic approach that leverages natural language processing and machine learning to identify decoy documents that have a high chance of deceiving targeted users [7].

CEA-ADS adopts a cognitive analysis approach to unravel the intricate linguistic nuances embedded within email communications. Through the integration of advanced Natural Language Processing (NLP) techniques, including sentiment analysis, syntactic parsing, and entity recognition, the mechanism aims to decode the emotional tone, semantic context, and relational structure of email content. By dissecting the linguistic fabric of each message, CEA-ADS strives to identify linguistic patterns indicative of potential BEC threats.

The following code could be used for sentiment analysis using Natural Language Toolkit (NLTK):

```
from nltk.sentiment import
    SentimentIntensityAnalyzer

sia = SentimentIntensityAnalyzer()
```

```
email_text = "Suspicious activity
    detected. Urgent response required."
sentiment_scores = sia.polarity_scores(
    email_text)
print(sentiment_scores)
```

The following code could be used for syntactic parsing using spaCy:

```
import spacy

nlp = spacy.load("en_core_web_sm")
email_text = "Please send the funds to
    this account."
parsed_email = nlp(email_text)
for token in parsed_email:
    print(token.text, token.pos_, token.
        dep_)
```

Sentiment analysis, a fundamental NLP technique, aids in gauging the emotional sentiment conveyed in emails. Beyond the realm of simple positive or negative sentiment classification, CEA-ADS seeks to decipher nuanced emotional nuances that might be exploited in BEC attacks. Syntactic parsing enables the identification of grammatical constructs and syntactic relationships, which are often manipulated by attackers to craft persuasive messages. Additionally, entity recognition assists in identifying key entities such as names, organizations, and monetary figures, which are crucial in detecting BEC-related information.

C. Automated Decision Support System

According to existing research, it is possible to harness AI-driven decision support to enhance the accuracy of BEC threat detection and provide contextually relevant recommendations to users. As previously mentioned, a systematic review of BEC phishing detection techniques, including the use of machine learning algorithms and features for detection was presented [5]. An automated phishing-advice tool, PhishEd, has been proposed [8], which allows people to report malicious emails and get automatically generated advice in response that is contextual to the suspicious email. Researchers in [9] focus on threat detection in the Internet of Things. Their publication highlights the importance of threat detection as a preventive measure against malware threats, ransomware, and attacks, which can be applied to BEC threat detection. Overall, researchers suggest that AI-driven decision support can be used to enhance the accuracy of BEC threat detection and provide contextually relevant recommendations to users.

The core innovation of CEA-ADS lies in its automated decision support system, designed to amplify users' ability to make informed decisions regarding potential BEC threats. Building upon the cognitive analysis phase, the system processes linguistic insights and contextual cues to generate contextually relevant recommendations.

The following sample code could be used for generating AI-driven recommendations:

979-8-3503-8368-3/23 $31.00 © 2023 IEEE

```python
class AIDecisionSupport:
    def generate_recommendation(self,
        linguistic_insights):
        # AI-based decision-making logic
        recommendation = "No immediate
            threat detected."
        return recommendation

# Instantiate the decision support system
ai_decision_support = AIDecisionSupport()
email_linguistic_insights = {"sentiment":
    "negative", "entities": ["urgent", "
    suspicious"]}
recommendation = ai_decision_support.
    generate_recommendation(
    email_linguistic_insights)
print("Recommendation:", recommendation)
```

This process involves the integration of advanced machine learning algorithms that learn from historical data and user interactions. As users engage with the system's recommendations, feedback loops enable the system to refine its decision-making process. The synergy between linguistic analysis and AI-driven recommendation systems empowers users by presenting actionable insights. By bridging the gap between the complexities of linguistic patterns and user intuition, CEA-ADS seeks to minimize false positives and equip users with the tools to respond effectively to potential BEC threats.

D. Integration of AI and Blockchain

An intriguing avenue explored by CEA-ADS involves the integration of artificial intelligence and blockchain technologies to enhance security and reliability. Researchers in [10] discuss how blockchain and AI can be used to secure IoT and IIoT applications. In [11], a blockchain-based system dynamic model to enhance cybersecurity in large corporations is proposed. Researchers have also reviewed recent studies on the convergence of AI and blockchain to secure IoT networks, suggesting future research directions [12]. Another recent research highlights the potential of AI and blockchain in cybersecurity applications, including securing cyber-physical systems [13].

The immutable nature of blockchain could be leveraged to securely store and authenticate the cognitive analysis results. This integration could serve to validate the accuracy of detected linguistic patterns and contextual insights, ensuring the integrity of the decision support system's outputs. While still in its conceptual stage, this integration introduces an additional layer of trust and transparency to CEA-ADS, aligning with the emerging trend of secure and verifiable AI-driven solutions.

E. Limitations and Challenges

The methodologies outlined in this section lay the theoretical foundation for CEA-ADS; however, it is essential to acknowledge the challenges and limitations that accompany its implementation. Empirical validation is paramount, as the effectiveness of CEA-ADS is predicated on the availability of diverse and high-quality training data. Additionally, the complexities of linguistic subtleties pose challenges in accurately detecting evolving attack tactics.

Ethical considerations are inherent in the application of AI-driven decision support systems. Addressing issues related to data privacy, bias in AI models, and the transparency of automated recommendations will be instrumental in building a responsible and effective solution. Furthermore, while the integration of AI and blockchain holds conceptual promise, practical implementation complexities and potential trade-offs must be carefully evaluated.

As we move toward future work and empirical validation, these limitations and challenges will guide the refinement and optimization of CEA-ADS, reinforcing its potential to revolutionize BEC prevention strategies.

III. DISCUSSION

The implementation of the CEA-ADS mechanism could yield notable outcomes in the domain of BEC prevention. While the empirical validation of CEA-ADS is slated for future research, this section provides an initial exploration of the potential implications of its integration.

A. Performance Evaluation

In a simulated environment, CEA-ADS could demonstrate promising capabilities in accurately detecting potential BEC threats. Leveraging sentiment analysis, syntactic parsing, and entity recognition, CEA-ADS could effectively decode the emotional tone, syntactic structures, and relevant entities within email communications.

The automated decision support system of CEA-ADS could generate contextually relevant recommendations based on the linguistic insights garnered. In an hypothetical scenario, a decision support system's recommendations could be compared with expert opinions, revealing its potential to aid users in making informed decisions regarding potential BEC threats.

B. Interpretation and Significance

The outcomes could underscore the significance of integrating cognitive analysis techniques and AI-driven decision support within BEC prevention strategies. CEA-ADS's ability to decipher linguistic subtleties and provide contextually aligned recommendations introduces a proactive layer of defense against BEC attacks. By mitigating the reliance on rule-based filtering and traditional machine learning methods, CEA-ADS elevates the precision and efficacy of BEC threat detection.

Furthermore, the integration of Python programming solutions may facilitate the translation of theoretical constructs into functional components. The sentiment analysis and syntactic parsing code snippets demonstrates the potential application of NLP techniques within CEA-ADS. The AI-driven decision support system's Python implementation showcases the synergy between linguistic insights and practical decision-making.

979-8-3503-8368-3/23 $31.00 © 2023 IEEE

C. Future Directions

Acknowledging the limitations of this initial exploration is essential. The actual performance of CEA-ADS in real-world contexts remains to be validated. The availability of diverse and representative training data will play a pivotal role in achieving robust accuracy.

Future work will encompass the empirical validation of CEA-ADS using real-world datasets and the refinement of AI models to accommodate the ever-evolving linguistic strategies employed by attackers. Additionally, the conceptual avenue of blockchain integration, while promising, requires rigorous investigation to address technical complexities and security considerations.

IV. CONCLUSION

The culmination of this study unveils the innovative potential of the CEA-ADS mechanism in the domain of BEC prevention. By synergizing cognitive analysis, AI-driven decision support, and programming solutions, CEA-ADS introduces a proactive and context-aware defense against the intricate challenges posed by BEC attacks.

A. Key Findings and Conclusions

The exploration of CEA-ADS yields several key findings that underscore its significance in BEC prevention. The integration of sentiment analysis, syntactic parsing, and entity recognition empowers CEA-ADS to decipher the emotional nuances, syntactic structures, and relevant entities within email communications. The AI-driven decision support system enhances users' decision-making capabilities by generating contextually aligned recommendations based on linguistic insights.

Theoretical constructs aligned with practical programming solutions can demonstrate the technical feasibility of CEA-ADS. The Python code snippets illustrate the application of natural language processing techniques and machine learning algorithms, anchoring the conceptual ideas in tangible implementations.

B. Benefits and Shortcomings

The benefits of CEA-ADS are manifest in its potential to revolutionize BEC prevention strategies. By focusing on cognitive analysis and AI-driven decision support, CEA-ADS transcends traditional rule-based filtering and static machine learning approaches. It augments users' capacity to interpret linguistic subtleties and empowers them to make well-informed decisions regarding potential BEC threats. Additionally, the conceptual exploration of blockchain integration introduces a layer of trust and transparency that aligns with emerging trends in secure AI-driven solutions.

However, it is important to recognize the limitations inherent in this study. The outcomes presented are based on an hypothetical scenario, and the actual performance of CEA-ADS in real-world environments remains to be validated. Furthermore, ethical considerations, including data privacy, bias mitigation, and interpretability of AI-driven recommendations, necessitate meticulous attention in the implementation of CEA-ADS.

C. Future Research Directions

The potential implications of CEA-ADS are far-reaching, extending beyond BEC prevention. The integration of cognitive analysis and AI-driven decision support can be extrapolated to address other cybersecurity challenges, emphasizing the versatility of the proposed mechanism.

Future research directions encompass the empirical validation of CEA-ADS using diverse and representative datasets. The refinement of AI models to adapt to evolving linguistic strategies employed by attackers is paramount. Furthermore, the exploration of blockchain integration demands thorough investigation into technical feasibility, security considerations, and practical implementation.

ACKNOWLEDGMENT

We acknowledge support of this work from the project "Immersive Virtual, Augmented and Mixed Reality Center of Epirus" (MIS 5047221) which is implemented under the Action "Reinforcement of the Research and Innovation Infrastructure", funded by the Operational Programme "Competitiveness, Entrepreneurship and Innovation" (NSRF 2014-2020) and co-financed by Greece and the European Union (European Regional Development Fund).

REFERENCES

[1] A. Papathanasiou, G. Liontos, V. Liagkou, and E. Glavas, 'Business Email Compromise (BEC) Attacks: Threats, Vulnerabilities and Countermeasures—A Perspective on the Greek Landscape', Journal of Cybersecurity and Privacy, vol. 3, no. 3, pp. 610–637, 2023.

[2] N. Saud Al-Musib, F. Mohammad Al-Serhani, M. Humayun, and N. Z. Jhanjhi, 'Business email compromise (BEC) attacks', Materials Today: Proceedings, vol. 81, pp. 497–503, 2023.

[3] A. Agazzi, 'Business Email Compromise (BEC) and Cyberpsychology'. 2020. arXiv preprint arXiv:2007.02415

[4] N. T N, D. Bakari, and C. Shukla, 'Business E-mail Compromise — Techniques and Countermeasures', in 2021 International Conference on Advance Computing and Innovative Technologies in Engineering (ICACITE), 2021.

[5] H. F. Atlam and O. Oluwatimilehin, 'Business Email Compromise Phishing Detection Based on Machine Learning: A Systematic Literature Review', Electronics, vol. 12, no. 1, p. 42, Dec. 2022.

[6] A. Kashapov, T. Wu, S. Abuadbba, and C. Rudolph, 'Email Summarization to Assist Users in Phishing Identification', in Proceedings of the 2022 ACM on Asia Conference on Computer and Communications Security, 2022.

[7] Sun, B., Ban, T., Han, C., Takahashi, T., Yoshioka, K., Takeuchi, J., Sarrafzadeh, A., Qiu, M. & Inoue, D. Leveraging machine learning techniques to identify deceptive decoy documents associated with targeted email attacks. *IEEE Access.* **9** pp. 87962-87971 (2021)

[8] Adam D. G. Jenkins, Nadin Kokciyan, and Kami Vaniea, 'PhishED: Automated contextual feedback for reported Phishing'. In 18th Symposium on Usable Privacy and Security. Usenix. 2022.

[9] N. Soltani, A. M. Rahmani, M. Bohlouli, and M. Hosseinzadeh, 'Artificial intelligence empowered threat detection in the Internet of Things: A systematic review', Concurrency and Computation: Practice and Experience, vol. 34, no. 22, Mar. 2022.

[10] M. A. Ferrag, L. Maglaras, and M. Benbouzid, 'Blockchain and Artificial Intelligence as Enablers of Cyber Security in the Era of IoT and IIoT Applications', Journal of Sensor and Actuator Networks, vol. 12, no. 3, p. 40, May 2023.

[11] S. Maesaroh, H. J. Permana, P. Dirgayusa Febrianaga, Noviyanti, and R. A. Pardosi, 'Blockchain Technology in the Future of Enterprise Security System from Cybercrime', Blockchain Frontier Technology, vol. 2, no. 1, pp. 1–8, Jun. 2022.

[12] S. Alharbi, A. Attiah, and D. Alghazzawi, 'Integrating Blockchain with Artificial Intelligence to Secure IoT Networks: Future Trends', Sustainability, vol. 14, no. 23, p. 16002, Nov. 2022.

[13] F. Muheidat and L. Tawalbeh, 'Artificial Intelligence and Blockchain for Cybersecurity Applications', in Studies in Big Data, Springer International Publishing, 2021, pp. 3–29.

Higher Order Probabilistic Analysis Of Network Trajectories Of Intelligent Agents In Thespian

Georgios Drakopoulos
Ionian University
c16drak@ionio.gr
0000-0002-0975-1877

Phivos Mylonas
University of West Attica
mylonasf@uniwa.gr
0000-0002-6916-3129

Abstract—Intelligent agents (IAs) are autonomous pieces of software designed to be deployed to and operate on infrastructure, whether physical or digital, and perform various tasks on them ranging from integrity check and structure discovery to functionality monitoring and information collection. Their task performing capability has been considerably increased with the recent advent of machine learning. An important parameter of IAs operating on networks such as the Web is their trajectory, which in many engineering scenarios depends heavily on random outcomes taking place at each vertex visited by the IA. In order to study the probabilistic properties of the trajectory length, said outcomes instead of being modeled or simulated are computed as the result as a game taking place between the vertex and the IA, developed in the Thespian framework for Python, and the vertex. The latter selects a random but fixed strategy, whereas the IA can adapt to learn this strategy either by observing the entropy of the choices of its opponent. If IA loses, then it backtracks, otherwise it chooses its next destination with a preferential attachment scheme. The mean, variance, skewness, and kurtosis of the trajectories of IAs operating on three scale-free graphs generated by NetworkX. Emphasis was placed on proper Pythonic code as many major Python modules such as threads, for generating game instances, Counter instances from collections to keep track of player choices, and functools for map/reduce functionality. Results indicate that IAs learning the opponent strategy have longer and richer network trajectories in terms of vertices, indicating the importance of learning.

Index Terms—intelligent agents, random walks, network trajectory, Thespian, NetworkX, strategy estimation, entropy, higher order data, collections, threading, functools, argparse, globals

I. INTRODUCTION

Intelligent agents (IAs) are autonomous pieces software designed to perform a broad spectrum of tasks including but not limited to structural integrity, link prediction, and even negotiation with other IAs. This functionality increases the digital awareness of their operators by collecting information about the environment they operate in. Interestingly, as IAs keep functioning therein, they generate information of their own, in addition to that they collect regularly, which can be mined. A major such piece of information for IAs operating in networks is the length of their trajectory expressed in the number of edges crossed. By probabilistically mining it, knowledge about the underlying network and about how easily the IA can cross it can be obtained.

The IAs presented and examined here traverse the network in order to visit each vertex. However, an opponent at each vertex the IA is currently visiting challenges the latter to a rock/paper/scissors (RPS) game whose outcome determines the next IA move. Specifically, if the IA emerges as a winner, then it chooses the next vertex to visit based on a preferential attachment policy, otherwise the IA backtracks to the vertex it came from. The RPS game was selected as it is straightforward to collect its results. Moreover, the IA moves based on the outcomes of actually played games instead of computing or simulating the probability of IA victory in each game.

The twofold primary research objective of this conference paper is to collect and perform a higher order probabilistic analysis of the lengths of the IA trajectories including the computation of skewness and kurtosis as well as to analyze the difference in performance of IAs relying on strategy estimation in each game from IAs which do not. As a secondary objective, heavy emphasis was placed on developing Pythonic code in order to achieve the desired functionality as well as to translate probability theoretic concepts to programming ones as determined by PEP8[1] complete with docstrings as stated in PEP 257[2]. Additionally, the IA was implemented in the Thespian[3] framework and the three synthetic benchmark datasets were obtained from the NetworkX[4] graph library. The methodology presented here can be applied to the study of network propagation phenomena such as meme diffusion is social media, SIR models, and the spread of biological virus.

The remainder of this work is structured as follows. In section II the recent scientific literature regarding IAs, Python modules, and parameter estimation is briefly reviewed. In section III the IA components and mechanics and the probablistic strategy estimation are explained, whereas in section IV the RPS source code is described. Results is explained in section V. Possible future research directions are given in section VI. Capital calligraphic letters symbolize random variables (rvs) and normal lowercase scalars. Technical acronyms are explained the first time they are encountered in text. Finally, the notation of this work is summarized in table I.

---

[1]https://peps.python.org/pep-0008
[2]https://peps.python.org/pep-0257
[3]https://github.com/thespianpy/Thespian
[4]https://networkx.org

979-8-3503-8368-3/23 $31.00 © 2023 IEEE

TABLE I
NOTATION SYNOPSIS.

Symbol	Meaning	First in
$\triangleq$	Definition or equality by definition	Eq. (1)
$\{s_1, \ldots, s_n\}$	Set with elements $s_1, \ldots, s_n$	Eq. (8)
$\langle s_k \rangle$	Sequence with elements s_k	Eq. (6)
$(t_1, \ldots, t_n)$	Tuple with elements $t_1, \ldots, t_n$	Eq. (12)
$\lvert \cdot \rvert$	Set or sequence cardinality functional	Eq. (6)
$\hat{y}$	Estimator of quantity y	Eq. (8)
$\text{prob}\{\Omega\}$	Probability of event Ω occurring	Eq. (2)
$\text{E}[\mathcal{X}]$	Expected value of random variable $\mathcal{X}$	Eq. (3)
$\text{Var}[\mathcal{X}]$	Variance of random variable $\mathcal{X}$	Eq. (3)
$\deg(v)$	Degree of vertex v	Eq. (2)
$\Gamma(v)$	Neighborhood of vertex v	Eq. (2)

II. PREVIOUS WORK

IAs are designed to operate autonomously in infrastructure [1], whether physical [2] or digital [3], to perform through machine learning (ML) a wide range of critical tasks such as link prediction [4], secure communications in unmanned aerial vehicles (UAVs) [5], and cooperation in autonomous vehicles [6]. In network environments such as social media [7], which can be represented by ordinary [8] or fuzzy graphs [9], or smart cities [10] IAs can be particularly effective as the modular and commuity-based graph structure [11] allows them to roam in order to collect information [12] about the network state such as expansion potential [13] and increase digital awareness [14] through state estimation [15] and swarm intelligence [16]. In such environments multi-objective optimization [17] and trajectory prediction in multi-agent engineering scenarios [18] are key problems. In settings like recommender systems [19] [20] the user acceptance of IAs is critical [21]. IAs can also be deployed in blockchains to negotiate smart contracts [22], collect state information [23], and verify consensus [24]. A survey for IA applications is [25].

Parameter estimation is an established field in signal processing [26] where a scalar parameter or a vector of parameters need to be inferred from a number of possibly corrupted measurements [27]. In classical estimation said parameters are considered fixed and they are estimated through probasilictic techniques like least minimum variance estimation [28] or model-based techniques like deterministic least mean squares [29] and maximum likelihood [30]. On the contrary, Bayesian estimation is also model-based but it treates the parameters as realizations of a random process [31]. Hence, a prior estimation is additionally required [32].

Python modules cover a broad spectrum of functionality such as behavioral analysis [33] and JSON parsing [34]. Functional programming [35] is an established programming paradigm where priority is placed on functions, considered first-class citizens, and their composition [36]. Besides purely functional languages such as Haskell, functional elements, mainly in the form of lambda expressions, can be found in Scala [37] and Java [38]. Functional data structures are described in [39] and functional aspects include map, filter, and reduce [40] [41].

III. INTELLIGENT AGENT DESIGN

A. Random Trajectories

In order to successfully carry out their task, the IAs typically have an architecture consisting of the following components:

- Sensors in order to perceive its environment.
- Actuators to act on its environment.
- A decision making scheme fed by sensor information.
- A state vector where the IA status is stored.

The trajectory an IA makes while traversing a network can be mined through statistical analysis or ML for patterns in order to extract working knowledge about the underlying network structure in addition to the information IA collects. In the context of this work the IA objective is to visit every vertex, which is a typical assignment. However, in order for it to be allowed to move, the IA must win an RPS game. The bot opponent, for brevity termed *bot* or *opponent*, of the IA selects independently and randomly one of the strategies of table II. The IA can be configured to either estimate the bot strategy or to select a strategy at random from the same table.

A number of N runs, shown in table VI, was executed. In such run the IA was first configured to estimate the bot strategy and then to select a strategy at random and the trajectory length, denoted respectively as L_e and L_r, was recorded and the ratio L of equation (1) was formed.

$$L \triangleq \frac{L_r}{L_e} \qquad (1)$$

When IA in a vertex u beats its opponent, then it can move to one of the neighbors of u. The probability of IA moving from vertex u to a neighboring v is computed by a preferential attachment policy shown in equation (2).

$$\text{prob}\{u \to v\} \propto \frac{\deg(v)}{\sum_{s \in \Gamma(v)} \deg(s)} \qquad (2)$$

The values of the trajectory length ratio of equation (1) can be considered as samples of an rv $\mathcal{L}$. Once its probabilistic mean and variance, respectively denoted as $\text{E}[\mathcal{L}]$ and $\text{Var}[\mathcal{L}]$, can be computed from the respective sample counterparts under mild ergodicity assumptions, then higer order probabilistic metrics can be computed. A common third order metric is the skewness κ_3 defined as in equation (3). On the condition that κ_3 comes from a unimodal distribution, when κ_3 is positive, then the tail of the distribution is to its right, while when it is positive, then its tail is to its left.

$$\kappa_3 \triangleq \frac{\text{E}\left[(\mathcal{L} - \text{E}[\mathcal{L}])^3\right]}{\text{Var}[\mathcal{L}]^{3/2}} \qquad (3)$$

Another higher order probabilistic metric is kurtosis, which is a fourth order metric evaluating the concentration around the mean and defined as in equation (4).

$$\kappa_4 \triangleq \frac{\text{E}\left[(\mathcal{L} - \text{E}[\mathcal{L}])^4\right]}{\text{Var}[\mathcal{L}]^2} \qquad (4)$$

979-8-3503-8368-3/23 $31.00 © 2023 IEEE

B. Strategy Estimation

In this game the bot player is configured such that at each round of each duel it makes a move which is independent of the past ones and according to a preselected stationary policy. In other words, the distribution of the next move made by the bot player is given in general by equation (5). Observe that said distribution generates categorical data. Moreover, decision independence precludes Markov chain type analysis.

$$s_k \triangleq \begin{cases} \textbf{rock}, & \text{prob}\{s_k = \textbf{rock}\} = p_r \\ \textbf{paper}, & \text{prob}\{s_k = \textbf{paper}\} = p_p \\ \textbf{scissors}, & \text{prob}\{s_k = \textbf{scissors}\} = p_s \end{cases} \quad (5)$$

A move sequence $\langle s_k \rangle$ of length n follows the trinomial distribution shown in equation (6):

$$\text{prob}\{s\} \triangleq \binom{n}{n_r}\binom{n - n_r}{n_p} p_r^{n_r} p_p^{n_p} p_s^{n_s}, \quad |s| = n \quad (6)$$

The distribution of (6) holds under the constrtaint pair of (7). This ensures that the distribution adds up to one.

$$\begin{aligned} n_r + n_p + n_s &= n \\ p_r + p_p + p_s &= 1 \end{aligned} \quad (7)$$

An estimation $\hat{p}_x$ of each of the three probabilities p_x of equation (6) can be constructed as a function $g_x(\cdot)$ of the frequencies of appearance, or equivalently the number of appearances, of rock, paper, and scissors denoted respectively as n_r, n_p, and n_s as shown in equation (8).

$$\hat{p}_x \triangleq g_x(n_r, n_p, n_s), \quad x \in \{r, p, s\} \quad (8)$$

One estimator originating directly from equation (6) but applied to a more generic context is equation (9), which coincides with the frequentist approach coined by Laplace.

$$\hat{p}_x \triangleq \frac{n_x}{n_r + n_p + n_s}, \quad x \in \{r, p, s\} \quad (9)$$

Based on the probability estimation the entropy of the opponent can be computed. In table II the entropy for the bot player strategies is shown. Equation (10) was used. Therein the base of logarithm b determines the base in which information is measured. Given that each logarithm is of the same order of magnitude, the selection of b does not critically influence the value of H. In this case b equals two and information is represented in bits.

$$H \triangleq \sum_k p_k \log_b \frac{1}{p_k} = -\sum_k p_k \log_b p_k \quad (10)$$

Quite often reduce is combined with **lambda**. For instance, in the following code segment the value of entropy is updated if and only if the value of argument p is not zero as otherwise a numerical error would occur.

```python
import functools as f
import numpy as np

vals = d.values()
H = f.reduce(lambda H,p: \
    H-p*np.log2(p) if p else H, vals, 0)
```

TABLE II
STRATEGIES AND THEIR ENTROPY.

Strategy	Distribution	Entropy (bits)
uniform	all moves 1/3	$\log_2 3$ (max)
last	play opponent's last move	between 0 and $\log_2 3$
rockx3	60% rock, 20% paper, scissors	1.3709
rockx2	50% rock, 25% paper, scissors	1.5
rockx0	50% paper, 50% scissors	1
rockx1	100% rock	0 (min)

Lambda expressions, functional elements, or even functional programming as a paradigm is supported in prominent programming languages such as Java[5] and C++[6]. Moreover, in newer languages such as Julia[7] the lambda expressions, therein called *anonymous functions*, are an integral part of the original language specifications[8]. Finally, in purely functional languages such as Haskell[9] and Scala[10], lambada functions not only are available, but they are *first class citizens*, namely the primary means of computation.

IV. RPS CODE OVERVIEW

A. Overview

The source code comprises of the files shown in table III. Therein is also shown a brief description of the respective file and the subsection which describes their functionality. The game relies heavily on the players exchanging messages in order to monitor RPS progress and keep track of moves.

TABLE III
SOURCE CODE FILES.

Name	Description	Shown in
bg_client_app.py	Client application	Sec. IV-B
bg_client.py	Client configuration	Sec. IV-C
bg_client_player.py	Basic player class	Sec. IV-D
bg_client_custom_players.py	Opponent player class	Sec. IV-E
bg_client_my_player.py	IA player classes	Sec. IV-F
bg_server_app.py	Server application	Sec. IV-G
bg_server.py	Main server file	Sec. IV-H
bg_server_worker.py	RPS game logic	Sec. IV-I
bg_messages.py	Server/client messages	Sec. IV-J

The event driven nature of the game means a non-linear source code execution. Moreover, the IA and the bot need not be on the same computer, in which case the network overhead should be taken into consideration.

Notice that in order to keep **from ... import** commands simple, all source files have been moved to the same directory. In this way some initial problems stemming from directory dependencies were automatically resolved.

Before a print operation takes place, a lock is obtained to ensure that only one thread displays a message. Moreover, all messages are sent to the standard stream sys.stderr,

---

[5]https://www.infoworld.com/article/2078836/love-and-hate-for-java-8.htm
[6]https://en.cppreference.com/w/cpp/language/lambda
[7]www.julialang.org
[8]https://docs.julialang.org/en/v1/manual/functions
[9]https://wiki.haskell.org/Anonymous_function
[10]https://docs.scala-lang.org/scala3/book/fun-anonymnous-functions.html

which is the proper stream for diagnostic and error messages. Moreover, for each such operation the flush option is enabled, meaning that each diagnostic message is guaranteed to appear on the screen the time the programmer intends to. Otherwise, given the buffered nature of I/O operations, there is a non-negligible probability that a program may crash before a message appears on screen, giving thus the wrong idea about the code crashing point. When flushing is enabed, however, all outbound messages have been properly displayed before the exception stack is unwound. Therefore, the developer has the correct perception about where code execution has reached and can look for the right causes. In order to ensure this, the partial method creates a restricted version of the generic print which sends its input to the desired stream with flushing. Recall that in order for this to work, the target function must support a keyword-pair structure of arguments.

```
import functools as f

ffprint = f.partial(print,\
    file=sys.stderr, flush=True)
```

B. File bg_client_app.by

This is the main client file, namely the file IA has to start in order for its player class to be initiated and for the connection to be set up. This file creates a client configuration dictionary and passes it to the main function of the *bg_client.py*. Note that the fields silent and verbose are set to **None** since they will be filled by the values of the server configuration dictionary.

C. File bg_client.py

This is the module containing the main client function, which is responsible for creating the various client players according to the appropriate command line arguments.

The globals() function is used early in this file in order to return a dictionary with every global variable[11] and then instantiate a player class, either an IA or a bot one depending on who is calling the rps_client_main() function. The latter is determined by the class name stored in the field class of the client configuration dictionary. As the bot player initialization function __init__() accepts different arguments than its counterparts of the IA classes, care is taken before invoking it. The variable klass() is an alias of the right initiator.

One the client is connetect to the server network socket, an initial message from the client is sent. Notice that in order to do so, a lock is obtained so that other threads may not print at the same time a message as well. The pair of functions acquire() and release() form a safe context manager, similar to the well-known **with** ... **as** one.

Note that a general technique after receiving over the network and deserializing a message using the pickle library is to check whether it is of the expected type. This is accomplished with as **assert** condition.

Since the bot player class does not have methods for keeping track of who the opponent is and for monitoring the opponent

---

[11] https://docs.python.org/3/faq/programming.html

moves, a check is made to determine whether the player class currently active has an attribute of the right name through the hasattr function[12] followed by a second check through the callable method to determine whether this attribute appears to be callable. The latter means that even if callable returns **True**, a function call with that attribute may still fail –a case not happening here by design, whereas if it returns **False**, then a function call with this attribute is bound to fail[13]. Because of the short-circuit evaluation[14] of logical **and** in Python, if the first check fails, then the second one is not even executed.

D. File bg_client_player.py

The rudimentary player class is stored in this short source file. This class is used to build the standard bot and IA classes through simple inheritance, even though Python is one of the few modern programming languages supporting multiple inheritance[15] like C++.

Another highlight is that since this is an elementary class without a fully fledged functionality, the method game_result() is just a placeholder. To emphasize this, a NotImplementedError is exception raised if this method is called instead of just putting a **pass** instruction in its definition.

E. File bg_client_custom_players.py

In this source file the standard bot player class can be found. If the __init__() is called with an argument of **None**, then a strategy out of the possible ones is picked at random. Otherwise, the prespecified strategy is selected and played out throughout the entire game.

Bias can be inserted in the method next_move() of the bot class by removing from or inserting to the initial list the *rock* option one or two times. Additionally, the bot can be instructed to play only *rock* as a move. Of course there is nothing special with this move, as taking the same steps with any of the other two moves would create similar distributions. Notice that in the selection of the next move the copy() method is needed, since the append() and remove() functions operate *in-place* and, hence, their return value as a result is **None**.

F. File bg_client_my_player.py

In this module the IA player class is defined. The latter inherits the majority of its functionality from the former, with the single exception of the function determining its next move. In this way, code is reused and the differences between the two classes become clearly visible. In table IV the methods of the IA class are explained.

At the heart of the IA player class lies a composite data structure where:

- A dictionary with keys the opponent ids as returned by the game server.
- Each corresponding value is a list with two elements:

---

[12] https://docs.python.org/3/library/functions.html#hasattr
[13] https://docs.python.org/3/library/typing.html
[14] https://www.geeksforgeeks.org/short-circuiting-techniques-python
[15] https://realpython.com/inheritance-composition-python

979-8-3503-8368-3/23 $31.00 © 2023 IEEE

TABLE IV
IA PLAYER METHODS.

Method	Description
__init__	Initializer
__str__	String conversion
__repr__	Object representation
update_moves	Keep track of opponent moves
update_opponent	Prepare for a new opponent
update_winner	Keep track of duel results
get_encounters	Return duel result history
rec_freq	Recommendation based on moves
rec_strategy	Recommendation based on entropy
next_move	Get next move

- A dictionary with each possible move as keys and an integer counter as value.
- A bool flag indicating whether the player is the IA.

Recall that int and str types are hashable and they can be dictionary keys. When this structure is initialized or when a new opponent in encountered, then a new entry is created using the dictionary update() method, which works on both existing and new structures. Should this measure had not been taken, a KeyError exception would have been raised.

G. File bg_server_app.py

This module must be started in order to set up the server controlling the game and spawning any bot players necessary. Since this file controls a considerable amount of functions, it also accepts a number of command line arguments through the argparse module in order to configure this functionality.

H. File bg_server.py

This is the main server file where the connections to players, whether IAs or vertex opponents, are established and the game is executed. Moreover, the main server function is an infinite **while** loop which can be only terminated when an exception occurs. In the exception hadling code care is taken to close the socket before exiting the program.

I. File bg_server_worker.py

This file contains the code for the actual execution of the game once the establishment of network sockets, the initial communication, and the game setup are done. Moreover, the game logic is implemented here.

J. File bg_messages.py

All possible messages sent from the server to client and vice versa are defined in file bg_messages.py. Messages with originally three arguments or more were converted to accept a dictionary with the individuals arguments as fields. Moreover, in order to secure message integrity, every initiator has an **assert** clause. Table V contains the messages as well as basic information about them.

V. RESULTS

The configuration parameters of the RPS game as well as the synopsis of the three synthetic benchmark datasets are shown in VI. Implementation was done in Python 3.12, the latest version currently available. The Thespian framework implements the Actor model in Python and relies heavily on message passing between the entities. The NetworkX library offers a wide array of functionality pertaining to graphs, including the generation of scale-free graphs with a predetermined number of vertices and density. NetworkX defines the density ρ for a graph with n_v vertices and n_e edges as in (11)[16].

$$\rho \triangleq \frac{2n_e}{n_v(n_v - 1)} \tag{11}$$

Therefore ρ is defined as the ratio of the number of edges of the graph to the number of edges of a complete graph with the same number of vertices. For directed graphs ρ is defined as half of that in (11). The synthetic benchmark graphs used here are undirected. Also please note that this definition of ρ differs from the one frequently used in the scientific literature.

From a graph structure perspective, the graph density ρ essentially determines on the average the number of choices the IA has when it wins an RPS game. Along the same line of reasoning, the graph diameter roughly determines how difficult is to reach the outermost vertices, which in turn may lead to longer trajectories if it is sufficiently high.

In table VII the results of the code execution with D duels of R rounds each per vertex are shown. The total number of games equals the number of vertices visited by the IA. Each table entry has the structure of equation (12). Since the game results are totally independent from the underlying graph topology, the IA and bot victory probabilities have been averaged over all three benchmarks. Also, the tie probability is not shown, as it is redundant information. If a tie occurs, then a new game starts on the same vertex.

$$(\text{prob}\{\text{IA wins}\}, \text{prob}\{\text{bot wins}\}) \tag{12}$$

Table VII should be interpreted as follows: The last row has the probabilities of victory, defeat, and tie for IA when it is configured to estimate bot strategy, whereas the remaining of the table has the same statistics when both the IA and the bot each randomly select a strategy. Each column corresponds to a bot strategy, while each row to an IA strategy. From the entries of table VII the following conclusions can be drawn:

- Each strategy against itself practically lead to tie.
- There was a symmetry between strategies played by bots, namely that the streategy pair (S_1, S_2) had roughly the same results as that of (S_2, S_1).
- The most effective strategy is *uniform*, which is the most unpredictable, followed by *rockx2*. This is expected given the entropies of these strategies.
- The worst one is *rockx1*, which is totally predictable. Moreover, it can be easily countered by an opponent constantly playing paper.

[16]https://networkx.org/documentation/stable/reference/generated/networkx.classes.function.density.html

979-8-3503-8368-3/23 $31.00 © 2023 IEEE

TABLE V
MESSAGE TABLE.

Class	Sender	Meaning	Parameters
ServerMsgHello	Server	Configuration message	Silent and verbose flags, rounds, duels
ServerMsgDuelStart	Server	Duel notification	Player ids, player and IA flags
ServerMsgRoundStart	Server	Round notification	Round id
ServerMsgRoundWinner	Server	Round winner	Player ids, player moves, winner id
ServerMsgDuelWinner	Server	Duel winner	Winner id
ClientMsgHello	Client	Connection message	Name, version, IA flag
ClientMsgOk	Client	General acknowledgement	Message
CientMsgRoundMove	Client	Move in round	Move, player id

TABLE VI
CONFIGURATION PARAMETERS.

Parameter	Value
Number of runs N	100000
Number of RPS duels D	101
Number of RPS rounds R	201
Number of vertices n_v	10000 / 10000 / 10000
Number of edges n_e	90000 / 70000 / 60000
Density ρ	0.00009 / 0.00007 / 0.00006
Diameter δ	11 / 12 / 14

- The *last* strategy is a peculiar case as the player employing it is totally dependent on the other. Still, it leads to a lot of ties, especially when the other plays *rockx1*.

In table VIII the probabilistic metrics of the trajectory length ratio $\mathcal{L}$ of (1) are shown. From the entries of this table the following conclusions can be drawn:

- The mean of $\mathcal{L}$ shows that on average the IA has to cross considerably more times the number of vertices when the bot strategy estimation is disabled.
- The relatively large value of the variance means that $\mathcal{L}$ is not very concentrated around its mean.
- The positive skewness indicates that the majority of the values, namely the tail of the distribution, of $\mathcal{L}$ is to the right of its mean.
- The kurtosis also seems to support the case that $\mathcal{L}$ flustuates. This could be possibly attributed to the random nature of the RPS game.
- The sparser and the more difficult to traverse the graph is, the higher the mean value of $\mathcal{L}$ and its fluctuations are, indicating more variability.

Entries from table VIII tend to support the hypothesis that enabling the IA to estimate the strategy of the bot in each vertex through the entropy of the distribution of the decisions of the latter reduces considerably the number of hops the IA must do in order to visit every vertex of the underlying graph.

VI. CONCLUSIONS AND FUTURE WORK

The focus of this conference paper is to perform higher order probabilistic analysis of the random trajectories of an intelligent agent (IA) operating on network infrastructures. At each vertex a bot opponent challenges the IA to a game of rock/paper/scissors (RPS). Depending on the outcome, the IA selects a neighboring vertex based on a preferential attachment rule if it is victorious or the IA backtracks to the previous vertex. Two scenarios were tested, depending on whether the IA could estimate the entropy of the decision distribution of its opponent or not. Results indicate that when that was true, the IA trajectory length was considerably shorter. The IA was implemented in the Python Thespian framework and the RPS game in Python utilizing modules like argparse, collections, threads, and functools. The RPS game was chosen such that the probability of the IA successfully advancing from one vertex to one of its neighboring ones to be determined by an actual game instead of being computed or simulated. This can be beneficial in the analysis of network propagation phenomena such as meme diffusion or virus propagation.

This work can be expanded in a number of ways. First and foremost, strategy estimators such as *chi square test* and *run test* can be implemented, both of which rely on the analysis of long sequences. Additionally, the proposed technique can be applied to larger benchmark networks to test its scalability.

ACKNOWLEDGMENT

This research was funded by the European Union and Greece (Partnership Agreement for the Development Framework 2014-2020) under the Regional Operational Programme Ionian Islands 2014-2020, project title: "Indirect costs for project "TRaditional corfU Music PresErvation through digiTal innovation", project number: 5030952.

REFERENCES

[1] S. Moussawi, M. Koufaris, and R. Benbunan-Fich, "The role of user perceptions of intelligence, anthropomorphism, and self-extension on continuance of use of personal intelligent agents," *European Journal of Information Systems*, vol. 32, no. 3, pp. 601–622, 2023.

[2] U.-H. Kim, J.-M. Park, T.-J. Song, and J.-H. Kim, "3-D scene graph: A sparse and semantic representation of physical environments for intelligent agents," *IEEE Transactions on cybernetics*, vol. 50, no. 12, pp. 4921–4933, 2019.

[3] J. Maurio *et al.*, "Agile services and analysis framework for autonomous and autonomic critical infrastructure," *Innovations in Systems and Software Engineering*, vol. 19, no. 2, pp. 145–156, 2023.

[4] M. Nikolaou, G. Drakopoulos, P. Mylonas, and S. Sioutas, "Intelligent agents with graph mining for link prediction over Neo4j," in *WEBIST*. SCITEPRESS, 2023.

[5] V. F. Sangeetha Francelin, J. Daniel, and S. Velliangiri, "Intelligent agent and optimization-based deep residual network to secure communication in UAV network," *International Journal of Intelligent Systems*, vol. 37, no. 9, pp. 5508–5529, 2022.

[6] H. Yang, Z. Wei, Z. Feng, X. Chen, Y. Li, and P. Zhang, "Intelligent computation offloading for MEC-based cooperative vehicle infrastructure system: A deep reinforcement learning approach," *IEEE Transactions on Vehicular Technology*, vol. 71, no. 7, pp. 7665–7679, 2022.

TABLE VII

IA VS BOT RESULTS (AVERAGE OVER THE THREE BENCHMARKS).

bot →	uniform	last	rockx3	rockx2	rockx0	rockx1
uniform	(0.3184, 0.3781)	(0.2189, 0)	(0.5621, 0.2238)	(0.4676, 0.2388)	(0.4527, 0.1293)	(0.955, 0)
last	(0.2686, 0.3681)	(0.199, 0.0099)	(0.6019, 0.1741)	(0.4975, 0.2587)	(0.4726, 0.4577)	(0.995, 0)
rockx3	(0.3532, 0.3034)	(0.2039, 0)	(0.592, 0.1691)	(0.4726, 0.2388)	(0.4676, 0.0746)	(0.998, 0)
rockx2	(0.3482, 0.3034)	(0.2039, 0)	(0.6218, 0.1641)	(0.5074, 0.2537)	(0.4427, 0.1839)	(0.995, 0)
rockx0	(0.2736, 0.3731)	(0.2039, 0)	(0.5971, 0.1939)	(0.4726, 0.2487)	(0.5771, 0)	(1, 0)
rockx1	(0.3333, 0.2338)	(0, 0.0049)	(0.5823, 0.1942)	(0.5472, 0.2089)	(0.4477, 0.3233)	(0, 0)
entropy	(0.4079, 0.194)	(0.4477, 0.1044)	(0.4378, 0.1144)	(0.4179, 0.099)	(0.4527, 0.1243)	(0.995, 0)

TABLE VIII

TRAJECTORY LENGTH RATIO STATISTICS.

Metric	$E[\mathcal{L}]$	$Var[\mathcal{L}]$	κ_3	κ_4
Benchmark 1	3.7819	3.2154	1.4471	3.1333
Benchmark 2	4.1373	3.7998	1.6147	3.7711
Benchmark 3	4.6333	4.0011	1.8533	4.9347

[7] G. Drakopoulos, E. Kafeza, P. Mylonas, and L. Iliadis, "Transform-based graph topology similarity metrics," *NCAA*, vol. 33, no. 23, pp. 16 363–16 375, 2021.

[8] G. Drakopoulos, I. Giannoukou, P. Mylonas, and S. Sioutas, "A graph neural network for assessing the affective coherence of Twitter graphs," in *IEEE Big Data*. IEEE, 2020, pp. 3618–3627.

[9] G. Drakopoulos, E. Kafeza, P. Mylonas, and S. Sioutas, "A graph neural network for fuzzy Twitter graphs," in *CIKM companion volume*, G. Cong and M. Ramanath, Eds., vol. 3052. CEUR-WS.org, 2021.

[10] Z. Tong, F. Ye, M. Yan, H. Liu, and S. Basodi, "A survey on algorithms for intelligent computing and smart city applications," *Big Data Mining and Analytics*, vol. 4, no. 3, pp. 155–172, 2021.

[11] G. Drakopoulos and P. Mylonas, "A genetic algorithm for Boolean semiring matrix factorization with applications to graph mining," in *Big Data*. IEEE, 2022.

[12] H. Li, Q. Zhang, and D. Zhao, "Deep reinforcement learning-based automatic exploration for navigation in unknown environment," *IEEE Transactions on neural networks and learning systems*, vol. 31, no. 6, pp. 2064–2076, 2019.

[13] G. Drakopoulos, E. Kafeza, P. Mylonas, and S. Sioutas, "Approximate high dimensional graph mining with matrix polar factorization: A Twitter application," in *IEEE Big Data*. IEEE, 2021, pp. 4441–4449.

[14] B. Shneiderman, "Design lessons from AI's two grand goals: Human emulation and useful applications," *IEEE Transactions on Technology and Society*, vol. 1, no. 2, pp. 73–82, 2020.

[15] A. Anand, E. Racah, S. Ozair, Y. Bengio, M.-A. Côté, and R. D. Hjelm, "Unsupervised state representation learning in atari," *Advances in neural information processing systems*, vol. 32, 2019.

[16] M. Schranz, G. A. Di Caro, T. Schmickl, W. Elmenreich, F. Arvin, A. Şekercioğlu, and M. Sende, "Swarm intelligence and cyber-physical systems: Concepts, challenges and future trends," *Swarm and Evolutionary Computation*, vol. 60, 2021.

[17] N. Yang, L. Han, R. Liu, Z. Wei, H. Liu, and C. Xiang, "Multi-objective intelligent energy management for hybrid electric vehicles based on multi-agent reinforcement learning," *IEEE Transactions on Transportation Electrification*, 2023.

[18] X. Mo, Z. Huang, Y. Xing, and C. Lv, "Multi-agent trajectory prediction with heterogeneous edge-enhanced graph attention network," *IEEE Transactions on Intelligent Transportation Systems*, vol. 23, no. 7, pp. 9554–9567, 2022.

[19] D. Roy and M. Dutta, "A systematic review and research perspective on recommender systems," *Journal of Big Data*, vol. 9, no. 1, p. 59, 2022.

[20] G. Drakopoulos, I. Giannoukou, S. Sioutas, and P. Mylonas, "Self organizing maps for cultural content delivery," *NCAA*, vol. 31, no. 7, 2022.

[21] E. Elshan, N. Zierau, C. Engel, A. Janson, and J. M. Leimeister, "Understanding the design elements affecting user acceptance of intelligent agents: Past, present and future," *Information Systems Frontiers*, vol. 24, no. 3, pp. 699–730, 2022.

[22] D. Kirli *et al.*, "Smart contracts in energy systems: A systematic review of fundamental approaches and implementations," *Renewable and Sustainable Energy Reviews*, vol. 158, 2022.

[23] S. Swain and M. R. Patra, "A distributed agent-oriented framework for blockchain-enabled supply chain management," in *ICBDS*. IEEE, 2022, pp. 1–7.

[24] G. Drakopoulos, E. Kafeza, and H. Al Katheeri, "Proof systems in blockchains: A survey," in *SEEDA-CECNSM*. IEEE, 2019.

[25] F. De la Prieta, S. Rodríguez-González, P. Chamoso, J. M. Corchado, and J. Bajo, "Survey of agent-based cloud computing applications," *Future generation computer systems*, vol. 100, pp. 223–236, 2019.

[26] X. Yi and C. Zhong, "Deep learning for joint channel estimation and signal detection in OFDM systems," *IEEE Communications Letters*, vol. 24, no. 12, pp. 2780–2784, 2020.

[27] K. Abratkiewicz, P. Samczyński, and K. Czarnecki, "Radar signal parameters estimation using phase accelerogram in the time-frequency domain," *IEEE Sensors Journal*, vol. 19, no. 13, pp. 5078–5085, 2019.

[28] T. L. Jensen and E. De Carvalho, "An optimal channel estimation scheme for intelligent reflecting surfaces based on a minimum variance unbiased estimator," in *ICASSP*. IEEE, 2020, pp. 5000–5004.

[29] S. Naz, R. Sultan, K. Zaman, A. M. Aldakhil, A. A. Nassani, and M. M. Q. Abro, "Moderating and mediating role of renewable energy consumption, FDI inflows, and economic growth on carbon dioxide emissions: Evidence from robust least square estimator," *Environmental Science and Pollution Research*, vol. 26, pp. 2806–2819, 2019.

[30] M. Li and X. Liu, "Maximum likelihood least squares based iterative estimation for a class of bilinear systems using the data filtering technique," *International Journal of Control, Automation and Systems*, vol. 18, no. 6, pp. 1581–1592, 2020.

[31] K. R. Mestav, J. Luengo-Rozas, and L. Tong, "Bayesian state estimation for unobservable distribution systems via deep learning," *IEEE Transactions on Power Systems*, vol. 34, no. 6, pp. 4910–4920, 2019.

[32] J. E. Bernhard, J. S. Moreland, and S. A. Bass, "Bayesian estimation of the specific shear and bulk viscosity of quark–gluon plasma," *Nature Physics*, vol. 15, no. 11, pp. 1113–1117, 2019.

[33] G. Drakopoulos, Y. Voutos, P. Mylonas, and S. Sioutas, "Motivating item annotations in cultural portals with UI/UX based on behavioral economics," in *IISA*. IEEE, 2021.

[34] G. Drakopoulos, E. Spyrou, Y. Voutos, and P. Mylonas, "A semantically annotated JSON metadata structure for open linked cultural data in Neo4j," in *PCI*. ACM, 2019.

[35] S. Chellappan and D. Ganesan, "Scala: Functional programming aspects," in *Practical Apache Spark*. Springer, 2018, pp. 1–37.

[36] P. Hudak, "Conception, evolution, and application of functional programming languages," *ACM CSUR*, vol. 21, no. 3, pp. 359–411, 1989.

[37] D. Pollak, V. Layka, and A. Sacco, "Functional programming," in *Beginning Scala 3*. Springer, 2022, pp. 79–109.

[38] M. Hartmann and R. Shevchenko, *Professional Scala: Combine object-oriented and functional programming to build high-performance applications*. Packt Publishing Ltd, 2018.

[39] C. Okasaki, *Purely functional data structures*. Cambridge University Press, 1999.

[40] S. F. Lott, *Functional Python programming: Discover the power of functional programming, generator functions, lazy evaluation, the built-in itertools library, and monads*. Packt Publishing Ltd, 2018.

[41] R. Dyer and J. Chauhan, "An exploratory study on the predominant programming paradigms in Python code," in *Joint European Software Engineering Conference and Symposium on the Foundations of Software Engineering*. ACM, 2022, pp. 684–695.

979-8-3503-8368-3/23 $31.00 © 2023 IEEE

Stimulating primary school students' interest in Data Structures through ESA's Astro Pi challenges

Athanasios Karakostas
Department of Information and Communication
Systems Engineering (ICSD)
University of the Aegean
Karlovasi, Greece
athkarakostas@aegean.gr

Akrivi Vlachou
Department of Information and Communication
Systems Engineering (ICSD)
University of the Aegean
Karlovasi, Greece
avlachou@aegean.gr

Abstract—In this paper we show how to stimulate the interest of primary school students of the Greek education system in simple data structures through activities carried out during our participation in European Space Agency's Astro Pi Mission Zero and Astro Pi Mission Space Lab projects. In our research, we focus on primary school students and show that it is possible for students to comprehend and use simple data structures such as single- and two-dimensional arrays in the context of simple tasks in the above-mentioned projects.

We conducted an in-class evaluation of the proposed approach using groups of students of two different age groups who had no previous contact with the subject of data structures and minimal programming experience. Our results shows that our approach managed to stimulate the interest of the majority of the students, while many of them recommend learning about data structures to future students of the same age group.

I. INTRODUCTION

In our previous work [1] we have stressed the fact that, while Information and Communications Technology (ICT) is a subject that has been taught in Greek primary schools for more than 10 years now, the currently implemented syllabus as mentioned in Ministerial Decision F.20/130336/D1/22-8-2019 clearly shows that the subject is focused on the use of computers as everyday tools to create and to communicate and less on the cultivation of algorithmic thinking and problem decomposition and solving capabilities. In our past work we showed that it is possible to introduce the students to simple algorithms such as sequential and binary search.

In order to be able to properly teach computing algorithms, a necessary foundation of knowledge is needed. For example, sorting algorithms are usually applied to a suitable data structure, such as a one- or multi-dimensional array, or lists or similar data structures. Before diving into teaching these algorithms to a younger audience, such as primary school students, it is therefore necessary first to stimulate interest in these data structures as a stepping stone before progressing further.

In this paper, we report activities that have been taught in the classroom merged with activities offered by the European Space Agency(ESA). ESA has been running for the past years two distinct challenges aimed at students aged up to 18 years of age, but suitable also for students of primary school [2] [3] [4]. This is one of the several active attempts from both private and public institutions and organizations to stimulate interest in computer programming and sciences in schools, as there exists clearly a skills gap between the needs of the industry and the workforce seen in computer science and other technology-related areas [5][6]. Generating interest and introducing to the classroom more complex topics of computer science such as programming, data structures, algorithms and similar topics is an ongoing topic with significant academic interest both in Greece and internationally.

In this paper we first provide a literature review on the subject, especially regarding the use of algorithms in primary education in other subjects (Section II). Then, we introduce ESA's Astro Pi challenges and provide an overview of our approach (Section III). In Section IV, we provide suitable activities for introducing students of primary education to computer science algorithm-related concepts. In Section V we evaluate our approach and present our results based on students' questionnaires. Finally, we conclude our work in Section VI and present our future work in this area.

II. RELATED WORK

Concerning the teaching of data structures and algorithms applied to them in the context of a Computer Science subject taught in education, the consensus is overwhelmingly positive. The majority of these works published refers to either secondary or university-level students, with a few welcome exceptions that also refer to primary education. The majority of work relates to teaching algorithms assuming the underlying data structures are well-known, which is not the case in primary education in Greece.

Dagdilelis et al. [7] argue that algorithms and computer programming should be included in basic education curricula and propose didactic scenarios towards this purpose, noting that the educators themselves need to be also suitably trained to successfully perform in this area.

Kordaki et al. [8] propose a web-based environment for teaching sorting algorithms to secondary education students, allowing also for a student-based interactive sorting procedure, allowing students to interpret their experience in free-text and pseudo-code prior to being introduced to typical sorting algorithms used in computer science.

979-8-3503-8368-3/23 $31.00 © 2023 IEEE

Boticki et al. [9] present a study on using a purpose-built mobile phone application to assist in teaching algorithms to university-level students, incorporating an interactive module to enhance the students' understanding of sorting algorithms, observing that the use of animation is beneficial in learning sorting algorithms and noting that the students that made use of the application had better understanding and academic results than those that did not use it.

Grivokostopoulou et al. [10] present an educational game based on Pacman for teaching search algorithms in university-level students of an Artificial Intelligence course, suggesting that educational games are an effective way of enhancing the engagement and deeper understanding of advanced (such as blind and heuristic) search algorithms.

Silva et al. [11] conclude that the search for effective pedagogical approaches for teaching algorithms and data structures in university-level students is of great concern for educators globally, while highlighting the need for more tools to support this purpose. Peer Instruction, Pair Programming, Problem-Based Learning, Project-Based Learning and Gamification are all suitable pedagogical strategies that can be used in the context of teaching algorithms to learners.

Yusupov et al. [12] propose a graph-based structuring of the content of a data structures course for university-level students in a sequence that allows for the mastering of each of the elements of the course and improves teaching effectiveness.

Muntean [13] proposes the model of a flipped classroom for teaching sorting and searching algorithms in university-level students, where the students study the related theory at home and then practice in the classroom to help increase in-class hands-on hours.

Tziortzioti et al. [14] propose a pedagogical approach that further bridges the gap between school and out-of-school science by undertaking an environmental water project involving IoT technologies and Arduino micro-controller for monitoring water-related variables via sensors. Although their proposal is very advanced for primary school education, the introduction of micro-controllers in primary education classrooms, as seen for example also in the case of the BBC Microbit for robotics contests [15] or sensors on ESA's RaspberryPi-based AstroPi module [2], also provide potential starting points for our activities.

Velaora and Kakarountas [16] support the idea of Game-Based Learning in university-level teaching, while Seralidou and Douligeris [17] suggest that even in the case of remote teaching (e.g. due to Covid19 or other circumstances) it is possible to use game creation activities when teaching Computer Science to secondary education classes to stimulate interest in areas such as Object Oriented programming, while helping in understanding the related concepts and building confidence.

Gibson [18] comments on the importance of problem-based learning, where algorithms form the solution to puzzles or games, and suggests an incremental approach to teach Graph data structures and related algorithms to students of all ages, starting from 5 years old.

Bernat [19] presents methods for teaching sorting algorithms in primary and secondary education, particularly aiming to help the students understand, practice, but also realise the efficiency of different sorting algorithms by counting the comparison steps involved in each case. He concludes that pictures are the most important tool in realizing the essence of algorithms, but animations help better comprehend the details of each algorithm.

Giordano and Maiorana [20] describe a didactic experience using visual language (BYOB) to implement a sorting algorithm, in a manner that can easily be adopted in the Scratch environment which is widely used in the Greek primary education system.

Moschella [21] explores the elements that compose computational thinking skills and their correlation with primary school learners and concludes that computational thinking is possible from early childhood when specific tools and activities are involved.

Finally, Seralidou and Douligeris [22] suggest using game-based learning techniques and game creation in order to cover parts of the curricula of the computer science subjects also in primary education, while teaching basic programming skills. They further note that the lack of textbook for computer science courses in primary education leads the teaching staff into either creating their own activities or to cover parts of the curricula of the computer science subjects.

III. PROPOSED TEACHING APPROACH

In our approach we are proposing using an enhanced version of ESA's Astro Pi Mission Zero [3] and Astro Pi Mission Space Lab [4], as well as an online tool for generating the RGB values for any selected colour, W3C School's HTML Color Picker [23], and Trinket's adapted emulator specifically designed for the AstroPi Mission Zero project [24] to run our code.

ESA's Astro Pi challenges are both aimed at helping students of primary and secondary education to learn how to code using Python that runs on a specialised Raspberry Pi computer. Two distinct challenges run every year, Mission Zero and Mission Space Lab. Mission zero is the simpler of the two, aiming to help students write a simple program that takes a reading from one of the computer's sensors (e.g. temperature, humidity etc) and displays this value on an 8x8 pixel matrix display, followed by a drawing displayed on the same 8x8 pixel matrix. Mission Space Lab is more complex, aiming to help students write more complex code that aids in performing a scientific experiment in space. Again in this case the 8x8 pixel matrix is present and can be used, while sensor measurements can be stored on file for further processing on Earth after the experiment is run.

The above-mentioned projects are clearly aimed at bringing science to the classroom to students of ages up to 18 years of age. Mission Zero is suitable for beginners and in our approach we will utilize this project to students of fifth grade (age 10) in a Greek primary school. Mission Space Lab is more complex in its deliverables and includes a scientific portion through an

979-8-3503-8368-3/23 $31.00 © 2023 IEEE

experiment proposal. We will utilize this project on students of sixth grade (age 11). Only the altered sections of these challenges that are linked to our aim of stimulating interest in data structures are presented in this paper, and not our complete approach on ESA's challenges.

The activities involved require a computer having an internet connection so that the related web-based tools can be utilised, however the initial steps can be generated also on a simple spreadsheet without requiring a computer connection or even in paper form until the simulation tool is needed. We utilised a hybrid approach, starting on a spreadsheet without needing internet access (this part could also be performed on a simple page of paper) and then continued on with the part of our tasks requiring online tools in order to be completed.

Knowledge of how colour is represented on a computer screen is essential, and while for the programming part the guidelines offered by ESA are rather straightforward, some compilation errors may be really tricky to spot for those not familiar with computer programming in Python. We provide step-by-step instructions for the teachers involved, however, especially for the Mission Space Lab project, some familiarity with topics related to Physics and an inventive step in the projects' proposal is also required. Finally, a questionnaire aimed at the students of both age groups is provided.

In the following section we will describe a representative lesson plan for introducing data structures to young students through these two ESA projects.

IV. LESSON PLAN FOR INTRODUCING DATA STRUCTURES

A. Step 1

In the first session the teacher welcomes the students and starts by showing a video of ESA's Raspberry Pi as positioned inside the International Space Station. The teacher explains that this is a special kind of computer, designed to perform scientific experiments on space that have been proposed and developed by school children all around Europe. The teacher stresses the fact that the small 8x8 pixel screen on Raspberry Pi's SenseHAT module is fully programmable and that every pixel of it can be assigned any colour desired in order to display an image, which can be our own creation. In the case of Mission Space Lab the teacher welcomes any questions about the nature of the experiments that can be performed, our motivation in order to take part in such a project. The timeline for each of the two projects is presented to the corresponding students. The main aim of this step is to generate interest amongst the students and a strong desire to participate and have their drawings displayed in space.

B. Step 2

The teacher asks the students to draw their own drawing in an 8x8 pixel-style canvas, which can be represented either in a spreadsheet where the cells are square-shaped by adjusting the height and width of the first 8 rows and columns to an identical size (e.g. 1 centimeter), or on an A4 page with a blank canvas pre-printed. The students proceed to generate their drawing by giving each cell a single colour, effectively producing pixel art

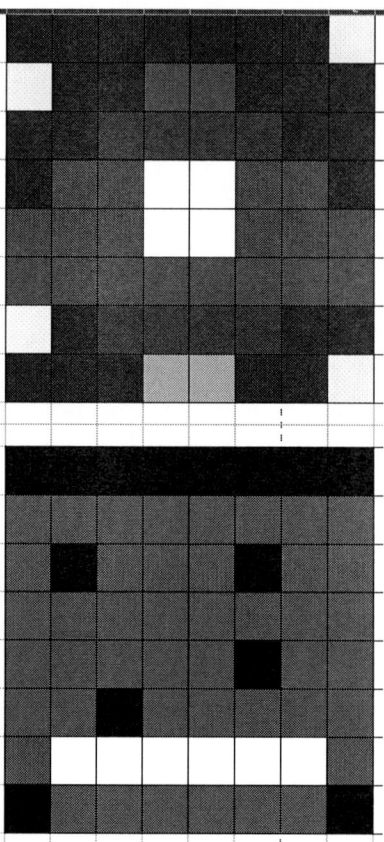

Fig. 1. Examples of student-proposed drawings

that fits the 8x8 pixel screen of the SenseHAT module. Two examples of such student-produced drawings can be seen in figure IV-B.

C. Step 3

This is the step where a very first introduction to data structures is given. The teacher explains the RGB colour scale to the students noting that three distinct values are needed to represent the Red, Blue and Green part of each coloured pixel on a screen. These three values are inseparable and all three together represent a single pixel's colour. In our Python each pixel's RBG values are stored in a tuple of three elements, the first data structure the students are exposed to. The students then are asked as an activity to identify all the colours that were used in their drawing, to determine the RGB values of the colours involved and to note down these values next to their drawing. In our case we used W3School's HTML Color Picker [23] which is presented to our students as a "Colour Factory". For the two drawings of figure IV-B the corresponding colours in RGB scale are shown in the two tables below:

D. Step 4

In this phase, the teacher engages the students to discuss about data structures. It is noted again that the three values for R, G and B are inseparable if one is to describe a specific colour. By writing the values in the board next to each other

TABLE I
RGB COLOURS FOR THE FIRST DRAWING

Colour	Red	Green	Blue
red	179	0	0
white	255	255	255
blue	0	71	179
dark blue	0	0	102
light yellow	255	255	128
dark yellow	230	184	0

TABLE II
RGB COLOURS FOR THE SECOND DRAWING

Colour	Red	Green	Blue
red	230	0	0
white	255	255	255
black	0	0	0
green	0	128	0

as a group, the teacher can then draw a rectangle surrounding them to denote their grouping together. This in essence is a one-dimensional array of size 3 (in Python coded as a tuple), with one value to record each colour's R, G or B segment. The teacher mentions that this grouping of 3 numbers could be applied to other problems of different size and asks the students to propose areas where a group of numbers can be treated as a whole in order to denote something. A common response could be for example that a similar approach can be taken for the grade sheet that is given to the parents of each student each term, where ten or so different grades for the different subjects are recorded in order to describe a student's performance for that term. As an activity, the students are asked to create variables in Python that define the colours of their drawing. For the second drawing shown, the corresponding result would be:

```
red = (230,0,0)
green = (0,128,0)
white = (255,255,255)
black = (0,0,0)
```

E. Step 5

At this stage the students are requested to alter an example offered by the ESA's guidelines that displays a drawing to the screen of the SenseHAT module or simulator. Describing in code an image for displaying on the SenseHAT module is done by defining a list of 64 pixels.

Having already calculated their colours in the previous stage, the task simply requires replacing 64 pixel values of the sample drawing with the correct colours of each student's drawing. Note that each element of this list is a tuple that describes an RGB colour as explained previously. While the list is one-dimensional of 64 elements, in essence it describes the 64 pixels of the 8x8 display matrix in sequence, so by writing our code so that every line of the list definition contains 8 elements it is easy to introduce the concept of 2-dimensional arrays of rows and columns of data. Students can easily answer the question "What is the colour of the 6th pixel of the 3rd

row of our image?" and "how can I turn it from black to red?". The fact that the internal R,G,B values for a colour can be ignored in this stage, as they were set earlier and now they are used inseparable for each pixel, is a good example of abstraction/generalisation, where unnecessary detail on the formation of each colour can now be ignored when generating an image, while the detail is still present and can be obtained if needed. For example we can see this image as a 1-dimensional array of 64 pixels, or as a 2-dimensional array of 8 rows and 8 columns of pixels. We can dive to 3-dimensional array logic by asking "What is the Blue component of the 6th pixel in the third row?". In Python we actually code it as a one-dimensional list of one-dimensional tuples, but it can be visualised by the students as a three-dimensional array with ease and these questions were answered in all four classes.

A completed code for the second drawing provided is given below. This can be run on the simulator module of Trinket's Mission Zero page [24].

```
from sense_hat import SenseHat
sense = SenseHat()
sense.set_rotation(270)

red = (230,0,0)
green = (0,128,0)
white = (255,255,255)
black = (0,0,0)

picture = [
    black, black, black, black, black, black, black, black,
    green, red, red, red, red, red, red, green,
    green, black, red, red, red, black, red, green,
    green, red, red, red, red, red, red, green,
    green, red, red, red, red, black, red, green,
    green, red, black, red, red, red, red, green,
    green, white, white, white, white, white, white, green,
    black, green, green, green, green, green, green, black
    ]

sense.set_pixels(picture)
```

The teacher mentions that this grouping of data could be re-written in rows and columns in a spreadsheet, where each row could be a pixel from 1 to 64 and each column would be the corresponding colour element in the RGB scale. It is again mentioned that this visualization could be applied to other problems of different size and asks the students to propose areas where a group of numbers can be treated in a similar fashion. A common response could be for example that a similar approach can be taken for the grade sheet that is given to the parents of each student at the end of the year, which for each of the ten or so different subjects contains a table of equalling number of rows and (in the Greek education system) 4 columns: One for each of the three different semesters and a final one for the final grade for the year for that subject deduced from the rounded-up average of the three grades of the terms.

F. Step 6

At this final stage the students are requested to complete a questionnaire which contains questions aiming to determine the level of understanding gained and whether the activities involved were found to be interesting and enjoyable to the

979-8-3503-8368-3/23 $31.00 © 2023 IEEE

students involved. The results in our case are provided in the following section.

V. EVALUATION

We evaluated our approach against a group of primary school students of four different classes, two classes of fifth-graders (E1 and E2) and two classes of sixth-graders (St1 and St2), during the school year 2021-2022. The distribution of the students is shown in Table III.

TABLE III
DISTRIBUTION OF STUDENTS

Class	Number of participants	Boys	Girls
E1	16	5	11
E2	15	5	10
St1	19	9	10
St2	22	10	12

A. Results of the student questionnaire

The students were split in two groups depending on their age and were given two questionnaires differing depending on the nature of the project undertaken (Mission Zero vs Space Lab). All questionnaires were anonymous and only grouped by age group, not by gender, school grades or other factors.

There were common questions in the two questionnaires relating to this study. All of the questions required choosing one of the following options: "Strongly Agree", "Agree", "Neither agree nor disagree", "Disagree", "Strongly Disagree". These answers were then weighted as shown in table IV to generate a score for each answer.

TABLE IV
DID I NEED THE TEACHER'S HELP?

Response	Weighting
Strongly Agree	2
Agree	1
Neither agree nor disagree	0
Disagree	-1
Strongly Disagree	-2

The first question was "Did you participate actively in the lesson?" and all fifth grade students answered Strongly Agree / Agree, averaging 1.69, while the sixth graders had all but five students answering also Strongly Agree/Agree averaging 1.21. This shows that not only the lesson plan was interesting but in addition the students felt that they were a part of the course.

The second question was "Did I co-operate with my desk-mate?". The result here are even higher, with scores of 1.85 and 1.61 respectively for fifth and sixth grade. The results show that during these lesson the cooperation of the students is strengthened and their joint creativity is enhanced.

The next question was "Did I successfully complete the part of this activity relating to our drawing", scoring 1.77 and 1.68, again showing successful completion of the specific task.

The next questions relate to their understanding of the first data structure-related task, identifying colours and storing them in an RGB-coded 3-element one-dimensional array. The questions were "I used the colour factory successfully" scoring 1.11 and 1.64 respectively, "I understand how the RGB scale works and how to identify these values for a colour" scoring 1.19 and 0.92, and finally "I understand how to describe a drawing/image pixel by pixel on a computer program" scoring 1.03 and 1.14. These responses are circling around the "Agree" rather than the "Strongly Agree" response, indicating that the students believe they have gained the knowledge but perhaps need additional activities to fully grasp the concepts involved.

The following question "I had a lot of questions during these lessons" averages negative scores for both age groups, -0.12 and -0.54 respectively, showing that at the time of teaching the concepts were clear to the students and that all students felt comfortable with the lesson material.

In the final set of questions, the purpose was to see if the students found the lesson interesting but were also satisfied with the knowledge they acquired. The questions asked were "I am pleased with this project", scoring 1.73 and 1.75, "I found the lessons relating to this project interesting". scoring 1.35 and 1.57, and "I would recommend this project to school students that attend this class next year", scoring 1.35 and 1.64. These responses further support the argument that the specific lessons were successful in introducing our students to the concept of data structures through a creative and enjoyable activity.

B. Results based on the students' feedback

Building on or previous work of introducing our students to computational thinking and algorithms in education, our aim through this work was to stimulate the interest of students to the world of data structures so that we can further build in subsequent studies in the areas of elementary sorting and cryptography algorithms. Our evaluation has shown that this goal has been achieved and we feel confident to proceed to the next step.

The activities managed to contribute to the understanding of data structures by children of fifth and sixth grade. The students were not familiar with these concepts beforehand, yet they were able to make use of simple data structures in a project that they found to be creative and enjoyable and that they strongly recommend to future students.

The active participation of students and their strong co-operation with their desk-mate, as reflected in the answers obtained from the questionnaire, enabled them to demonstrate the knowledge obtained and to describe through the computer code written complex data in the form of an image drawn on an 8x8 pixel screen.

VI. CONCLUSIONS AND FUTURE WORK

In this paper, we presented our approach to introduce primary school students to basic data structures through activities carried out under ESA's Mission Zero and Mission Space Lab challenges. We have prepared material and lesson plans that aim to introduce the students to the specific subject, through creative, interesting and pleasant activities.

In summary we would like to point out that, by inspecting the answers of our questionnaires, we conclude that by taking a group of fifth and sixth grade students who had no previous contact with the subject of data structures we managed to stimulate the interest of the majority of the students. We believe that further activities are required to build on this success and further boost the students' understanding and application of data structures in similar projects.

In conclusion, our goal has been achieved to a very satisfactory degree. The students had a positive attitude towards these activities and we strongly believe that further related activities will be met with similar response.

Following our initial investigation in teaching simple search algorithms in primary education and this further step of introducing the concept of simple data structures to our students, we aim to further build on this and proceed with activities aiming to teach basic algorithmic concepts of computer science to primary education students. As a next step we aim to introduce our students to elementary sorting and cryptography algorithms. Our aim is to get the students to re-invent well-known algorithms without prior teaching and evaluate their results against groups of students that have been taught algorithms in a more traditional way, as well as perform statistical analysis of the two different groups.

ACKNOWLEDGMENT

We would like to thank the students of the 35th Primary School of Peristeri that participated in our study. Moreover, we would like to thank the ESA Education programme, which in collaboration with the Raspberry Pi Foundation gave us the opportunity to participate to the the European AstroPi Challenge.

REFERENCES

[1] A. Karakostas, V. Liarokapis, and A. Vlachou, "Introducing algorithmic thinking in primary education in greece," in *2022 7th South-East Europe Design Automation, Computer Engineering, Computer Networks and Social Media Conference (SEEDA-CECNSM)*. IEEE, 2022, pp. 1–6.

[2] [Online]. Available: https://astro-pi.org/

[3] "Astro pi: Mission zero step-by-step guide." [Online]. Available: https://projects.raspberrypi.org/el-GR/projects/astro-pi-mission-zero

[4] "Astro pi mission space lab phase 2 guide." [Online]. Available: https://projects.raspberrypi.org/en/projects/code-for-your-astro-pi-mission-space-lab-experiment

[5] R. Singh Dubey, J. Paul, and V. Tewari, "The soft skills gap: a bottleneck in the talent supply in emerging economies," *The International Journal of Human Resource Management*, pp. 1–32, 2021.

[6] I. Vasileiou, "The cyber skills gap," in *Cybersecurity Issues in Emerging Technologies*. CRC Press, 2021, pp. 185–198.

[7] V. Dagdilelis, M. Satratzemi, and G. Evangelidis, "Introducing secondary education students to algorithms and programming," *Education and Information Technologies*, vol. 9, pp. 159–173, 06 2004.

[8] M. Kordaki, M. Miatidis, and G. Kapsampelis, "A computer environment for beginners' learning of sorting algorithms: Design and pilot evaluation," *Computers & Education*, vol. 51, no. 2, pp. 708–723, 2008.

[9] I. Boticki, A. Barisic, S. Martin, and N. Drljevic, "Teaching and learning computer science sorting algorithms with mobile devices: A case study," *Computer Applications in Engineering Education*, vol. 21, no. S1, pp. E41–E50, 2013.

[10] F. Grivokostopoulou, I. Perikos, and I. Hatzilygeroudis, "An educational game for teaching search algorithms," in *International Conference on Computer Supported Education*, vol. 3. SCITEPRESS, 2016, pp. 129–136.

[11] D. B. Silva, R. de Lima Aguiar, D. S. Dvconlo, and C. N. Silla, "Recent studies about teaching algorithms (cs1) and data structures (cs2) for computer science students," in *2019 IEEE Frontiers in Education Conference (FIE)*. IEEE, 2019, pp. 1–8.

[12] F. Yusupov, I. Shamuratova, D. Yusupov, and T. Khudayberganov, "Improving the effectiveness of teaching the course "data structure and algorithms" based on structuration and integration of the discipline," in *2019 International Conference on Information Science and Communications Technologies (ICISCT)*. IEEE, 2019, pp. 1–4.

[13] C. Muntean, "Teaching tip: Flipping the class to engage students in learning programming algorithms," in *Society for Information Technology & Teacher Education International Conference*. Association for the Advancement of Computing in Education (AACE), 2019, pp. 2320–2325.

[14] C. Tziortzioti, E. Mavrommati, I. Chatzigiannakis, and V. Komis, "Bridging the gap between school and out-of-school science: A making pedagogical approach," in *2020 5th South-East Europe Design Automation, Computer Engineering, Computer Networks and Social Media Conference (SEEDA-CECNSM)*. IEEE, 2020, pp. 1–6.

[15] [Online]. Available: https://wrohellas.gr/panellinios-diagonismos-ekpaideutikis-rompotikis/

[16] C. Velaora and A. Kakarountas, "Game-based learning for engineering education," in *2021 6th South-East Europe Design Automation, Computer Engineering, Computer Networks and Social Media Conference (SEEDA-CECNSM)*. IEEE, 2021, pp. 1–6.

[17] E. Seralidou and C. Douligeris, "Motivating students in distance programming learning using games," in *2021 6th South-East Europe Design Automation, Computer Engineering, Computer Networks and Social Media Conference (SEEDA-CECNSM)*. IEEE, 2021, pp. 1–7.

[18] J. P. Gibson, "Teaching graph algorithms to children of all ages," in *Proceedings of the 17th ACM annual conference on Innovation and technology in computer science education*, 2012, pp. 34–39.

[19] P. Bernát, "The methods and goals of teaching sorting algorithms in public education," 2014.

[20] D. Giordano and F. Maiorana, "Teaching algorithms: Visual language vs flowchart vs textual language," in *2015 IEEE Global Engineering Education Conference (EDUCON)*. IEEE, 2015, pp. 499–504.

[21] M. Moschella, "Observable computational thinking skills in primary school children: How and when teachers can discern abstraction, decomposition and use of algorithms." 03 2019, pp. 6259–6267.

[22] E. Seralidou and C. Douligeris, "Creating and using digital games for learning in elementary and secondary education," in *2020 5th South-East Europe Design Automation, Computer Engineering, Computer Networks and Social Media Conference (SEEDA-CECNSM)*. IEEE, 2020, pp. 1–8.

[23] "Html color picker." [Online]. Available: https://www.w3schools.com/colors/colors_picker.asp

[24] "Trinket - mission zero." [Online]. Available: https://trinket.io/mission-zero

Teachers' Evaluation of 3D Design and 3D Printing Activities in Secondary Education

Eleni Seralidou
Department of Informatics
University of Piraeus
Piraeus, Greece
eseralid@unipi.gr

Theodoros Karvounidis
Department of Informatics
University of Piraeus
Piraeus, Greece
tkarv@unipi.gr

Christos Douligeris
Department of Informatics
University of Piraeus
Piraeus, Greece
cdoulig@unipi.gr

Abstract— **Three-Dimensional (3D) printing is a modern, burgeoning technology with numerous practical applications in everyday scenarios. When integrated with 3D design methods, it becomes a potent educational tool, fostering engagement and igniting curiosity. In this paper, 3D design and 3D printing educational activities for secondary education are presented. These activities use educational worksheets and they are evaluated by secondary education teachers. The ultimate goal is to offer valuable educational material suggestions to educators and gather their opinions about the implementation of 3D design and 3D printing in education. The teachers applied the educational activities in an organized workshop and offered their opinions. According to their opinions, these activities are not only engaging but also beneficial and can be adopted by different specialties.**

Keywords—3D Printing, 3D Design, Secondary Education, Design Thinking

I. INTRODUCTION

Three-Dimentional (3D) Printing, which is also known as Additive Manufacturing (AM), is a radical contemporary manufacturing process that is currently used in numerous sectors including industries, art, healthcare and education [1]. The emergence and rapid development of 3D printing technologies emphasizes the imperative to not only improve 3D skills in the industry, but also to provide opportunities for new educational practices that will facilitate the adoption of these new technologies across a range of educational subjects [2]. The technical evolution and use of applications have made 3D printers easily accessible by the educational community.

The academic communities have shown substantial interest in 3D printing technology because of its capacity to produce intricate structures using customizable materials and features [3]. 3D printers have found applications in diverse educational settings, spanning research institutions, libraries, and high schools. Their adoption enhances STEM education through a range of learning methods and experiences [4][5]. Some of the offered advantages encompass increased student engagement, improved comprehension of theoretical concepts through visualization, and the integration of both practical and theoretical skills [6].

Additionally, 3D design serves as the creative blueprint for transforming digital concepts into tangible objects through the process of 3D printing. Creating 3D models through exploiting 3D design techniques, serves as a crucial preliminary step for 3D printing [7]. 3D design involves the generation of 3D models using digital tools, and software, as well as block programming techniques [8]. It is a versatile field with uses spanning across diverse industries such as gaming, animation, architecture, and product design [9]. 3D design has arisen as a transformative educational tool, enhancing the learning journey across multiple subjects. Through the incorporation of 3D design in classrooms, students not only acquire valuable technical expertise but foster critical and creative thinking as well [10]. This innovative method promotes experiential learning, empowering students to conceptualize, model, and visualize intricate ideas and concepts. Whether applied in the realm of science, art, or engineering, 3D design empowers students to interact dynamically and immerse themselves in their subjects, equipping them for the demands of a technology-driven world [11].

Within the educational settings, students have the opportunity to employ 3D design software for the purpose of conceiving and producing their own educational tools, models, and prototypes [12]. This dynamic approach not only fosters creativity and problem-solving skills but also empowers learners to turn their ideas into physical objects through 3D printing [7]. No matter the object produced, this combination sparks curiosity, innovation, and a deeper understanding of subjects, making education more immersive and inspiring. Thus, the synergy between 3D design and 3D printing is a game-changer in education, since it allows students to transcend traditional textbooks and participate in unprecedented hands-on learning experiences.

Nonetheless, the pace of technological advancement in 3D printing outpaces that of education. Many schools have not yet fully implemented this type of technology into their educational programs. In order for that to happen curricula and teaching methodologies must constantly evolve to match the rapid pace of development [13]. Moreover, the implementation of 3D printing technology into education remains relatively limited. This absence of sufficient technology-focused education is regarded as a significant impediment to its widespread adoption. Additionally, existing teacher education programs have not adequately equipped trainees with the essential knowledge and skills

979-8-3503-8368-3/23 $31.00 © 2023 IEEE

required for effectively utilizing digital fabrication within the classroom [14].

Moreover, there is a shortage of research examining precise pedagogical approaches for educating using 3D printing technology [15]. Even though the 3D printing technology has been shown to positively affect learning, it is has not yet been widely adopted in schools [16]. Unfortunately, in the educational environment, especially in public schools, there is a huge lack of resources which are essential for the educational development of children. Hence, to a large extent, the use of contemporary technologies, as 3D printing, have not yet reached these schools [17]. Moreover, while education related to 3D printing is progressively gaining prevalence across diverse domains, studies reveal that the accompanying educational resources, including appropriate books and materials, are lacking in adequacy [18].

Additionally, the teachers need to develop general knowledge, skills and abilities and master the achievements of pedagogy, psychology, methodological sciences, modern techniques and advanced technologies at the level where they will be able to integrate contemporary technologies into the educational settings [19]. Furthermore, there is a scarcity of instructional materials designed for integrating 3D printers into standard curricula, and teachers often face challenges due to the absence of comprehensive professional training.

Consequently, in order to effectively integrate 3D printers into educational settings, it is essential to allocate resources not only for budgetary support but also for teacher training and for the creation of appropriate teaching and learning materials [20]. Keeping the above in mind, in order to effectively incorporate 3D printing into the K-12 curriculum, educators need to acquire the skills of designing 3D models, operating 3D printers, and troubleshooting printing issues [14]. Thus, to cover the previous needs, adequate educational material that is designed based on pedagogical frameworks in order to make it suitable for education, is necessary.

In this paper, we present five educational worksheets for the secondary education. These worksheets cover basic concepts of 3D design and 3D printing following the principles of the "Design Thinking" pedagogical framework for integrating 3D design and 3D printing in education, as it is described in the next paragraphs. The activities included in the worksheets focus on exploiting the features of digital 3D modeling environments, which include creating 3D objects with 3D design and with block programming techniques, and the benefits of using 3D printers in the educational process. These worksheets are addressed to all specialties.

The contribution of this paper includes the exploitation of contemporary technology in educational settings and the creation of educational material to support the integration of this technology. Additionally, the opinions and suggestions of teachers in secondary education, that were gathered through the workshop that we organized in order to demonstrate the proposed educational material, gave us valuable insights. The workshop took place in a vocational high school in Greece with the participation of teachers from various specialties.

The rest of the paper is organized as follows: In Section II the theoretical foundations are presented followed by the presentation of the 3D design and 3D printing activities in Section III. In section IV, the research methodology and the research results are included, while in Section V a discussion of the results is outlined. Finally, the conclusion and future steps are presented in Section VI.

II. THEORETICAL FOUNDATIONS

In the rapidly evolving realm of modern technology, 3D design and 3D printing have emerged as transformative forces. These intricately connected fields demonstrate the power to convert digital ideas into tangible forms. In the context of education, the combination of 3D design and 3D printing goes beyond conventional learning boundaries, providing students and educators with an exceptional canvas for exploration and innovation.

A. 3D Design in Education

3D design alludes to the process of creating and manipulating 3D digital models using specialized software [21]. Unlike 2D design, which deals with flat images and drawings, 3D design entails the construction of virtual objects possessing depth, and volume, as well as the ability to be observed from multiple perspectives. These models serve a multitude of purposes, encompassing visualization, prototyping, simulation, and even production via technologies such as 3D printing. In fields like architecture, product design, animation, video games, and engineering, 3D design plays a pivotal role, allowing designers to manifest their concepts in a more authentic and tangible manner.

The integration of 3D design into education is progressively becoming essential, providing students with a dynamic platform to nurture their creative and technical abilities [14]. Utilizing specialized software, students can create intricate 3D models, promoting a deeper comprehension of spatial connections and design principles. This interactive method goes beyond conventional learning approaches, as it allows students to envision abstract concepts and prototypes. The participation in 3D design projects nurtures students' problem-solving skills, digital literacy, and fosters an innovative mindset, equipping them for contemporary careers that demand expertise in technology and design-oriented thinking [22]. Moreover, integrating 3D design into educational curricula promotes collaborative teamwork and empowers students to materialize their ideas into concrete visualizations, bridging the divide between imagination and practical applications in the real world [23].

B. 3D Printing in Education

3D printing is a manufacturing technique that entails the construction of 3D objects by gradually adding material, guided by a digital model [24]. This differs from conventional subtractive manufacturing techniques, in which material is subtracted from a solid block to shape the end product. 3D printing constructs the object progressively, layer by layer, employing diverse technologies and materials such as plastics, metals, ceramics, and even biological substances. This procedure enables the fabrication of intricate geometries, personalized designs, and complex internal structures, which could be difficult or unattainable using traditional manufacturing techniques [25].

From an educational point of view, 3D printing is transforming education by offering hands-on learning

979-8-3503-8368-3/23 $31.00 © 2023 IEEE

experiences. It allows students to create physical objects from digital designs, enhancing their understanding of complex concepts in subjects like science, technology, engineering, art, and math (STEAM) [26]. Through creativity and design projects, students develop problem-solving skills and gain exposure to modern technologies relevant to various industries [27]. This approach fosters collaboration, critical thinking, and digital literacy, preparing students for future careers while making learning engaging and interactive [28]. Simultaneously, young individuals play a role as innovators in the realm of science and technology by engaging in the 3D printing process, which entails ideation, 3D modeling, printing, reflection, and refinement, ultimately leading to the creation of advanced products [28].

While students express their appreciation for the technology in terms of engagement and enhanced learning effectiveness, it is noteworthy to note that professionals in the field highlight practical constraints when it comes to its application [18]. Presently, the primary impediment to the broad adoption of 3D printing technology is its cost, yet ongoing innovations in this domain are poised to render it more attainable in the foreseeable future [29].

C. A Pedagogical Framework for Integrading 3D Design and 3D Printing in Education

The "Design Thinking" approach is an effective pedagogical framework for integrating 3D design and 3D printing into education [30]. Design Thinking is a human-centered problem-solving methodology that encourages creativity, collaboration, and practical application. It can be applied to 3D design and 3D printing in education by taking into account the following steps [31] [32]:

- **Empathize:** Understand the needs, interests, and goals of the students. Determine educational goals and domains in which 3D design and printing can enrich the learning experience.

- **Define:** Clearly define the learning objectives and project goals. Ascertain the ways in which 3D design and printing can support these goals, whether through concept visualization, prototyping, or crafting functional models.

- **Ideate:** Encourage students to brainstorm and generate various ideas for their 3D design projects. Cultivate an environment conducive to nurturing creativity, enabling students to delve into various design opportunities.

- **Prototype:** Have students translate their ideas into 3D digital models using design software. Highlight the cyclic aspect of the process, motivating them to enhance and refine their designs through feedback.

- **Test:** Create prototypes via 3D printing and assess their performance, visual austhetics, and alignment with the educational goals. Students can acquire practical experience in assessing the real-world results of their creations.

- **Iterate:** Offer students guidance in making crucial design modifications based on test results and feedback. This phase reinforces the value of critical thinking and the significance of ongoing enhancement.

- **Implement:** Implement the completed 3D designs using a range of methods, whether it involves showcasing them to the class, integrating them into projects, or producing functional objects for specific uses.

- **Reflect:** Encourage students to reflect on their design and printing experiences. Engage in conversations about their acquired knowledge, the obstacles they faced, and the competencies they honed.

Through the utilization of the "Design Thinking" framework, educators can establish a structured yet adaptable method that fosters creativity, problem-solving, and collaboration. This approach aligns 3D design and 3D printing with the educational goals, nurturing a comprehensive learning experience that equips students for the dynamic challenges of the contemporary world.

III. THE 3D DESIGN – 3D PRINTING ACTIVITIES

The "Design Thinking" framework, as it was described in the previous section, was used for the design and creation of the worksheets activities for the secondary education. The activities were organized to cover six didactic hours, combining the 3D design, by modelling and block programming, with the 3D printing of 3D objects. The activities are designed not to address a single specialty, but to offer an introductory way to basic 3D design and 3D printing principles to all who are interested in learning. This direction was followed because of the fact that the integration of 3D design and 3D printing activities is suggested in the curriculum of several courses in elementary and secondary education. Teachers are encouraged to implement contemporary technologies, like 3D printers, into their lessons [38].

The activities were created following the stages of the "Design Thinking" framework. Hence, in the stage of Empathize through our experience as educators we witness the need of students to move over and beyond the traditional learning methods and engage with educational methods that combine theory with practice. In the stage of Define we set the learning objectives which included the introduction to students of 3D modeling and 3D printing techniques. Hence, in the educational worksheets activities, which focused on the creation of 3D shapes through design and also block programming, the actual creation of the designed objects through the use of a 3D printer during the activity was included.

During the stage of Ideate, we included tasks where students are able to alter the suggested shapes according to their ideas, while into the stage of Prototype following the activities instructions the students create their own objects. During the stage of test the students actually print the selected objects by adjusting the printer and the object adequately. Into the stage of Iterate, the students have the opportunity to improve the original designs and reprint them if they wish to.

During the stage of Implement, the students are called to think for what purpose their objects can be used or what other objects they can create for a purpose they want to fulfill. Finally, during the Reflect stage, a discussion over the produce objects in the students' ideas is encouraged.

979-8-3503-8368-3/23 $31.00 © 2023 IEEE

More specifically, the first worksheet with the title "Creating a simple 3D shape" covers one didactic hour and includes the creation and printing of an object by the combination and properties' adjustment of two 3D shapes. The 3D object is created initially with 3D design techniques and then using block programming by placing the right blocks into the right order. The second worksheet with the title "Transforming a two dimension image into a three dimension image" covers one didactic hour and includes the creation and printing of a 3D object by the transformation of a 2D shape.

Fig. 1. A 3D Object sample in worksheet 1.

The third worksheet has the title "Create spiral shapes" which includes the creation and properties adjustment of different shapes by applying 3D techniques as well as block programming, for one didactic hour. The printing of the produced objects is included as well. The fourth worksheet, with the title "Creating a 3D clock", includes the creation and printing of a 3D clock, and it also covers one didactic hour using mainly 3D techniques.

.

Fig. 2. A 3D Clock sample in worksheet 4.

Lastly, the fifth worksheet, with the title "Creation of 3D repeated shapes", covers two didactic hours and includes the creation of similar in the appearance houses only using block based programming and includes the printing of the produced objects as well.

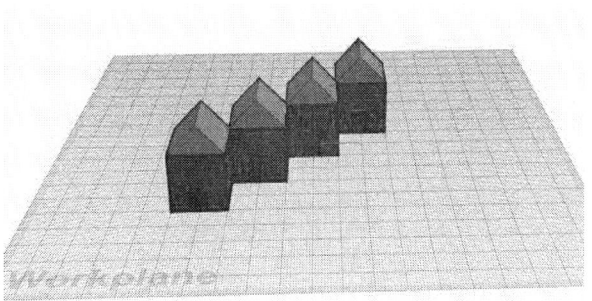

Fig. 3. A 3D repeated shapes sample in worksheet 5.

In all the worksheets, instructions for the creation of printable files are given. Also time for adjusting properly the dimensions, in order for the printing process to be short in time, and set up instructions for the 3D printing device is included.

IV. THE RESEARCH APPROACH

An important step in our study was the workshop organized for teachers of secondary education. This workshop helped us gather important opinions and suggestions for the improvement of our original work. The workshop was organized in March 2023, lasted for two-hours and took place in a vocational high school in Greece with the participation of eighteen teachers of various specialties namely Informatics, Mechanics, Electronics and Mathematics. The workshop covered basic principles of 3D design and 3D printing with the use of appropriate educational tools and learning material in order to support and enhance the learning and teaching process.

A. Methodology

A presentation of the Tinkercad online software [34], the 3D printers' functionality, the 3D printers' slicing software and the suggested learning material were presented in order for the teachers to become familiar with the activities' content and the tools to be used. The teachers implemented selected activities from the learning material and in the end they gave their opinion though filling a questionnaire, which followed a 5-point Likert scale (1= Strongly Disagree – 5= Strongly Agree). The questionnaire was divided in two parts; the first part included general demographic questions while the second investigated the teaching and learning experience. The purpose of this study was to gather the teachers' opinions about the suggested material for 3D design and 3D printing.

More specifically, in the two-hour workshop 3D printer technologies were presented, focusing on the way a 3D printer works and how it can be adjusted. Next, different 3D printer types of slicing software were presented in order for the teachers to be informed about the way 3D objects can be prepared and adjusted for 3D printing. After that, a short display of the Tinkercad online software was performed. During that display the basic functions of 3D design and 3D design through programming were

979-8-3503-8368-3/23 $31.00 © 2023 IEEE

introduced to the teachers. These presentations lasted for one hour.

During the second hour, the activities worksheets were distributed to the teachers, which worked in groups using the Tinkercad online software. In parallel, the 3D printer was used to print some of the produced 3D objects.

At the end of the two-hour workshop, a questionnaire was distributed to the teachers who filled it giving their views about the activities they performed during the workshop.

B. Results

Eighteen teachers participated in the workshop, seven women and eleven men. The majority of them were over 40 years old and had previous experience in programming. Almost half of the teachers (55.5%) had also previous experience in 3D design, but the majority (66.7%) had no previous experience in 3D printing techniques.

In the learning and teaching experience part of the questionnaire, in question 1 about the clarity in what was going to be taught through the worksheets activities the majority of the teachers' answers were 5-Strongly agree (72.2%). In question 2, about the clarity in the understanding of the content of the worksheets activities, again the majority of the teachers selected option 5-Strongly agree (77.8%). In the next question, about having problems in the organization and implementation of the activities, the 77.8% strongly agreed that there were no problems encountered.

In the fourth question, about the clarity and conciseness of the activities' wording, the teachers strongly agreed that there was no problem as well (77.7%). In the fifth question, about the interest over the activities' content, the 83.3%, strongly agreed that the activities were interesting. In the next question, about the given instructions for the better understanding of the activities, a 61,1% strongly agreed and a 38.9% agreed that the given instructions were helpful.

In question 7, about better understanding of the activities through visual examples, again 61,1% strongly agreed and 38.9% agreed that there was enough visual help in the worksheets. In question 8, all the teachers strongly agreed or agreed that by applying the activities an understanding of the basic concepts of 3D design and 3D printing can be achieved.

In question 9, about how interesting is the using of block programming to create 3D designs, the majority of the teachers strongly agreed or agreed (in total 88.9%) that using programming to create 3D objects is interesting. In question 10, the teachers strongly disagreed (50%) or disagreed (38.9%) that they had to know 3D design techniques before in order to participate in the activities. In question 11, the teachers also strongly disagreed or agreed (in total 88.9%) that they had to know programming with blocks techniques previously in order to implement the activities.

For a better insight, we present in figure 4 the mean values of the teachers' answers for questions 1 to 11.

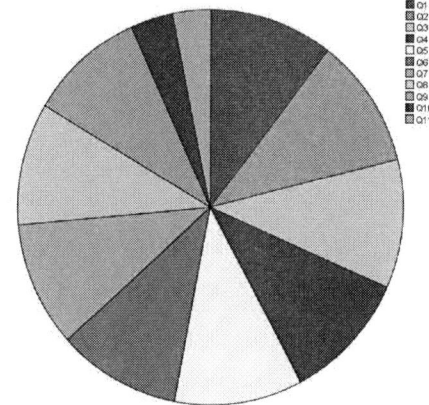

Fig. 4. Mean values of the answers to questions 1 to 11.

Lastly, in question 12, the teachers were asked about suggestions for improvement. Four of them thought that improvements are needed regarding the planned time of the activities and in one case the activities' content progression.

V. DISCUSSION

According to the results obtained in the workshop, the participating teachers seem to have a positive opinion about the 3D design and 3D printing activities' implementation [36]. More specifically, the teachers think that the subject to be taught is clear in the activities and the concepts that are included follow one another in a clear and understandable way. Also, the teachers think that the activities were well organized, clear and interesting and there was no problem in their implementation.

Additionally, the instructions given through the worksheets were helpful in order for one to understand basic concepts of 3D design and 3D printing. Furthermore, the visual examples included in the activities helped in the better understanding of the activities.

Moreover, the teachers think that by implementing the activities they understood basic concepts of 3D design and 3D printing and found interesting the creation of 3D designs using block programming [37].

Finally yet importantly, in the teachers' opinion, it was not prerequisite to know principles of 3D design and block programming beforehand in order to participate in the activities' implementation.

VI. CONCLUSIONS AND FURTHER RESEARCH

3D printing is an emerging technology within educational institutions, gaining popularity among both educators and students [35]. As a result, there is a necessity to explore novel approaches for its integration into educational environments [33]. In this paper, we presented educational activities for 3D design and 3D printing and their implementation by educators in the context of a two-hour workshop. In order to evaluate the produced activities we firstly addressed teachers of secondary education, before implementing them into actual classroom settings.

The teachers' feedback was very positive about the content and the structure of the suggested activities. According to their opinion, the activities are suitable for teaching and learning 3D design and 3D printing principles and also enhance design and programming skills. Additionally, we witnessed the teachers' positive attitude towards the implementation of 3D design and 3D printing activities into the teaching process.

In the future, we intend to expand the activities content in such a way as to include more subjects organized in a way that is relevant to the teachers' specialties. In addition, in some cases we are going to reconsider the planned time for some of the activities.

At the school level, it is important to include 3D technology into the curriculum in order to cultivate the necessary proficiency over the subject of 3D design and 3D printing. Consequently, it is of paramount importance to conduct research on the inclusion of 3D printing courses in forthcoming education curricula [15].

Taking into account the imperative that derives from all the previously mentioned, we are planning to take our research one step further by implementing the proposed activities into the educational process and by asking the teachers and the students to give their opinion in order to test and improve them further.

ACKNOWLEDGMENT

This work has been partially supported by COSMOTE through a PEDION24 grant.

REFERENCES

[1] Powar, K. P., & Patil, S. D. (2022). Promoting Technology-Enhanced Project-Based Learning through Application of 3D Printing Technology for Mechanical Engineering Education. Journal of Engineering Education Transformations, 35(Special Issue 1).

[2] Ford, S., & Minshall, T. (2019). Invited review article: Where and how 3D printing is used in teaching and education. Additive Manufacturing, 25, 131-150.

[3] Govil, K., Kumar, V., Pandey, D. P., Praneeth, R., & Sharma, A. (2019). Additive Manufacturing and 3D Printing: A Perspective. Advances in Engineering Design: Select Proceedings of FLAME 2018, 321-334.

[4] Pinger, C. W., Geiger, M. K., & Spence, D. M. (2019). Applications of 3D-printing for improving chemistry education. Journal of Chemical Education, 97(1), 112-117.

[5] Güleryüz, H., & Dilber, R. (2022). Robotic coding and 3D printer with STEM activities; the effect of science teacher candidates on STEM awareness and STEM self-efficacy. Education and Information Technologies, 1-21.

[6] Assante, D., Cennamo, G. M., & Placidi, L. (2020, April). 3D Printing in Education: an European perspective. In 2020 IEEE Global Engineering Education Conference (EDUCON), 1133-1138.

[7] Ford, S., & Minshall, T. (2019). Invited review article: Where and how 3D printing is used in teaching and education. Additive Manufacturing, 25, 131-150.

[8] Küçük, M., Talan, T., & Demirbilek, M. (2023). The Effect of Creating 3D Objects with Block Codes on Spatial and Computational Thinking Skills. Informatics in Education.

[9] Mou, T. Y. (2020). Students' evaluation of their experiences with project-based learning in a 3D design class. The Asia-Pacific Education Researcher, 29(2), 159-170.

[10] Dere, H. E., & Kalelioglu, F. (2020). The effects of using web-based 3D design environment on spatial visualisation and mental rotation abilities of secondary school students. Informatics in Education, 19(3), 399-424.

[11] Ng, O. L., & Chan, T. (2019). Learning as Making: Using 3D computer-aided design to enhance the learning of shape and space in STEM-integrated ways. British Journal of Educational Technology, 50(1), 294-308.

[12] Anđić, B., Ulbrich, E., Dana-Picard, T. N., & Laviza, Z. (2022). Usability of 3D modelling and printing in STEAM education: primary school teachers perspective. In Twelfth Congress of the European Society of Research in Mathematics Education (CERME12). Available at: https://hal.science/hal-03745402

[13] Pikkarainen, A., & Piili, H. (2020). Implementing 3D Printing Education Through Technical Pedagogy and Curriculum Development. Int. J. Eng. Pedagog., 10(6), 95-119.

[14] Üçgül, M., & Altıok, S. (2023). The perceptions of prospective ICT teachers towards the integration of 3D printing into education and their views on the 3D modeling and printing course. Education and Information Technologies, 1-31.

[15] Chun, H. (2021). A Study on the Impact of 3D Printing and Artificial Intelligence on Education and Learning Process. Scientific Programming, 2021, 1-5.

[16] Trust, T., Woodruff, N., Checrallah, M., & Whalen, J. (2021). Educators' interests, prior knowledge and questions regarding augmented reality, virtual reality and 3D printing and modeling. TechTrends, 65, 548-561.

[17] Batocchio, A., Silva, S. S., Franco, M., & Minatogawa, V. L. (2022, February). 3D Printer: An Application in Teaching. In Product Lifecycle Management. Green and Blue Technologies to Support Smart and Sustainable Organizations: 18th IFIP WG 5.1 International Conference, PLM 2021, Curitiba, Brazil, July 11–14, 2021, Revised Selected Papers, Part II (pp. 327-340). Cham: Springer International Publishing. https://link.springer.com/chapter/10.1007/978-3-030-94399-8_24

[18] Kim, S., Shin, Y., Park, J., Lee, S. W., & An, K. (2021). Exploring the potential of 3D printing technology in landscape design process. Land, 10(3), 259.

[19] Karimov, O. (2021). The importance of the use of 3D printers in the field of technological education. Mental Enlightenment Scientific-Methodological Journal, 172-178.

[20] Pai, S., Gourish, B., Moger, P., & Mahale, P. (2018). Application of 3D printing in education. International Journal of Computer Applications Technology and Research, 7(7), 278-280.

[21] Garstki, K. (2017). Virtual representation: the production of 3D digital artifacts. Journal of Archaeological Method and Theory, 24, 726-750.

[22] Li, T., & Zhan, Z. (2022). A systematic review on design thinking Integrated Learning in K-12 education. Applied Sciences, 12(16), 8077.

[23] Chien, Y. H., & Chu, P. Y. (2018). The different learning outcomes of high school and college students on a 3D-printing STEAM engineering design curriculum. International Journal of Science and Mathematics Education, 16, 1047-1064.

[24] Arvanitidi, E., Drosos, C., Theocharis, E., & Papoutsidakis, M. (2019). 3D Printing and Education. International Journal of Computer Applications, 177(24), 55-59.

[25] Attaran, M. (2017). The rise of 3-D printing: The advantages of additive manufacturing over traditional manufacturing. Business horizons, 60(5), 677-688.

[26] Sun, Y., & Li, Q. (2018). The application of 3D printing in STEM education. In 2018 IEEE international conference on applied system invention (ICASI), 1115-1118.

[27] Trust, T., & Maloy, R. W. (2017). Why 3D print? The 21st-century skills students develop while engaging in 3D printing projects. Computers in the Schools, 34(4), 253-266.

[28] Chen, J., & Cheng, L. (2021). The influence of 3D printing on the education of primary and secondary school students. In Journal of Physics: Conference Series (Vol. 1976, No. 1, p. 012072). IOP Publishing.

[29] Wilk, R., Likus, W., Hudecki, A., Syguła, M., Różycka-Nechoritis, A., & Nechoritis, K. (2020). What would you like to print? Students' opinions on the use of 3D printing technology in medicine. PloS one, 15(4), e0230851.

[30] Smith, R. C., Iversen, O. S., & Hjorth, M. (2015). Design thinking for digital fabrication in education. International Journal of Child-Computer Interaction, 5, 20-28.

[31] Tramonti, M., Dochshanov, A. M., & Zhumabayeva, A. S. (2023). Design Thinking as an Auxiliary Tool for Educational Robotics Classes. Applied Sciences, 13(2), 858. https://www.mdpi.com/2057492

[32] Brenner, W., & Uebernickel, F. (2016). Design thinking for innovation. Research and practice.

[33] Tejera, M., El Bedewy, S., Galván, G., & Lavicza, Z. (2022). 3D Printing and GeoGebra as artefacts in the process of studying mathematics through architectural modelling. Mathematics Education in Digital Age 3, 276. https://hal.science/hal-03925304/document#page=288

[34] Autodesk Tinkercad (2023). Available at: https://www.tinkercad.com/

[35] Assante, D., Cennamo, G. M., & Placidi, L. (2020). 3D printing in Education: an European perspective. In 2020 IEEE Global Engineering Education Conference (EDUCON), 1133-1138.

[36] Yildirim, G. (2018). Teachers' Opinions on Instructional Use of 3D Printers: A Case Study. International Online Journal of Educational Sciences, 10(4), 304-320.

[37] Chytas, C., Diethelm, I., & Tsilingiris, A. (2018). Learning programming through design: An analysis of parametric design projects in digital fabrication labs and an online makerspace. In 2018 IEEE Global Engineering Education Conference (EDUCON), 1978-1987.

[38] Instructions for teaching the high school Informatics course for the school year 2023-2024, Available at: https://iep.edu.gr/IEP/pliroforiki_2023_24.pdf

Nanotechnology as a Tool for Computational Thinking Skills using Open Hardware, Embedded Systems and Repository Platform, in Industry 4.0 Era

Konstantinos Kalovrektis
Dept. of Digital Systems
University of Thessaly
Larisa, Greece
kkalovr@uth.gr

Ioannis Dimos
Dept. of Computer Science and Biomedical Informatics
University of Thessaly
Lamia, Greece
ioadimos@uth.gr

Apostolos Xanakis
Dept. of Digital Systems
University of Thessaly
Larisa, Greece
axenakis@uth.gr

Athanasios Kakarountas
Dept. of Computer Science and Biomedical Informatics
University of Thessaly
Lamia, Greece
kakarountas@uth.gr

Abstract—The advancement in Microelectromechanical System (MEMS) technology resulted in accurate and high-performance miniature device systems. These devices are so tiny that they are not noticeable by the human eye and exhibit excellent feasibility in miniaturization sensors due to their small dimensions, low power consumption, and superior performance. The area of science and engineering where MEMS are developed (dimensions in the manometer scale) is called Nanotechnology. Nanotechnology is one of the fastest growing scientific research related to Industry 4.0. Nanotechnology may introduce industrial skills deficits as well as opportunities for new teaching practices in several subjects and educational frameworks. In the present work, we investigate the attitude of STEM (i.e. technology/engineering) and non STEM - related instructors, regarding the integration of Nanotechnology applications in Higher education curricula. Their opinions, concerning the applied teaching method, the learning content material and expected student skills, should always be taken in to account, as they may boost any reformations proposed. Moreover, we propose a repository platform, with which instructors may interact with 3D designs and MEMs material to built their didactic plans. This work's findings is critical for the design and innovative training material and computational thinking (CT) activities, which will prepare student with skills related to Industry 4.0 demand.

Index Terms—Nanotechnology, MEMs, Industry 4.0, Computational Thinking, Open Hardware, Repository Platform

I. INTRODUCTION AND BACKGROUND

A. Nanotechnology in Education

In the past few decades, advances in microelectronic device fabrication technologies have produced compelling, accurate, and high-performance device systems. Technology has been squeezed to the point where we can make devices so tiny that they are not noticeable by the human eye. Microelectromechanical systems (MEMS) involve the innovation of tiny devices that can represent the models as sensors or actuators

[1]–[3] and convey data from the nanoscale to the macroscopic scale [10]. In recent times, MEMS technology has grown significantly in acknowledging different sorts of natural sensors and actuators. Besides, it has been utilized in miniaturized sensor manufacturing in a large number of applications due to low power ratings, quick response, ease, cheapness, and better sensitivity [4]. As a result, smarter consumer electronics have opened up new possibilities for citizens in terms of communication, sports, industry, entertainment, etc. The impart of such innovative devices and upcoming revolutionary developments on everyday life, both present and challenge, science and technology education researchers, to incorporate this cutting - edge fields in Higher education contexts [5], [6]. However, several worth noticing issues which emerge and need to be considered, are: [7]:

- To which extend teachers acknowledge the educational and technological significance of Nanotechnology's applications inclusion, within innovative curricula ?
- Do they have all necessary and up - to date teaching material and means to support their STEM activities ?
- Are they well - trained in order to teach modules related to Nanotechnology ?
- Is there a complete learning framework to promote Nanotechnology, as a part into the problem solving process, within the STEM approach ?

In this work we give a quantitative and qualitative explanation regarding the aforementioned issues. Research findings in [8], [9], indicate that teachers' perspectives and attitudes, should be carefully taken into account, in any attempt for curricula change and innovation. Additionally, their opinions regarding teaching methods, learning context and expected

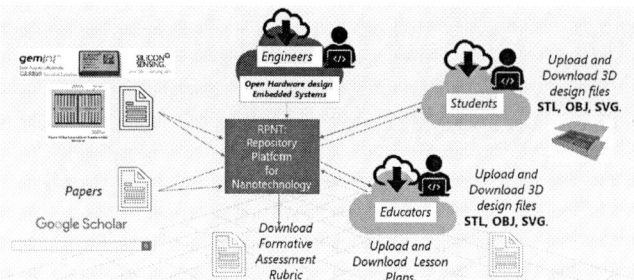

Fig. 1: RPNT: Repository Platform for Nanotechnology

learning outcomes, should be take into account as well, as teachers are considered as the *agents of change* for any potential reformation and innovation [11]. Therefore, the contribution of this study aims to record the STEM and non - STEM related instructors regarding their attitude towards the educational significance of Nanotechnology's inclusion within the new curricula. To do so, we use a five Likert scale questionnaire [25] and we conduct Mann - Whitney U test, as a non parametric test, to investigate the differences in the rank - distribution of the samples. To the best of our knowledge, no similar study relates to such inquiry [7]. Finally, we design and propose and open software - open hardware repository of embedded systems, as a learning design framework, to promote Nanotechnology problems, as a part into the problem solving STEM process approach.

B. Computational Thinking and STEM

Computational Thinking (CT) is undoubtedly considered a fundamental skill as reading, writing, and arithmetic in the 21st century [16] [26], [27]. Janette Wing, in a continuing effort to improve her initial definition of CT, expressed that CT refers to the *mental processes involved in formulating a problem and expressing its solution(s) in such a way that a computer—human or machine—can carry out the task effectively* [17]. Currently, CT has become commonly accepted as a problem-solving method [18] which includes a set of concepts such as *abstraction, decomposition, generalization, algorithmic thinking, evaluation, simulation, verification and predictions* [18], [19], [27]. These concepts mainly arises from fundamental Computer Science and Computing Science practices. According to [26], the dimensions of CT are as follows:

- The ability to think algorithmically
- The ability to decompose an initial problem to smaller problems and to try to firstly solve the smaller ones
- The ability to draw conclusions (i.e. generalize) and to use patterns
- The ability to evaluate a model
- The ability to think abstractly

The majority of European (EU) and non - EU countries, realize the importance and center role of CT in educational activities and adopt CT training into their curricula [11]. Nanotechnology is a interdisciplinary scientific and engineering field, devoted to designing, producing and using structures and systems, by controlling molecules and atoms at nano - scales [20]. Nanotechnology education and didactic activities have to do with understanding analogies, revealing patterns and projecting the results to large scale applications [28]. To this end, CT dimensions are evident to nanotechnology education activities. The applications of nanotechnology are very beneficial to society, ranging from smart materials to information and communication technology, energy technology, and medicines. This means that nanotechnology has already been embraced by many Industry 4.0 sectors.

Currently, there are plenty of learning and teaching materials related to CT, which propose activities to develop CT competencies, available in various formats [14]. These teaching materials are accessible by both STEM and non-STEM related instructors [12]. An example of this type of educational material is available in the project CS Unplugged [13], which explains how the concepts of CT (algorithmic thinking, abstraction, decomposition, generalization and patterns, logic, and evaluation) can be applied to each pedagogical activity.

Additionally, the development of emerging technologies, such as Virtual Reality (VR), Robotics, Augmented Reality (AR), etc. has led to the adoption of new didactic material and tools, such as STEM courses and experiments utilizing Arduino controllers [31] [33]. These materials require students' familiarization with Computer Science principles and Computational thinking skills in order to be utilized properly. A characteristic example is shown in [32] where the Arduino board is used in Primary education for teaching STEM following the problem-based learning (PBL) methodology.

Although in [14], [15], various approaches relate educational activities to CT skills cultivation, to the best of our knowledge, there are no similar materials, to encourage and support instructors in utilizing Nanotechnology withing their STEM (or non - STEM) training activities [7]. On the contrary, the majority of existing Nanotechnology learning material is mainly technical, focusing on Biomedical, Biology, Chemistry, Materials domains [20], [21]. There are very limited (or no) works on STEM activities with a focus on the correlation of understanding Nanotechnology, scales and applications, by engaging with CT activities.

To deal with the aforementioned gap, in this work, we investigate teacher's attitude towards integration of Nanotechnology activities, within STEM related and CT framework, as well as their willingness to follow appropriate training, in order to understand, design and produce innovative learning material and nanotechnology related activities. We also propose and acquire a free online learning platform, available to the teaching community, which works as a repository with ready 3D designs for Nano - structures, nano - sensors, along with ready lesson plans and STEM rubrics. Our main goal is to promote Nanotechnology, within this learning design framework, as a part of the problem solving STEM approach.

979-8-3503-8368-3/23 $31.00 © 2023 IEEE

II. RESEARCH METHODOLOGY

A. Questionnaire

Surveys in Greece and Cyprus [6], [23], conducted for schooling education teachers, show that there is no consensus on whether Nanotechnology need to be included in the curriculum [21]. According to [24], the majority of learning curricula is related to technical domains for chemistry, nanomedicine, nanoelectronics, smart materials etc., but there is a gap for learning activities, introductory to nanotechnology applications and projections among nano and macro scales. As there is no (or very limited) investigation about the importance of teaching Nanotechnology basic ideas and its correlation with CT skills as *problem-solving, decomposition, and abstraction (CT concepts)*, we aim to provide a learning framework to close this gap.

In [7], authors underline the necessity for research community to propose and promote ways to contribute to a comprehensive curricula for teaching Nanotechnology in high schools and higher education. The lack of properly designed, STEM - based and practical hands - on material, may cause teachers to become averse to teaching Nanotechnology's background concepts.

In [25], we design and use a five scale Likert survey questionnaire, based on the following categories (C):

C1 Teachers' beliefs and perceptions of Nanotechnology in education ([Q1], [Q2], [Q3])

C2 Teachers' beliefs and perceptions about the relationship between Nanotechnology and Computational Thinking skills development ([Q4], [Q5], [Q6])

C3 Teachers' beliefs and perceptions about career perspectives in the Nanotechnology field ([Q7], [Q8], [Q9])

C4 Teachers' beliefs and perceptions about the existing teaching material concerning Nano literacy ([Q10], [Q11], [Q12])

The questionnaire is anonymous and the target is STEM and non - STEM related teachers, instructors and professors, which are informed by a promo video. This video explains what is Nanotechnology, which skills are related to this scientific field and why is it worth including to new curricula, to support the skills for Industry 4.0 application. The questionnaire has two parts. In the 1st part, we collect demographic information, regarding participants' field area, years of working experience and their education level. In the 2nd part, the survey focuses on questions related to the aforementioned categories. In Table I, we give all questions per category.

B. Proposed Learning Framework and Repository

Apart from investigating the teachers' attitude towards the significance and potential inclusion of Nanotechnology activities in innovative STEM curricula, we envision and propose a learning framework, based on the *Engineering Design Process (EDP)* [29]. EDP is a systematic and iterative approach, used by engineers, to solve problems and develop new products and systems. It is a contemporary teaching method, appropriate to implement STEM education scenarios [30]. According to Fig.

TABLE I: Research Questions

	Questions
Q1	Nanotechnology is an important tool for understanding STEM
Q2	Nanotechnology should be taught in compulsory education
Q3	Teaching Nanotechnology orients students to new research and technology opportunities and helps them enhance their ambitions
Q4	Nanotechnology has contributed greatly to fixing problems in the world
Q5	Nanotechnology develops evaluation skill (Embedded Digital Twin)
Q6	Nanotechnology prompts both teachers and students to better model a problem
Q7	I can easily participate in a discussion for Nanotechnology
Q8	I'm certified in teaching Nanotechnology
Q9	I understand the career opportunities in nanotechnology
Q10	I can easily find useful teaching Nanotechnology material on the web
Q11	I am familiar with the usage of Nanotechnology teaching material in the classroom
Q12	I need didactic material for boosting CT concepts through Nanotechnology teaching

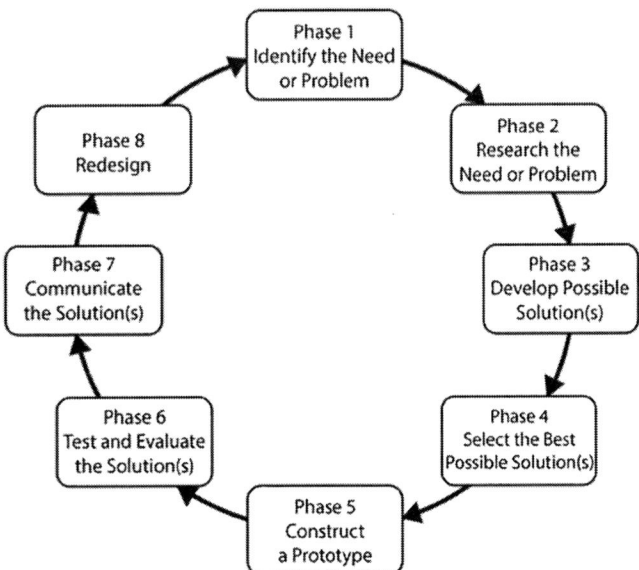

Fig. 2: Phases of the Engineering Design Process (EDS)

2, it involves a series of eight steps, that guides students (i.e. working as *engineers*), from identifying a problem, to creating a prototype and implementing a solution for a specific STEM problem. EDP process, is not always conducted in a strictly linear and circular way, and students may revisit earlier steps, gather more information, giving feedback to the prototype or encounter new challenges along the way.

In Fig. 1, the System Model for *RPNT* Repository Platform for Nanotechnology is shown. In essence, RPNT interacts with both students and instructors, collects and categorizes all necessary 3D designs and didactic plans gives advice and constructs rubrics, according to the learning objectives. Students may upload and download 3D designs (i.e. .STL, .OBJ, .SVG) files, mainly designed in *TinkerCAD*, as shown in Fig. 3 Teachers also interact with RPNT by uploading and downloading lesson plans, activities and 3D design files. The

979-8-3503-8368-3/23 $31.00 © 2023 IEEE

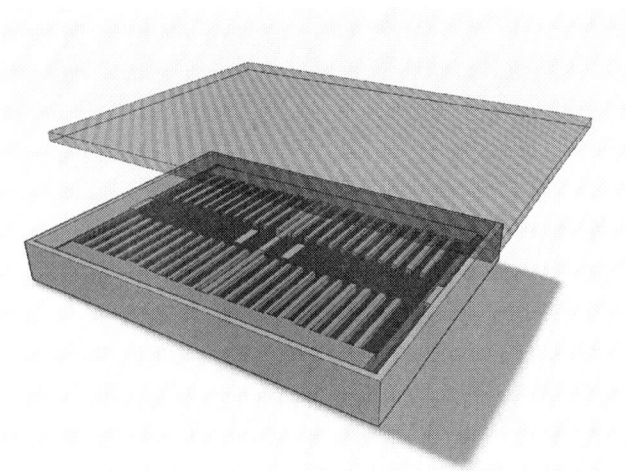

Fig. 3: Dual-Axis Accelerometer design in TinkerCAD

TABLE II: Scientific field Distribution

Levels	Answers	Total %	Cumulative %
STEM (Technology/Engineer/Informatics)	39	81.3%	81.3%
non-STEM	9	18.8%	100.0%

TABLE III: Years Of Experience Frequencies

Years	Answers	Total %	Cumulative %
0 - 5	10	20.8%	20.8%
6 - 15	17	35.4%	56.3%
16 - 25	7	14.6%	70.8%
26+	14	29.2%	100.0%

heart of RPNT is an expert system, which produces as an output, assessment rubrics, according to the input learning objectives.

Focusing on the technical characteristics of RPNT, the educational material relates to:

1) 3D designs of Micro-Electro-Mechanical Systems (MEMS) such as:
   - Accelerometers
   - Gyroscopes
   - Humidity sensors
   - Temperature sensors
   - 3D accelerometer and 3D magnetometer
   - MEM Microphone
2) Lesson plans and tutorials
3) Course evaluation rubrics

The repository platform is built on WordPress 6.3.2, with MariaDB as the database, while the lesson plans are constructed according to WordPress Tutor LMS and each lesson material also relates to a video course, along with quizzes, with h5p.org Web 2.0 tools.

III. RESULTS AND DISCUSSION

The anonymous study, is conducted during September 2023, and includes N = 48 answers from STEM and non - STEM related teachers (i.e. instructors). Their distribution, according their scientific field, working experience and education level, is the following:

- *Scientific field:* 39 teach STEM related course (i.e. Maths, Science, Computers, Engineering and 9 teach non - STEM courses, according to TableII
- *Working experience*: 10 with 1 - 5 year, 17 with 6 - 15 years, 7 with 16 - 25 years and 17 with 26 - 35 years, according to Table III
- *Education level*: 8 with a Bachelor degree, 25 with Master's, 8 with a PhD and 7 with a PostDoc experience

The results, concerning descriptive statistical parameters such as *Median, Mode*, and *Min, Max*, according to questionnaire's Category 1 (C1), are shown in Table IV. Following, the answers frequencies, related to questions *Q1, Q2, Q3*, are shown in Fig. 4. We observe that the majority replies as *Agree* or *Strong Agree*. However, it is worth further analyzing how the answer to *Neutral* option, affects or not, the interpretation of the results.

To do so, we apply the *Mann - Whitney U* test, also known as the *Wilcoxon rank-sum test*, which is a non - parametric statistical test, used to compare two independent groups, when the dependent variable is ordinal or continous, but not normally distributed. To this end, we investigate the differences in the rank distribution of the samples, between the *independent* variable *Scientific Field* of the answer samples. As long as value $p \geq 0.05$, there is no significant statistical difference, so the *Scientific field* does not affect the participants' answers. The p - *value* for questions Q1, Q2, Q3, as shown in V are

(a) Answers for Q1

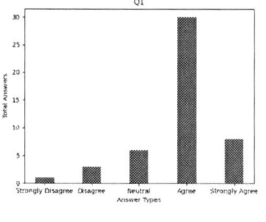

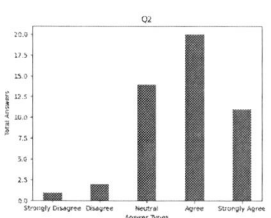

(c) Answers for Q3

(b) Answers for Q2

Fig. 4: Answer Frequencies for Category 1 Questions

TABLE IV: Teachers' beliefs and perceptions of Nanotechnology in education

Descriptives						
	N	Missing	Median	Mode	Min	Max
Q1	48	0	4.00	4.00	1	5
Q2	48	0	4.00	4.00	1	5
Q3	48	0	4.00	4.00	2	5

TABLE V: Mann-Whitney U statistics for Scientific field

		Static	p
Q1	Mann-Whitney U	142	0.307
Q2	Mann-Whitney U	108	0.061
Q3	Mann-Whitney U	137	0.277

greater than 0.05

Following, in order to investigate differences in the rank - distribution of the samples for each question Q1,Q2,Q3, we examine the non - parametric *Kruskal - Wallis* test, between the independent variable *Education Level* of the sample and we find that there is not statistically significant difference, as according to *p - values* of V. This result states that *Education Level* did not affect the participants' answers.

In the sequel, in order to investigate differences in the rank - distribution of the samples (Rank) for questions Q1,Q2,Q3, we examine the non - parametric *Kruskal-Wallis test* between the independent variable *Working Experience*, and we find that there is no statistically significant difference, as *p - value* is greater than 0.05 in all cases, according to VII

Finally, for the results of *Wilcoxon by* H_a, $\mu \geq 3$ (neural), for Category 1 questions bundle (i.e. textitTeachers' beliefs and perceptions of Nanotechnology in education), we find that there is a statistically significant difference present to a positive attitude, towards Nanotechnology use in Education, as *p - value* indicates $p \leq 0.01$, according to VIII

Additionally, we perform similar statistical analysis for the remaining categories entitled: C2: *Teachers' beliefs and perceptions about the relationship between Nanotechnology and Computational Thinking skills development*, C3: *Teachers' beliefs and perceptions about career perspectives in the Nanotechnology field* and C4: *Teachers' beliefs and perceptions about the existing teaching material concerning Nano literacy.* According to the tests, regarding the parameters of *years of*

TABLE VI: One-Way AOVA (Non-parametric) Kruskal-Wallis (Education Level)

	x^2	df	p
Q1	3.95	3	0.267
Q2	3.51	3	0.319
Q3	7.40	3	0.060

TABLE VII: One-Way AOVA (Non-parametric) Kruskal-Wallis (Working Experience)

	x^2	df	p
Q1	0.867	3	0.833
Q2	1.692	3	0.639
Q3	1.912	3	0.591

TABLE VIII: Test of Wilcoxo W

		Static	p
Q1	Wilcoxon W	814	<.001
Q2	Wilcoxon W	544	<.001
Q3	Wilcoxon W	757	<.001
Note. H_a $\mu > 3$			

working experience and *education level*, we find that there is a statistically significant positive attitude towards Nanotechnology to support Computational Thinking skills development and to contribute positively in career perspectives, related to Nanotechnology field.

IV. CONCLUSIONS

According to this work, the research emphasizes the need for teachers' support in teaching and preparing material for basic Nanotechnology concepts and principles, within the STEM framework. We statistically investigate instructors' attitude towards Nanotechnology applications and curricula enhancements and we observe that the majority supports the inclusion of MEMs technology to be included in their didactic materials. Additionally, we design and propose an innovative learning platform, with which teachers and students interact, exchange ideas, upload and download new material etc. In essence, the platform will work as a repository for material and good practices for Nanotechnology. Apart from their working experience and level of education, the majority of teachers believe that there is a necessity for the existence of such platform, which will improve the quality of education and make STEM lessons more attractive.

Moreover, the challenge posed by the proposed platform, is the inclusion and offering of learning activities, which boost CT skills and the evaluation of both teaching and learning outcomes by custom - based automated Nanotechnology skills - based rubrics.

REFERENCES

[1] M. Versaci, A. Jannelli, and G. Angiulli, "Electrostatic Micro-Electro-Mechanical-Systems (MEMS) Devices: A Comparison Among Numerical Techniques for Recovering the Membrane Profile," IEEE Access, vol. 8, pp. 125874–125886, 2020, doi: 10.1109/ACCESS.2020.3008332.

[2] Y. Li, D. Gu, S. Xu, X. Zhou, K. Yuan, and Y. Jiang, "A Monoclinic V1-x-yTixRuyO2 Thin Film with Enhanced Thermal-Sensitive Performance," Nanoscale Res. Lett., vol. 15, no. 1, p. 92, Dec. 2020, doi: 10.1186/s11671-020-03322-z.

[3] R. De Oliveira Hansen et al., "Magnetic films for electromagnetic actuation in MEMS switches," Microsyst. Technol., vol. 24, no. 4, pp. 1987–1994, Apr. 2018, doi: 10.1007/s00542-017-3595-2.

[4] W. Tong, Y. Wang, Y. Bian, A. Wang, N. Han, and Y. Chen, "Sensitive Cross-Linked SnO2:NiO Networks for MEMS Compatible Ethanol Gas Sensors," Nanoscale Res. Lett., vol. 15, no. 1, p. 35, Dec. 2020, doi: 10.1186/s11671-020-3269-3.

[5] Huffman, D.; Ristvey, J.; Morrow, C.; Deal, M. Integrating Nanoscience in High School Science: Curriculum Models and Instructional Approaches. In 21st Century Nanoscience—A Handbook: Public Policy, Education, and Global Trends; Sattler, K., Ed.; CRC Press: Boca Raton, FL, USA, 2020; Volume 10, p. 10. Available online: https://bookshelf.vitalsource.com/ (accessed on 16 August 2021).

[6] Mandrikas, A.; Michailidi, E.; Stavrou, D. In-service Teachers' Needs and Mentor's Practices in Applying a Teaching–Learning Sequence on Nanotechnology and Plastics in Primary Education. J. Sci. Educ. Technol. 2021, 30, 630–641.

979-8-3503-8368-3/23 $31.00 © 2023 IEEE

[7] Hingant, B.B.; Albe, V. Nanosciences and nanotechnologies learning and teaching in secondary education. Stud. Sci. Educ. 2010, 46, 121–152.

[8] G. H. Roehrig, R. A. Kruse, and A. Kern, "Teacher and school characteristics and their influence on curriculum implementation," J. Res. Sci. Teach., vol. 44, no. 7, pp. 883–907, Sep. 2007, doi: 10.1002/tea.20180.

[9] Laherto, A. An Analysis of the Educational Significance of Nanoscience and Nanotechnology in Scientific and Technological Literacy. Sci. Educ. Int. 2010, 21, 160–175.

[10] V. L. Kalyani and P. Kumawat, "Recent Advancement in Nanosensors with Special Reference To Biomedical Applications," Jul. 2020, doi: 10.5281/ZENODO.3926187.

[11] European Commission. Joint Research Centre., Developing computational thinking in compulsory education: implications for policy and practice. LU: Publications Office, 2016. Accessed: Mar. 27, 2022. [Online]. Available: https://data.europa.eu/doi/10.2791/792158

[12] I. Dimos, C. Velaora, K. Louvaris, A. Kakarountas, and A. Antonarakou, "How a Rubric Score Application Empowers Teachers' Attitudes over Computational Thinking Leverage," Information, vol. 14, no. 2, p. 118, Feb. 2023, doi: 10.3390/info14020118.

[13] Bell, T.; Vahrenhold, J. CS Unplugged-How Is It Used, and Does It Work? In Adventures Between Lower Bounds and Higher Altitudes; Lecture Notes in Computer Science; Springer: Cham, Switzerland, 2018; Volume 11011.

[14] de Araujo, A.L.S.O.; Andrade, W.L.; Guerrero, D.D.S. A systematic mapping study on assessing computational thinking abilities. In Proceedings of the 2016 IEEE Frontiers in Education Conference (FIE), Eire, PA, USA, 12–15 October 2016; pp. 1–9.

[15] Araujo, A.L.; Andrade, W.; Guerrero, D. Um mapeamento sistemático sobre a avaliaçao do pensamento computacional no brasil. Anais dos Workshops do Congresso Brasileiro de Informática na Educação 2016, 5, 1147.

[16] Wing, J. Computational thinking's influence on research and education for all. Ital. J. Educ. Technol. 2017, 25, 7–14.

[17] Wing, J. Research notebook: Computational thinking—What and why. Link Mag. 2011, 6, 20–23.

[18] Avila, C.O.; Foss, L.; Bordini, A.; Debacco, M.S.; da Costa Cavalheiro, S.A. Evaluation rubric for computational thinking concepts. In Proceedings of the 2019 IEEE 19th International Conference on Advanced Learning Technologies (ICALT), Alagoas, Brazil, 15–18 July 2019; Volume 2161, pp. 279–281.

[19] Martin, F. Rethinking Computational Thinking. Commun. ACM 2018, 59, 8

[20] Y. Dong, X. Wu, X. Chen, P. Zhou, F. Xu, and W. Liang, "Nanotechnology shaping stem cell therapy: Recent advances, application, challenges, and future outlook," Biomed. Pharmacother., vol. 137, p. 111236, May 2021, doi: 10.1016/j.biopha.2021.111236.

[21] A.Spyrtou, L. Manou, and G. Peikos, "Educational Significance of Nanoscience–Nanotechnology:Primary School Teachers' and Students' Voices after a Training Program," Educ.Sci., vol. 11, no. 11, p. 724, Nov. 2021, doi: 10.3390/educsci11110724.

[22] J. L. Hess, A. Chase, D. Minner, M. Rizkalla and M. Agarwal, "An evaluation of a research experience for teachers in nanotechnology," 2017 IEEE Frontiers in Education Conference (FIE), Indianapolis, IN, USA, 2017, pp. 1-7, doi: 10.1109/FIE.2017.8190701.

[23] Papanastasiou, E. C., & Angeli, C. (2008). Evaluating the Use of ICT in Education: Psychometric Properties of the Survey of Factors Affecting Teachers Teaching with Technology (SFA-T3). Educational Technology & Society, 11 (1), 69-86.

[24] Achilleas Mandrikas, Emily Michailidi & Dimitris Stavrou (2020) Teaching nanotechnology in primary education, Research in Science & Technological Education, 38:4, 377-395, DOI: 10.1080/02635143.2019.1631783

[25] Survey Five Scale Likert Questionnaire, {https://shorturl.at/acCY3}

[26] Psycharis, S., Kalovrektis, K., and Xenakis, A. (2020). A conceptual framework for computational pedagogy in STEAM education: determinants and perspectives. Hellenic J. STEM Educ. 1, 17–32. doi: 10.51724/hjstemed.v1i1.4

[27] Spyridon Kourtis, Apostolos Xenakis, Konstantinos Kalovrektis, Antonios Plageras, and Ioanna Chalvantzi. 2022. An Exploratory Teaching Proposal of Greek History Independence Events based on STEAM Epistemology, Educational Robotics and Smart Learning Technologies. In Proceedings of the 2021 European Symposium on Software Engineering (ESSE '21). Association for Computing Machinery, New York, NY, USA, 120–128. https://doi.org/10.1145/3501774.3501792

[28] Buse, Karsten H and Havenith, Martina, and Hofmann, Oliver, 2019, he Nanoworld in the Classroom: A Comprehensive and Hands-On Approach to Teaching Nanoscience, Journal of Chemical Education, 96:5,932-939, https://10.1021/acs.jchemed.9b00020

[29] P. Baligar, R. Kandakatla, G. Joshi and A. Shettar, Integrating cooperative learning principles into the engineering design process: A mixed-methods study at first-year undergraduate engineering, 2021 IEEE Frontiers in Education Conference (FIE), Lincoln, NE, USA, 2021, pp. 1-5, doi: https://10.1109/FIE49875.2021.9637161.

[30] Nguyen Quang Linh and Le Thi Thu Huong, Engineering design process in STEM education: an illustration with the topic wind energy engineers, Journal of Physics: Conference Series, Volume 1835, 2nd International Annual Meeting on STEM education (I AM STEM) 2019 27-29 September 2019, DOI 10.1088/1742-6596/1835/1/012051

[31] M. Oprea and M. Mocanu, "Bluetooth Communications in Educational Robotics," 2021 23rd International Conference on Control Systems and Computer Science (CSCS), Bucharest, Romania, 2021, pp. 408-413, doi: 10.1109/CSCS52396.2021.00073.

[32] García-Tudela, Pedro & Marín-Marín, José-Antonio. (2023). Use of Arduino in Primary Education: A Systematic Review. Education Sciences. 13. 134. 10.3390/educsci13020134.

[33] A. Xenakis, K. Kalovrektis, K. Theodoropoulou, A. Karampelas, G. Giannakas, D. Sotiropoulos, D. Vavougios, "Using Sensors and Digital Data Collection/Analysis Technologies in K–12 Physics Education Under the STEM Perspective", The International Handbook of Physics Education Research: Teaching Physics, Mehmet Fatih Taşar, Paula R. L. Heron

"Athens Museum Explorer": A pilot application of cultural proposals for a smart city

Anastasia Gasidou
MSc, Archaeologist
Digital Culture, Smart Cities,
IoT and Advanced Digital
Technologies, Department of
Informatics, University of Piraeus,
Greece,
anastasia.gkasidou@yahoo.com

Dimitrios Kotsifakos
Post-Doc, PhD, MSc,
Electronic Engineering,
Digital Culture, Smart Cities, IoT and
Advanced Digital Technologies,
Department of Informatics, University
of Piraeus,
Greece,
kotsifakos@unipi.gr

Christos Douligeris
Professor
Digital Culture, Smart Cities, IoT and
Advanced Digital Technologies,
Department of Informatics, University
of Piraeus, Greece,
cdoulig@unipi.gr

Abstract— In this paper, we present a pilot design and development study of a culture application, which can be part of a smart city. The article analyses the pilot study concerning the creation of the Athens Museum Explorer mobile application and highlights the functions of the application, and the techniques, tools, and technologies used to develop the application. In addition, the accessibility aspects of the application are presented, recognizing their importance in ensuring an inclusive user experience. The results of its implementation and the difficulties encountered during the app's implementation are presented.

Keywords— Pilot web application, cultural proposals, smart city

I. INTRODUCTION

This paper presents a comprehensive and in-depth "pilot study", which explores the crucial role of a digital mobile application in the context of smart city development. This "pilot study" presents a construction of the implementation of a mobile application designed to provide cultural information services about the cultural life of citizens in the city. This pilot study focuses on the development of a digital application for mobile devices called "Athens Museum Explorer" (Fig. 1). The study is part of the overall framework of the development of smart applications for smart cities. Focusing on the city of Athens in Greece, the development of the paper delves into the design of the application, its functions, and its impact on enhancing the city's cultural landscape, tourism potential, and overall smart city goals. More specifically, the idea for the creation of this application came from the need for citizens to easily find and evaluate museums.

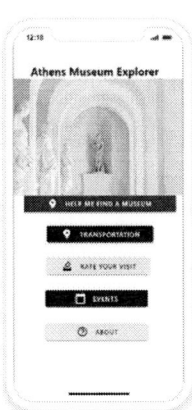

Fig. 1. The mobile application "Athens Museum Explorer".

The implementation of the construction is part of the planning and development initiatives of Athens as a smart city and focuses on preserving and promoting cultural heritage. As the smart city concept revolves around the use of technological developments to enhance urban life and sustainability [1], the creation of a "pilot study" as a starting point for a series of similar smart applications was deemed urgent and necessary. Through a comprehensive analysis, in this article, we present the functions of the pilot application, its impact on different aspects of the city, and finally, how the difficulties encountered during its implementation were addressed. The results of the study provide valuable insights into the potential of mobile device technology for supporting applications for smart cities.

The research questions for this project can be outlined as follows:

- How can the implementation of the Athens Museum Explorer app contribute to the advancement of Athens as a smart city, particularly in the context of preserving and promoting its cultural heritage?

- What are the key functions and features of the pilot application, and how do they impact various aspects of the city, including cultural exploration, tourism, and sustainability?

- What technological difficulties were encountered during the app's implementation, and how were these challenges addressed to ensure the successful development of the application?

In this "Introduction", we presented the context of this project, how it fits in with other research on the topic of smart cities and what is the research questions about the application. In section 2 (Methods), we present the scientific areas and methods we use for the construction. In section 3 (Results) we focus on the result of the project, the screenshots, and the accessibility features of the application. and in section 4 (Discussion / Conclusions) we present the project's conclusions.

II. METHODS

A. Techniques, tools, and technologies

In creating a cultural app for museums in Athens, various techniques, tools, and technologies were used to enhance the user experience and provide an exciting exploration of the city's rich cultural heritage. By using various technologically advanced techniques, tools, and technologies, the cultural app for museums in Athens can provide an enriched and enjoyable experience for visitors, promoting a deeper connection with the city's living cultural heritage and historical treasures.

979-8-3503-8368-3/23 $31.00 © 2023 IEEE

More specifically, in creating the "Athens Museum Explorer" app for a smart city, two fundamental techniques that play a key role in its success are user research and content curation. User research involves a comprehensive study of the target audience, including both residents and tourists visiting Athens. Through surveys, interviews, and field observations, valuable insights are gained into the preferences, needs, and behaviors of users. This ensures that the features and content of the application meet their expectations.

Content curation is equally critical, as it involves the careful selection and organization of attractive and informative material related to the museums and their exhibits. By curating high-quality content such as images, locations, and museum operation information, the Athens Museum Explorer app provides users with an immersive and engaging experience. The curated content highlights the cultural significance of each cultural site. In addition, the content curation process includes collaboration with museum experts, historians, and curators to ensure accuracy and authenticity. By combining user research knowledge with expertly curated content, the Athens Museum Explorer app provides a seamless and enriching journey through the city's cultural heritage, promoting a deeper appreciation and understanding of Athens' remarkable history.

B. Delimitation of the research frameworks when creating the application

The following research frameworks were defined during the development of the application:

Definition of the product application: The objectives of the application and the scope of its capabilities and functions were clearly defined. At the same time, key objectives were identified, such as enhancing the visitor experience (User Experience Design, UX Design), promoting the museum's collections, or offering educational content [2].

Audience survey: A brief survey was conducted with potential users, to understand their preferences and needs [3]. Further, demographics, user preferences, and technological proficiency were analyzed to shape the application accordingly. The survey is ongoing through a questionnaire via the embedded use of Google Forms in the application (https://tinyurl.com/3dfs48np).

Data and content collection: Research is ongoing to collect accurate and relevant historical and cultural information. Collection and interactive content for the application. It is worth noting that it has been ensured that all content complies with copyright and intellectual property laws [4].

User experience design and User Interface (UI) design: focus groups and usability testing were initially conducted to design an intuitive and user-friendly interface. The goal was to optimally visualize the layout and flow of the application. Some designs and layouts needed to be tweaked based on user feedback to achieve optimal usage.

Choice of technology and platform: The platform was selected based on the target audience and the requirements of the application. Furthermore, factors such as scalability, security, and integration with existing systems were considered. Finally, it was confirmed that the application would be accessible to all users while respecting accessibility standards.

When developing the Athens Museum Explorer app for a smart city, the use of tools such as Adalo and other complementary technologies can improve the development process of the app and enhance its functionality. Adalo, a code-free app development platform, allows designers and developers to create mobile apps without extensive coding knowledge, significantly reducing development time and costs. The in-app drag-and-drop interface allows the developer to quickly prototype and re-design the app, ensuring a user-friendly and visually appealing interface for seamless navigation. Adalo's flexibility allows for the integration of various functionalities such as interactive maps, multimedia content, and user profiles, facilitating a comprehensive exploration of Athens' museums.

In summary, the successful development of a cultural application requires the thoughtful integration of various techniques, tools, and technologies. It is worth noting that user research provides designers with valuable information about the preferences of the target audience, ensuring that the application meets their needs and expectations. Content curation plays a key role in curating engaging and informative materials that highlight the cultural significance of exhibits and historical objects, creating a compelling narrative for users to explore. By integrating these techniques, tools, and technologies, cultural applications can bridge the gap between the past and the present, providing an engaging and educational experience that fosters a deeper connection with cultural heritage in the digital age. However, the real essence of its impact lies in the user experience, as evidenced by meticulously designed working screens, an unwavering commitment to accessibility, and unparalleled efficiency [5]. These aspects serve as the pillars on which the app's success is based, making it an indispensable platform for culture lovers who wish to embark on a seamless and immersive journey through Athens' rich and diverse heritage.

III. RESULTS

A. Working screens, accessibility features, and the effectiveness of the application

In this, we present working screens of the application "Athens Museum Explorer", its accessibility features, and its effectiveness. The purpose is to provide an in-depth analysis of the application's user interface, its compliance with accessibility standards, and its overall performance in improving the user experience. Specifically, the Athens Museum Explorer app has a well-designed User Interface (UI) that aims to provide a seamless and intuitive experience for its visitors [6]. The app features a clean and visually appealing layout, using a combination of icons, images, and text to guide users through the various functions. The app's working screens contain useful data and information about the museums of the city of Athens. The home screen serves as the central hub of the app, providing quick access to basic functions such as searching for museums, information on transportation to museums, visitor experience ratings, exploring events, and general information related to the app.

The home screen of this cultural app combines sleek aesthetics and thoughtful features, encapsulating the essence of cultural exploration and appreciation [7]. By incorporating buttons such as "Help me find a museum", "Transfer", "Rate your visit", "Events" and "About", the app begins a transformative journey, inspiring users to embark on exciting cultural discoveries, connect with like-minded enthusiasts and become active members in preserving and celebrating the world's rich cultural heritage. As a result, you are committed to accessibility, which is highlighted through a robust set of features designed to provide a seamless and enriching experience for users of all abilities.

The options proposed to the user are oriented to the general principles and rules of accessibility. First, the option "Help me find a museum" lists directions for accessible exploration in the city. The user-friendly interface includes clear and concise descriptions and high-contrast visuals to accommodate the visually impaired. The "Transport" button further reinforces the app's commitment to inclusivity by incorporating features to support users with different mobility requirements. The app references public transport systems to display accessibility information for each route, such as the availability of wheelchair-accessible vehicles and stations with lifts. This integrated approach ensures that all users can access cultural destinations safely and independently without facing barriers to transport.

The app's main menu acts as a gateway to a rich and exciting world of heritage, art, and traditions, designed to promote cultural exploration and appreciation in the digital age. Upon launching the app, users are greeted with a visually striking and dynamic interface that seamlessly blends contemporary aesthetics with the charm of tradition. The app's logo, a symbol representative of the diverse cultural mosaic it encompasses, anchors elegantly in the top corner, exuding a sense of identity and unity. Beneath this, some options categorize the app's content. When navigating the app, users are met with an intuitive and user-friendly interface, carefully designed to cater to a diverse audience of all ages and cultural backgrounds. The app user is provided with basic information for a tour of the city's museum "web" through appropriate "Help me find a museum", "Transportation", and "Events" options. Each option is accompanied by small icons that reflect the cultural theme of the app and invite users to delve deeper into it.

B. Maintaining the Application Specifications

When selecting "Help me find a museum" (Fig.2) the user is supported to select a museum that belongs to his/her immediate cultural interests. The option acts as a "compass", guiding users in a search to discover cultural institutions and museums in Athens. Utilizing advanced technology, the app provides a user-friendly map interface of the museums, with comprehensive details about their collections, opening hours, and special exhibitions.

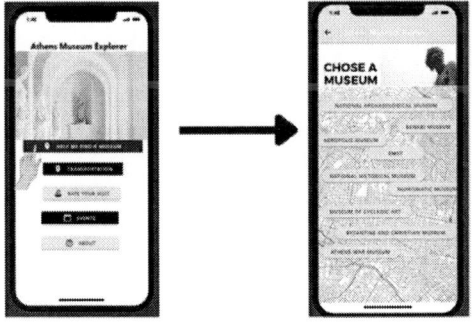

Fig. 2. The "Help me find a museum" option.

The selection points the way to accessible exploration, as the design of the application follows the principles of

universal accessibility. The user-friendly interface includes clear and concise descriptions and high-contrast visuals to accommodate the visually impaired. The 'Transfer' button further reinforces the app's commitment to inclusivity by incorporating features to support users with different mobility requirements. The app references public transport systems to display accessibility information for each route, such as the availability of wheelchair-accessible vehicles and stations with lifts. This integrated approach ensures that all users can access cultural destinations safely and independently without facing barriers to transport.

Then, the "Transportation" option (Fig. 3) provides seamless integration with various transport services, simplifying travel by public transport to cultural destinations. Through this option, users can obtain real-time information about the lines serving the museums in the application, eliminating the hassle of logistics and allowing for effortless cultural tours.

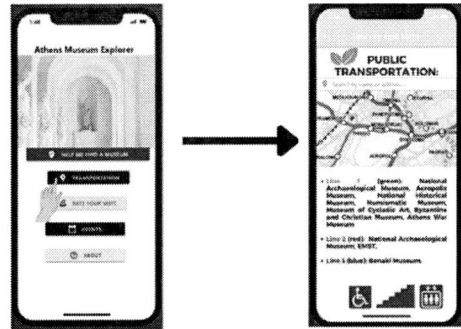

Fig. 3. The "Transportation" option.

The "Events" option (Fig. 4) reveals a calendar full of cultural events, workshops, exhibitions, performances, and much more which is updated by the app's creator as new events occur. Users can explore a wide range of upcoming events, locations, or themes and note their attendance. In addition, the app offers event reminders and ticket purchase options, ensuring users never miss out on the great cultural events around them.

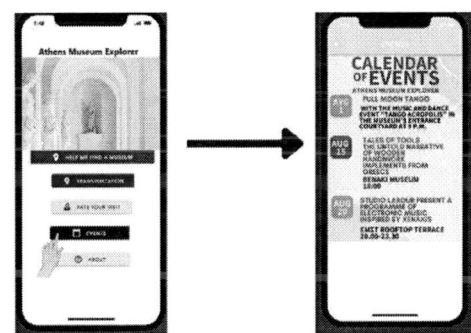

Fig. 4. The "Events" option.

Finally, the "About" option serves as a gateway for the user to discover the mission, vision, and team behind the creation of the application (Figure 7.8). Users can learn about the app's commitment to promoting cultural appreciation and understanding, its partnerships with cultural institutions, and its dedication to embracing diversity and inclusivity.

The home screen of this cultural app combines elegant aesthetics and thoughtful features, encapsulating the essence of cultural exploration and appreciation. By developing and integrating diverse options, the app begins a transformative journey, inspiring users to embark on exciting cultural discoveries, connect with like-minded enthusiasts, and become active members in preserving and celebrating the world's rich cultural heritage [8]. As a result, a powerful set of features emerges that aim to provide a seamless and enriched experience for users of all abilities. For the user, the application options are oriented with the general principles and rules of accessibility.

The "Athens Museum Explorer" app seeks efficiency for cultural exploration and appreciation. The "Help me find a museum" button is a testament to the app's effectiveness, offering users a seamless and intuitive interface to discover the wealth of cultural institutions Athens has to offer.

Fig. 5. The "About" option.

The app quickly generates personalized recommendations based on user's preferences and interests, presenting them with a curated list of museums and historic sites that match their cultural curiosities. To achieve this, the app can utilize machine learning algorithms that analyze user data, such as historical interests and location, to swiftly curate a tailored list of museums and historic sites aligning with their cultural preferences. In addition, the list of museums listed in the app includes detailed information about each venue, from opening hours to ticket prices, enabling users to plan their visits efficiently and maximize their cultural experiences. By streamlining the process of discovering museums and putting valuable information at their fingertips, the "Help me find a museum" button is at the heart of the app's effectiveness, revolutionizing the way users engage with Athens' cultural heritage.

The "Athens Museum Explorer" application is a product of technological innovation, responding to modern culture lovers with its seamless screens, unwavering commitment to accessibility, and undeniable efficiency. Incorporating a series of interactive working screens, the app encourages users to embark on an exciting journey into the diverse world of Athens' art, history, and cultural heritage. The considered accessibility features exemplify a transformative approach, ensuring that people of all abilities can participate in cultural exploration, promoting inclusivity and diversity in the app's vibrant community.

Moreover, the app's effectiveness in simplifying museum discovery and streamlining transportation facilitates a seamless and enjoyable cultural experience, maximizing the time users can spend immersing themselves in Athens' rich cultural mosaic. As a powerful tool to promote cultural

appreciation and understanding, the Athens Museum Explorer app exemplifies the potential of technology to connect individuals with the treasures of the past, inspiring a generation of cultural enthusiasts to embrace and celebrate the timeless beauty and significance of our shared heritage.

Fig. 6. The "Choose a Museum" option.

In conclusion, the Athens Museum Explorer app contributes significantly to the city's broader initiatives for its development as a smart city, as it aligns perfectly with the vision of creating a technologically advanced, sustainable, and visitor-friendly urban environment [9]. Integrating the application of cutting-edge technology with the promotion of cultural heritage enhances the overall smart city experience, meeting the needs of both residents and tourists. Firstly, the app supports sustainability by reducing dependence on physical guides and maps. Traditional paper guides and maps produce significant waste and contribute to environmental pollution. By providing integrated virtual tours and interactive content through the app, tourists can access all the necessary information without the need for printed material [10]. This move towards digitization not only reduces paper waste but also supports the city's commitment to becoming more environmentally friendly.

Finally, on all screens we find an option that returns users to the home menu, making the application easy to use. This functionality is complemented by personalized user profiles that track the progress and preferences of each user, promoting a customized cognitive experience, as well as it has also incorporated a multi-language support system, making it accessible to visitors from abroad and promoting the intercultural commitment of Greek tourism policy [11]. To achieve this, the app can employ data analytics and machine learning to create personalized user profiles, enabling progress tracking and tailored recommendations, while also integrating a comprehensive multi-language support system to enhance accessibility for international visitors, aligning with the goals of promoting the intercultural commitment of Greek tourism policy. In addition, a comprehensive events calendar informs users about upcoming exhibitions, workshops, and cultural events organized by the museum, encouraging repeat visits and a lasting interest in Athens' cultural heritage.

The proposed mobile app, the Athens Museum Explorer, represents a commendable and transformative solution for cultural preservation and tourism in Athens. Its development methodology, which heavily relies on user research and content curation, exemplifies a user-centric approach that caters to the diverse preferences and needs of both residents and tourists. This emphasis on accessibility, inclusivity, and user-friendliness is a notable strength of the app. It enhances the user experience by providing valuable information about museums, events, and transportation, ensuring cultural exploration is both convenient and engaging for everyone. The app's commitment to accessibility features is particularly noteworthy, as it ensures that individuals with varying abilities can participate in cultural exploration seamlessly.

Furthermore, the app's impact extends beyond cultural preservation. It plays a pivotal role in driving tourism and economic development in Greece. By offering a comprehensive, immersive, and informative platform for tourists, it attracts and retains visitors, contributing to increased revenue for the local economy. The app promotes sustainable tourism by encouraging eco-friendly transportation options and minimizing the use of printed materials. It aligns seamlessly with Athens' vision of becoming a smart and sustainable city.

IV. DISCUSSION / CONLUSIONS

The application "Athens Museum Explorer" was developed to facilitate cultural exploration in the city of Athens. The app provides citizens and tourists of the city with an interactive platform to discover the cultural heritage of the city while contributing to its broader objectives. This user-friendly app acts as a personal guide, offering a wealth of information about the city's renowned museums. The app and its potential for promoting cultural tourism and sustainable urban development [12] are integrated into the vision of the smart city of Athens. The "Athens Museum Explorer" app was designed to enhance the visitor's experience and promote their visit to Athens' renowned museums. The key feature of the "Athens Museum Explorer" app is the interactive map for museums, which acts as a personal curator for each visitor. Through the app's user-friendly interface [13], visitors can access a comprehensive map that highlights the various options for museums in Athens. The interactive guide offers detailed descriptions and informative text for each museum, providing information on museum type, hours, ticket prices, and location to enrich the visitor's understanding.

The "Athens Museum Explorer" app records user satisfaction and popular destinations as users are asked to vote for their favorite museum. The "Rate your visit" option (Figure 7.4) combined with access to the content of each museum encourages users to engage in a meaningful dialogue with the cultural app community [14], sharing their first-hand experiences and knowledge after visiting museums and events. This interactive feature enables users to leave detailed feedback, provide ratings, and offer valuable recommendations, cultivating a dynamic and vibrant platform for culture lovers to connect, collaborate, and grow together. As users engage with cultural institutions through the "Rate your visit" option, the app inclusively defines the visitor community by providing accessible feedback mechanisms [15]. The app uses satisfaction survey forms with easy-to-understand questions, avoiding jargon and technical language that can be confusing to some users. The app also encourages museums and event organizers to actively seek feedback from users with disabilities, facilitating a continuous process of improving and better adapting cultural experiences to different needs [16]. By prioritizing accessibility at every level, from the presentation of information to user feedback, the cultural app exemplifies a transformative approach to inclusivity,

seeking to empower and enrich the lives of all individuals, regardless of their abilities.

The "Athens Museum Explorer" app combines modern technology with historical richness, enabling users to embark on an educational and exciting journey through the museums of Athens. Combining immersive features such as virtual tours with interactive learning tools and event notifications, this app upgrades the traditional museum visit to a holistic and enriching cultural exploration. Incorporating the principles of the circular economy into our cultural app not only promotes sustainability but also encourages users to re-imagine and reuse cultural content, fostering a more environmentally conscious and creative digital ecosystem [17].

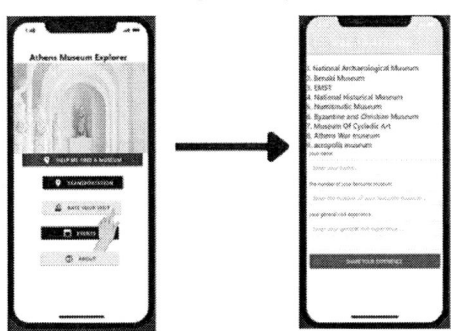

Fig. 7. The "Rate your visit" option.

As a flexible tool, it meets the interests of different categories of users, from tourists looking for an informative experience to students who want to enrich their historical knowledge. With its user-oriented design and emphasis on education and accessibility, the app is a prime example of how cultural apps can foster a deeper appreciation for cultural heritage while embracing the potential of modern technology [18].

The Athens Museum Explorer app acts as a powerful catalyst for the promotion and preservation of cultural heritage in Athens. Through its innovative and interactive features, the app has a profound impact on the promotion of cultural heritage, fostering a deeper understanding of the city's historical significance and artifacts to both local and global audiences. First, the app leverages technology to overcome physical barriers, allowing users from around the world to engage with Athens' cultural treasures before they visit. This accessibility helps democratize cultural heritage, ensuring that knowledge about Athens' museums and historical sites reaches a wider and diverse audience, including those who do not normally have access to such resources. In addition, the role of the app in promoting cultural events and facilitating community participation strengthens the link between museums and their audiences, ensuring the continued appreciation and preservation of Athens' rich cultural heritage. As a pioneer in the promotion of cultural heritage, the Athens Museum Explorer app is a prime example of how technology can revitalize and preserve the legacies of ancient civilizations for future generations. It is noteworthy that the development of this application aims to contribute to tourism and the economic development of the country in the following ways:

1] The ability of the application to improve the overall tourist experience in Athens plays a key role in attracting and retaining visitors [19]. By providing tourists with a seamless and interactive platform to explore the museum's exhibits, objects, and historical landmarks, the app significantly enriches their cultural journey. The integration of virtual tours allows tourists to immerse themselves in the historical grandeur of the city even before they arrive in Greece. Such immersive experiences not only pique curiosity and anticipation but also encourage tourists to extend their stay and explore more of Athens' cultural offerings. The app acts as a comprehensive guide, offering valuable information about the museum's collections, historical context, and significance, creating a more comprehensive and educational experience for visitors, and

2] The positive impact of the application on tourism translates directly into economic growth for Greece. A thriving tourism industry not only generates revenue for the museum but also boosts the entire local economy. Increased tourist visitation leads to higher demand for hospitality services, including hotels, restaurants, and transportation, providing an important source of income and employment opportunities for local businesses and communities [20]. In addition, the promotion from the implementation of cultural events and exhibitions organized by the museum contributes to the growth of ancillary industries such as event management, catering, and souvenir sales. Moreover, as tourists extend their stay to explore more cultural sites facilitated by the app, they are likely to spend more money on accommodation, food, shopping, and entertainment, further contributing to Greece's overall economic growth.

In general, an application dedicated to the museums of Athens is a powerful tool for the promotion of tourism and economic development in Greece. By enhancing the tourist experience and providing access to rich historical and cultural content, the app attracts travelers to immerse themselves in Athens' heritage. As the app attracts more visitors and encourages longer stays, it stimulates economic activities in various sectors, helping to create jobs, increase revenues, and strengthen the local economy [21]. In this symbiotic relationship, the application not only preserves and promotes Athens' cultural heritage but also acts as a driving force for the sustainable tourism and economic prosperity of Greece.

In addition, the app encourages environmentally friendly transport options, helping to reduce carbon dioxide emissions and air pollution. The integration of interactive maps and navigation functions guides tourists through efficient and sustainable transport routes. By promoting walking and public transport options, the app reduces dependence on private vehicles, leading to a greener and more sustainable urban environment. Athens' smart city goals include promoting sustainable transport solutions and the Athens Museum Explorer app actively supports this goal by encouraging users to make environmentally conscious choices.

To further strengthen the Athens Museum Explorer app, it would be beneficial to provide a more explicit and articulated link between the app's technological aspects and its broader cultural and societal impact. While the presentation discusses the app's features and functionalities, it could delve deeper into how the application leverages technology to revolutionize cultural exploration and its profound implications for society. Emphasizing how the app enhances cultural understanding, bridges generational gaps, or fosters intercultural appreciation through technology would make a compelling case for its significance. A clearer connection between the technological innovations and their direct influence on cultural preservation, visitor experiences, and economic growth would underscore

the app's transformative potential, rendering it an even more indispensable asset for Athens and its residents and visitors.

"Athens Museum Explorer" stands as a pioneering exemplar of the transformative power of technology within the context of cultural exploration in a smart city. This pilot application embodies the fusion of cultural heritage with innovation, offering personalized user profiles that dynamically evolve with user preferences, ensuring a cognitive experience that is both enlightening and immersive. Its integration of a robust multi-language support system not only fosters accessibility for global tourists but also serves as a testament to the commitment of Greek tourism policy to fostering intercultural exchange. As we navigate the digital age, "Athens Museum Explorer" not only exemplifies the potential for technology to breathe new life into our historical and cultural treasures but also underscores the role of smart city initiatives in harnessing technology for the betterment of society, enriching our understanding of the past, and forging connections across cultures. In the grand tapestry of urban innovation, this application stands as a shining thread, weaving together history, technology, and a vibrant future.

Finally, the Athens Museum Explorer app is a prime example of how smart city initiatives can be seamlessly integrated into cultural tourism. The app's digital approach reduces waste, optimizes tourist flows, and encourages environmentally friendly transport options, all of which contribute to the city's wider sustainability efforts. By leveraging technology to enhance the cultural experience while promoting responsible tourism practices, the app is perfectly aligned with Athens' vision of becoming a sustainable and digitally advanced city. Through such innovative solutions, the Athens Museum Explorer app sets a positive precedent for other smart city projects, highlighting how technology [22] can promote sustainable development while preserving and promoting cultural heritage.

Acknowledgment

This work has been partly supported by COSMOTE under a PEDION24 grant. (University of Piraeus Research Center).

References

[1] D. M. El-Sherif, and E. E. Khalil, Policy instruments for facilitating smart city governance. Chapter 22 at J. R. Vacca, Smart Cities Policies and Financing. Elsevier. 2022, Pp. 305–317.

[2] R. Unger. A project guide to UX design: For user experience designers in the field or in the making (Second edition). New Riders.2012.

[3] H. M. Kim, S. Sabri, and A. Kent, Chapter 2—Smart cities as a platform for technological and social innovation in productivity, sustainability, and livability: A conceptual framework. at H. M. Kim, S. Sabri, & A. Kent, Smart Cities for Technological and Social Innovation. Academic Press. 2022, pp 9-28.

[4] L. G. Anthopoulos. Understanding smart cities: A tool for smart government or an industrial trick? Cham, Switzerland: Springer International Publishing, 2017, p. 293.

[5] S. Barile, M. V. Ciasullo, F. Iandolo, and G. C. Landi, The city role in the sharing economy: Toward an integrated framework of practices and governance models. Cities, 2021, pp. 119.

[6] S. Y. Christensen, J. Dickinson, K. Machac, and H. Cline. Define UX Design. 2020 Intermountain Engineering, Technology and Computing (IETC), 2020, pp. 1–5.

[7] Diamantaki, Katerina, et al. 'Evaluating the User Experience of a Mobile User in a Smart City Context'. International Journal of Intelligent Engineering Informatics, vol. 3, no. 2/3, 2015, p. 120.

[8] Ferrer-Mavárez, María De Los Ángeles, et al. 'Applicability of the User Experience Methodology: Communication and Employment Web

Portal for Older Adults'. Media and Communication, vol. 11, no. 3, June 2023.

[9] C. F. Fratini, S. Georg and M. S. Jørgensen. Exploring circular economy imaginaries in European cities: A research agenda for the governance of urban sustainability transitions. Journal of Cleaner Production, 228, 2019, 974–989.

[10] Gasco-Hernandez, M., Nasi, G., Cucciniello, M., & Hiedemann, A. M. The role of organizational capacity to foster digital transformation in local governments: The case of three European smart cities. Urban Governance.2022.

[11] Heinlein, Alexander, et al. 'Grounding Human-Object Interaction to Affordance Behavior in Multimodal Datasets'. Frontiers in Artificial Intelligence, vol. 6, Jan. 2023, p. 1084740.

[12] Jedliński, Mariusz. 'The Position of Green Logistics in Sustainable Development of a Smart Green City'. *Procedia - Social and Behavioral Sciences*, vol. 151, 2014, pp. 102–11.

[13] Karr, Ashley. 'UX Research vs. UX Design'. *Interactions*, vol. 22, no. 6, Oct. 2015, pp. 7–7.

[14] Komninos, N., Kakderi, C., Collado, A., Papadaki, I., & Panori, A. (2022). Digital transformation of city ecosystems: Platforms shaping engagement and externalities across vertical markets. In Sustainable Smart City Transitions (pp. 91-112). Routledge.

[15] Krupiy, T. T. A vulnerability analysis: Theorising the impact of artificial intelligence decision-making processes on individuals, society and human diversity from a social justice perspective. Computer law & security review, 38, 2020, 105429.

[16] Landsbergen, David, et al. 'Governance Rules for Managing Smart City Information'. *Urban Governance*, vol. 2, no. 1, 2022, pp. 221–31.

[17] Li, Xiaoting, et al. 'Development of Circular Economy in Smart Cities Based on FPGA and Wireless Sensors'. Microprocessors and Microsystems, vol. 80, 2021, p. 103, 600.

[18] Lim, C., Kim, K. J., & Maglio, P. P. Smart cities with big data: Reference models, challenges, and considerations. Cities, 82, 86-99.Marginson, S. (2022). What drives global science? The four competing narratives. Studies in Higher Education, 47(8), 2018, 1566-1584.

[19] Moore, Robert J., et al. 'Conversational UX Design'. Proceedings of the 2017 CHI Conference Extended Abstracts on Human Factors in Computing Systems, ACM, 2017, pp. 492–97.

[20] Nicolas-Agustin, A., Jiménez-Jiménez, D., & Maeso-Fernandez, F. (2022). The role of human resource practices in the implementation of digital transformation. International Journal of Manpower, 43(2), 395-410.

[21] Park, Joon, and Seungho Yoo. 'Evolution of the Smart City: Three Extensions to Governance, Sustainability, and Decent Urbanisation from an ICT-Based Urban Solution'. International Journal of Urban Sciences, vol. 27, no. sup1, 2023, pp. 10–28.

[22] Smith, Adrian, and Pedro Prieto Martin. Going Beyond the Smart City? Implementing Technopolitical Platforms for Urban Democracy in Madrid and Barcelona. 2022, p. 20.

Precision education through the eyes of humanistic teaching in the L2 classroom

Ioanna Moustaka
Department of Informatics
Ionian University
Corfu, Greece
imoustaka@hotmail.com

Spyridon Doukakis
Department of Informatics
Ionian University
Corfu, Greece
sdoukakis@ionio.gr

Marina Mattheoudakis
School of English
Aristotle University of Thessaloniki
Thessaloniki, Greece
marmat@enl.auth.gr

Abstract—In the fast-paced lifestyle of the 21st century, individuals are increasingly reliant on a variety of technological devices, particularly mobile devices, as integral parts of their everyday routines. Likewise, the realm of education is actively working to incorporate and leverage the swiftly advancing technology to enrich the process of learning. This has given rise to the concept of mobile learning (m-learning), where the educational sector seeks to harness the potential of mobile devices to enhance student's learning experience. Upon these premises, precision education emerges; the optimal learning subject is transformed into a collection of quantifiable measures, capable of being computed, improved, influenced and even forecast. This command over shaping a future-ready educational subject stems from insights generated on digital platforms, coupled with data originating from behavioral and life sciences, with a specific emphasis on each student's individual learning style and preferences. Notwithstanding, there are challenges inherently linked with the use of MALL for educative purposes, and that is what we intend to examine. Following a similar train of thought, our endeavor aims to guarantee that second language (L2) learners extract the utmost educational advantages by capitalizing on personalised technological advancements (precision education) and methodologies. Simultaneously, the teaching and learning process will be enriched and completely re-imagined as a positive journey from a humanistic standpoint, since all the potential challenges will be highlighted and spotted.

Index Terms—Precision Education, Humanistic Teaching, Mobile Learning, Second Language Teaching

I. INTRODUCTION

The need for English L2 learners with developed communication skills in the target language has become more acute over the past years mainly due to globalization [1]. The global count of English language learners (ELL) has surged significantly on a worldwide scale. With more than 1 billion learners universally, as stated by McKay in 2012 [2], English holds the title of the most extensively instructed foreign language, solidifying its position as the primary global communication medium. Concurrently, technology has revolutionized all facets of human existence, including language learning and acquisition. Having introduced a fresh phase in education, technology has undoubtedly rendered learning highly creative, even more adaptable and exploratory. Halverson and Smith contend that technology enhances effectiveness and efficiency in the learning processes [3]. This is true since applications and mobile devices, seamlessly integrated into our daily routines, serve various functions, incorporating educational aims as well. Consequently, they play a vital role in the journey of acquiring foreign language skills [4]. According to [5], the field of second language (L2) instruction has witnessed significant interest in mobile learning (often abbreviated as m-learning) due to the latest advancements in mobile technology. That being the case, utilizing mobile devices for L2 instruction has emerged as a prominent trend in technology-driven language learning research and practice [6].

Any instance in which students employ a computer resulting in the enhancement of language skills is categorized as Computer Assisted Language Learning or simply CALL [7]. This term gained prominence in language education during the early 1980's, primarily involving desktop computers with rudimentary software programs. Over time, this domain has expanded to encompass applications, vlogs (video blogs) and online blogs, online courses, virtual learning environments and so many more.

However defining mobile learning, which is a broader concept that includes all forms of learning that occur through mobile devices, presents challenges due to the inherent ambiguity surrounding the specific technological instruments involved such as laptops, tablets, computers and the like, and the contexts in which and times at which they are deployed. Broadly speaking, mobile learning, often referred to as m-learning, pertains to the utilization of portable mobile devices such as smartphones, tablets, laptop computers, and wearable technology and encompasses a more comprehensive range of educational activities and subjects conducted through mobile devices [8]. These devices play a facilitating role in the learning process, supporting a wide range of educational activities conducted not only within traditional classroom settings but also beyond them. The emergence and widespread usage of mobile devices has given rise to a novel field known as Mobile-Assisted Language Learning (MALL), which is utilized in both formal and informal second language learning [9]. MALL made its maiden appearance around 2005, as [10] tells us, when several universities in the United States started distributing complimentary mobile devices to their students. The global recognition of this trend gained traction around the year 2009, marked by the British

979-8-3503-8368-3/23 $31.00 © 2023 IEEE

Council's creation of mobile applications (apps) designed for language learning [11]. The significant contribution of major publishers in English language teaching (ELT), who produced independent applications or those associated with coursebooks, played a pivotal role in hastening the worldwide expansion of Mobile-Assisted Language Learning (MALL) [12].

Unlike CALL, MALL is characterized by personal use and the utilization of portable devices, enabling novel modes of learning for the second language (L2) student. It places emphasis on mobility; this allows for spatial and temporal shifts that create expanded learning opportunities, even in the language learning classroom, compared to traditional face-to-face language learning and paper and pencil method. Although personal portable devices have unarguably been in use for quite a while [13], MALL remains a research domain with great potential and capacity.

A crucial insight, evident across all the published MALL research, is that students universally hold a highly positive perception of utilizing mobile technology due to its convenient "anywhere, anytime" accessibility. As stated in [14], the favorable student outlook on mobile technology has significantly fueled endeavors to incorporate MALL. Especially among the student demographic, the widespread ownership of mobile phones has established them as the preferred technological medium for MALL developers.

Scholars over the years have explored novel methodologies so as to embody computer-assisted programs into second language learning [15]. Golonka et al. assessed diverse technological modalities and their consequent efficacy in second language learning [16]. Their review highlighted the effectiveness of computer-assisted pronunciation training in enhancing students' pronunciation, while offering valuable feedback at the same time.

Authors in [17] conducted a thorough examination of research related to Mobile Assisted Language Learning (MALL), with a specific focus on speaking and listening aspects. Their review encompassed various categories of mobile devices, such as mobile phones, tablet PCs, MP3 players and other similar technologies.

In a parallel fashion, authors in [18] undertook a study examining the impact of mobile technology and texting on second language learning (L2), revealing that students derived enjoyment while developing their vocabulary through their mobile phones. They actually "learned new words with the help of their mobile phones!" (p. 78).

Through the utilization of authentic texts on the Internet and multimedia-enhanced computer-assisted language learning platforms, language learners are presented with unparalleled prospects to cultivate second language literacy skills and foster intercultural comprehension. Bearing this context in mind, the author in [19] conducted a meta-analysis of 11 studies investigating the impact of computer-mediated glosses in second language reading comprehension and vocabulary learning. Based on the outcomes of this meta-analysis and considering the general attributes of gloss-related research, he put forth suggestions for forthcoming investigations. These recommendations encompass the replication of studies and the methodical exploration of reading variables and learner-specific disparities within multimedia learning settings featuring authentic texts.

Wang et al. introduced a situated computer game into English language classes for sixth graders, evaluating the efficacy of various guidance strategies in bolstering English vocabulary [20]. Addressing reading comprehension issues, Liu et al. developed context-aware learning environments on mobile platforms to enhance second language learners' reading comprehension abilities [21].

Along similar lines in their study, authors in [22] assessed the utilization of an innovative MALL-P (Mobile Aided Learning - Portal) in Philippine tenth graders for learning English as L2. The feedback they received from the students regarding the use of the portal was quite positive; they characterized it as "helpful", "easy", "understandable", "fun" and "enjoyable" and further asserted they had a "meaningful learning experience".

With the growing sophistication and accessibility of mobile technologies, scholars like Hwang and Wu propose that employing mobile apps to support English language learning proves to be a suitable and beneficial approach [23].

So apart from the conventional use of mobile phones for verbal communication, the contemporary versatile mobile technology empowers users to seamlessly connect to the Internet for educational purposes. This accessibility facilitates tasks like completing homework and studying, watching educational videos and e-book reading (read alouds). This newfound mobility has additionally unlocked the potential for learning beyond physical confines and time constraints, enabling second language learning (L2) and education beyond classroom settings [24].

In summary, the collective results indicated a positive impact of Mobile-Assisted Language Learning (MALL) on students' English language learning. Every study reached the consensus that the utilization of diverse MALL applications and devices produced favorable outcomes. The researchers noted an enhancement in students' motivation, academic progress, active participation and focused learning time. Earlier research had also hinted at the potential of technology integration for targeted student development, and the above mentioned findings provided substantial support for that notion.

II. AFFECTIVE FILTER HYPOTHESIS AND MOTIVATION IN SERVICE OF PRECISION EDUCATION

The emulation of interactive processes within an application stands as a pivotal element in arousing learner engagement. This form of stimulation holds the capacity to captivate learners, compelling them to interact with contextualised input actively. Hence, a crucial consideration lies in evaluating whether an app incorporates mechanisms for furnishing constructive feedback or enabling users to autonomously self-correct their responses within the learning tasks featured in the app.

Drawing from Krashen's affective filter hypothesis [25], emotional factors can divert the attention of language learners

979-8-3503-8368-3/23 $31.00 © 2023 IEEE

during language acquisition. Krashen asserts that factors like self-esteem issues, overwhelming anxiety and low levels of motivation hinder the utilization of comprehensible input for acquisition. To phrase it differently, when the filter is raised, it obstructs the progress of language acquisition. Conversely, a positive emotional state is a prerequisite for acquisition, though not solely sufficient. Consequently, a well-crafted mobile learning (m-learning) application should diminish the impact of the affective filter, enabling users to actively engage in assigned tasks. These interactions have the potential to decrease learners' anxiety since they work singularly stimulated by the responses they receive from the learner who is under scrutiny and the center, the attention of the whole process. Since the learning tasks will be personalised and adapted to each individual learner, there are high chances for their motivation and self esteem levels to be boosted. Dealing with the environment issues that could additionally cause anxiety and uneasiness to the students, successful outcomes could be produced. Hence, it becomes imperative to consider motivation as a crucial element while assessing mobile learning (m-learning) apps designed for language learning.

Motivation is definitely a notion difficult to examine since it is multifaceted and it refers to a complex set of constructs that relate to social alongside with psychological issues and perceptions [26]. It is the desire that initiates the longing to learn an L2, accompanied by the effort we put to sustain it. But for motivation to be fruitful, it needs to stem from the heart, in other words it needs to be intrinsic. Students that are intrinsically motivated according to [27] are associated with higher levels of academic achievement. They are stimulated and knowledge-driven, they experience joy through learning the language which stems from a genuine desire to become acculturated and be more like their target language group members.

So recapitalising, it is evident that learners who possess higher levels of motivation exhibit more favorable receptivity to knowledge acquisition. This underscores the significance of tailored pedagogical approaches, wherein the provision of personalized and progressively challenging assignments can be characterized as a pivotal facet. This approach, if characterised as customised, engenders a sense of motivation among students. Consequently, the execution of personalised and tiered tasks not only ensures their successful accomplishment but also mitigates the likelihood of encountering disillusionment or sentiments of inadequacy. The cumulative effect of these strategies not only nurtures and sustains high motivational levels but also propels learners toward the attainment of their scholastic objectives in mastering the target second language (L2).

III. Does MALL have a Pedagogical Dimension?

Mobile applications are crafted to deliver content to their users, necessitating the evaluation of such applications to be firmly grounded in instructional design theories and frameworks. Reeves [28] delineated fourteen dimensions for appraising diverse forms of computer-based education which encompass the following sectors: motivation, pedagogical philosophy, underlying psychology, experiential value, epistemology, teacher role, program flexibility, error handling significance, accommodation of individual differences, learner control, user engagement, cooperative learning, goal orientation and cultural sensitivity.

Over the last decades, this evaluative structure has set the basis for researchers and instructional designers to observe certain guiding principles and to scrutinize and create instructional materials for computer-based programs. This approach has also been adapted for more specialised domains, including distance education [29], instructional applications [30], educational software and programs (Rodríguez et. al, 2012) and Massive Online Open Courseware [31].

As emphasized in [32], a thorough examination of the instructional material or program's content, tasks and context is very crucial. Therefore, the assessment of content quality in applications holds paramount importance. Considering the learning nature of these apps and simultaneously evaluating the pedagogical consistency of language skills within the learning activities, becomes imperative.

Given that apps are software integrated into mobile devices, evaluating their usability, customization options and sharing capabilities becomes a valuable aspect of assessment. In summation, seven key elements emerge for the evaluation of language-learning mobile apps: content excellence, pedagogical consistency in language skills, feedback and self-correction mechanisms, motivation, usability, customization potential and sharing functionalities.

Despite the reported benefits of MALL (Mobile-Assisted Language Learning) in delivering effective English language instruction, it is important to acknowledge that its adoption has not been uniform across all contexts. Besides the so many and valuable additions it provides to education, research has highlighted various issues, challenges and limitations associated with the integration of MALL in teaching the English language. For instance, authors in [33] identified several challenges encountered when employing MALL for teaching English, encompassing difficulties equally faced by educators and learners such as cultural and ethical concerns, challenges tied to mobile devices and applications, as well as obstacles linked to specific regional conditions within Indonesia. These challenges encompass factors such as educators' and students' limited digital literacy, reluctance to embrace technological tools and financial constraints related to acquiring mobile devices.

In addition to observations in [33], further investigations have shed light on the limitations associated with the implementation of MALL. Kukulska et al. also brought to light some adverse aspects, including concerns about distractions, safety issues, feelings of uncertainty and technological difficulties [34]. These affective factors underscore the multifaceted nature of challenges related to the use of MALL in education.

Expanding on the challenges, authors in [35] contribute additional insights, highlighting psychological constraints, pedagogical shortcomings, technical restrictions and the issue of screen size. Without doubt, the limitations of small screen sizes and the challenges of text input methods restrict the quantity of data that can be presented to users along with the extent of learner responses that can be anticipated [36]. Beyond the school environment, there's a propensity for students to lean towards utilizing computers with educational software and fast internet connectivity for learning purposes, as opposed to mobile phones, which are predominantly employed for interpersonal communication rather than educational pursuits. Educators encounter challenges when it comes to conducting assessments. The mode of interaction is intrinsically tied to the dimensions of the interactive communication device's screen, signifying its pivotal role.

Additionally, research conducted in [37] revealed that there was an absence of consistent motivation, self-monitoring and self-management by learners, which inevitably led to limited utilization of applications. Referring to similar challenges, Jones et al. indicated that the size of the screen used by learners has an impact on the speed of internet search tasks [38]. Furthermore, the effectiveness of learning through mobile devices is significantly shaped by the availability of internet connectivity, which can again present challenges like unstable links or application malfunctions, as noted in [39].

IV. THROUGH THE EYES OF HUMANISM

However as technology popularity grew, classroom attention spans have started to decrease. According to [6], while mobile learning is definitely not a novel concept, the introduction of new devices with advanced capabilities has significantly heightened interest, especially among language educators.

In sight of all the above mentioned, and taking into consideration the holistic aspect of teaching, a step concerning a more humanistic approach towards educating students nowadays is precision education.

A humanistic-oriented management of education necessitates an approach that unlocks the potential of each individual within an evolving world. This educational evolution, aiming to actualise individuals' potential, is encapsulated by the concept of precision education [40], [41]. Precision education represents an emerging paradigm that emphasises individually tailored, highly efficient and personalised approaches to teaching and learning [42], [43], [44]. By pledging scientific, evidence-based, customised and individualised learning, precision education emerges as a comprehensive remedy to the perceived limitations and challenges associated with conventional schooling and education.

Drawing inspiration from the foundational principles of precision medicine that seeks treatment for ailments that consider patients' behavioral, emotional, neurological, biological and genetic makeup [40], precision education shares a parallel aim of providing personalised teaching according to each student's uniquely individual idiosyncrasy. Consequently, the essence of precision education can be articulated by adapting the definition of precision medicine proposed by Ferryman and Pitcan (as cited in [44]); it involves the systematic collection, integration and analysis of a diverse array of genetic and non-genetic data. Employing techniques derived from big data analysis and machine learning, precision education seeks to uncover insights pertaining to teaching, education and learning that are precisely tailored to the unique characteristics of each individual.

Notwithstanding, by integrating knowledge and conclusions related to psychological and biological parameters into teaching methods, which arise from the brain's transformations during learning, new horizons are opened.

Chen's (2016) evaluative research employs a theory-driven rubric to appraise the capabilities of language learning applications tailored for adult L2 learners. The findings indicate that a universal language-learning application capable of catering to the diverse and unique needs of adult L2 learners does not exist. Nevertheless, the assessment underscores that mobile learning applications furnish numerous avenues and forms through which adult learners can better refine their language abilities. By skillful instructional planning and applying precision educative methods, these mobile apps can be seamlessly integrated into curricula or language-learning modules, offering students an opportunity to elevate their language skills and making the whole educative process more empowering; with perspective and driven by motivation.

In a similar vein, the engagement of students in decision-making about their education poses a challenge to an inherently authoritative system where education is overseen by educators, politicians and other figures. As reported in [45], this situation leads to an examination of opposing interests that imbue significance into and shape student involvement. A lens of humanistic critical viewpoint serves as the means to delve into this exploration.

The notion that humans possess an innate creativity suggests that they naturally manifest this creativity through their actions, unless such resourceful capacity is hindered or eradicated. The act of generating value is intrinsic to our human essence. When we commend individuals for their 'resilience and moral fiber,' we are essentially recognizing their exceptional aptitude for generating value [45].

This is exactly why precision education that empowers each person to view learning within the framework of their unique characteristics is so powerful.

V. CONCLUSIONS

The rapid expansion of technological devices and information technologies, along with their widespread application across various domains, serves as a catalyst and source of inspiration for educators. This encouragement prompts them to integrate these tools into their instructional designs, aiming to elevate the learning process through enhanced educational methods. Shaping the future of education requires policy-making grounded in "objective" and "evidence-based" knowledge.

Despite the proven efficacy of mobile learning for content delivery, instructional applications in Mobile-Assisted Language Learning (MALL) still encounter the educational hurdle of delivering feedback and tracking learners' progress [14]. The feedback for responses typically remains confined to presenting a correct answer or offering a "Right/Wrong" assessment. Except for retrospective tests of effectiveness, the monitoring of students' performance by instructors has been nearly absent, and this is a sector we could apply ourselves with as future work.

Several factors contribute to the predominant one-way flow of information from teacher to learner in the majority of tutorial Mobile-Assisted Language Learning (MALL) applications. However, the primary reason for this dynamic is the absence of the necessary infrastructure, both in terms of technology and personnel, to facilitate two-way communication between learners and instructors within a mobile setting. Consequently, only a limited number of initiatives have made an effort to tackle the challenge of establishing systematic feedback mechanisms and comprehensive student monitoring [14].

While neuroscience might not offer immediately implementable recommendations for enhancing classroom instruction, it has the potential to indirectly impact and shape the field of education [46]. Chang et al. [47], however, believe that by incorporating insights and findings from psychological and biological factors into instructional approaches, which stem from the adaptations that occur in the brain during the learning process, new opportunities are unveiled.

Due to the widespread ownership of mobile phones, tablets, laptops, PCs and the like globally and the continuous enhancement of their features along with their suitability for educational purposes, Mobile-Assisted Language Learning (MALL) is expected to play a progressively vital role in aiding the teaching of foreign languages in the imminent future. It is anticipated that many educators will contribute to this future development, as the effectiveness of technology relies heavily on the quality of the teaching methods. In this regard, the knowledge and innovative approaches of language educators cannot be replaced, emphasizing their irreplaceable role in shaping successful pedagogical technology.

Learning should refrain from being coercive or obligatory. Instead of forcefully imparting information to students, there should be a focus on nurturing logical thinking and fostering moral and aesthetic values, humanistic aspects of the educative procedure that is, starting from the early stages of education up to the very college level [45]. What we need to change in the educative approach is not to exclusively produce future employees or managers but human game changers, students who can think differently. These values are inevitably enforced by the practice of the humanistic aspect of precision education, since it is holistic in its core, tailor-made and singularly targeted towards each student individually.

REFERENCES

[1] X. Chen, "Evaluating language-learning mobile apps for second-language learners," *Journal of Educational Technology Development and Exchange (JETDE)*, vol. 9, no. 2, p. 3, 2016.

[2] S. L. McKay, "Teaching materials for english as an international language," *Principles and Practices of Teaching English as an International Language*, vol. 3, no. 9, pp. 70–83, 2012.

[3] R. Halverson and A. Smith, "How new technologies have (and have not) changed teaching and learning in schools," *Journal of Computing in Teacher Education*, vol. 26, no. 2, pp. 49–54, 2009.

[4] R. Gafni, D. B. Achituv, and G. J. Rachmani, "Learning foreign languages using mobile applications," *Journal of Information Technology Education: Research*, vol. 16, p. 301, 2017.

[5] G. Stockwell, "Using mobile phones for vocabulary activities: Examining the effect of platform," *Language Learning and Technology*, vol. 14, no. 2, pp. 95–110, 2010.

[6] R. Godwin-Jones, "Emerging technologies: Mobile apps for language learning," *Language Learning & Technology*, vol. 15, pp. 2–11, 2011.

[7] H. Jarvis and M. Achilleos, "From computer assisted language learning (CALL) to mobile assisted language use (MALU)," *TESL-EJ*, vol. 16, no. 4, p. 4, 2013.

[8] I. Moustaka, S. Doukakis, and M. Mattheoudakis, "The use of mobile applications in the l2 learning classroom: Is it worth the while?" in *International Conference on Interactive Mobile Communication, Technologies and Learning (IMCL)*, 2023.

[9] X.-B. Chen, "Tablets for informal language learning: Student usage and attitudes," *Language Learning and Technology*, vol. 17, pp. 20–36, 2013.

[10] G. M. Chinnery, "Going to the mall: Mobile assisted language learning," *Language Learning & Technology*, vol. 10, no. 1, pp. 9–16, 2006.

[11] N. Hockly, "Mobile learning," *ELT Journal*, vol. 67, no. 1, pp. 80–84, 2013.

[12] G. Dudeney and N. Hockly, "Ict in elt: how did we get here and where are we going?" *ELT Journal*, vol. 66, no. 4, pp. 533–542, 2012.

[13] R. Gafni, "Measuring quality of m-learning systems," *Informing Science Press*, 2009.

[14] J. Burston, "Exploiting the pedagogical potential of mall," *Mobile Learning as the Future of Education*, 2011.

[15] C. A. Chapelle, "The relationship between second language acquisition theory and computer-assisted language learning," *The Modern Language Journal*, vol. 93, pp. 741–753, 2009.

[16] E. M. Golonka, A. R. Bowles, V. M. Frank, D. L. Richardson, and S. Freynik, "Technologies for foreign language learning: A review of technology types and their effectiveness," *Computer Assisted Language Learning*, vol. 27, no. 1, pp. 70–105, 2014.

[17] A. Kukulska-Hulme and L. Shield, "An overview of mobile assisted language learning: From content delivery to supported collaboration and interaction," *ReCALL*, vol. 20, no. 3, pp. 271–289, 2008.

[18] N. Cavus and D. Ibrahim, "m-learning: An experiment in using SMS to support learning new english language words," *British Journal of Educational Technology (BJET)*, vol. 40, no. 1, pp. 78–91, 2009.

[19] L. B. Abraham, "Computer-mediated glosses in second language reading comprehension and vocabulary learning: A meta-analysis," *Computer Assisted Language Learning*, vol. 21, no. 3, pp. 199–226, 2008.

[20] Z. Wang, G. Hwang, Z. Yin, and Y. Ma, "A contribution-oriented self-directed mobile learning ecology approach to improving EFL students' vocabulary retention and second language motivation," *Journal of Educational Technology & Society*, vol. 23, no. 1, pp. 16–29, 2020.

[21] G. Liu, G. Hwang, Y. Kuo, and C. Lee, "Designing dynamic english: A creative reading system in a context-aware fitness centre using a smart phone and QR codes," *Digital Creativity*, vol. 25, no. 2, pp. 169–186, 2014.

[22] T. M. Mengorio and R. Dumlao, "The effect of integrating mobile application in language learning: An experimental study," *Journal of English Teaching*, vol. 5, no. 1, pp. 50–62, 2019.

[23] G. Hwang and P. Wu, "Applications, impacts and trends of mobile technology-enhanced learning: A review of 2008-2012 publications in selected SSCI journals," *International Journal of Mobile Learning and Organisation*, vol. 8, no. 2, pp. 83–95, 2014.

[24] J. Yang, "Mobile assisted language learning: Review of the recent applications of emerging mobile technologies," *English Language Teaching*, vol. 6, no. 7, pp. 19–25, 2013.

[25] S. Krashen, "Principles and practice in second language acquisition," 1982.

[26] E. R. Lai, "Motivation: A literature review," *Person Research's Report*, vol. 6, pp. 40–41, 2011.

[27] M. Vansteenkiste, W. Lens, and E. L. Deci, "Intrinsic versus extrinsic goal contents in self-determination theory: Another look at the quality

of academic motivation," *Educational Psychologist*, vol. 41, no. 1, pp. 19–31, 2006.

[28] T. Reeves, "Evaluating what really matters in computer-based education," *Computer Education: New Perspectives*, pp. 219–246, 1994.

[29] M. T. Eskey and H. Roehrich, "A faculty observation model for online instructors: Observing faculty members in the online classroom," *Online Journal of Distance Learning Administration*, vol. 16, no. 2, pp. 1–14, 2013.

[30] T. S. Chener, C. Lee, A. Fegely, and L. A. Santaniello, "A detailed rubric for assessing the quality of teacher resource apps," *Journal of Information Technology Education Innovations in Practice*, vol. 15, pp. 117–143, 2016.

[31] W. Admiraal, B. Huisman, and O. Pilli, "Assessment in massive open online courses," *Electronic Journal of E-learning*, vol. 13, no. 4, pp. 207–216, 2015.

[32] P. L. Smith and T. J. Ragan, *Instructional Design*. John Wiley & Sons, 2004.

[33] S. Solihin, "Using mobile assisted language learning (mall) to teach english in indonesian context: Opportunities and challenges," *VELES (Voices of English Language Education Society)*, vol. 5, no. 2, pp. 95–106, 2021.

[34] A. Kukulska-Hulme and O. Viberg, "Mobile collaborative language learning: State of the art," *British Journal of Educational Technology*, vol. 49, no. 2, pp. 207–218, 2018.

[35] D. P. Sam and R. Shalini, "Limitations and advantages in implementing mall in the tertiary esl classrooms: A review," *International Journal of Recent Technology and Engineering*, vol. 9, no. 5, pp. 27–32, 2021.

[36] E. Brown, "Mobile learning explorations at the stanford learning lab," *Speaking of Computers*, vol. 55, pp. 112–120, 2001.

[37] G. G. Botero, F. Questier, and C. Zhu, "Self-directed language learning in a mobile-assisted, out-of-class context: Do students walk the talk?" *Computer Assisted Language Learning*, vol. 32, no. 1-2, pp. 71–97, 2019.

[38] M. Jones, G. Buchanan, and H. W. Thimbleby, "Improving web search on small screen devices," *Interacting with Computers*, vol. 15, no. 4, pp. 479–495, 2003.

[39] Z. Zaitun, M. S. Hadi, and E. D. Indriani, "Tiktok as a media to enhancing the speaking skills of efl student's," *Jurnal Studi Guru Dan Pembelajaran*, vol. 4, no. 1, pp. 89–94, 2021.

[40] S. A. Hart, "Precision education initiative: Moving toward personalized education," *Mind, Brain, and Education*, vol. 10, no. 4, pp. 209–211, 2016.

[41] P. K. Kuhl, S.-S. Lim, S. Guerriero, and D. van Damme, "Developing minds in the digital age: Towards a science of learning for 21st century education," *Educational Research and Innovation*, 2019.

[42] K. Brunila, J. Honkasilta, E. Ikävalko, M. Lanas, A. Masoud, K. Mertanen, and K. Mäkelä, "The cultivation of subjectivity of young people in youth support systems," *The Routledge International Handbook of Global Therapeutic Cultures*, 2020.

[43] K. J. Saltman, "Artificial intelligence and the technological turn of public education privatization: In defence of democratic education," *London Review of Education*, vol. 18, no. 2, pp. 196–208, 2020.

[44] B. Williamson, "Digital policy sociology: Software and science in data-intensive precision education," *Critical Studies in Education*, vol. 62, no. 3, pp. 354–370, 2021.

[45] R. N. Swain and B. C. M. Patnaik, "Dying out b schools and their survival through value education," *A Journal of Economics and Management*, vol. 2, no. 3, 2013.

[46] E. Lekati and S. Doukakis, "Neuroeducation and mathematics: The formation of new educational practices," in *Worldwide Congress on "Genetics, Geriatrics and Neurodegenerative Diseases Research"*, 2022, pp. 91–96.

[47] Z. Chang, M. S. Schwartz, V. Hinesley, and J. M. Dubinsky3, "Neuroscience concepts changed teachers' views of pedagogy and students," *Frontiers in Psychology*, vol. 12, p. 685856, 2021.

979-8-3503-8368-3/23 $31.00 © 2023 IEEE

Cybersecurity and Democracy: A Review

Fotios I. Roumpies, Phd candidate
Computer Science and Biomedical Informatics
University of Thessaly
Lamia, Greece

Athanasios P. Kakarountas, Assosiate Professor
Computer Science and Biomedical Informatics
University of Thessaly
Lamia, Greece

Abstract—The threats presented in Cyberspace, through the ever-increasing human networking in it, result in the conduct of Cyberattacks by state and non-state actors on the vital infrastructure of society, but also the manipulation of public opinion through disinformation of citizens. This research paper aims to substantiate the reasons why Cybersecurity is an important factor, both for the proper functioning of the internet and its dependent activities and material, as well as for avoiding the alteration of democratic institutions and digital freedoms of the individual.

Index Terms—Malware, Cyberspace, Cybersecurity, Democracy, Disinformation.

I. INTRODUCTION

The Internet has a tremendous impact on culture, education, and trade and has introduced communication via e-mail, blogs, and social networking services. It has also contributed to the creation of a digital space known as Cyberspace, while enabling various actors to develop multiple Cyber threats. Private and public infrastructures constitute targets for cyber actors, which threaten National Security, Public Order, and Democratic Institutions.

This paper is divided into five distinct parts. The first part contains a brief, presentation of Cyberspace. The second part focuses on Cybersecurity, cyberattacks, techniques, actors, and their impact on individuals. The third part focuses on altering democratic functions through digital disinformation. It mainly analyses the mechanism of influencing public opinion and, democracy, through digital sources of infotainment. The fourth part focuses on response measures, and legislative framework, to enhance cybersecurity and confront online disinformation. The fifth part presents conclusions and proposals.

II. CYBERSECURITY

A. Cyberthreats, Cyberattacks and Malware

A methodical approach to the related terms happened through the study of Susan W. Brenner, on data collected during a malicious cyber action that took place during a 2006 attack on computer systems of a sensitive U.S. Department of Commerce. In her article, analyses the data collected after the investigations. Finally, a group of websites originating from Chinese internet service providers was detected as the origin of the attack. Still, it was impossible to identify those responsible for the attack, as much as the nature of the attack. This incident projected anonymity and impunity of the attackers in cyberspace. In addition, questions were raised to specify the kind of attack (act of Cybercrime, Cyberterrorism, Cyberwar, or something else) In the article, cyber threat was defined as the use of computer technology to engage them in activities that stimulate society's ability to maintain its internal and external order. On one hand, the internal order of a society is maintained by establishing, and implementing rules so that any deviant behaviour is punishable by law and prosecuted by law enforcement agencies. On the other hand, external order is practically shielded by the Armed Forces and interstate agreements.

This dual state of internal-external threat, and the choice between law enforcement and military power, determines the dynamic response of the state to any impending attack. This hypothesis is based on the fact that every society occupies a geographical area and that the concepts of state entity, its "territory", but also its "sovereignty" are inseparable. Also, threats to the social order are defined as internal in case of crime-terrorist threat or as external in case of a war act. However, the use of computer systems erodes the above binary approach, as it makes the concepts of state-territory indistinguishable. Therefore, various actors (State-sponsored or not) that seek to undermine society's capacity to uphold order and perplex governmental systems on the assignment of action to the right services, may launch cyberattacks [4].

From another approach, cyberattack means any type of malicious activity that attempts to collect, disrupt, deny, degrade, or destroy information system resources or information. Cyber intruders mainly use malware, which injects corrupted code into existing software and eventually forces PCs to perform actions or processes unintentionally by their operators. There are several types of malware, each with its mode of operation, the most representative being the following [46]:

(a) Virus and Worm, spread "infected" programs, damaging useful files [44].

(b) Spyware, monitors users' activity and transmits their confidential data to the attackers without the users' knowledge or agreement (e.g.adware, mobile spyware, Cookie trackers, etc) [25].

(c) Trojan Horse, perform harmful actions in the background, in the form of screensavers, useful programs or files etc [1].

(d) Ransomware is used for extortion. When attacking a device, it encrypts the data stored on the disk, and then a demand of ransom is sent to the user [28].

979-8-3503-8368-3/23 $31.00 © 2023 IEEE

1) Purpose of Cyberattacks, main Categories and actors : The CIA triad (loss of Confidentiality, Integrity, Availability) is a model that organizations use to assess their security capabilities and risks from cyberattacks. Firstly, confidentiality refers to the protection of confidential data from unauthorized acquisition by intruders who scan the web for them. Nevertheless, we also need to be cautious of "social engineering attacks," that involve coercing someone into divulging private data, like e-mail Phishing, Pharming, Spoofing etc [22]. Continuously, ensuring data integrity involves stopping malicious or unintentional actions from improperly changing data, which leads to fraud or bad decision-making, like man-in-the-middle and salami attacks. Lastly, the availability of IT systems is against the loss of computer network functionality and efficacy, like DoS attacks [1].

Generally speaking, there are four main categories of cyberattacks: disruption, interception, modification, and fabrication. All of the CIA triad's aforementioned elements may be impacted by a digital assault. Furthermore, it might be difficult to distinguish between different assault patterns and the precise impacts they might produce, because an assault should fall under more than one category or have different kinds of effects.

In cyberspace, cyber attackers have different educational backgrounds, skills, and resources (state-sponsored or not) and engage in activities of questionable legitimacy. They are generally grouped into categories, according to the purpose they serve. Primary, professional hackers and activists (hacktivists), act to gain unauthorized access to PCs and extract information, cause damage, etc, due to financial motives and socio-political agenda) [8]. Additionally, cybercriminals have a personal agenda and/or financial gain, targeting data, which they illegally remove until they get ransom or exploit it for personal-financial gain (e.g Industrial spies) [43]. Furthermore, cyberterrorists are politically motivated hackers who can attack a wide range of targets to inflict enough damage and cause panic [49]. Last, but not least, are internal actors, usually disgruntled employees within the target firm, like Edward Joseph Snowden, and vandals who cause damage without any specific personal benefit. [5].

B. Cybercrime, Cyberwarfare, and Cyberterrorism

Cybercrime refers to criminal acts committed online using electronic communications networks and information systems. However, there is a flood of definitions of what exactly Cybercrime is, with a wide range from simple (a crime committed on a computer network - Dictionary.com) to very specialized ones, which cover every small category of Cybercrime. Also, many types of crimes, including child sexual abuse, human and drug trafficking, have moved online or are facilitated online [19]. In addition, relevant bodies of European Commission, report that investigations of almost all forms of crime have a digital footprint in the year 2019 [10]. A typical example of Cybercrime is the Case of "NotPetya", in Ukraine, where a series of cyberattacks using ransomware began in 2017, flooding the websites of Ukrainian financial institutions, government

structures, media, and energy providers. "NotPetya" encrypted all files on infected PCs and the cybercriminals demanded ransom to restore them [27].

There are no precise definitions of Cyberwarfare and Cyberterrorism, accepted by the EU or UN, and so far there is no evidence that cyberassaults can cause loss of life or injury to a person. Cyberterrorism could be defined as illegal attacks and threats of attack against PCs, their networks, and the information stored on them, in order to intimidate or coerce governments and people into accepting the promotion of their socio-political agenda [12]. Thus, the concept of Cyberwarfare refers to a conflict between states that takes place in Cyberspace, or how a cyberattack constitutes an armed attack based on "jus ad bellum" [38].

In 2007, Estonia's conflict with Russia over the transfer of the Tallinn Bronze Soldier is a typical example of situations that have the features of cyberwarfare events against important public and private facilities. Although the attacker's identity has not been made clear, it is widely believed that the services of an opposing State were involved. Subsequently, the Estonian government, law enforcement authorities, banks, media, and internet infrastructure, suffered three weeks of DDoS cyberattacks. Diplomatic interest in this cyberattack was high due to the possible reinterpretation of NATO's Article 5. In the end, no one was able to find any proof of the official involvement of an enemy State Eventually, NATO carried out an internal process, through which it assessed the level of cybersecurity in its infrastructure and created the Cooperative Centre of Excellence in Cyber Defence (CCDCOE), and developed the Tallinn Handbook on CyberWar [24].

C. Impact of cyberattacks on society

Although cyber assaults have not caused harm to anyone's physical integrity yet, the question arises whether cyberattacks pose a threat to individuals and whether they have similar psychological effects on people as conventional attacks. A series of studies conducted by the Israeli University of Haifa, from 2013 to 2016, related to the psychological effects of Cyberterrorism on individuals' emotions, and political behaviors", provided answers to the above-mentioned questions.

In summary, the scientific methodology of the research was based on the collection of data from adults, who were exposed to simulated cyber-terrorist attacks, and measurable conclusions were drawn, about their impact on the civilians' sense of security and emotions. Also, among others, conclusions were drawn about the perception of the threat these cyberattacks present in relation to incidents of conventional terrorism and how they affect political actions. Particularly, in Table I and II, results of the research are shown (measurements in State-Trait Anxiety Inventory -STAI) [45].

The results, derived from the evaluation of the data, showed that, the exposure of individuals to cyberattack incidents shares many negative emotions with exposure to conventional terrorist attacks. Data also show that announcing the existence of a potential cyberthreat can exacerbate sadness and stress,

TABLE I
MEASUREMENT OF STRESS PRODUCED AFTER EXPERIMENTAL CYBERTERRORIST ATTACKS IN A SAMPLE OF PEOPLE. [SCALE: 1 (NO STRESS) TO 4 (HIGH ANXIETY)].

Perpetrator Treatment group	Unidentified	Terrorist
Control: no terrorism	2.3	2.7
Cyberterrorism, non-lethal	3.5	3.4
Cyberterrorism, lethal: deaths-injuries	3.6	3.6
Conventional terrorism, lethal: deaths-injuries	-	4.0

TABLE II
THREAT PERCEPTION MEASURES FOLLOWING EXPERIMENTAL CYBERTERROR ATTACKS. [SCALE: 1 (LOW) TO 4 (HIGH)].

Experiment perpetrator Treatment Team	Unidentified	Terrorist
Control: no terrorism	2.9	3.1
Cyberterrorism, non-lethal, loss of funds	-	3.4
Cyberterrorism, non-lethal, asset and data loss	3.4	-
Cyberterrorism, lethal: deaths and injuries	3.5	3.6
Conventional terrorism, lethal: deaths-injuries	-	3,8

and encourage people to conform to limits on their constitutional freedoms, individuality, civil liberties and adopt radical political beliefs. Likewise, conventional terrorism reinforces radical political views and advocates for more stringent state control and Internet rules. [20].

D. The Cyberthreat Landscape in the EU

The EU Agency for Cybersecurity (ENISA), in the 10th edition of its Report, provides an overview of the cyber threat landscape between July 2021 and July 2022 in the EU, based on a collection of various open-source data.

The report revealed that cybersecurity threats are multiplying, particularly during the second half of 2021 and 2022. The threats identified during the reporting period to have the highest frequency of presence are ransomware, malware, social engineering threats, data threats, availability threats, supply chain attacks, and disinformation. Also, showed that the Russia-Ukraine crisis defined a new era for cyberattacks and hacktivism, as well as their impact on armed conflicts. Destabilized geopolitical situations can trigger potentially damaging cyberattacks in parallel implementation of conventional military actions, marking a new type of warfare that will affect international rules in cyberspace.

In addition, it is noted that disinformation campaigns continue to increase, driven by the widespread use of digital social media platforms and other online media. Social media and even search engines are now the main sources of information for many people in the Eurozone. Disinformation campaigns are often used to reduce the overall perception of trust in the institutions that provide the security environment to communities [17].

III. DEMOCRACY AND CYBERSPACE

A. Digital rights

Although the rapid digital transformation offers to increase government organizations' performance, at the same time raises threats to democratic processes, especially in relation to the security of digital information, confidentiality, and unauthorized monitoring. Digital rights are basic human rights. The rights to online privacy and freedom of expression, are truly extensions of the equal and inalienable rights set out in the United Nations Universal Declaration of Human Rights. According to the UN, disconnecting people from the internet violates these rights and is contrary to international law [47].

The European Parliament, the Council, and the Commission have formally submitted to the EU a Joint Declaration on Digital Liberties and values for the next ten digital years. Important topics covered in the draft include placing people and their rights at the heart of things, encouraging collaboration and diversity, guaranteeing internet liberty, encouraging involvement in online environments, enhancing privacy, and advancing the viability of the digital tomorrow. European digital rights and principles will complement existing rights such as data protection, online privacy, and the Charter of Fundamental Rights. [36].

B. Disinformation in Cyberspace

Disinformation is being intentionally disseminated widely and systematically on the internet, posing a growing threat to democratic institutions. Inaccurate or deceptive material that is produced, presented, and distributed for financial advantage or with the intent to deceive the general population and maybe create public damage to them is what the EU refers to as disinformation.

1) Types of disinformation: Today, the Council of Europe categorizes Three categories of informative issues to distinguish among genuine and fraudulent communications, as well as which ones are generated or disseminated by individuals with malicious intent and which do not. These types are the following. (a) Disinformation refers to erroneous or deceptive material that is produced, displayed, and shared with the intention of deceiving people or generating financial advantage, which may result in damage for the community.(b) Misinformation is defined as false information, but the person disseminating it believes it's true. The direct dissemination of fake news that has not been verified if it is correct takes place without malicious purposes, in the majority of cases. (c) Malinformation is information that is based in reality but is used to cause harm to an individual, organization, or country [9].

2) Instigators of disinformation - misinformation and its dissemination: Disinformation can be produced and disseminated when using a social media account that is owned by individuals, businesses, state authorities, etc. UNESCO's Working Group on Freedom of Expression and Countering Disinformation makes a distinction amongst those who disseminate content (Actors) and those who create misinformation (Abettors) when examining the players accountable for disinformation. [3].

Significant pervasive dangers to civil rights and democratic procedures come from organized efforts to conduct synchronized social media activities across several platforms. A Meta's report on Foreign Interference and Coordinated Authentic Conduct shows that States, political groups, the military, and/or advisory companies that provide services to these organizations are frequently associated with extensive misinformation operations. A foreign state-sponsored agitator or disinformation agent could breach public international law's of non-interference norm [18].

Concerns over foreign influence campaigns including misinformation have increased since State-sponsored organizations spread disinformation via the internet under fictitious identities. For example, was made public that Russian IRA agency purportedly influenced the 2016 US presidential election. The society's observation demonstrates that State-sponsored meddling in other States' democratic processes includes meddling in voting processes via organized disinformation efforts, encouraging democratic regression [13].

Also, in 2020, a Parliamentary body of the Council of Europe (PACE) raised worries regarding the extent of information contamination in a society that is more divided and linked digitally. Further, noted the propagation of disinformation operations intended to sway public perceptions, with the use of patterns regarding foreign meddling and alteration in elections, as well as harassing conduct and an increasing level of hate speech on social platforms [23].

3) The Impact of Disinformation on the Individual - Global online Survey: Disinformation has the power to mislead and deceive voters, foster mistrust of global laws, or democratically approved policies, as well as sabotage elections. Comprehending the motivations behind disinformation in any field of concern is crucial in order to counteract it.

The global online survey of the research company Ipsos for the Center of Innovation and International Governance (CIGI), conducted from 21/12/2018 until 10/02/2019, in 25 major economies and 25,229 Internet users (1000+ people per country), in collaboration with the Internet Society and the United Nations Conference on Trade and Development (UNCTAD), provided a representative insight into Internet users' views on a wide range of topics, including online privacy, social media and fake news. Below, are presented the most important results of the research, that concern the areas of the present paper.

As shown in Figure 1, the majority of people, feel that social media platforms have made it easier for their users to access information.

Also, as shown in Figure 2, on Facebook, social media, and the Internet in general, false information recollection seems to be most common. Few people say they have never come across false information in a conventional media source, which contributes to the perception that false information is considerably less common in conventional media providers.

Lastly, as shown in Figure 3, fewer than half of global citizens express at least some degree of confidence that any of the algorithms they are using are unbiased, in any context.

From the combined study of the above figures, the following key points are derived:

1) Majorities argue that social media has made it simpler to obtain data.
2) Fake news is considered less common in traditional media and increasingly common in digital media.
3) Fewer than half of the people who took part in the survey have a bare minimum of faith in the objectivity of everyday algorithms. The most frequent causes of trust issues in algorithms are objectivity, their perceived exploitative nature, a lack of transparency, and the absence of human judgment in the decision-making process. On the contrary, those who show faith in the objectivity of algorithms, highlight the absence of human influence, and emotion in decision-making [41].

C. Alteration of the electoral process

Elections intervention is the term used to describe illegal and inappropriate methods of swaying people's opinions and decisions in order to limit their capacity to fulfil their right to vote. This implies that the exercise of their political rights must occur both in the absence of hateful language while not interfering with the liberties of expression, and thinking. This assertion has been debunked by the widespread use of misinformation by organizations. Other states might not do a good enough job of defending this freedom for their citizens, even if they do not openly utilize misinformation during election periods. Virtual misinformation may also be used by non-state and foreign entities to affect and sabotage elections. [11].

Using a cyberattack to manipulate the results of the vote, as was done in 1994 when an unauthorized program at first lowered Nelson Mandela's (South Africa) win, is likewise not creative. [40]Furthermore, different State and non-state actors along with a range of illicit organizations, terrorist groups, and hacker activists, employ a variety of strategies to erode public confidence in election systems, such as threatening voters or winning support for specific candidates by unleashing potentially fabricated information. These practices make use of modern technology and affect the transparency of democratic processes, with the ultimate goal of undermining trust in elections. [40].

Independent news environments have emerged, as public confidence in conventional media has decreased. The concept of online platforms promotes click-generated content, and this has increased polarization. This favors the creation of a more homogeneous audience, which in turn undermines tolerance of alternative views. For example, Russia's Internet Research Agency (IRA) bought about 3,400 ads on Facebook and Instagram during the 2016 U.S. presidential campaign, according to a 2019 U.S. Department of Justice report [33].

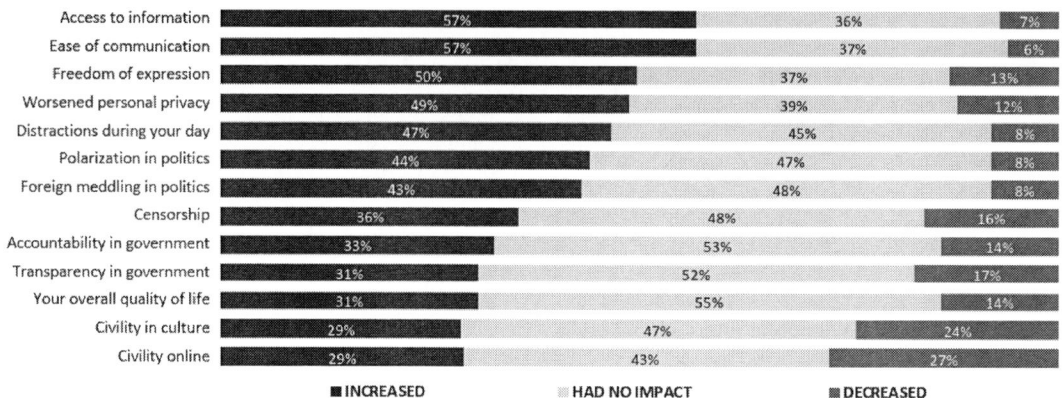

Fig. 1. Contribution of social media to information[Source:https://www. ipsos. com/en/2019-cigi-ipsos-global-survey-internet-security-and-trust]

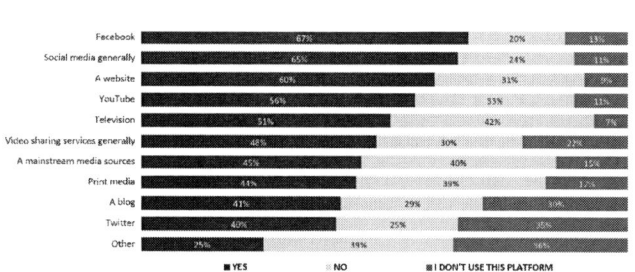

Fig. 2. Fake news on social media, Internet, and mainstream media.[Source:simpson2019cigi]

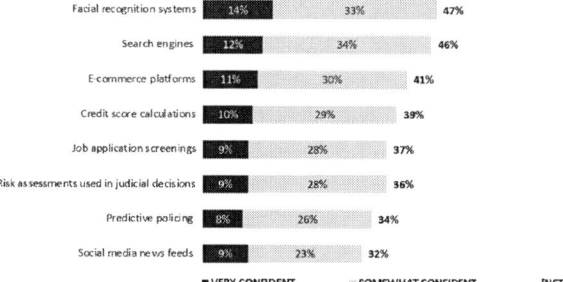

Fig. 3. Degree of confidence people express to algorithms used in daily life and believe that they are unbiased, in any context.[Source:simpson2019cigi]

A high standard of democratic public conversation requires information and common perspectives as foundations. The European Commission's virtual future policy also takes into account the fact that misinformation undermines public confidence in organizations, alongside with conventional and online media, and therefore damages democracies by impeding individuals' capacity to make reasoned decisions. Additionally, it issues a warning that misinformation will exacerbate already existing conflicts and threaten democratic foundations like elections, by polarizing political views. Finally, the UN, specifically stated that It is the duty of elected officials to guarantee that voters can form opinions independently, without violence or threat of violence, coercion, incitement, or manipulative interference of any kind (Article 21 and 25 UDHR) [9].

D. Digital repression

Digital repression refers to the oppressive utilization of communications technologies (ICT) by an authority to demonstrate dominion not only over possible but also over present challenges and contestations. It also encompasses a range of strategies that governments might employ to track and regulate the online behavior of its population, such as cyber tracking, as well as misinformation operations etc [16].

However, as extreme as the above may sound to each reader, they are already applied in countries with increased state surveillance of their citizens. For example, Chinese authorities

in the Xinjiang Uyghur Autonomous Region, collect data from a variety of sources by force, due to a predictive law enforcement initiative that allows them to detect possible dangers [48].

Some governments frequently impose internet blackouts and other limits on the internet on their citizens; many of these limitations are driven by concerns for public safety or national security. [50]. These shutdowns ultimately do not restore order, but instead caused fear and confusion among citizens. Internet outages not only harm national security through the suppression of free speech, but they can also cause panic and raise public health concerns. By restricting access to digital communication tools, emergency services, and reliable public safety data, outages jeopardize citizens' physical safety. Such violations also undermine the norms based on the international framework of Internet administration [35].

In many cases, cyber "militias" and "troll farms" are used to stifle the opposition, by accusing them of promoting lies or being enemies of society. It is therefore understood that the creation of online content aimed at attacking the opposition, as well as the discrediting of critical voices of dissent towards the views and actions of a government, bypasses the fundamental guidelines for democratic dialogue, which must be transparent in all democratic societies.

979-8-3503-8368-3/23 $31.00 © 2023 IEEE

IV. ADDRESSING CYBER THREATS IN THE CONTEXT OF SAFEGUARDING DEMOCRACY

A. IT Security architectures

The current basic structure of an organization's IT systems is highly complex. Its basic features cover a wide landscape, which is difficult to control in terms of management, processing, and exchange of an Institution's sensitive information, while the network territory is not demarcated anymore. These take place in a digitally connected world, unsafe to suspicious activity, where networks, IT systems, the number of devices, as well as cloud and multi-cloud ecosystems complications, are increasing. It is therefore understood that security requirements are increasing to protect the critical data of the Institution from cloud and multi-cloud ecosystem complications, Illicit leaks, intentional modification etc. The National Cybersecurity Authority of the Greek State provides stakeholders with a cybersecurity guidebook with administrative risk reduction strategies for IT systems. Regarding effective defense against constantly evolving Cyber threats, two architecture models are mainly proposed (min digital.gr).

1) "Defense in depth" Architecture: This model presents the adoption of sequential layer security measures across the network and data of an organization, where all layers together face a multitude of threats. If one threat discovers a way to bypass one layer, then it will face the defense of the next, as shown below in Figure 4. A successful defense-in-depth plan consists of technical means, administrative strategies (liberties limitation, compelled data etc.), network security (firewalls, invasion identification mechanisms, VPNs, etc.), guarding of devices – applications – data (anti-malware programs, data backup, encryption, etc.) [42].

2) 'Zero trust' Architecture: The security model called "zero trust," is based on the scenario that every threat exists inside and outside the traditionally delineated perimeters of networks,as shown below in Figure 5. Specifically, the underlying assumptions of the framework include the following.

- Never trust, always verify. With this assumption, the flow of data on each device, the device itself but also the user himself or even the device administrator are not considered inspiring trust, so to connect to the Organization's network they must be authenticated and then explicitly authorized with minimum privilege requirements.
- Consider there is a collapse. With this assumption, it is understood that an aggressive action was carried out, which aims to damage the devices and network of the Entity. Then, the "deny by default" instruction is followed when there is any request to access a device or data feed application, etc. Access is then granted, preceded by a thorough examination of several parameters (e.g. username, name, and location of the device, etc.) [34].

B. Legislation- Regulatory Framework and Initiatives to Safeguard Cybersecurity and Democracy

The existing legal framework does not include cyber-attacks among the acts that can be considered aggressive and therefore such actions cannot trigger the activation of Article 51 of the UN Charter. In addition, the problem of accountability contributes negatively to the identification of the perpetrator of attacks, despite the development of technology, and weakens deterrence. However, preparation at the regulatory and legal level is fundamental to avoid surprises, but also to provide a suitable ground for intercepting and deterring future more sophisticated cyberattacks.

1) European Union: Below are listed the main regulatory provisions related to the Internet and Cybersecurity, as well as indicatively some of the most important Initiatives of its institutions.

- Budapest Convention on Cybercrime (Directive 2013/40/EU), aims to tackle cybercrime and strengthen the security of IT systems, [31].
- Permanent Structured Cooperation (PESCO) aims to strengthen defense cooperation between the Member States of the Eurozone,(CFSP) 2018/340 [29].
- The European Agency for Cybersecurity - ENISA, contributes to the formulation of EU policy in the field of Cybersecurity, Regulation (EU) 2019/881 [6].
- NIS Directive of the European Parliament and of the Council. Directive (EU) 2016/1148 of the European Parliament and of the Council (NISD) [15].
- General Data Protection Regulation (GDPR). Regulation (EU) 2016/679, protection of natural persons against the processing of personal data and the unwilling distribution of them [32].
- Law Enforcement Data Protection,Directive (EU) 2016/680, on the protection of natural persons against irregular processing of personal data for the prosecution of criminal offenses [26].
- EU cybersecurity strategy for the Digital Decade, 22/03/2021 (JOIN-2020- 18 final) [37].
- The CDDG is the intergovernmental forum of the Council of Europe where Member States (M.S) leaders convene to create European standards [14].
- Strategic agenda of the European Council (SC). The European Council's strategic agenda for 2019-2024 [2].
- EU Cyber Diplomacy toolbox, Council Decision (CFSP) 2020/1127 amending Decision (CFSP) 2019/797 concerning restrictive measures against cyber-attacks threatening EU [39].

2) NATO and UN: Cyber threats are characterized as complex, and will potentially become catastrophic, so NATO adapts new strategies in the evolving cyber threat landscape while working closely with EU [7].

United Nations Initiatives Against Disinformation. In 2011 the UN Human Rights Council unanimously adopted the UN Guiding Principles on Business and Human Rights [30] [21].

V. CONCLUSIONS – PROPOSALS

A. Conclusions

The technological revolution transforms human society into a gradually linked cyberspace ecosystem. However, apart from

---

979-8-3503-8368-3/23 $31.00 © 2023 IEEE

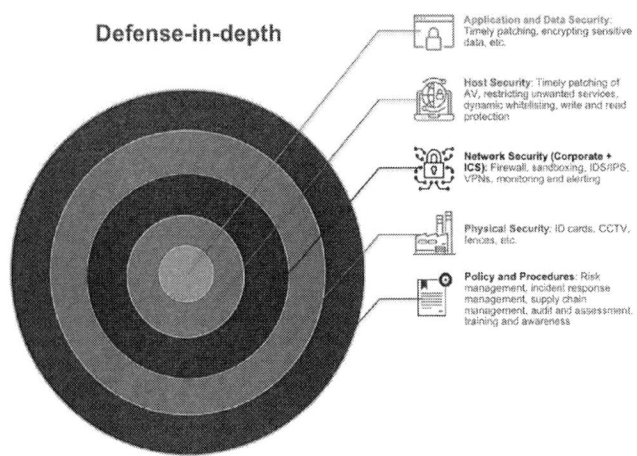

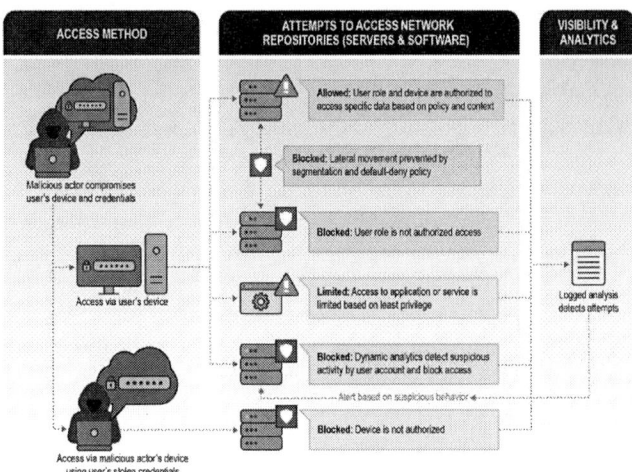

Fig. 4. Defence in depth [Source:https://modernciso.com]

Fig. 5. Zero Trust [Source:https://mindigital.gr]

the undeniable advantages provided by the digital world to citizens, there are also several risks for state and private bodies, as well as the individual in general. The fact that this virtual space transcends geographical boundaries, and lowers barriers to cyberthreats and political propaganda while providing anonymity and lack of accountability to actors, creates clear opportunities for malicious actions on a global scale.

Cyber threats exploit cyber vulnerabilities and take place with various types and techniques, aiming the confidentiality, integrity, and availability of systems connected to the internet. Cyberattacks are carried out by various actors, who have personal, economic, ideological, or even geopolitical motives. As a result of these actions, critical infrastructure is affected, creating confusion and anxiety for the individual, but also influencing public opinion.

Due to the unstable geopolitical status in some regions of the earth, it is expected to occur more cyberattacks. Beyond that, social media platforms tend to be the main source of information, but also the main tool of manipulation. It is therefore understood that conventional and non-conventional forces, soldiers and civilians, sabotage of IT systems, and disinformation interact with each other and "play" with the psychology of individuals in a new theatre of operations.

Furthermore, the attempt to manipulate democratic institutions, whether successful or not, affects voters' views and choices and harms democracy, since it raises doubts about whether democratic institutions work well, in terms of reflecting citizens' choices. Meanwhile, authoritarian states have used the pretext of cracking down on the spread of fake news to hurt the opposition, while restricting freedom of expression and media freedom through digital repression to maintain power.

The responsibility for a safer Cyberspace reflects on all involved Bodies and Organizations. The EU seeks to promote an open and free cyber ecosystem, enforced with adequate legislative framework and related initiatives. The core values

of human dignity, freedom, democracy, equality, human rights, and, by extension, digital rights, are commitments that the EU wishes to safeguard through its regulatory work.

B. Proposals

(a) Proposals for improving Cybersecurity

- The awareness and responsibility of each user, to safeguard their personal and professional data, contributing to the formation of a safe online environment.
- The exchange of know-how between the relevant Bodies in the EU, including the Private Sector and Academia, so as to develop international Cybersecurity standards, resilient critical infrastructures, and supply chains.
- The strengthening of international cooperation between Law Enforcement Agencies and Judicial authorities in the fight against Cybercrime, including the exchange of best practices.
- The update of the already implemented framework for imposing sanctions on cyber-attacks against the EU or Member States.
- The creation of the European Network of Cybersecurity Centres which will work closely with similar centres in third countries.
- The development of strong encryptions to protect the fundamental rights and digital security of users, businesses, and governments. Also, the use of artificial intelligence (AI) and machine learning (ML) by cybersecurity professionals to shrink the scope of cyberattacks, instead of wasting time and resources on constantly pursuing malicious activity.

(b) Proposals on combating online disinformation and reinforcing the security of democratic institutions.

- The development of a UN Convention on Universal Digital Human Rights, as well as the strengthening of existing relevant UN resolutions.

979-8-3503-8368-3/23 $31.00 © 2023 IEEE   108

- The EU should use its diplomatic influence to push other governments to implement transparency standards and restrict disinformation in large digital companies.
- Civil society needs to fight disinformation and protect human rights online by creating a digital barrier to the spread of fake news.

REFERENCES

[1] Jason Andress. *The basics of information security: understanding the fundamentals of InfoSec in theory and practice.* Syngress, 2014.

[2] Suzana Anghel and Ralf Drachenberg. Eprs— european parliamentary research service. 2019.

[3] Kalina Bontcheva, Julie Posetti, Denis Teyssou, Trisha Meyer, Sam Gregory, Clara Hanot, and Diana Maynard. Balancing act: Countering digital disinformation while respecting freedom of expression. *Geneva, Switzerland: United Nations Educational, Scientific and Cultural Organization*, 2020.

[4] Susan W Brenner. At light speed: Attribution and response to cyber-crime/terrorism/warfare. *J. Crim. L. & Criminology*, 97:379, 2006.

[5] Bryan Burrough, Sarah Ellison, and Suzanna Andrews. The snowden saga: A shadowland of secrets and light. *Vanity Fair*, 23, 2014.

[6] Federica Casarosa. Cybersecurity certification of artificial intelligence: a missed opportunity to coordinate between the artificial intelligence act and the cybersecurity act. *International Cybersecurity Law Review*, 3(1):115–130, 2022.

[7] Jeffrey S Caso. The rules of engagement for cyber-warfare and the tallinn manual: A case study. In *The 4th Annual IEEE International Conference on Cyber Technology in Automation, Control and Intelligent*, pages 252–257. IEEE, 2014.

[8] Samuel Chng, Han Yu Lu, Ayush Kumar, and David Yau. Hacker types, motivations and strategies: A comprehensive framework. *Computers in Human Behavior Reports*, 5:100167, 2022.

[9] Carme Colomina, Héctor Sánchez Margalef, Richard Youngs, and Kate Jones. The impact of disinformation on democratic processes and human rights in the world. *Brussels: European Parliament*, 2021.

[10] European Commission. Joint communication to the european parliament and the council–the eu's cybersecurity strategy for the digital decade. *JOIN (2020) 18 Final*, 2020.

[11] United Nations Human Rights Committee et al. General comment no. 25: The right to participate in public affairs, voting rights and the right of equal access to public service (art. 25).

[12] Dorothy E Denning. Cyberterrorism: The logic bomb versus the truck bomb. *Global Dialogue*, 2(4):29, 2000.

[13] Larry Diamond. The democratic rollback-the resurgence of the predatory state. *Foreign Aff.*, 87:36, 2008.

[14] Ardita Driza Maurer, Melanie Volkamer, and Robert Krimmer. Council of europe guidelines on the use of ict in electoral processes. In *European Symposium on Research in Computer Security*, pages 585–599. Springer, 2022.

[15] Unión Europea. Directive (eu) 2016/1148 of the european parliament and of the council of 6 july 2016 concerning measures for a high common level of security of network and information systems across the union. *Official Journal of the European Union*, 19, 2016.

[16] Steven Feldstein. How artificial intelligence is reshaping repression. *J. Democracy*, 30:40, 2019.

[17] European Union Agency for Cybersecurity (ENISA). Enisa threat landscape 2021. 2022.

[18] Nathaniel Gleicher. Removing coordinated inauthentic behavior. *Retrieved August*, 22:2022, 2020.

[19] Sarah Gordon and Richard Ford. On the definition and classification of cybercrime. *Journal in computer virology*, 2:13–20, 2006.

[20] Michael L Gross, Daphna Canetti, and Dana R Vashdi. The psychological effects of cyber terrorism. *Bulletin of the Atomic Scientists*, 72(5):284–291, 2016.

[21] António Guterres. Roadmap for digital cooperation. *United Nations. June*, 2020.

[22] Gary Hinson. Social engineering techniques, risks, and controls. *ED-PAC: The EDP Audit, Control, and Security Newsletter*, 37(4-5):32–46, 2008.

[23] Alicja Jaskiernia. Information pollution in a digital and polarized world as a challenge to human rights protection-the council of europe's approach. *Rev. Eur. & Comp. L.*, 46:7, 2021.

[24] Eric Talbot Jensen. The tallinn manual 2.0: Highlights and insights. *Geo. J. Int'l L.*, 48:735, 2016.

[25] Engin Kirda, Christopher Kruegel, Greg Banks, Giovanni Vigna, and Richard Kemmerer. Behavior-based spyware detection. In *Usenix Security Symposium*, page 694, 2006.

[26] Mark Leiser and Bart Custers. The law enforcement directive: Conceptual challenges of eu directive 2016/680. *Eur. Data Prot. L. Rev.*, 5:367, 2019.

[27] Reyner Aranta Lika, Danushyaa Murugiah, Sarfraz Nawaz Brohi, and Daksha Ramasamy. Notpetya: cyber attack prevention through awareness via gamification. In *2018 International Conference on Smart Computing and Electronic Enterprise (ICSCEE)*, pages 1–6. IEEE, 2018.

[28] Apostolos Malatras, Zoran Stanic, Ifigeneia Lella, Ricardo De Sousa Figueiredo, Eleni Tsekmezoglou, Marianthi Theocharidou, Rossen Naydenov, and Anastasios Drougkas. Enisa threat landscape: Transport sector (january 2021 to october 2022). 2023.

[29] Benjamin Martill and Carmen Gebhard. Combined differentiation in european defense: tailoring permanent structured cooperation (pesco) to strategic and political complexity. *Contemporary Security Policy*, 44(1):97–124, 2023.

[30] Janne Mende. The united nations guiding principles on business and human rights (2011). *Sources of the History of Human Rights*, 2018.

[31] Adrian Cristian Moise. Analysis of directive 2013/40/eu on attacks against information systems in the context of approximation of law at the european level. *Special Issue JL & Admin. Sci.*, page 374, 2015.

[32] Christopher F Mondschein and Cosimo Monda. The eu's general data protection regulation (gdpr) in a research context. *Fundamentals of clinical data science*, pages 55–71, 2019.

[33] Robert S Mueller et al. *The Mueller Report*. e-artnow, 2019.

[34] RESILIENCE OF NETWORK. Hellenic republic ministry of digital governance national cybersecurity authority. 2021.

[35] Ted Piccone. Democracy and digital technology: The unique challenges that digital technology is presenting to democratic governments and how they, together with civil society, need to respond. *Sur: Revista Internacional de Direitos Humanos*, 15(27), 2018.

[36] CAR POLONA. European declaration on digital rights and principles. 2022.

[37] Margarita Robles-Carrillo. The european union strategy for cybersecurity margarita robles-carrillo. *The Legal Challenges of the Fourth Industrial Revolution: The European Union's Digital Strategy*, 57:173, 2023.

[38] Marco Roscini. World wide warfare-'jus ad bellum'and the use of cyber force. *Max Planck Yearbook of United Nations Law*, 14:85–130, 2010.

[39] Vera Rusinova and Ekaterina Martynova. Fighting cyber attacks with sanctions: Digital threats, economic responses. *Israel Law Review*, pages 1–40, 2023.

[40] Scott J Shackelford, Angie Raymond, Abbey Stemler, and Cyanne Loyle. Defending democracy: taking stock of the global fight against digital repression, disinformation, and election insecurity. *Wash. & Lee L. Rev.*, 77:1747, 2020.

[41] S Simpson. Cigi-ipsos global survey on internet security and trust. *Published by Centre for International Governance Innovation. https://www. ipsos. com/en/2019-cigi-ipsos-global-survey-internet-security-and-trust*, 2019.

[42] Clifton L Smith. Understanding concepts in the defence in depth strategy. In *IEEE 37th Annual 2003 International Carnahan Conference onSecurity Technology, 2003. Proceedings.*, pages 8–16. IEEE, 2003.

[43] Russell Smith, Peter Grabosky, and Gregor Urbas. Cyber criminals on trial. *Criminal Justice Matters*, 58(1):22–23, 2004.

[44] Eugene H Spafford. Computer viruses–a form of artificial life? 1990.

[45] CD Spielberger, RL Gorsuch, R Lushene, PR Vagg, and GA Jacobs. Manual for the state-trait anxiety inventory; palo alto, ca, ed. *Palo Alto: Spielberger*, 1983.

[46] Tim Stevens. Cyberweapons: an emerging global governance architecture. *Palgrave Communications*, 3(1):1–6, 2017.

[47] Gaye Tuchman. The currency of democracy: Politics, the media, and corporate control, 2013.

[48] Human Rights Watch. China: Big data fuels crackdown in minority region. 2018.

[49] Gabriel Weimann. Cyberterrorism: The sum of all fears? *Studies in Conflict & Terrorism*, 28(2):129–149, 2005.

[50] Darrell M West. Internet shutdowns cost countries $2.4 billion last year. *Center for Technological Innovation at Brookings, Washington, DC*, 2016.

Access to personal data is still tempting for mobile apps even after the GDPR implementation

Gerasimos S. Magoulas
Department of
Digital Media
and Communication
Ionian University
Kefalonia, Greece
gmagoulas@ionio.gr

Spyros E. Polykalas
Department of
Digital Media
and Communication
Ionian University
Kefalonia, Greece
s.polykalas@ionio.gr

Abstract—**Mobile phones have become a vital part of our daily lives with users relying on them for a variety of activities. Users don't hesitate to provide their personal data to free or paid mobile applications, in order to use them with no interruption. However, the amount of personal data being exchanged has raised concerns, resulting in the introduction of regulations for protecting personal data at many countries. A regulation in this direction that was enacted in Europe is the General Data Protection Regulation known as GDPR that came into effect in May 2018. This study examines the impact of GDPR on Google Play applications by comparing a dataset of metrics collected in 2017 to a post-GDPR dataset gathered in 2022. The focus of this research is to analyze the effect of GDPR on mobile app usage and its associated metrics by comparing the datasets of the two above periods. Our analysis shows that progress has been made towards achieving GDPR compliance, but there is still a considerable distance to be covered before reaching full compliance. A key observation is that the number of permissions used by apps has not been reduced even after the GDPR implementation.**

Keywords—Google Play Store, Mobile Apps, App Permissions, GDPR, data privacy, personal data

I. INTRODUCTION

In today's world, the mobile phone has become an essential accessory that controls many aspects of our daily life. Unlike in the 90s, when citizens primarily used mobile phones for communication, they now serve as an essential tool for communication via voice and messaging, as well as for diverse activities such as education, payments, security, relationships, entertainment, and more. As a result, it has earned the name "smartphone" due to the intelligent applications that users install on it, either free or paid, from different categories based on their usefulness.

However, the interaction between users and these applications on mobile phones involves the exchange of personal data. The increasing use of personal data by these apps over the years raises questions about whether such information is genuinely necessary for the apps to serve their intended purpose.

To address these concerns, the EU introduced a regulation on personal data in 2016, which came into effect in May 2018. The regulation requires that data processors inform users for the reason their personal data is being collected and obtain their consent before processing it. The application can only proceed once users' data have been processed. Failure to comply with the GDPR can result in severe penalties for data processors. Given these regulations, it is essential to examine whether companies have taken the necessary steps to comply with them fully.

In the mobile application spectrum, Google Play and App Store are the most popular application repositories for Android and iOS operating systems, respectively[1] during the period of our survey. The scope of this research paper is to focus on the impact of GDPR on the length / amount of access to personal data requested by mobile apps. In particular a comparison between two different datasets is presented. The first dataset in 2017 was collected for the Google Play repository. We will compare our pre-GDPR research in 2017 to a new study conducted after the GDPR's implementation in the first quarter of 2022.

II. LITERATURE REVIEW

There is rich research done on data privacy issues of the mobile apps examining different aspects. It is a topic that has been extensively researched and analyzed but there are still more points of view that can be thoroughly analyzed. In [1], user reviews are considered, and they are analyzed using app and review crawlers to explore the correlation between reviews related to security and privacy concerns and the subsequent app updates addressing these issues. It shows that developers take into consideration such reviews since data privacy and security issues concern users. In the same direction [2] proposes a framework that identifies functionality relevant user reviews that describe the actual permissions that apps use rather than the ones described in the description of the apps. The research in [3] shows the factors that can influence app users for giving high rates despite the quality of the app. In [4] the main research point is similar to our research question if GDPR has affected the apps. The research focuses on the so-called "dangerous" permissions and examines pre-GDPR and post GDPR versions of the apps and concludes that there is a decrease in the number of permissions apps request for, in the post GDRP period. This conclusion is slightly different from our research findings which shows that the percentage of number of permissions wasn't reduced in general but there was a small decrease in only one permission category. The compliance with the

---

[1] https://www.statista.com/statistics/276623/number-of-apps-available-in-leading-app-stores/ (access date: May 2023)

GDPR is also examined in [5] that shows that most apps describe the permissions they use in their privacy policy but there is a number of apps that use third party libraries that already use personal data before getting the consent from the user. Finally, the research work in [7] examines if apps have a privacy policy and quantifies the associations between privacy policy and other apps' features i.e., if the developer address is in Europe, it is more likely to have a privacy policy.

Our research is a comparative analysis of specific metrics between two datasets gathered before and after the implementation of GDPR and examines if this regulation has positively affected the developers..

III. ANALYSIS

Our research is a comparative study of the permissions that Google Play applications request from the users in order to operate fully, between two periods. These two periods are the pre GDPR and post GDPR period. The data for the pre GDPR period were taken from [6] and pertains to 529 apps.

The post GDPR data were collected in the Q1 of 2022, when Google Play app store had 3.3 million apps available according to statista [2]. This period, Google Play classified the applications in three main categories which were: General, Family and Games. These three main categories were classified furtherly in 31 subcategories in total. Specifically, the category General contained 27 subcategories, the category Family contained 2 subcategories and Games 2 subcategories.

Our research is focused on the permissions that these applications request from the users in order to operate fully. We have collected data for 610 apps in January 2022 with the tag "Recommended for you" of all 31 different subcategories, for 30 different user profiles. This tag refers to applications that are either popular in the area of the user visiting Google Play, or are similar to other apps that are installed by the user.

The distribution of our downloaded applications per category compared to the research in 2017 is shown in Fig. 1.

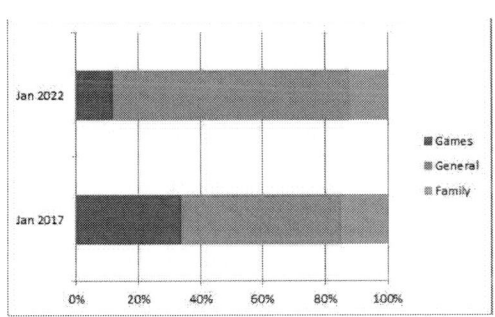

Fig. 1. Comparison of distribution of mobile apps per category between 2017 and 2022 dataset

As implemented in [6] for each mobile app in this research we have collected the same data which are: the relevant category/subcategory, the number and rate of reviews, the number of installations, the user data in which the mobile application required access known as permissions.

The first subject of research is the rating of the mobile apps by the users. The rating is a value between 0 and 5. The comparison of the ratings of the two datasets is depicted in Fig. 2.

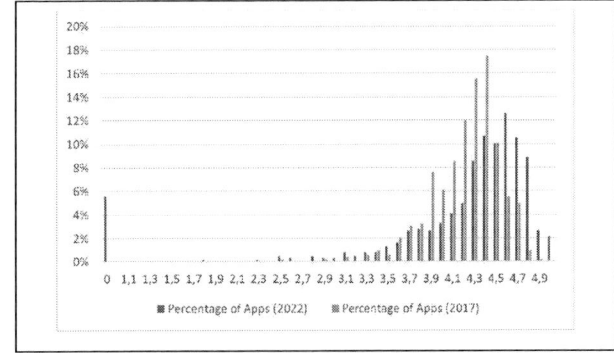

Fig. 2. Comparison of user rating distribution between datasets of 2017 and 2022

The ratings for the mobile apps in the 2017 research shows that the majority of mobile apps have high ratings between 3.9 and 4.7. The explanation for this fact was that the apps chosen were the most popular. The distribution of the new research based on apps with the tag "Recommened for you" have also high ratings which means that most applications try to meet high standards even if they are not the most popular. In this analysis there is also a percentage of 5.57% with no rating which means that there were very new apps shown under this tag, not having a rate yet.

The number of downloads in Google Play show the unique number of downloads by accounts no matter how many devices have installed the app. The comparison between the two datasets is shown in Fig. 3.

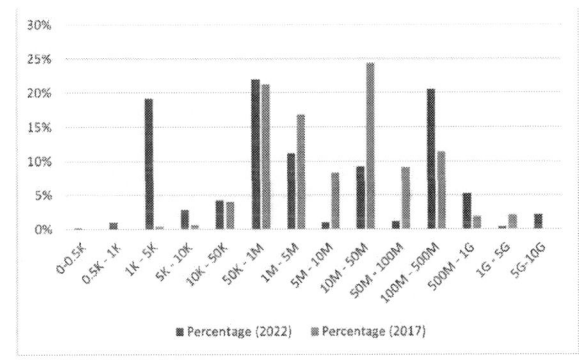

Fig. 3. Comparison of Distribution of number of installations between 2017 and 2022 datasets

Since the download intervals were slightly different between the year 2017 and 2022, we have adapted the research of 2017 with the download intervals of 2022. In the 2022 dataset, we can see that almost 20% of applications were downloaded between 1000 and 5000 times since the tag of these apps was "Recommended for you" and the app could

---

[2] https://www.statista.com/statistics/289418/number-of-available-apps-in-the-google-play-store-quarter/ (access date: May 2023)

belong to any downloading range. The popularity criteria which are examined in 2017 dataset shows that most apps have large downloading ranges more than 50.000 downloads.

The main part of this research is the part that refers to the personal data that the application requires to access in order to operate fully, either they are relevant to the purpose of the application, or not.

The applications during the installation or operation use different parts of the devices and also have access to personal data of the user. The developer declares which parts of the device and personal data are used in the privacy policy of the application. The required access to components of the device or personal data are called permissions [3]. When the user downloads the app and installs it, it is assumed that he/she has read the privacy policy and knows which permissions are needed in order for the app to operate correctly. Google Play has classified the permissions in categories based on the resources they use and the personal data they use. Our research has recorded the permissions and permission categories of the above applications, which are presented in Table 1.

TABLE I.   TYPE OF REQUIRED ACCESS (PERMISSIONS)

Permission Category	Permission
Photos/Media/Files	read the contents of your USB storage
	modify or delete the contents of your USB storage
	access USB storage filesystem
Location	approximate location (network-based)
	precise location (GPS and network-based)
Storage	read the contents of your USB storage
	modify or delete the contents of your USB storage
Camera	take pictures and videos
Wi-Fi connection information	view Wi-Fi connections
Contacts	read your contacts
	find accounts on the device
Identity	find accounts on the device
	add or remove accounts
	read your own contact card
Phone	read phone status and identity
	directly call phone numbers
Microphone	record audio
Calendar	read calendar events plus confidential information
Device & app history	read sensitive log data
Wearable sensors/Activity data	body sensors (like heart rate monitors)
Device ID & call information	read phone status and identity
Device & app history	retrieve running apps

Other	send sticky broadcast
	disable your screen lock
	control Near Field Communication
	create accounts and set passwords
	modify system settings
	read Home settings and shortcuts
	control flashlight
	install shortcuts
	close other apps
	read sync statistics
	Access all system downloads
	expand/collapse status bar
	Google Play license check
	write Home settings and shortcuts
	set an alarm
	read sync settings
	toggle sync on and off
	send SMS messages
	uninstall shortcuts
	measure app storage space
	add words to user-defined dictionary
	read TV channel/program information
	write TV channel/program information
	read terms you added to the dictionary
	receive data from Internet
	view network connections
	run at startup
	prevent device from sleeping
	full network access
	control vibration
	access Bluetooth settings
	pair with Bluetooth devices
	download files without notification
	read Google service configuration
	connect and disconnect from Wi-Fi
	allow Wi-Fi Multicast reception
	change your audio settings
	draw over other apps
	use accounts on the device
	change network connectivity

Each Permission Category contains one or more permissions that may record personal data or have access to device components that have for sure nothing to do with personal data such as the flashlight. In the datasets, the number of applications per permission category represents the number

---

3
https://support.google.com/googleplay/answer/11416267?hl=en&co=GENI
E.Platform%3DAndroid#zippy=%2Cunderstand-app-permissions%2Ccontrol-app-permissions-data-collection-after-download (access date: May 2023)

of applications that access at least one permission belonging to that category. Even if an application uses more than one permission from a category, we still count it as one application in that permission category.

As a result, the comparison between the datasets of 2017 and 2022 are depicted in Fig. 4.

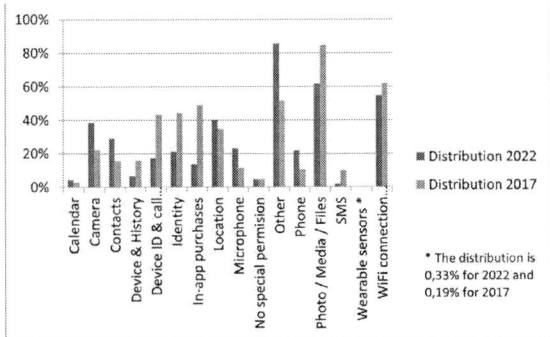

Fig. 4. Type of permission required by mobile apps in dataset 2017 and 2022

In order to conclude which period was the most demanding in permissions we estimated the average distribution per year which is 27% in 2022 and 29% in 2017 which means that the most demanding period was 2017.

In the post GDPR period, we can see that Google uses the same description for the Permission category that is not detailed in order to explain particularly which personal data are accessed in the Photo/Media/Files or what are the contacts used for, by an app that needs access to them.

The comparison between the datasets has the following results:

1) Even if the datasets contain a different percentage of subcategories of apps the 2 most demanded permission category is "Photo/Media/Files" and "Other"

2) The percentage of apps that don't require any special permission are the same

3) There is a big difference in the permission "Device and History" between the two datasets and specifically in 2022 the percentage has been reduced which means that applications avoid to use the device history which is personal data.

4) There are some contradicotry results: Access in some types has been increased between the two periods such as access to: Calendar, Camera, Contacts, Location, Microphone, Other and Phone. On the other hand access to the following types of permissions has been decreased between the two monitoring periods: Device History, Identity, In-app purchases, Photo / Media / Files, SMS, WiFi connection.

It should be mentioned that access to personal data that are crucial for marketing issues such as location remains high in both periods presenting a slightly increase. In addition, the high difference in some types of access such as in app purchase requires further analysis and justifications. One interesting fact that could have influenced these differences is the data safety form that Google mandated all developers to complete by July 2022 [4]. This form gathers information about privacy concerns and security practices of applications and must be filled out by both existing and new developers. The form was introduced in late 2021 [5], with developers being notified to provide the necessary details before the July 2022 deadline. This requirement might have prompted developers to begin considering privacy issues during the data collection process, which took place in the first quarter of 2022.

Another interesting point is the number of permissions categories that each app has access to which is depicted in Fig. 5 for the two periods.

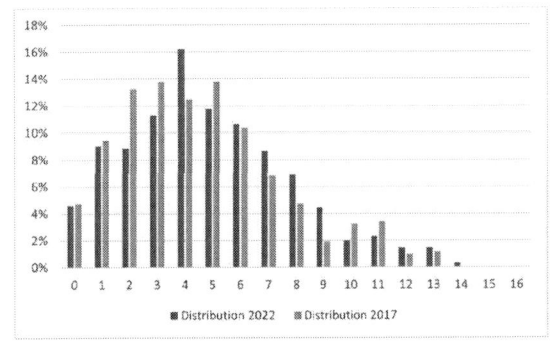

Fig. 5. Number of permissions per mobile app

As we can see from the graph most applications require access to 3 to 7 access types. There is a close similarity between the datasets which means that the number of permissions did not significant reduce as expected in the post GDPR period.

The number of applications that remain approximately the same and have the same permissions in the two datasets are no permissions with approximately 4%, 1 permission with approximately 8%, 6 permissions with approximately 10%.

It should be noted that between the two datasets only 14 apps were the same (2,5% of the sample).

The correlation between the number of installations, the number of reviews, the rate of reviews and the number of required types of data access based on the dataset of [6] in 2017 showed that the only correlation, is between the number of installations (Ninstalls) and the number of reviews (Nreviews) which proves the common sense that the number of reviews are high when the number of installations are high (the correlation index was 0.8). The method used was all examined apps were grouped in three groups based on the number of installations. The intervals of number of installations that Google Play provides for each app has been normalized to the relevant middle number of the range.

The strength of this assumption is not confirmed in the 2022 research as Table 2 shows.

---

[4] https://support.google.com/googleplay/android-developer/answer/10787469?hl=en (access date: May 2023)

[5] https://developers.google.com/admob/android/privacy/play-data-disclosure (access date: May 2023)

TABLE II. CORRELATION MATRIX

		Ratings	Ninstalls	NReviews	Naccess
ALL	Ratings	1			
	Ninstalls	0,0551654	1		
	NReviews	0,0100893	0,5877433	1	
	Naccess	0,1029746	0,2821053	0,2910813	1

		Ratings	Ninstalls	NReviews	Naccess
>1M	Ratings	1			
	Ninstalls	0,1552502	1		
	NReviews	0,0812903	0,5878125	1	
	Naccess	0,6158042	0,3151546	0,2923218	1

		Ratings	Ninstalls	NReviews	Naccess
1M<X<50M	Ratings	1			
	Ninstalls	0,6287502	1		
	NReviews	0,4631499	0,6162166	1	
	Naccess	0,7359144	0,5348799	0,4608046	1

		Ratings	Ninstalls	NReviews	Naccess
>100M	Ratings	1			
	Ninstalls	0,5117728	1		
	NReviews	0,2825239	0,5899621	1	
	Naccess	0,8380364	0,5845846	0,4629778	1

As the table shows there is a slight correlation between the number of installations and the number of reviews (correlation index 0.59). In this research the correlation that emerges is between the rating and the number of required type of access.

Regarding the data privacy issues that is the main purpose of this research we can see that there is a strong correlation between the rating and the number of required type of access (permissions) in the more than 100 million installation group which are the most popular apps (correlation index 0.84). This is contradictory to the common perception that the higher number of permissions would make the user more cautious to give a high rating to the app. It seems that the users still don't pay the attention required to their personal data and the number of permissions doesn't affect at all their rating.

In Fig. 6 the distribution of review rate is depicted for all four categories of apps grouped by the number of installations also used in the above correlation. In this Figure we can see that the distributions in all categories follow almost identical trends just as in the 2017 dataset. The higher percentage of mobile apps in the "All" category shows a review rate between 4.5 and 4.7.

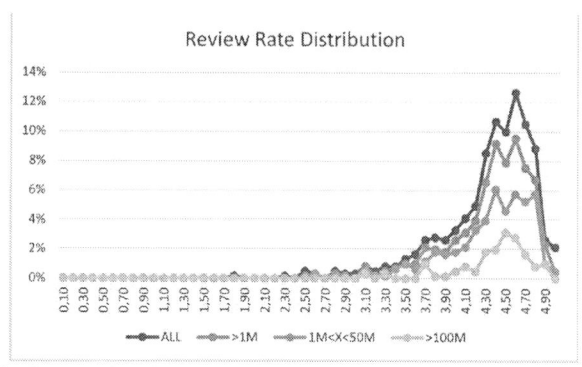

Fig. 6. Rate review distribution

The last part of this research is presented in Table 3, where there is a summary of the mean values of the review rate, the number of installations, the number of reviews, the number of data access types and also the relation between the number of reviews and the number of installations for all the above categories of apps grouped by the number of installations for the 2022 dataset.

TABLE III. MEAN VALUES PER MOBILE APP CATEGORY

Category	Number of Mobile Apps	Mean Review Rate	Mean number of installations	Mean number of Reviews	Mean Number of data access types	Relation reviews / installations
ALL	610	4	230995852	1568940	5	0.68%
>1M	444	4.2	317209460	2150191	5	0.68%
1M<X<50M	307	4.2	15605863	1410997	5	0.90%
>100M	105	4.3	1272857143	8308943	7	0.65%

In this table we can see that the mean review rate for all four categories is between 4 and 4.3. The mean number of data access types is between 5 and 7. The relation between the number of reviews and number of installations is between 0.68% and 0.9%.

In the 2017 dataset the mean review rate for all categories was very close to the 2022 dataset since it was between 4.2 and 4.3. The mean number of data access types was between 4 and 7 which is very close too. The relation between the reviews and installations was between 2% and 3%. This relation is greater than the respective 0.68% and 0.9% of the 2022 dataset which is expected since the number of reviews in the 2017 dataset are greater than the 2022 dataset in the respective categories. This is explained because the apps in the 2017 dataset were the most popular apps in each category but the 2022 dataset had the tag "Recommended for you" as explained above.

IV. CONCLUSIONS

This comparative study between the pre GDPR and post GDPR period was implemented to monitor the progress of the compliance of the mobile applications in the Google Play ecosystem with the GDPR. The results of this research are based on datasets collected from Google Play in the periods of 2017 published in [6] and 2022.

Although GDPR imposes the principle of personal data minimization our research indicates that mobile apps developers require access to personal data of end users to the same extent as the period before GDPR implementation. That means that either mobile apps developers had already followed the principle of data minimazation before GDPR, or that they have not yet complied with principles contained in the GDPR, in particular with the principles of data minimization and scope elimination. Since the number of mobile apps access permissions remains high, it allows us to conclude that the current policy of mobile apps developers is not fully in compliance with GDPR.

Although the two datasets do not contain exactly the same sets of mobile apps the high number of examined applications in both periods allow us to make comparisons. The comparison of the datasets could be done since the size of the samples were equal and the categories of the application repository were still the same.

The installation process is still the same between the two periods. The data privacy section of the apps informs the users about the permissions that the application needs on components and personal data of the mobile device. The explanations of the permissions are not detailed and in some situations are very general so the user isn't sure about which particular personal data are used.

In both datasets the majority of the mobile apps had high ratings even if there was a difference in the tags of the mobile apps collected. The 2017 dataset were the most popular in each category but the 2022 dataset were the "Recommended for you" applications. This means that even these apps might not be the most popular in each category the developers try to meet high standards for their apps.

In both periods the two most accessed permissions are "Photo/Media/Files" and "Other". Both categories have a big amount of personal data and especially in the Photo/Media/Files category the description is very general since it doesn't describe what data are accessed in the USB storage filesystem. There is also a percentage of apps that don't require of any special permission which is the same at both datasets.

Observing the permission distribution of the mobile apps, there is an improvement since the permission's percentage "Device and History" in 2022 has been reduced which means that applications avoid to use the device history which is personal data.

In general, the permission categories are 15 and in both datasets the percentage of permission categories used by apps, are quite the same (3 to 7).

Examining the correlation that might exist among the number of installations, the number of reviews, the rate of reviews and the number of required types of data access the only correlation existed was between the review rate and the number of required type of access in the more than 1million installation group which are the more popular apps. This result shows that the users don't pay the attention required to their personal data and the number of permissions doesn't affect at all their rating when the app is popular.

This work has limitations since the number of applications examined is low but it is comparative with the dataset of 2017. Another issue is that the applications compared are neither the same nor their distribution per category is the same. Futhermore, this research is conducted using the apps within the Google Play ecosystem, which poses a limitation in achieving more broadly applicable conclusions for mobile applications.

The future work should take into consideration these issues and the number of examined datasets should be increased.

ACKNOWLEDGMENT

This research was funded and conducted in the framework of MSc "New Media Communication and Digital Marketing" offered by the Department of Digital Media and Communication of the Ionian University.

REFERENCES

[1] D. C. Nguyen, E. Derr, M. Backes and S. Bugiel, "Short Text, Large Effect: Measuring the Impact of User Reviews on Android App Security & Privacy," *2019 IEEE Symposium on Security and Privacy (SP)*, San Francisco, CA, USA, 2019, pp. 555-569, doi: 10.1109/SP.2019.00012.

[2] R. Wang, Z. Wang, B. Tang, L. Zhao and L. Wang, "SmartPI: Understanding Permission Implications of Android Apps from User Reviews," in *IEEE Transactions on Mobile Computing*, vol. 19, no. 12, pp. 2933-2945, 1 Dec. 2020, doi: 10.1109/TMC.2019.2934441.

[3] Mahmood, A. Identifying the influence of various factor of apps on google play apps ratings. *J. of Data, Inf. and Manag.* **2**, 15–23 (2020). , doi: 10.1007/s42488-019-00015-w

[4] N. Momen, M. Hatamian and L. Fritsch, "Did App Privacy Improve After the GDPR?," in *IEEE Security & Privacy*, vol. 17, no. 6, pp. 10-20, Nov.-Dec. 2019, doi: 10.1109/MSEC.2019.2938445.

[5] Kaijun Liu, Guoai Xu, Xiaomei Zhang, Guosheng Xu, Zhangjie Zhao, "Evaluating the Privacy Policy of Android Apps: A Privacy Policy Compliance Study for Popular Apps in China and Europe", *Scientific Programming*, vol. 2022, Article ID 2508690, 15 pages, 2022. , doi: 10.1155/2022/2508690

[6] S. E. Polykalas, G. N. Prezerakos, F. D. Chrysidou and E. D. Pylarinou, "Mobile apps and data privacy: When the service is free, the product is your data," *2017 8th International Conference on Information, Intelligence, Systems & Applications (IISA)*, Larnaca, Cyprus, 2017, pp. 1-5, doi: 10.1109/IISA.2017.8316392.

[7] Story, P., Zimmeck, S., Sadeh, N. (2018). Which Apps Have Privacy Policies?. In: Medina, M., Mitrakas, A., Rannenberg, K., Schweighofer, E., Tsouroulas, N. (eds) Privacy Technologies and Policy. APF 2018. Lecture Notes in Computer Science(), vol 11079. Springer, Cham., doi: 10.1007/978-3-030-02547-2_1

Reliability Analysis of Fault Tolerant Memory Systems

Yagmur Yigit*, Leandros Maglaras*, Mohamed Amine Ferrag†, Naghmeh Moradpoor*, Georgios Lambropoulos‡

* School of Computing, Engineering and The Build Environment, Edinburgh Napier University, UK
† Technology Innovation Institute, United Arab Emirates, UAE
‡ University of Piraeus, Department of Informatics, Piraeus Greece

Email: yagmur.yigit@napier.ac.uk, L.Maglaras@napier.ac.uk,
mohamed.ferrag@tii.ae, N.Moradpoor@napier.ac.uk, ; george@lambropoulos.com

Abstract—This paper delves into a comprehensive analysis of fault-tolerant memory systems, focusing on recovery techniques modelled using Markov chains to address transient errors. The study revolves around the application of scrubbing methods in conjunction with Single Error Correction and Double Error Detection (SEC-DED) codes. It explores three primary models: 1) Exponentially distributed scrubbing, involving periodic checks of memory words within exponentially distributed time intervals; 2) Deterministic scrubbing, featuring regular, periodic word checks; and 3) Mixed scrubbing, which combines both probabilistic and deterministic scrubbing approaches. The research encompasses the estimation of reliability and Mean Time to Failure (MTTF) values for each model. Notably, the findings highlight the superior performance of mixed scrubbing over simpler scrubbing methods in terms of reliability and MTTF.

Index Terms—Error Correction, RAM, Memory System Reliability, Scrubbing Techniques.

I. INTRODUCTION

When designing a fault-tolerant memory system, one encounters a plethora of choices. A robust fault-tolerant system should possess the capability to detect, diagnose, isolate, mask, compensate for, and recover from errors and faults. In the models examined in this article, SEC-DED (Single Error Correction-Double Error Detection) codes are employed [1], as they strike a balance between minimal overhead (attributed to extra parity bits) and highly efficient error detection. Additionally, memory scrubbing proves to be a highly effective approach for recovering from transient faults caused by environmental disruptions and intermittent faults arising from inherent weaknesses within the circuit [2], [3]. It is a technique that reads words, checks their correctness, and rewrites the corrected data in its initial position [4], [5]. A previous study analyzed the reliability of scrubbing recovery techniques for memory systems, emphasizing the use of SEC-DED codes [6]. This research examined two models: exponentially distributed scrubbing and deterministic scrubbing. The authors derived reliability and mean-time-to-failure (MTTF) equations and compared the results with memory systems lacking redundancy and utilizing only SEC-DED codes. Probabilistic scrubbing is based on the fact that every time the program in execution processes a word, it is read and checked for correctness. This technique offers a good solution for environments where all words are addressed at the same rate by the program [7]. But in real systems, it is almost certain that some words are more often used than others, making this technique inadequate [8]. Deterministic scrubbing, on the other hand, depends on a mechanism that reads every word and checks for its correctness periodically, thus improving system reliability and MTTF [9], [10]. This technique achieves better reliability than the probabilistic one but demands more complex architecture, especially when a small scrubbing interval is required. Mixed scrubbing is finally based on the fact that the system uses exponentially distributed scrubbing with the addition of a mechanism that reads every word. This model achieves better reliability than the first one and achieves better reliability from the second one with a much simpler architecture (scrubbing interval required). The main contribution of this paper is that it thoroughly analyses the three techniques and computes and compares the reliability and MTTF of RAM of different word bit sizes.

The rest of this paper is organized as follows: Section 2 presents the terminology, assumptions, and notation used in this article. Section 3 presents the reliability analysis of the memory system. Section 4 presents the experimental results, and Section 5 concludes the article.

II. TERMINOLOGY, NOTATIONS, ASSUMPTIONS

Table I: Notations

Notation	Definition
λ	Failure Rate per Memory Bit
w	Number of Bits per Word
c	Number of Check Bits per Word
L	Error Rate of One Word
M	Memory Size in Number of Words
$r(t)$	Reliability of One Word
$R(t)$	Reliability of the Memory System
T	Interval of Deterministic Scrubbing
n	Number of Intervals of Deterministic Scrubbing
μ	Rate of Probabilistic Scrubbing
MB	Megabyte
$SEC\text{-}DED$	Single Error Correction, Double Error Detection Process in Memory Word
$Memory\ System$	Computer Hardware Associated with Data Storage and Retrieval
$Memory\ Word$	Group of Bits Capable of Storing Data
$MTTF$	Mean Time to Failure

979-8-3503-8368-3/23 $31.00 © 2023 IEEE

The notations are presented in Table I, and our underlying assumptions can be summarized as follows:

1) Transient faults manifest with a Poisson distribution.
2) Failures of individual bits are statistically uncorrelated.
3) The control and correction mechanisms integrated within the memory operate flawlessly.
4) Each word is treated as a statistically independent entity.
5) It is possible for an erroneous bit to recover due to the presence of another error.

III. RELIABILITY ANALYSIS OF MEMORY SYSTEMS

In this section, we present a reliability analysis for memory systems that use deterministic, probabilistic, or mixed scrubbing techniques.

A. Exponentially Distributed Scrubbing

The scrubbing mechanism employed in this method relies on the premise that every time a memory location is accessed, it undergoes a check and correction process if necessary. The system utilizes a Single Error Correction-Double Error Detection (SEC-DED) code to detect and rectify errors. However, it's important to note that this code may fail when a memory word contains a sufficient number of multiple-bit errors [11]. We represent with S0, S1, and S2 the states of having zero, one or two memory errors in an SEC-DEC-protected memory word, respectively. The interval between two consecutive accesses to the same memory location follows an exponential distribution with a rate denoted as μ [12]. While this assumption may not hold true for all memory systems, the operating system can enforce the use of exponential scrubbing.

A Markov chain is employed to model the recovery scenario, which features as many states as the error correction code being used [13]. Fig. 1 depicts the Markov chain representing this model.

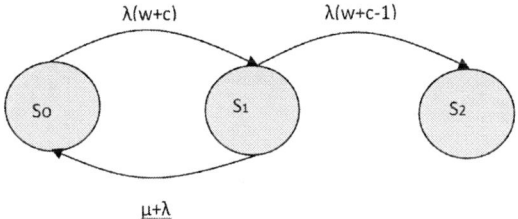

$$\lambda(w+c) \qquad \lambda(w+c-1)$$

$$\mu+\lambda$$

Figure 1: State transition rate diagram for probabilistic scrubbing

Assuming that the error rate of a memory word with w data and c check bits is $\lambda(w + c)$, assuming all bits in the word fail statistically independently with a failure rate λ. The error rate of a memory word when being in state a is:

$$L = \lambda + \lambda + \lambda + \lambda + ... + \lambda = \lambda(w + c) \quad (1)$$

The transition from state S_0 to state S_1 occurs when:

- The scrubbing mechanism is activated (the word is addressed by the program in execution) [rate μ].

- An error occurs in the faulty bit cancelling the initial one [rate λ].

When the word is in state S_1, the occurrence of an error in any bit other than the faulty leads the memory word to state S_2 [rate $\lambda(w + c - 1)$].

The differential equations that describe the system are:

$$dP_0(t)/dt = -\lambda(w + c)P_0(t) + (\lambda + \mu)P_1(t) \quad (2)$$

$$dP_1(t)/dt = \lambda(w + c)P_0(t) - (\lambda(w + c) + \mu)P_1(t) \quad (3)$$

$$dP_2(t)/dt = (\lambda(w + c - 1))P_1(t) \quad (4)$$

The solution to the aforementioned equations can be derived by taking their Laplace transforms into account [14] while also considering their respective initial conditions:

$$P_0(0) = 1, P_1(0) = 0, P_2(0) = 0$$

The final form of the reliability is:

$$r(t) = \frac{a_{1p}}{a_{1p} - a_{2p}} exp(-a_{2p}t) + \frac{a_{2p}}{a_{2p} - a_{1p}} exp(-a_{1p}t) \quad (5)$$

The reliability of the memory system with M words is:

$$R(t) = r(t)^M \quad (6)$$

The MTTF is:

$$MTTF = \int_0^\infty R(t)\,dt \quad (7)$$

The final form of MTTF is:

$$MTTF = \frac{4(\mu + \lambda)}{M(\lambda(2w + 2c - 1))^2} \quad (8)$$

B. Deterministic Scrubbing

The deterministic scrubbing used in this model is based on the fact that one or more processors cycle through the memory system, scrub the memory words, and correct any transient errors [15]. The cycling time is T, and checking the word is executed in parallel with the normal operations when this part of the memory is idle. The memory system fails every time that two errors co-exist in the same memory word. The time interval T can be reduced depending on the hardware mechanism that is used.

Deterministic scrubbing ensures that all the memory words are checked and corrected, if possible, at the same rate, improving the reliability and MTTF of the memory system. The complexity of the hardware, on the other hand, is a disadvantage of this technique that must be used only in very noisy environments when transient errors occur at a high rate [16].

The behaviour of a memory word in the time interval $0 < t < T$ can be modelled using the Markov chain, as can be seen in Fig. 2.

Assuming that the error rate of every bit is λ and that the different bits are statistically independent, the error rate of a word when being in state a is:

979-8-3503-8368-3/23 $31.00 © 2023 IEEE

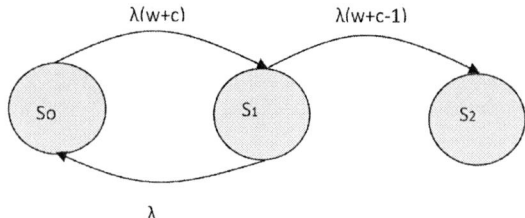

Figure 2: State transition rate diagram for deterministic scrubbing

$$L = \lambda + \lambda + \lambda + \lambda + ... + \lambda = \lambda(w + c) \quad (9)$$

The transition from state S_1 to state S_0 occurs when an error occurs in the faulty bit, cancelling the initial one [rate λ].

When the word is in state S_1, the occurrence of an error in any bit other than the faulty leads the memory word to state S_2 [rate $\lambda(w + c - 1)$].

The differential equations that describe the system are:

$$dP_0(t)/dt = -\lambda(w + c)P_0(t) + (\lambda)P_1(t) \quad (10)$$

$$dP_1(t)/dt = \lambda(w + c)P_0(t) - (\lambda(w + c))P_1(t) \quad (11)$$

$$dP_2(t)/dt = (\lambda(w + c - 1))P_1(t) \quad (12)$$

By using the Laplace transforms of the aforementioned equations and keeping in mind that the following initial conditions apply:

$$P_0(0) = 1, P_1(0) = 0, P_2(0) = 0$$

The reliability of one word in the time interval (0,t) is given by the type:

$$r_0(t) = \frac{a_{1d}}{a_{1d} - a_{2d}}exp(-a_{2d}t) + \frac{a_{2d}}{a_{2d} - a_{1d}}exp(-a_{1d}t) \quad (13)$$

The reliability of a word at a specific moment is established by the multiplication of two probabilities: the probability of its survival throughout all the preceding scrubbing intervals and the probability of enduring the additional time until the present interval.

$$r(t) = [r_0(t)]^n r_o(x), t - nT + x, 0 \le x \le T \quad (14)$$

At time nT, the reliability of a word is determined by the probability that it has endured without more than one error co-existing in the word during all the preceding n time intervals. Consequently, it can be expressed as the reliability at time T raised to the power of n. Since all memory words are considered to be statistically independent, the overall reliability of the memory system can be calculated as follows:

$$R(t) = r(t)^M \quad (15)$$

The MTTF is:

$$MTTF = \int_0^\infty R(t)\, dt \quad (16)$$

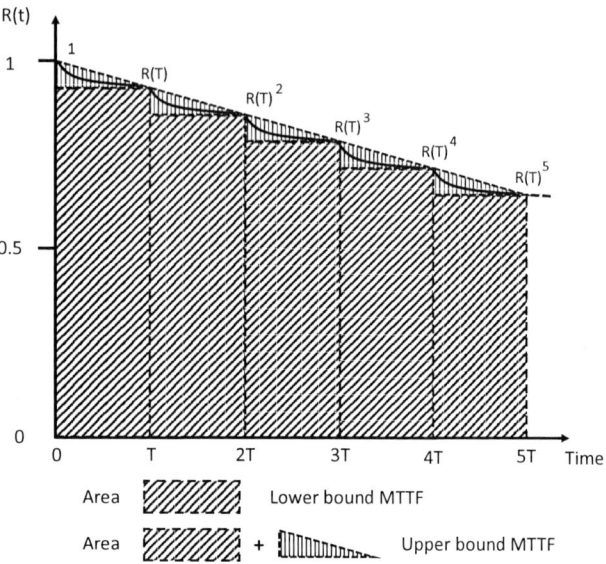

Figure 3: MTTF for deterministic scrubbing

The calculation of the MTTF can be achieved using the following approximation. Fig. 3 shows that the MTTF has an upper (MTTFu) and lower bound (MTTFl). Using that, we can say that $MTTFl < MTTF < MTTFu$.

$$MTTFl = \frac{TR(T)}{1 - R(T)} \quad (17)$$

$$MTTFu = \frac{T[1 + R(T)]}{2[1 - R(T)]} \quad (18)$$

So the MTTF bounds are:

$$\frac{TR(T)}{1 - R(T)} < MTTF < \frac{T[1 + R(T)]}{2[1 - R(T)]} \quad (19)$$

C. Mixed Scrubbing

The model described in this section combines probabilistic and deterministic scrubbing. Whenever a word is addressed, it is checked for correctness and rewritten in its initial condition. The rate of probabilistic scrubbing is μ. In order to enhance the system's reliability and the MTTF, a specialized hardware mechanism is introduced to systematically traverse through the memory and scrub each word [17]. The time interval of the deterministic scrubbing is T.

The Markov chain that describes the behaviour of one memory word in the time interval (0, t) is the same as the probabilistic scrubbing [18]. The transition of the system from one state to the other follows the same rules as in the probabilistic scrubbing, and the final form of the reliability of one memory word in the time interval (0, t) is:

$$r_0(t) = \frac{a_{2m}}{a_{2m} - a_{1m}}exp(-a_{1m}t) + \frac{a_{1m}}{a_{1m} - a_{2m}}exp(-a_{2m}t) \quad (20)$$

Using the same reasoning as in the deterministic scrubbing method, we can find that the final form of the reliability of the memory word system at time t is:

979-8-3503-8368-3/23 $31.00 © 2023 IEEE

Table II: Summary of Three Systems

Memory System	Reliability R(t)	MTTF
Probabilistic Scrubbing	$[\frac{a_{1p}}{a_{1p}-a_{2p}}exp(-a_{2p}t) + \frac{a_{2p}}{a_{2p}-a_{1p}}exp(-a_{1p}t)]^M$	$\frac{4\mu}{M(2\lambda+2w-1)^2}$
Deterministic Scrubbing	$[R_d(T)]^n R_d(x), t=nT+x, 0 \leq x \leq T$	$\frac{T[1+R_d(T)]}{2[1-R_d(T)]}$
Mixed Scrubbing	$[R_m(T)]^n R_m(x), t=nT+x, 0 \leq x \leq T$	$\frac{T[1+R_m(T)]}{2[1-R_m(T)]}$

$$R(t) = r(t)^M \qquad (21)$$

Where:

$$r(t) = [r_0(t)]^n r_o(x), t - nT + x, 0 \leq x \leq T \qquad (22)$$

MTTF is also calculated in a similar way as in the deterministic method:

$$\frac{TR(T)}{1-R(T)} < MTTF < \frac{T[1+R(T)]}{2[1-R(T)]} \qquad (23)$$

The summary of the reliability and MTTF can be seen in Table II for the three systems described above. For deterministic and mixed scrubbing, the upper limit of MTTF (MTTFu) is used.

IV. EVALUATION

Table III: MTTF of memory systems with 32-bit words
$\lambda = 0,00001 upsets/bit/day$, w=32, c=7, T=10 sec, $\mu = 0,1sec$

Memory size	Probabilistic	Deterministic	Mixed
1 MB	3,6E+05	6,35E+05	1,02E+06
2 MB	1,8E+05	3,18E+05	5,08E+05
4 MB	9,0E+04	1,58E+05	2,54E+05
8 MB	4,5E+04	7,945E+04	1,27E+05
16 MB	2,2E+04	3,97E+04	6,35E+04
32 MB	1,12E+04	1,99E+04	3,18E+04
64 MB	5,62E+03	9,93E+03	1,59E+04
128 MB	2,81E+03	4,96E+03	7,94E+03

Table IV: MTTF of Memory Systems with 64-bit Words
$\lambda = 0,00001 upsets/bit/day$, w=64, c=8, T=10 sec, $\mu = 0,1sec$

Memory size	Probabilistic	Deterministic	Mixed
1 MB	1,66E+05	2,61E+05	4,47E+06
2 MB	8E+04	1,30E+05	2,23E+05
4 MB	4,15E+04	6,52E+04	1,12E+05
8 MB	2,08E+04	3,26E+04	5,58E+04
16 MB	1,04E+04	1,63E+04	2,79E+04
32 MB	5,19E+03	8,14E+03	1,40E+04
64 MB	2,59E+03	4,07E+03	6,98E+03
128 MB	1,30E+03	2,04E+03	3,19E+03

We conducted extensive simulated scenarios in order to evaluate the MTTF and reliability values for different memory systems with various error rates and memory sizes. The value λ represents the error rate per bit per day, and the values $1/\mu$, T are scrubbing intervals in seconds.

When we compare the values provided in Table III, Table IV, Table V, and Table VI, we observe that the ratio of the MTTF for deterministic scrubbing to the MTTF for

Table V: MTTF of Memory Systems with 32-bit Wwords
$\lambda = 0,0001 upsets/bit/day$, w=32, c=7, T=10 sec, $\mu = 0,1sec$

Memory size	Probabilistic	Deterministic	Mixed
1 MB	3,6E+03	7,22E+03	9,77E+03
2 MB	1,8E+03	3,61E+03	4,89E+03
4 MB	9E+02	1,80E+03	2,44E+03
8 MB	4,5E+02	9,02E+02	1,22E+03
16 MB	2,25E+02	4,51E+02	6,11E+02
32 MB	1,12E+02	2,26E+02	3,06E+02
64 MB	5,62E+01	1,13E+02	1,53E+02
128 MB	2,61E+01	5,64E+01	7,64E+01

Table VI: MTTF of memory systems with 64-bit words
$\lambda = 0,0001 upsets/bit/day$, w=34, c=8, T=10 sec, $\mu = 0,1sec$

Memory size	Probabilistic	Deterministic	Mixed
1 MB	1,66E+03	3,32E+03	4,52E+03
2 MB	8,30E+02	1,66E+03	2,26E+03
4 MB	4,15E+02	8,29E+02	1,13E+03
8 MB	2,08E+02	4,14E+02	5,64E+02
16 MB	1,045E+02	2,07E+02	2,82E+02
32 MB	5,19E+01	1,03E+02	1,41E+02
64 MB	2,59E+01	5,18E+01	7,05E+01
128 MB	1,30E+01	2,59E+01	3,53E+01

Table VII: The Comparison of MTTF of the Three Memory Systems

(T=10, μ=0.1), $1 < M < 128$	Ratio ($MTTF_a/MTTF_b$)
Deterministic / Probabilistic	1.6 - 2
Mixed / Deterministic	1.6 - 1.75
Mixed / Probabilistic	2.5 - 3.5

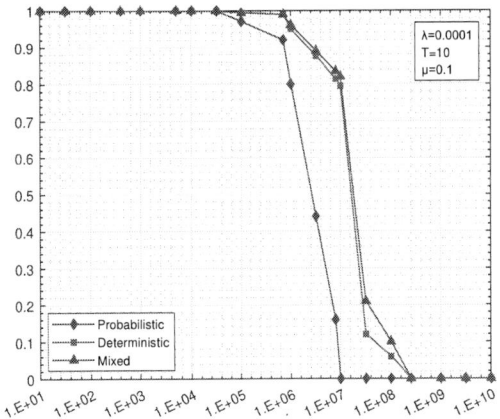

Figure 4: Reliability of memory systems of 128 MB of 32-bit words for various memory systems.

Table VIII: The Effect of Parameters T and μ to the Values of MTTF of the Memory Systems *[λ=0.00001, (w+c)=72, M=128MB]*

(a)	μ=0.1	μ=0.01	μ=0.001
Probabilistic	1300	130	13
Mixed	3490	2618	2528

(b)	T=10	T=100	T=1000
Deterministic	2400	260	26
Mixed	3490	1434	1300

(c)	T=10, μ=0.1	T=100, μ=0.01	T=1000, μ=0.001
Probabilistic	1300	130	13
Deterministic	2400	260	26
Mixed	3490	352	35

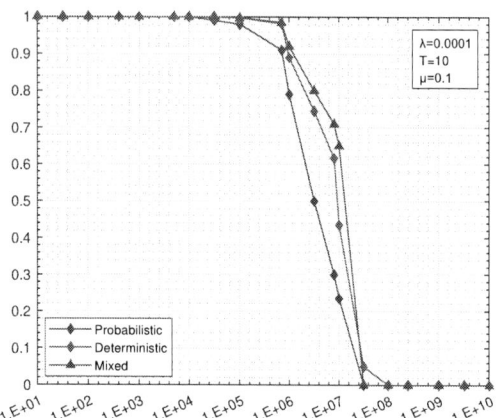

Figure 5: Reliability of memory systems of 128 MB of 64-bit words for various memory systems.

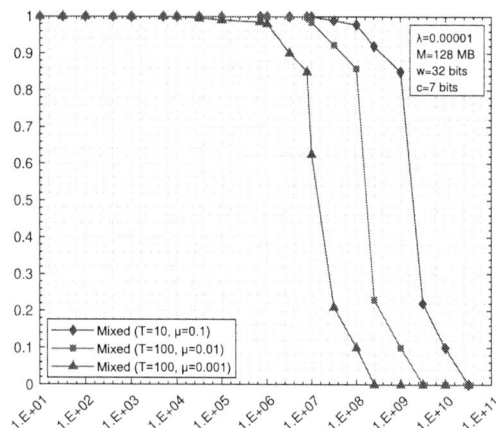

Figure 6: Reliability of memory systems with mixed scrubbing for different values of parameters T and μ.

probabilistic scrubbing falls in the range of 1.6 to 2. This comparison pertains to memory systems with identical scrubbing intervals, where the reciprocal of the mean time between faults equals the scrubbing interval ($1 / \mu = T$). Summarizing these findings, Table VII presents the MTTF for memory systems employing deterministic scrubbing and confirms the mentioned conclusions.

The MTTF of the memory system with mixed scrubbing is relatively less influenced by changes in the values of a single parameter, T or μ. This is attributed to the combined effect of deterministic and probabilistic scrubbing applied to the memory words. This distinct behaviour is evident when compared to memory systems employing either probabilistic or deterministic scrubbing alone, as illustrated in Table VIII.

In Table VIII-a, it is evident that for $\mu = 0.01$, the MTTF of the memory system utilizing probabilistic scrubbing is 130 days. In contrast, the MTTF for the system employing mixed scrubbing is significantly higher at 2618 days. This showcases that with a tenfold increase in the parameter $1/\mu$ (from 10 to 100), the MTTF of the first system reduces by tenfold. In contrast, the MTTF of the memory system with mixed scrubbing only experiences a 24% decrease. Further reducing the parameter μ by tenfold ($\mu = 0.001$), the MTTF of the first system reduces by tenfold, whereas the MTTF of the memory system with mixed scrubbing only decreases by 3%.

Table VIII-b supports similar conclusions, indicating that a tenfold increase in the parameter T (from 10 to 100) results in a 58% decrease in the MTTF of the memory system with deterministic scrubbing.

Lastly, Table VIII-c demonstrates that the MTTF of memory systems employing mixed scrubbing is affected similarly to systems with either deterministic or probabilistic scrubbing when both parameters T and $1/\mu$ are altered.

Fig. 4 and Fig. 5 illustrate the reliability of memory systems concerning error rates and memory size. These figures clearly demonstrate that mixed scrubbing yields the highest reliability, while probabilistic scrubbing results in the lowest reliability when T=$1/\mu$.

In Figure 6, we can observe the reliability of the three memory systems for various values of T and $1/\mu$. The systems being compared in these figures are configured with 32-bit words, a memory size of 128 MB, and an error rate of 0.00001.

Table IX: The MTTF of the Proposed Mixed Scrubbing Compared with Simple Deterministic or Probabilistic Scrubbing

	MTTF (days)	Complexity
Probabilistic ($1/\mu$=10)	2811	Simple
Deterministic (T=20 sec)	2836	Very High
Mixed (T=1 day)	2896	Simple

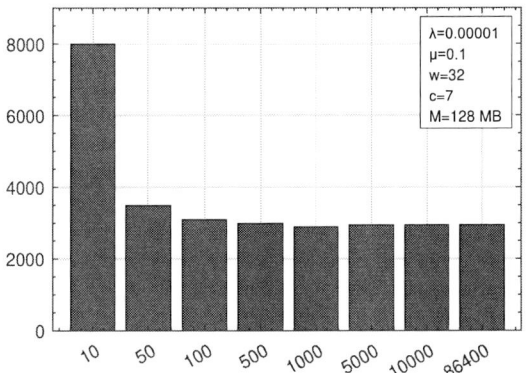

Figure 7: The MTTF of mixed scrubbing for various T.

Figure 7 illustrates the variation in the MTTF of the memory system utilizing mixed scrubbing for different values of the parameter T. It is evident from Figure 7 that the MTTF of the mixed system experiences a gradual decrease before eventually stabilizing at a value closely approximating that of the system employing probabilistic scrubbing, which is 2811 days. Considering the information presented above, we recommend the adoption of mixed scrubbing, particularly when deterministic scrubbing is performed on a daily basis. This adjustment results in reduced hardware complexity, as the T interval can be relatively large. The achieved MTTF is superior to that of probabilistic scrubbing with parameters μ=0.1 and deterministic scrubbing with a parameter T=20. The corresponding results are detailed in Table IX.

V. CONCLUSIONS

In this paper, we have explored and analyzed the reliability of fault-tolerant memory systems using three distinct scrubbing techniques: exponentially distributed scrubbing, deterministic scrubbing, and mixed scrubbing. These models employed Single-Error Correction and Double-Error Detection (SEC-DED) codes and aimed to ensure system reliability in the face of environmental disruptions and inherent circuit weaknesses. Our findings reveal that mixed scrubbing outperforms probabilistic and deterministic scrubbing methods in terms of reliability and MTTF. The mixed scrubbing model's ability to combine the strengths of probabilistic and deterministic approaches while maintaining a simpler architecture, particularly when utilizing a relatively large scrubbing interval, stands out as a key advantage. This approach not only enhances system reliability but also extends the MTTF, making it a compelling choice for memory systems.

The findings of this research suggest that mixed scrubbing is a promising approach for improving the reliability of fault-tolerant memory systems, particularly in scenarios where daily deterministic scrubbing is performed. It offers a robust solution for addressing transient errors and environmental disruptions, making it a valuable addition to the design of fault-tolerant memory systems.

REFERENCES

[1] J. Samanta, J. Bhaumik, and S. Barman, "Compact and Power Efficient SEC-DED Codec for Computer Memory," *Microsystem Technologies*, vol. 27, pp. 359–368, 2021.

[2] V. Vlagkoulis, A. Sari, G. Antonopoulos, M. Psarakis, A. Tavoularis, G. Furano, C. Boatella-Polo, C. Poivey, V. Ferlet-Cavrois, M. Kastriotou *et al.*, "Configuration Memory Scrubbing of SRAM-Based FPGAs Using a Mixed 2-D Coding Technique," *IEEE Transactions on Nuclear Science*, vol. 69, no. 4, pp. 871–882, 2022.

[3] A. Cook, A. Nicholson, H. Janicke, L. Maglaras, and R. Smith, "Attribution of Cyber Attacks on Industrial Control Systems," *EAI Endorsed Transactions on Industrial Networks and Intelligent Systems*, vol. 3, no. 7, p. e3, Apr. 2016. [Online]. Available: https://publications.eai.eu/index.php/inis/article/view/458

[4] T. Kwon, M. Imran, and J.-S. Yang, "Reliability enhanced heterogeneous phase change memory architecture for performance and energy efficiency," *IEEE Transactions on Computers*, vol. 70, no. 9, pp. 1388–1400, 2020.

[5] L. A. Maglaras, M. A. Ferrag, H. Janicke, N. Ayres, and L. Tassiulas, "Reliability, Security, and Privacy in Power Grids," *Computer*, vol. 55, no. 9, pp. 85–88, 2022.

[6] A. Saleh, J. Serrano, and J. Patel, "Reliability of Scrubbing Recovery Techniques for Memory Systems," *IEEE Transactions on Reliability*, vol. 39, no. 1, pp. 114–122, 1990.

[7] J.-C. Baraza-Calvo, J. Gracia-Morán, L.-J. Saiz-Adalid, D. Gil-Tomás, and P.-J. Gil-Vicente, "Proposal of an adaptive fault tolerance mechanism to tolerate intermittent faults in RAM," *Electronics*, vol. 9, no. 12, p. 2074, 2020.

[8] L. Maglaras, M. A. Ferrag, H. Janicke, W. Buchanan, and L. Tassiulas, "Bridging the Gap between Cybersecurity and Reliability for Critical National Infrastructures," in *THE BRIDGE*, vol. 119. The Magazine of IEEE-Eta Kappa Nu, 2023, pp. 14–19.

[9] S. Scargall and S. Scargall, "Reliability, Availability, and Serviceability (RAS)," *Programming Persistent Memory: A Comprehensive Guide for Developers*, pp. 333–346, 2020.

[10] L. Maglaras, H. Janicke, and M. A. Ferrag, "Combining security and reliability of critical infrastructures: The concept of securability," p. 10387, 2022.

[11] Y. Yigit, C. Chrysoulas, G. Yurdakul, L. Maglaras, and B. Canberk, "Digital Twin-Empowered Smart Attack Detection System for 6G Edge of Things Networks," in *2023 IEEE Globecom Workshops (GC Wkshps)*, 2023.

[12] Y. Yigit, K. Huseynov, H. Ahmadi, and B. Canberk, "YA-DA: YAng-Based DAta Model for Fine-Grained IIoT Air Quality Monitoring," in *2022 IEEE Globecom Workshops (GC Wkshps)*, 2022, pp. 438–443.

[13] Y. Yigit, L. D. Nguyen, M. Ozdem, O. K. Kinaci, T. Hoang, B. Canberk, and T. Q. Duong, "TwinPort: 5G Drone-assisted Data Collection with Digital Twin for Smart Seaports," *Scientific Reports*, vol. 13, p. 12310, 2023.

[14] M. A. Ferrag, L. Maglaras, H. Janicke, and R. Smith, "Deep Learning Techniques for Cyber Security Intrusion Detection: A Detailed Analysis," in *6th International Symposium for ICS & SCADA Cyber Security Research 2019 (ICS-CSR)*, 9 2019.

[15] M. Dagli, S. Keskin, Y. Yigit, and A. Kose, "Resiliency Analysis of ONOS and Opendaylight SDN Controllers Against Switch and Link Failures," in *2020 Fifth International Conference on Research in Computational Intelligence and Communication Networks (ICRCICN)*, 2020, pp. 149–153.

[16] G. Secinti, P. B. Darian, B. Canberk, and K. R. Chowdhury, "Resilient end-to-end Connectivity for Software-defined Unmanned Aerial Vehicular Networks," in *2017 IEEE 28th Annual International Symposium on Personal, Indoor, and Mobile Radio Communications (PIMRC)*, 2017, pp. 1–5.

[17] L. Maglaras, "From Mean Time to Failure to Mean Time to Attack/Compromise: Incorporating Reliability into Cybersecurity," p. 159, 2022.

[18] D. A. Santos, A. M. P. Mattos, D. R. Melo, and L. Dilillo, "Enhancing Fault Awareness and Reliability of a Fault-Tolerant RISC-V System-on-Chip," *Electronics*, vol. 12, no. 12, p. 2557, Jun 2023. [Online]. Available: http://dx.doi.org/10.3390/electronics12122557

Accelerate Processing of Image with the Keccak-512 Algorithm on Cryptoprocessor

Argyrios Sideris[1], Theodora Sanida[1], Maria Vasiliki Sanida[2], Michael Dossis[1] and Minas Dasygenis[1]

[1]*Department of Electrical & Computer Engineering, University of Western Macedonia, 50131, Kozani, Greece*
[2]*Department of Digital Systems, University of Piraeus 18534, Piraeus, Greece*
asideris@uowm.gr, thsanida@uowm.gr, sanidasilia@gmail.com, mdossis@uowm.gr, mdasyg@ieee.org

Abstract—**Image hashing is a crucial technique for content authentication and image retrieval in various applications, including multimedia databases and copyright protection. This study introduces a novel approach to image hashing that leverages the capabilities of FPGAs and harnesses the cryptographic strength and computational efficiency of the Keccak-512 algorithm. Our experimentations were conducted on the Intel Arria 10 GX FPGA board, and the versatile Nios II enhanced microprocessor. The results gleaned from these experiments underscore the tremendous potential of our FPGA-based image-hashing approach. It presents an alluring solution, merging efficiency with security, effectively addressing the burgeoning demands for content authentication and image integrity verification across diverse image-centric applications. This approach provides a powerful tool for addressing the increasing demand for image authentication and integrity verification in the digital age.**

Index Terms—**Cryptography, Keccak-512, NIOS II enhanced microprocessor, Image hashing, FPGA.**

I. INTRODUCTION

The rapidly evolving digital landscape has created an era where visual content authentication and integrity verification have become paramount. As a pivotal technique, image hashing is crucial in addressing these imperatives, permeating domains such as healthcare, robotics, defence, meteorology, commerce, and more. Consequently, the mission of the cryptographic community has evolved to encompass the development of specialized image hash features dedicated to fortifying the security of visual information [1], [2].

So, the demand for robust solutions in visual content authentication and integrity verification continues to soar to the ever-expanding digital frontier. As this transformative wave of digitalization sweeps across the globe, the cryptographic community stands at the forefront, shouldering the responsibility of innovation. Their mission is not merely to adapt but to lead as they steer the development of specialized image hash technologies meticulously designed to reinforce the fortifications around the integrity and authenticity of visual information. This evolution in cryptographic research and application mirrors the urgency with which our digital society seeks to preserve the trustworthiness of the visual content that increasingly defines our interactions and decision-making in this digital age [3]–[5].

Within the scope of this study, we have developed and integrated the distinguished Keccak-512 algorithm into the FPGA Intel Arria 10 GX board [6]. To assess the effectiveness of our approach, we have conducted comprehensive comparisons with alternative designs, employing conventional evaluation metrics such as throughput and efficiency. Through these evaluations, we aim to shed light on our approach's unique advantages and performance characteristics, contributing valuable insights to the broader field of research and technology development.

This article brings forth two substantial contributions to the field of image hashing:

- We introduce a pioneering approach for adapting the renowned Keccak image processing algorithm, mainly focusing on 512 x 512 pixels resolution images. We aim to significantly enhance FPGA devices' acceleration and overall performance in image hashing applications. Our selection of the Keccak variant with an output size of 512 bits is deliberate, as it offers heightened security levels and data safeguards.

- Within our novel technique, we introduce a methodological enhancement leading to a notably reduced cycle count, optimizing the overall evaluation metrics. This improvement is instrumental in enhancing the efficiency and throughput of our approach, making it ideal for image hashing and cryptographic applications.

The subsequent sections of this document follow this organizational structure: In the forthcoming Section II, we review related studies that share similarities with our work. In Section III, we delve into our image hashing implementation on an FPGA device. Section IV is dedicated to presenting a comprehensive evaluation of the performance achieved through our experimental work. Our study's outcomes and prospects for future investigation are outlined in Section V.

II. RELATED WORK

In this part, our focus shifts towards an in-depth exploration of works that resemble the scope and objectives of our research. The chosen articles were for their relevance to our work but also their intrinsic merit in pioneering innovative applications of the Keccak algorithm within the realm of image hashing [7]

The authors of [8] focus on the performance implementation of the Keccak-256 algorithm for hashing image sizes 256 x 256 pixels. The primary objective of the study is to produce

979-8-3503-8368-3/23 $31.00 © 2023 IEEE

high-throughput and high-area solutions. An Intel DE2-115 device was used in their study, and VHDL was the programming language of choice for the implementation. Their design for the Keccak-256 algorithm was able to attain a throughput rate of 27.821 Gbps, an efficiency rate of 10.62 Mbps/Slices, a maximum frequency rate of 378 MHz, and an area (slices) of 2624. They have, additionally, reached a maximum frequency of 378 MHz.

In the study [9], the authors set their sights on encrypting 256 x 256 grayscale images. Their approach was done utilizing VHDL on the Intel Arria 10 GX. The design proposed for the Keccak-256 algorithm attained a throughput rate of 35.593 Gbps. Furthermore, they earned the maximum operating frequency of 458 MHz. Regarding the hardware resource utilization aspect, the area occupation is 2.984 slices. Finally, they gained an efficiency rating of 11.92 Mbps/Slices.

In contrast to prior investigations, we implemented Keccak-512 designs using the Nios II/fast enhanced microprocessor focusing on images with a 512 x 512 pixels resolution. The method presented in this study offers a secure Keccak-512 technique and aims to significantly enhance FPGA devices' acceleration and overall performance in image hashing applications.

III. IMAGE HASHING IMPLEMENTATION ON ARRIA 10 GX FPGA DEVICE

In this part, we will break down the individual components of the technique employed for the Keccak-512 algorithm, focusing on the specific tools and hardware utilized in our experiments. Our chosen tools include Quartus II version 18.3 Standard Edition (SE) and the Arria 10 GX FPGA device, which form the foundation of our experimentation setup.

Quartus II version 18.3 SE serves as our primary design and development environment. Quartus II is a well-established software suite for designing and implementing digital systems on FPGA platforms. The Standard Edition offers comprehensive features and capabilities instrumental in creating and optimizing FPGA designs. It provides a user-friendly interface for hardware description language coding, synthesis, simulation, and place-and-route operations.

The choice of the Intel Arria 10 GX FPGA board as our hardware platform is pivotal to our experimentation. It offers a range of resources, including programmable logic cells, embedded memory, and high-speed transceivers, making it well-suited for demanding applications like cryptographic algorithms.

A. Nios II/fast enhanced microprocessor

The Nios II/fast enhanced microprocessor is designed to meet the demands of a wide range of embedded systems and real-time applications. What distinguishes the Nios II/fast enhanced microprocessor is its remarkable balance between performance and resource utilization, making it an exceptional choice for FPGA-based designs [10] One of the standout features of the Nios II/fast enhanced microprocessor is its configurability. This adaptability allows for creating highly optimized solutions, minimizing resource consumption and maximizing processing efficiency.

B. System design of the Keccak-512 core

Figure 1 visually represents our Keccak-512 approach design, providing a concise overview of its architectural components and their roles in the algorithm's execution. At the heart of this architecture lies a series of essential elements that collectively facilitate data's secure and efficient hashing.

First and foremost, we have the zero state, a fundamental component meticulously crafted to serve as the initial state for the Keccak-512 algorithm. This zero state sets the stage for the algorithm's iterative processing, acting as a blank canvas onto which data will be iteratively incorporated and transformed. It is important to note that the original zero states, stored within the zero state component, play a specific and pivotal role in the initial iteration of the Keccak algorithm.

The bitrate Xor input plays a pivotal role in the algorithm, serving as a channel through which external data is introduced into the computation. This input is XORed with the zero state, marking the beginning of the data absorption process crucial to Keccak's hashing mechanism.

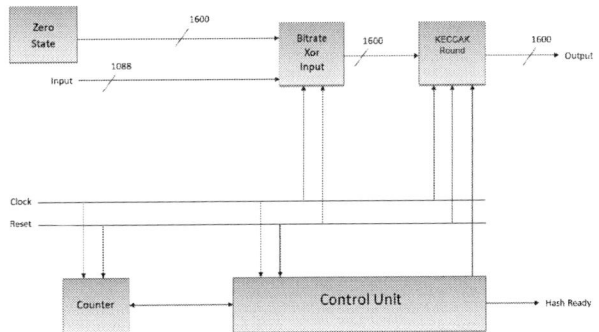

Fig. 1. Keccak-512 approach design overview.

A control unit is responsible for managing the various stages of the algorithm's operation. It ensures the orderly progression of data through the different components of the design, regulating the flow and ensuring the algorithm's integrity.

A counter is a crucial part of the architecture, responsible for keeping track of the number of iterations or rounds completed during the Keccak-512 computation. This iterative nature is inherent to the algorithm and is vital for its security properties.

Lastly, the Keccak round represents a fundamental processing step within the algorithm. This component embodies the core operations that manipulate and transform data as it moves through the algorithm's rounds, ultimately leading to the process of creating the final hash value.

C. The Keccak round design component for 512-bit output size

The Keccak cryptographic algorithm is characterized by 24 permutation rounds, each comprising five successive phases: theta, rho, pi, chi, and iota. These phases work together harmoniously to transform the state array as data moves through the algorithm. In the Keccak algorithm, the state array is the central data structure that is updated at every step as each permutation round is executed. This array serves as the canvas upon which the cryptographic operations are applied, and its transformation is integral to the algorithm's hashing process.

Figure 2 visually represents the 2-stage pipelined format inherent to the Keccak Round. This structure optimizes the algorithm's performance by efficiently managing data flow and processing. Notably, a 2-to-1 multiplexer is employed at the onset of the round to facilitate feedback. Within this pipelined approach, each register is input independently, ensuring efficient data handling.

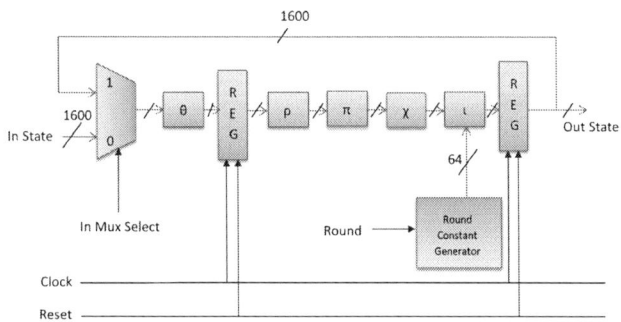

Fig. 2. The Keccak-512 round design component.

To minimize the critical path, the first register is strategically positioned between the theta and rho stages of the algorithm, a design choice that enhances the algorithm's processing speed. Additionally, as the round draws to a close, the second register is introduced slightly before the feedback part, allowing for efficient data capture and manipulation throughout the round's execution. These 2 registers are controlled by clock and reset signals, ensuring precise synchronization and functionality within the algorithm.

D. Integrative system method

Our work started with an original grayscale image boasting a pixel resolution of 512 by 512 pixels. The first step in our data flow involved storing this image data efficiently in SDRAM memory, which served as the designated storage medium for the input block of data. Subsequently, this stored data block was meticulously fed as input into the Keccak-512 core, a crucial component in our cryptographic algorithm. To ensure the reliability and precision of our system, we implemented each essential component of this architecture using VHDL, a powerful hardware description language. This approach allowed us to tailor and fine-tune each component to meet the specific requirements of our project.

To validate the correctness and functionality of our VHDL implementations, we conducted a battery of rigorous tests using various test benches. These test benches independently evaluate each VHDL file, ensuring that every element of our system operates as intended. The ModelSim 10.6d simulator, known for its robust simulation capabilities, was the tool of

choice for running these tests. We utilized authentic input examples supplied by the NIST for the Keccak-512 algorithm, guaranteeing that our tests were conducted under real-world conditions.

Furthermore, to provide a comprehensive overview of our project's architecture, we meticulously designed and documented block diagrams for each individual VHDL file within the top module. To solidify the validity of our results, we leveraged the precise input benchmarks provided by NIST for the Keccak-512 algorithm. These benchmarks were invaluable in rigorously assessing our system's performance and verifying its adherence to established cryptographic standards. Moreover, as part of our project's overarching goals, we embarked on developing and enhancing the Nios II microprocessor, ensuring that it seamlessly integrated with our cryptographic algorithm. The aim was to corroborate our simulation findings using ModelSim and ensure that our system operated effectively in the real world.

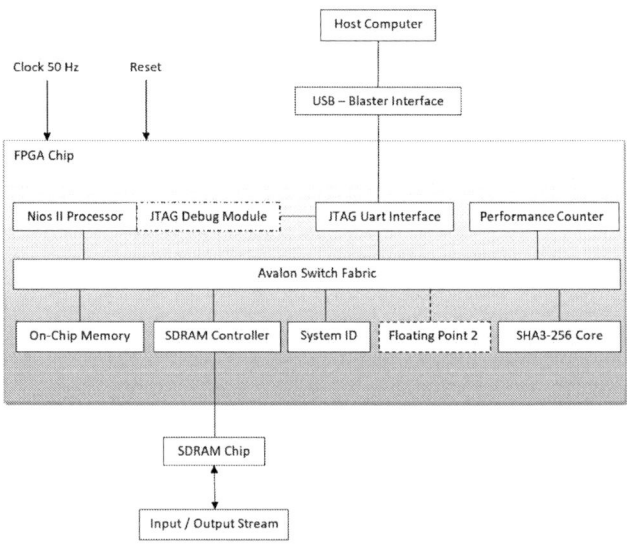

Fig. 3. Integrative system method on Intel Arria 10 GX FPGA board.

Several vital components were thoughtfully implemented within the microprocessor design to ensure seamless functionality and optimal performance. These components include a clock subsystem to maintain precise timing, a Phase-Locked Loop (PLL) for frequency synthesis, a DDR4 SDRAM controller for efficient memory access, a peripheral ID system for component identification, a performance counter to monitor and optimize design performance, on-chip RAM that performs as the essential functional memory for the Nios II-CPU, a JTAG-UART for debugging and communication, and the core of our cryptographic operations, the Keccak-512 module.

One notable feature of our design is the incorporation of on-chip RAM, strategically utilized as the primary functional memory for the Nios II-CPU. This on-chip memory ensures rapid data access and processing, contributing significantly to the design's overall performance. Figure 3 visually shows the comprehensive design of our method, demonstrating the inte-

979-8-3503-8368-3/23 $31.00 © 2023 IEEE

gration of the Nios II enhanced microcontroller into our work framework, and Figure 4 shows the experimental application.

IV. THE OUTCOMES OF IMAGE HASH

The calculation of throughput (TGT) is a fundamental metric in assessing the performance of our system, and it is determined using Equation (1). So, in Equation (1), the number of bits parameter corresponds to the bitrate size denoted as r. This indicates the size of the data blocks being processed by our system, reflecting the fundamental unit of information that our circuit handles. The frequency parameter represents the maximum frequency reported by our design tool, a crucial factor in assessing the speed at which our circuit operates. This frequency value directly influences the overall throughput of the system, as it determines how quickly data can be processed.

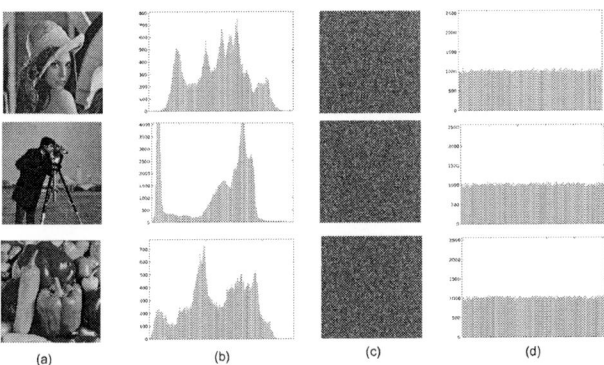

Fig. 4. (a) image, (b) histogram of the (a), (c) cipher and (d) histogram of the (c) on Intel Arria 10 GX FPGA board.

The number of clock cycles parameter signifies the latency of our circuit, which, in turn, reflects the number of clock cycles required for the 5 essential functions, to collectively generate the hash value. This metric is indicative of the efficiency and speed with which our cryptographic operations are executed.

$$TGT = \frac{Number\ of\ bits\ \star\ frequency}{Number\ of\ clock\ cycles} \quad (1)$$

The efficiency (EFC) is calculated by utilizing Equation (2).

$$EFC = \frac{TGT}{Area} \quad (2)$$

Our design takes up 2402 slices of the FPGA, a testament to its efficient use of hardware resources. Operating at a commendable frequency of 486 MHz, our system achieves a noteworthy throughput of 12.724 Gbps when handling a bitrate size of 576 bits. This data processing rate is critical to our system's ability to quickly and effectively hash data. What truly stands out in these results is the calculated efficiency of our design, which measures 5.297 Mbps per Slices. This figure signifies a balance that we have struck between resource utilization and data processing capabilities. It underscores the optimization of our design, ensuring that we make the

most efficient use of the FPGA's resources while delivering a competitive level of data processing performance.

V. CONCLUSIONS AND FUTURE WORK

This study showcases our extensive investigation into achieving optimal performance in the hashing of 512 x 512-pixel images through the utilization of the Keccak-512 algorithm, seamlessly integrated with the Nios II/fast enhanced microprocessor on the Intel Arria 10 GX FPGA board. Our comprehensive experiments centred around the innovative 2-staged pipelined design have yielded promising results, significantly enhancing the throughput and efficiency of the Keccak algorithm for image hashing applications. Our findings underscore the remarkable speed, performance, and security features inherent in the presented algorithm. Combining the computational power of the Nios II/fast enhanced microprocessor with the advanced capabilities of the Keccak-512 algorithm, we have forged a synergy that results in an exceptional solution tailored specifically for the task of hashing 512x512-pixel images. This approach enhances the processing speed, enabling swift and efficient hashing of image data and ensures robust data protection, a critical aspect in applications involving sensitive image data. Our analysis paves the way for enhanced security and efficiency in a variety of image-related applications, ranging from data integrity verification.

ACKNOWLEDGMENTS

This work was partially supported by the University of Western Macedonia.

REFERENCES

[1] W. Villegas-Ch, J. García-Ortiz, and J. Govea, "A comprehensive approach to image protection in digital environments," *Computers*, vol. 12, no. 8, p. 155, 2023.

[2] A. Sideris, T. Sanida, and M. Dasygenis, "Hardware acceleration design of the sha-3 for high throughput and low area on fpga," *Journal of Cryptographic Engineering*, pp. 1–13, 2023.

[3] M. Rana, Q. Mamun, and R. Islam, "Lightweight cryptography in iot networks: A survey," *Future Generation Computer Systems*, vol. 129, pp. 77–89, 2022.

[4] A. Sideris, T. Sanida, and M. Dasygenis, "A novel hardware architecture for enhancing the keccak hash function in fpga devices," *Information*, vol. 14, no. 9, p. 475, 2023.

[5] J. Ruiz-Rosero, G. Ramirez-Gonzalez, and R. Khanna, "Field programmable gate array applications—a scientometric review," *Computation*, vol. 7, no. 4, p. 63, 2019.

[6] A. Sideris, T. Sanida, and M. Dasygenis, "High throughput implementation of the keccak hash function using the nios-ii processor," *Technologies*, vol. 8, no. 1, p. 15, 2020.

[7] A. Sideris and M. Dasygenis, "Enhancing the hardware pipelining optimization technique of the sha-3 via fpga," *Computation*, vol. 11, no. 8, p. 152, 2023.

[8] A. Sideris, T. Sanida, A. Chatzisavvas, M. Dossis, and M. Dasygenis, "High throughput of image processing with keccak algorithm using microprocessor on fpga," in *2022 7th South-East Europe Design Automation, Computer Engineering, Computer Networks and Social Media Conference (SEEDA-CECNSM)*. IEEE, 2022, pp. 1–4.

[9] A. Sideris, T. Sanida, D. Tsiktsiris, and M. Dasygenis, "Image hashing based on sha-3 implemented on fpga," in *Recent Advances in Manufacturing Modelling and Optimization*. Springer, 2022, pp. 521–530.

[10] H. Cao and U. Meyer-Baese, "Xml-based automatic nios ii multiprocessor system generation for intel fpgas," *Electronics*, vol. 11, no. 18, p. 2840, 2022.

A Leveraging Matterport for Industry-Focused Mobile Applications: Augmented Reality Training for Vocational Education and Training

Dimitrios Kiriakos
*Department of Business Administration,
University of West Attica*
Attica, Athens, Greece
kiriakos@uniwa.gr
[0000-0002-2736-2074]

Dimitrios Kotsifakos
*Department of Informatics, School of
Information and Communication
Technologies,*
University of Piraeus, Piraeus, Greece
kotsifakos@unipi.gr
[0000-0002-6656-260X/print]

Yannis Psaromiligkos
*Department of Business Administration,
University of West Attica*
Attica, Athens, Greece
yannis.psaromiligkos@uniwa.gr
[0000-0002-8420-8663]

Abstract—Although the 4th Industrial Revolution has already progressed, traditional manufacturing, factoring plants, and repetitive tasks affect production schedules and increase final costs. Augmented Reality technology through mobile devices, with the display of Three-dimensional, designs in combination with the exploitation of the capabilities of Augmented Reality software scanning, measurement, but also interactions of virtual objects with notes that convey data and information about the real world and spatialization of a location. This paper explores the innovative use of Augmented Reality technology and the Matterport platform in Vocational Education and Training. The Vocational Education and Training laboratories employ Matterport's Augmented Reality capabilities, to train students in scanning and creating Augmented Reality applications, enabling them to develop industry-specific mobile applications for various sectors. In many specializations of Vocational Education and Training laboratories are utilized to create immersive virtual environments that students can explore and interact with using their mobile devices. The research focuses on the integration of the Matterport platform into the Vocational Education and Training curriculum, emphasizing its potential for enhancing vocational training. In our paper, we present the strategic planning for the enrichment of the application in the new Vocational Education and Training curricula and the first reporting elements from this development.

Keywords—Mobile Applications, Augmented Reality Training, Vocational Education and Training

I. INTRODUCTION. FOCUS ON A REAL PROBLEM: AN UP-TO-DATE EDUCATION OF ENGINEERS AND VOCATIONAL EDUCATION AND TRAINING GRADUATES

This article suggests a solution to a problem that is observed during the production processes, mainly in traditional factories of construction and organization of raw materials [1]. In some of these traditional factories, a lack of proper coordination between the planning of actions and the practical applications and implementations in the factory plan was observed. This in turn causes delays in production schedules, which increases the final cost. Thus, the significant discrepancy between the collection of data and the creation of plans and budgets related to the planning of the production of a project, with the outcome of the project, is observed. Both industrial units and traditional factories do not have an up-to-date digital infrastructure and as a result, many project plans are made with traditional two-dimensional (2D) drawings. Our proposal involves the crucial place of engineering education and Augmented Reality (AR) technology through mobile devices [2]. An accurate answer to that problem is to reorganize education and concentrate on the up-to-date education of engineers and Vocational Education and Training (VET) graduates. The basic research idea of the paper is to explore the innovative use of AR technology and the Matterport platform in the context of VET. The authors aim to address a real-world problem observed in traditional manufacturing and production processes, where a lack of proper coordination between planning and practical applications can lead to delays and increased costs. To tackle this issue, the paper suggests the integration of AR technology and Matterport's capabilities into the VET curriculum. Specifically, the research focuses on:

- Utilizing Matterport's AR capabilities to train students in scanning and creating AR applications.

- Enabling students to develop industry-specific mobile applications for various sectors.

- Highlighting projects undertaken by VET laboratories across industries such as the food industry and renewable energy systems.

- Emphasizing the potential for enhancing vocational training through AR technology and Matterport.

Discussing the benefits of integrating AR technology into the curriculum, such as enhanced student engagement, practical skill acquisition, and industry relevance. In summary, the research idea revolves around using AR and Matterport to improve vocational training by providing students with practical experience in creating AR applications and preparing them for industry-specific challenges. The research focuses on the integration of the Matterport platform [3], into the VET curriculum [4], emphasizing its potential for enhancing vocational training. In our paper, we present various projects and teaching scenarios undertaken by VET laboratories that employ Matterport's Augmented Reality (AR) capabilities. Our teaching scenario engages students in scanning and creating Augmented Reality applications, enabling them to develop industry-specific mobile applications for various sectors of VET's laboratories, which span across industries such as the Food industry and Renewable Energy systems (photovoltaic-wind energy). Students from different specializations collaborate to develop mobile applications that showcase their expertise in virtual environments related to these industries. Through the Matterport platform [5], AR applications deliver realistic and interactive experiences, enabling students to learn and engage with the content using their mobile devices. The research methodology involves an analysis of the training process at VET's laboratories, including the integration of Matterport technology, project development, and student feedback. The findings highlight the benefits of using augmented reality and the Matterport platform in VET, such as enhanced student engagement, practical skill acquisition, and industry relevance. In our paper we discuss the potential impact of AR training by combining theoretical knowledge with hands-on experience in creating AR applications, students develop a deep understanding of industry-specific concepts and gain valuable skills sought after by employers. With the display of Three-Dimensional (3D) designs combined with the exploitation of AR software capabilities [6].

By leveraging AR's capabilities, VET's institutions can provide students with a dynamic and effective learning experience, preparing them for the challenges and opportunities of modern industries. The involvement of AR has a catalytic effect on the measurements and calculations of the tasks during a project. Interactions of virtual objects with references and transfers of data and information containing representations from the real world support the completion of a real-world project based on the specific spatial planning of a location. We also explore the innovative use of AR technology and the Matterport platform in VET [7]. The VET laboratories [8], employ Matterport's AR capabilities [9], to train students in scanning and creating AR applications, enabling them to develop industry-specific mobile applications for various sectors. Specializations such as Web Design Development Technicians, Internal Architectural Decoration and Object Designers, Automation Technicians, and Renewable Energy Technicians, are utilized to create immersive virtual environments that students can explore and interact with using their mobile devices. The use of AR technology enables students to gain practical experience and skills in creating immersive digital environments that align with specific industries' needs.

The following sections of this paper delve into the details of how AR training using Matterport can revolutionize vocational education. It examines the approach taken in VET laboratories, highlighting projects undertaken by students from different specializations in sectors like the Food industry and Renewable Energy systems (photovoltaic-wind energy systems). The paper also explores the benefits of integrating Matterport technology into the curriculum, such as enhanced student engagement, practical skill acquisition, and industry relevance. In addition to the present Chapter 1, the paper is divided into four subsequent chapters. The second chapter is entitled Why VET, why now? The Feasibility of the Research" presents the limits of today's social and strategic transformations of the connection of technical education with the forces of labor. Also is presented in what ways the development of knowledge and skills of engineers and graduates of technical education affects the production process. The third chapter "Augmented Reality in VET" delves into the concept of AR and explores its applications in VET. This chapter provides a deeper understanding of how applications of AR and Matreport apply to the VET curriculum. The article closes with the "Conclusions" fourth chapter that presents the perspective, but also the difficulties of AR Training for VET.

II. EASE OF USE. WHY VET, WHY NOW? THE FEASIBILITY OF THE RESEARCH

This research is concentrated on the up-to-date education of Engineers and VET graduates. These are crucial for several reasons (Fig. 1) as Technological Advancements [10], Industry Relevance [11], Meeting Regulatory Requirements [12], Enhanced Problem-Solving Abilities [13], Career Growth and Opportunities [14], and finally Interdisciplinary Collaboration [15]. Engineering fields and vocational industries are constantly evolving due to technological advancements. New tools, techniques, and processes emerge, and professionals need to stay updated to effectively contribute to their respective fields. By receiving up-to-date education, engineers and VET graduates can acquire the latest knowledge and skills required to work with innovative technologies and remain competitive in the job market [16]. The key findings of that study are:

- The match between what people study and the jobs they get is high for technicians,

- The mismatch between intended and destination occupations reflects the generic aspect of VET, and finally,

- The waste of some skills relevant to their current occupation.

The integration of Matterport technology into VET involved several implementations aimed at enhancing vocational education through augmented reality. These implementations were rigorously assessed and evaluated to gauge their effectiveness and impact on students' learning experiences. The assessments were conducted across different time frames and student groups to ensure a comprehensive understanding of the outcomes. These trainings were part of the Erasmus+ KA210 program AR4Food (2022-1-EL01-KA210-VET-000081341).

- *Intensive 3-Week Program (July 2023)*

In July 2023, a 3-week intensive program was conducted, engaging 24 VET students. The objective was to provide an immersive learning experience using Matterport technology. A multifaceted evaluation approach was employed, combining quantitative and qualitative methods. Key assessment parameters included:

Student Engagement: Observations of student participation and interaction with AR-enhanced content.

Skill Acquisition: Evaluation of students' ability to use Matterport for creating AR applications and forming digital twins.

Feedback and Perception: Collection of student feedback through surveys and interviews to gauge their perception of the learning experience.

- *Short-Term Programs (May 2023)*

Two short-term programs were conducted in May 2023, each involving eight (8) VET students. These programs provided a snapshot of the immediate impact of Matterport technology on student learning. Assessments primarily focused on skill acquisition and engagement levels during the short timeframe.

- *Semester-Long Courses (Feb-June 2023)*

Over one semester, Matterport was integrated into the curriculum for two groups, each comprising twenty-five (25) VET students. These students participated in Matterport-enhanced vocational education for three hours per week. Comprehensive evaluations were conducted, including pre- and post-assessments to measure knowledge gain, skill development, and long-term retention. The results can be seen at Thematic Public Institute of Vocational Training Aigaleo [25].

979-8-3503-8368-3/23 $31.00 © 2023 IEEE

- *Teacher Training Course*

A one-week VET teacher training course was conducted to prepare instructors for effectively utilizing Matterport technology in their teaching. The assessments focused on the instructors' ability to adapt AR technology to vocational training, ensuring they could effectively transfer their knowledge to students.

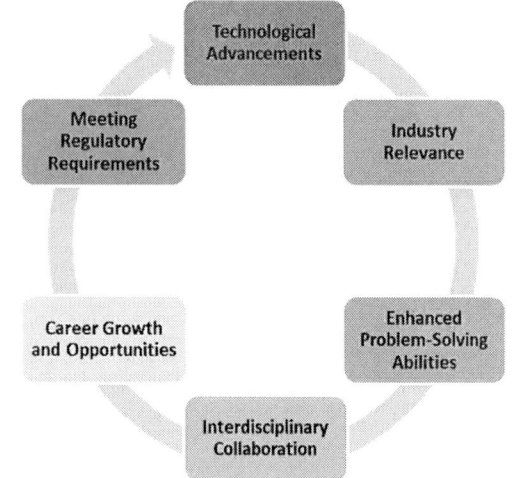

Fig. 1. The reasons for up-to-date education of engineers and VET graduates.

The roadmap for the implementation of our proposals is as follows:

1. Needs Assessment and Planning: Identify the specific vocational education and training programs or courses that can benefit from integrating AR and Matterport technology. Evaluate the existing curriculum and identify areas where AR-based learning can enhance student engagement and practical skill acquisition. Develop a clear plan and timeline for integrating AR and Matterport into the curriculum.

2. Infrastructure and Resources: Acquire the necessary hardware and software resources, including AR-capable devices (e.g., smartphones, tablets), Matterport cameras or scanners, and AR development tools. Ensure that students and instructors have access to the required technology and training to use these tools effectively.

3. Curriculum Design: Redesign or adapt the curriculum to include AR-enhanced learning experiences. Identify specific lessons, projects, or modules where AR can be integrated. Define learning objectives for AR-based activities and align them with the broader educational goals of the program.

4. Content Creation: Train instructors and students on how to create AR content using Matterport and AR development tools. Encourage students to collaborate on creating industry-specific AR applications related to their vocational specializations.

5. Implementation: Incorporate AR-enhanced learning experiences into the curriculum as planned. Ensure that students have access to the necessary AR devices and content. Monitor the progress of students and gather feedback on their AR learning experiences.

6. Evaluation and Assessment: AR integration by measuring student engagement, skill acquisition, and the alignment of learning outcomes with industry needs. Collect feedback from students and instructors on their experiences with AR-based learning. Use performance metrics and user surveys to evaluate the impact of AR on vocational training.

7. Continuous Improvement: Based on the evaluation results and feedback, make necessary adjustments to the AR-enhanced curriculum. Stay updated with advancements in AR technology and consider how new developments can further enhance vocational education. Share best practices and success stories within the educational community to inspire others to adopt similar approaches.

8. Scaling and Expansion: Consider expanding the use of AR and Matterport technology to other VET programs or institutions (e.g. Use Matterport in Health care – capturing an Intensive Care Unit - ICU space). Explore partnerships with industry stakeholders to ensure that AR content remains relevant to current industry needs. Seek opportunities to showcase the achievements of students and instructors in AR-based vocational training.

9. Research and Publication: Conduct research to assess the long-term impact of AR-enhanced vocational education on student career outcomes and industry relevance.

Share research findings and insights through academic publications and presentations to contribute to the broader educational community. By following this roadmap, you can effectively implement the proposals and leverage AR and Matterport technology to enhance VET

Industries rely on engineers and VET graduates to solve complex problems and drive innovation. By staying current with the latest trends and developments in their respective fields, professionals can ensure that their knowledge and skills align with industry needs. This relevance enables them to address industry-specific challenges and contribute meaningfully to their organizations or sectors [17]. Many engineering disciplines have regulatory bodies that set standards and guidelines to ensure safety, quality, and ethical practices. Staying updated with the latest regulations is crucial for engineers to design, implement, and maintain projects in compliance with legal requirements. Similarly, VET graduates working in regulated industries must stay abreast of any changes in regulations to perform their roles effectively and maintain compliance [18]. Technology advancements often bring new challenges that require engineers and VET graduates to think creatively and solve complex problems. Up-to-date education equips professionals with the knowledge and skills needed to analyze and address these challenges. They can apply the latest theories, methodologies, and best practices to develop innovative solutions and improve efficiency in their work [19]. The job market is highly competitive, and employers seek professionals who can contribute immediately and adapt to evolving circumstances. Up-to-date education demonstrates a commitment to continuous learning and professional growth, making engineers and VET graduates more attractive to employers. It opens doors to new career opportunities, promotions, and leadership roles within their fields [20]. Many engineering projects and vocational industries involve collaboration among professionals from various disciplines. By staying updated, engineers and VET graduates can effectively communicate and collaborate with experts from diverse backgrounds. They can understand and integrate ideas from multiple disciplines, leading to more comprehensive and innovative solutions [21].

III. AUGMENTED REALITY IN VET

AR has emerged as a transformative technology that has the potential to "revolutionize" the VET curriculum. By merging virtual elements with the real world, AR enhances the learning experience using digital twins, providing interactive and immersive environments for students. AR in education refers to the technology that overlays digital information, such as images, videos, or 3D models, onto the real-world environment in real time.

A. The Augmented Reality in the VET Curriculum

Unlike virtual reality, which immerses users in a completely computer-generated environment, AR enriches the physical environment with virtual elements. By using various devices such as smartphones, tablets, or smart glasses, users can interact with these virtual elements within their real-world context. The AR in the VET curriculum [22] Enhanced the Learning Experience, Contextualized the Real-world Stuff, Organized Experiential Learning, Personalized Adaptive Learning, and directed collaboration and Remote Learning (Fig 2). AR transforms the learning process from passive to active, enabling students to engage with the subject matter in a more interactive and immersive manner. By visualizing complex concepts and processes through virtual models and simulations, students gain a deeper understanding of the subject matter. AR also provides students with real-world context by overlaying digital information onto physical objects or environments. This contextualization helps students connect theoretical concepts with practical applications, bridging the gap between classroom learning and real-world scenarios. AR enables students to experience hands-on learning in a controlled and safe environment. Through virtual simulations and digital twins, students can practice skills, make informed decisions, and learn from mistakes without real-world consequences. This experiential learning approach fosters skill acquisition and competence development.

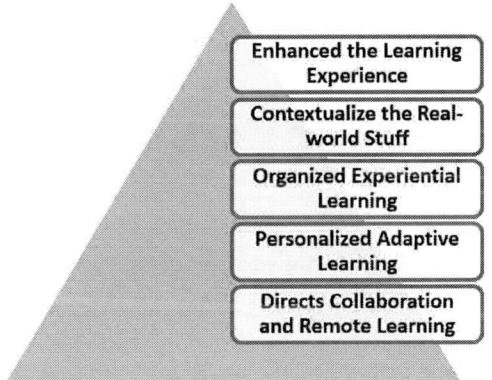

Fig. 2. The AR in the VET curriculum

Finally, AR technology can be tailored to meet individual student's needs and learning styles. Virtual content can be customized to provide targeted guidance, feedback, and additional resources based on each student's progress and performance. AR facilitates collaboration among students, allowing them to work together on virtual projects or simulations. Furthermore, AR enables remote learning by connecting students and instructors in different locations, breaking down geographical barriers, and expanding access to vocational education. AR technology has diverse applications in VET (Fig. 3) in various ways:

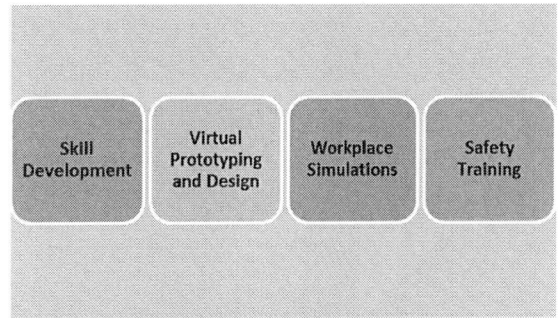

Fig. 3. AR technology has diverse applications in VET.

- Skill Development: AR can be utilized to train students in specific vocational skills, such as machinery operation, equipment maintenance, or technical procedures. By overlaying step-by-step instructions, safety guidelines, and interactive elements onto real-world objects, students can learn and practice these skills in a virtual setting.

- Virtual Prototyping and Design: AR enables students to design and visualize virtual prototypes of products or structures, using digital twins. By overlaying digital models onto physical objects or environments, students can assess the feasibility, functionality, and aesthetics of their designs, making iterative improvements before physical production.

- Workplace Simulations: AR can simulate real-world work environments, allowing students to gain practical experience in industry-specific settings. For instance, students pursuing careers in interior design can use AR to visualize and decorate virtual spaces, while automotive students can simulate repairs and maintenance tasks.

- Safety Training: AR can enhance safety training by creating virtual scenarios that simulate hazardous situations. Students can practice responding to emergencies, identifying risks, and implementing appropriate safety measures, all within a controlled virtual environment.

AR presents numerous opportunities in VET (Fig 4). Several challenges need to be addressed:

- Technological Infrastructure: Implementing AR requires adequate technological infrastructure, including devices, software, and connectivity. Institutions must ensure the accessibility and compatibility of AR tools for students to effectively engage in AR-based learning.

- Training and Support: Educators need proper training and support to effectively integrate AR into the curriculum. They must understand the pedagogical principles behind AR and possess the technical skills to create and deliver AR content.

- Cost and Scalability: AR technology can be costly to implement, considering the hardware, software, and content development expenses. Institutions must consider the cost implications and scalability of AR solutions to ensure long-term sustainability.

- Evaluation and Assessment: Developing effective methods to assess student learning and performance in AR environments is crucial. Institutions need to explore

979-8-3503-8368-3/23 $31.00 © 2023 IEEE

appropriate evaluation strategies that capture the unique aspects of AR-based learning experiences.

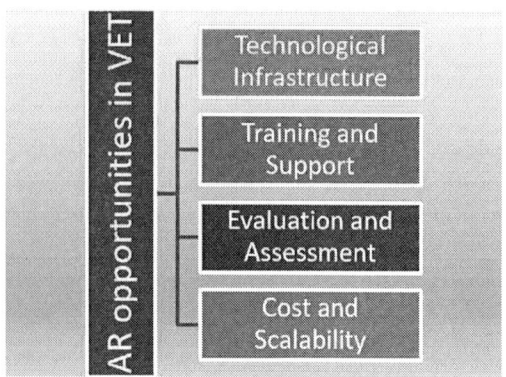

Fig. 4. AR opportunities in VET.

- Technological Infrastructure: Implementing AR requires adequate technological infrastructure, including devices, software, and connectivity. Institutions must ensure the accessibility and compatibility of AR tools for students to effectively engage in AR-based learning.

- Training and Support: Educators need proper training and support to effectively integrate AR into the curriculum. They must understand the pedagogical principles behind AR and possess the technical skills to create and deliver AR content.

- Cost and Scalability: AR technology can be costly to implement, considering the hardware, software, and content development expenses. Institutions must consider the cost implications and scalability of AR solutions to ensure long-term sustainability.

- Evaluation and Assessment: Developing effective methods to assess student learning and performance in AR environments is crucial. Institutions need to explore appropriate evaluation strategies that capture the unique aspects of AR-based learning experiences.

B. Matterport is a leading platform for Augmented Reality in VET

Matterport is the platform for overall superior performance and popularity in virtual tour applications [23]. Matterport is a leading platform that enables the creation, visualization, and interaction with immersive 3D virtual environments. It offers a range of tools and technologies that empower users to capture real-world spaces and transform them into interactive digital models. Matterport is the standard for 3D space capture. Our all-in-one platform transforms real-life spaces into immersive digital twin models. Matterport empowers people to capture and connect rooms to create truly interactive 3D models of spaces (https://matterport.com/). Matterport in education is used as a tool with high potential use. The use of these systems in technical education, particularly in the field between VET and industrial practice is essential [24]. In the context of VET Matterport plays a pivotal role in creating industry-specific AR applications and enhancing the learning experience. Matterport combines hardware and software solutions to capture and create 3D models of physical spaces. It utilizes specialized cameras and scanners to capture high-resolution imagery, depth data, and spatial information. These captured data are then processed through Matterport's software platform, where they are transformed into fully interactive and navigable virtual environments. The Matterport platform also provides an intuitive user interface that allows users to explore virtual spaces from various angles, navigate through rooms, and interact with objects within the virtual environment. Users can access these virtual spaces through web browsers, mobile devices, or Virtual Reality (VR) headsets, providing flexibility and accessibility. The virtual environment offers the students and trainees the ability to interact and experiment with items and constructs in a similar way they would in the real world [25]. Matterport in VET offers several key benefits (Fig. 5):

- Immersive Learning Environments: Matterport enables the creation of immersive and realistic virtual environments that replicate real-world settings. This immersive experience enhances student engagement and allows for a deeper understanding of industry-specific concepts and environments.

- Industry-Specific Applications: Matterport's AR capabilities can be leveraged to develop industry-specific applications in various vocational fields. Students can create virtual environments related to their specialization, such as Web Design Development Technicians, Internal Architectural Decoration & Object Designers, Automation Technicians, and Renewable Energy Technicians. These applications enable students to explore and interact with virtual representations of industry-specific scenarios, fostering practical skill acquisition.

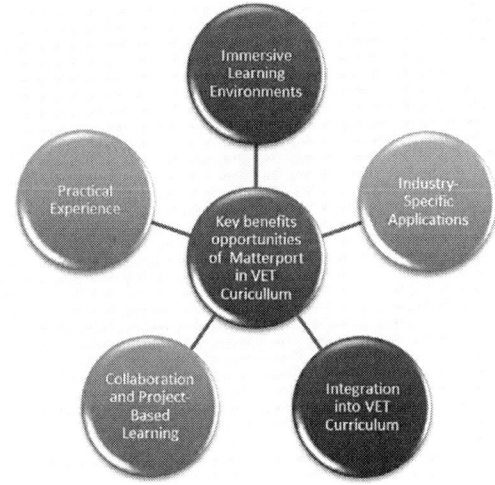

Fig. 5. Key benefits opportunities Matterport offers in VET.

- Practical Experience: Through Matterport, students can gain practical experience by working with virtual models and simulations. They can navigate through virtual spaces, interact with objects, and make informed decisions based on real-world contexts. This hands-on approach prepares students for the challenges they will encounter in their future careers.

- Collaboration and Project-Based Learning: Matterport facilitates collaboration among students by allowing them to work together on virtual projects. They can collaborate remotely, share their virtual environments, and collectively develop industry-specific applications. This promotes teamwork, communication, and project-based learning.

- Integration into VET Curriculum: Matterport can be integrated into the VET curriculum, aligning with specific learning outcomes and competencies. Instructors can design lessons and assignments that utilize Matterport's capabilities to enhance student learning and skill development.

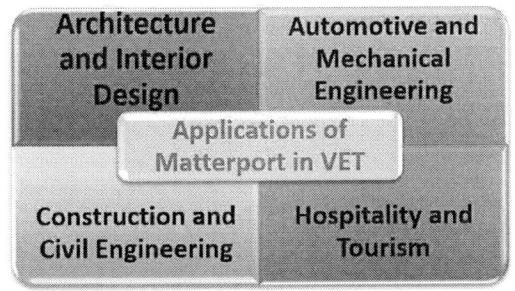

Fig. 6. Example Applications of Matterport in VET.

Matterport can be applied in various vocational domains (Fig. 6), including but not limited to:

- Architecture and Interior Design: Students can use Matterport to create virtual walkthroughs of architectural designs or interior spaces. They can experiment with different layouts, furniture arrangements, and material choices, gaining insights into spatial planning and aesthetics.

- Construction and Civil Engineering: Matterport can be employed to simulate construction sites or building projects. Students can explore virtual environments, understand construction processes, and identify potential design or safety issues before physical construction.

- Automotive and Mechanical Engineering: Matterport can aid in the virtual prototyping of automotive components or mechanical systems. Students can visualize and test different configurations, assess performance, and make design improvements.

- Hospitality and Tourism: Matterport can create virtual tours of hotels, resorts, or tourist attractions. Students can develop interactive experiences that showcase different amenities, layouts, and customer interactions.

IV. EVALUATIONS AND CONCLUSIONS

In our research, we focused on the integration of the Matterport platform into the VET's curriculum, and we emphasized the potential for enhancing vocational training. We present various projects undertaken by VET's laboratories, which span across industries. The generated 3D models will be used to carry out the development work in combination with the tools needed in each case. By combining theoretical knowledge with hands-on experience in creating AR applications, students can develop a deep understanding of industry-specific concepts and gain valuable skills sought after by employers. The mobile nature of the applications enables students to conveniently learn and interact with virtual environments, fostering immersive and interactive learning experiences. By continuously updating their knowledge and skills, professionals can thrive in their fields and contribute meaningfully to technological advancements and societal progress. Through the utilization of AR technology and the Matterport platform, we contribute to the advancement of VET and training, preparing students for the challenges and opportunities presented by the modern industrial landscape. The assessments and evaluations of these implementations consistently highlighted several positive outcomes:

- Enhanced Student Engagement: Matterport implementations significantly increased student engagement, as observed through active participation and interaction with AR-enhanced content.

- Practical Skill Development: Students demonstrated proficiency in scanning and creating AR applications using Matterport, indicating the practical skills gained during the programs.

- Positive Student Feedback: Student feedback was overwhelmingly positive, with many expressing enthusiasm for the immersive and interactive learning experiences facilitated by Matterport.

- Knowledge Gain: Semester-long courses showed substantial knowledge gain, as evidenced by improved performance in pre- and post-assessments.

- Instructor Preparedness: The teacher training course effectively equipped instructors with the skills and confidence to integrate Matterport into their teaching methods.

These assessments and evaluations collectively validate the effectiveness of Matterport technology in enhancing VET, emphasizing its potential for improving skill acquisition, knowledge retention, and student engagement.

The value added by this paper lies in several key aspects:

- Innovative Integration of AR: The paper introduces an innovative approach to integrating augmented reality (AR) technology, specifically through the Matterport platform, into VET. This integration goes beyond traditional teaching methods and demonstrates the potential of AR to enhance practical skill development and industry-specific training.

- Addressing Real-World Challenges: By addressing the challenges faced in traditional manufacturing and production processes, the paper recognizes the need for up-to-date education in engineering and VET. It proposes a solution that leverages AR to bridge the gap between theoretical knowledge and practical applications, directly addressing real-world problems in industrial settings.

- Multi-Industry Applications: The paper showcases the versatility of Matterport-enabled AR applications across various industries, such as the food industry and renewable energy systems. This breadth of application areas demonstrates the adaptability and relevance of the proposed approach to a wide range of vocational disciplines.

- Empirical Evidence: The paper provides empirical evidence of the impact of Matterport implementations in VET. Through detailed assessments and evaluations, it substantiates the positive outcomes, including increased student engagement, practical skill development, and knowledge gain.

- Teacher Training Component: The inclusion of a teacher training course highlights the practicality and scalability of the proposed approach. It emphasizes the importance of preparing instructors to effectively utilize AR technology, thereby ensuring its successful implementation in educational settings.

- Interdisciplinary Collaboration: The paper acknowledges the significance of interdisciplinary collaboration in addressing complex industrial challenges. It emphasizes

979-8-3503-8368-3/23 $31.00 © 2023 IEEE   131

how AR can facilitate collaboration among students from different specializations, promoting teamwork and communication skills.

- Industry Relevance: The integration of AR technology aligns education with industry needs, making graduates more competitive in the job market. The paper recognizes the importance of staying current with industry trends and regulations, which is crucial for professionals in engineering and VET fields.

- Interactive Learning Environments: Matterport-enabled AR applications create immersive and interactive learning environments. This enhances the learning experience by allowing students to visualize complex concepts, explore virtual environments, and practice real-world skills in a controlled setting.

Finally, this paper adds value by presenting a forward-looking approach to vocational education and training that leverages AR technology through the Matterport platform. It addresses practical challenges, provides empirical evidence of its effectiveness, and highlights its adaptability across industries. Furthermore, it underscores the importance of teacher training and interdisciplinary collaboration in preparing students for modern industrial settings, ultimately contributing to the advancement of education and skill development in engineering and VET. The initial functions consider the utilitarian principles of technical constructions, but also the experience of users. The overall gain from this educational extension allows students to conceive of space as a three-dimensional grid measured by the actual distances between objects. In addition, the interaction of objects in the space, the perspectives from their movement, and the creation of annotations in various parts of the virtual workspace become visible. But the most important fact is that it learns to adapt to the set of requirements that a particular workplace may have and discovers customized AR solutions. The result is the development of a set of AR tools that will provide different functions and adapt to each user. Our research includes the up-to-date education of engineers and VET graduates are essential for their professional growth, industry relevance, regulatory compliance, problem-solving abilities, career opportunities, and interdisciplinary collaboration. This enabled the leveraging of user knowledge of the situation and maximizing the strength of AR technology while solving a problem. The main part of the program was implemented at the Public Thematic Institute of Aigaleo, Athens, Greece.

REFERENCES

[1] Skilton, M., & Hovsepian, F.: The 4th industrial revolution. Springer Nature (2018).

[2] Nincarean, D., Alia, M. B., Halim, N. D. A., & Rahman, M. H. A.: Mobile augmented reality: The potential for education. Procedia-social and behavioral sciences, 103, 657-664 (2013).

[3] Kim, M. K., Hwang, D., & Park, D.: Analysis of Maintenance Techniques for a Three-Dimensional Digital Twin-Based Railway Facility with Tunnels. Platforms, 1(1), 5-17 (2023).

[4] Ortega-Gras, J. J., Gómez-Gómez, M. V., Bueno-Delgado, M. V., Garrido-Lova, J., & Cañavate-Cruzado, G.: Designing a Technological Pathway to Empower Vocational Education and Training in the Circular Wood and Furniture Sector through Extended Reality. Electronics, 12(10), 2328 (2023).

[5] Yadav, K., Ramrakhya, R., Ramakrishnan, S. K., Gervet, T., Turner, J., Gokaslan, A., ... & Chaplot, D. S.: Habitat-matter port 3d semantics dataset. In Proceedings of the IEEE/CVF Conference on Computer Vision and Pattern Recognition (pp. 4927-4936) (2023).

[6] Sarkar, P., & Dewangan, O.: Augmented Reality-Based Virtual Smartphone. Journal of Data Acquisition and Processing, 38(2), 1983-1990 (2023).

[7] Spöttl, G., & Windelband, L.: The 4th industrial revolution–its impact on vocational skills. Journal of Education and Work, 34(1), 29-52 (2021).

[8] Fomunyam, K. G.: Education and the Fourth Industrial Revolution: Challenges and possibilities for engineering education. International Journal of Mechanical Engineering and Technology, 10(8), 271-284 (2019).

[9] Fitria, T. N.: Augmented Reality (AR) and Virtual Reality (VR) Technology in Education: Media of Teaching and Learning: A Review. International Journal of Computer and Information System (IJCIS), 4(1), 14-25 (2023).

[10] Marzuki, M. A. B., & Yunus, J. N.: Industry-4.0 Based Teaching and Learning Technology: An Acceptance Investigation Among Mechanical Engineering Lecturers in Higher TVET Education Institution. Advances in Business Research International Journal, 8(3), 1-9 (2022).

[11] Lund, H. B., & Karlsen, A.: The importance of vocational education institutions in manufacturing regions: adding content to a broad definition of regional innovation systems. Industry and Innovation, 27(6), 660-679 (2020).

[12] Bendanillo, A. A., Arcadio, S. M. N., Yongco, J. M. A., Arcadio, R. D., & Arcadio, J. R. N.: Enhancing Technical Proficiency and Industry Readiness: A Comprehensive Diploma Program for Engineering Technology. European Journal of Innovation In Nonformal Education, 3(7), 20-44 (2023).

[13] Ma, J., Wang, S., & Huang, W.: Research on Finite Element Technology in the Development of Mechanics Course Resources. Scientific and Social Research, 5(6), 49-53 (2023).

[14] Frady, K.: Use of virtual labs to support demand-oriented engineering pedagogy in engineering technology and vocational education training programs: a systematic review of the literature. European Journal of Engineering Education, 1-20 (2022).

[15] Christiansen, L., Hvidsten, T. E., Kristensen, J. H., Gebhardt, J., Mahmood, K., Otto, T., ... & Laursen, E. S.: A Framework for Developing Educational Industry 4.0 Activities and Study Materials. Education Sciences, 12(10), 659 (2022).

[16] Oviawe, J. I., Uwameiye, R., & Uddin, P. S.: Bridging skill gap to meet technical, vocational education and training school-workplace collaboration in the 21st century. International Journal of vocational education and training research, 3(1), 7-14 (2017).

[17] Mohammed, U. K., Katken, K. K., Adamu, M. D., & Igwe, C. O.: Curriculum and Industrial Demand: A Tool for Industrial Efficiency. Proceedings of 7th International Conference of School of Science and Technology Education, Federal University of Technology, Minna (2019).

[18] Kaske, A., Torres, R., & Jeon, S.: Promoting work-based learning for vocational teachers. In Technical and Vocational Teacher Education and Training in International and Development Co-Operation: Models, Approaches, and Trends (pp. 67-82). Singapore: Springer Nature Singapore (2022).

[19] Subandi, M. S., Sudjimat, D. A., & Tuwoso, T.: TPACK and HOTS preparation for engineering education students with integrated project-based learning live MOOCs. In AIP Conference Proceedings (Vol. 2590, No. 1). AIP Publishing (2023, May).

[20] Pogatsnik, M.: Dual Training in Engineering Education. In 2023 IEEE 21st World Symposium on Applied Machine Intelligence and Informatics (SAMI) (pp. 000169-000174). IEEE (2023, January).

[21] Wrigley, C., & Mosely, G.: Design thinking pedagogy: facilitating innovation and impact in tertiary education. Taylor & Francis (2022).

[22] Ng, R. Y. K., & Lam, R. Y. S.: Using mobile and flexible technologies to enhance workplace learning in vocational education and training (VET). Innovations in open and flexible education, 85-95 (2018).

[23] Chi, H. Y., Sha, J., & Zhang, Y.: Bring Environments to People–A Case Study of Virtual Tours in Accessibility Assessment for People with Limited Mobility. In Proceedings of the 20th International Web for All Conference (pp. 96-103) (2023, April).

[24] Kuna, P., Hašková, A., & Borza, Ľ.: Creation of Virtual Reality for Education Purposes. Sustainability, 15(9), 7153 (2023).

[25] https://iekaigal.att.sch.gr/?page_id=328

Optimizing Player Engagement in an Educational Virtual Game through Fuzzy Logic-based Challenge Adaptation

Akrivi Krouska, Christos Troussas, Yorghos Voutos, Phivos Mylonas, Cleo Sgouropoulou
Department of Informatics and Computer Engineering
University of West Attica
Egaleo, Greece
{akrouska, ctrouss, gvoutos, mylonasf, csgouro}@uniwa.gr

Abstract—**Adaptive virtual learning environments provide an ideal foundation for enhancing personalized learning experience. Moreover, the incorporation of game elements enhances motivation levels, further enhancing the potential for improving learning outcomes. This paper explores the integration of fuzzy logic as a dynamic adaptivity tool in virtual educational games. By employing user-specific characteristics as input, including student progress, gaming time, and knowledge, the output of the fuzzy logic system represents the challenging level presented to the player in the educational game. Then, a set of rules is applied to tailor the complexity and pacing of challenges presented to learners. This adaptation extends to virtual character behavior, the learning path, and task complexity, aligning with learning objectives and proficiency levels. The study presents a practical implementation of this fuzzy logic-driven adaptivity mechanism in a virtual game environment designed for C++ programming education. The virtual game has been evaluated and the results of the evaluation showcase the potential of this approach in enhancing the learning process by tailoring the environment to learners' specific needs and preferences.**

Keywords—*virtual games, puzzle-based learning, intelligent tutoring system, adaptivity, personalization, fuzzy logic, game difficulty, virtual environment, game-based learning, programming education, adaptive virtual learning environments*

I. INTRODUCTION

Virtual games have primarily focused on entertainment purposes, but during the last decades they have been powerful tools for educational reasons as well [1]. Due to their inherent nature holding immersive and interactive characteristics, they provide an engaging environment for the learners and can support knowledge acquisition [2]. Indeed, the merging of education and entertainment has provoked the interest of many researchers as a means to revolutionize traditional teaching approaches. Needless to say, virtual games align with other technological approaches, such as social media, to provide innovative avenues for enhancing and modernizing education.

The blending of entertainment and education in virtual games environments is a task that hinders compelling opportunities or even obstacles [3]. Education does not simply involve the delivery of information to learners; its main objective is to create a fertile ground for learners to comprehend this information and at the same time to be motivated to learn. On the other hand, entertainment has the ability to engage users and as such it can help towards the creation of captivating learning environments. By combining elements of entertainment, virtual games can be used for education as platforms where learning is an enjoyable and immersive experience.

Due to the different background of users of virtual games, it is of great importance to adapt these environments to their learning needs and interests [4]. Adaptivity techniques stem from the field of artificial intelligence (AI) and are used to provide a personalized environment for the learners to support their evolving preferences. Adaptability is crucial to enhancing engagement and offer a challenging environment to learners that is tailored to their abilities.

There are several AI techniques that can offer adaptation to learning environments. Machine learning techniques, such as artificial neural networks or decision trees, are considered valuable tools for promoting adaptation in learning technology systems. These techniques have the ability to analyze a large pool of data towards identifying patterns among them and subsequently tailor their environment to the learners. Furthermore, reinforcement learning has also been utilized in ameliorating the learning experience in a personalized way. One other technique is fuzzy logic [5]. Despite the numerous adaptation techniques at our disposal, fuzzy logic remains unquestionably a standout choice. Fuzzy logic is a branch of artificial intelligence which is responsible for handling uncertain situations. Indeed, it can model the complexity of several characteristics, including the ones of users, and offer a sophisticated tool for decision making. Fuzzy logic is based on the concept of fuzzy sets, which allows for the representation of linguistic variables as well as gradual transitions between different states rather than strict binary distinctions.

Analyzing the related literature, it can be inferred that adaptivity plays a crucial role in building personalized learning environments. The same also applies in virtual educational environments, where researchers have explored various ways to tailor the learning experience to learners based on their requirements and preferences [6-15]. In more detail, in these works, the researchers have focused on enhancing the engagement, motivation and the learning outcomes through adaptivity. Moreover, within the concept of adaptivity, fuzzy logic has been utilized as a powerful tool for its advancement. Indeed, the researchers have employed fuzzy logic for representing and managing variables like student preferences, cognitive load and emotional states, allowing for a nuanced adaptation of content and challenges to optimize the learning process [16-25].

979-8-3503-8368-3/23 $31.00 © 2023 IEEE

In view of the above, this paper presents a novel adaptivity mechanism for tailoring the challenges presented to users in a virtual educational game. To achieve this, the proposed system uses a fuzzy control system to personalize an educational virtual game by adjusting content, challenges, and virtual character behavior. This system considers inputs like student progress, gaming time, and knowledge to maintain an engaging and adaptive learning experience. The fuzzy control system predicts the game's challenge level, ensuring tasks align with the player's abilities to maintain motivation. Then, several fuzzy rules are applied to provide the key concepts for challenges that will be delivered to users. This adaptation extends to the behavior of virtual characters, the learning path, and the complexity of puzzles and tasks, all designed to align with learning objectives and the students' proficiency level. As a testbed for our research, the virtual game is designed with a focus on facilitating the learning of C++ programming concepts, covering a spectrum from novice to more advanced topics.

II. FUZZY-BASED DETECTION OF CHALLENGING LEVEL

The proposed system provides an innovative way of adapting to student profile using a fuzzy control system. In particular, the developed educational virtual game adjusts its content and challenges to match the individual needs and progress of each student. This feature is a crucial for enhancing the effectiveness of the game by making learning more engaging, personalized, and efficient.

The fuzzy control system is applied to predict the level of challenge in the game. This parameter represents how demanding the game's tasks, puzzles, problems, or challenges are for the player to complete or solve. It encourages the player to stay engaged and motivated to overcome obstacles and learn new concepts. As such, game difficulty matches the player's abilities, resulting in an enjoyable and productive learning experience. Based on the challenging level, the developed game adjusts the following:

- The behavior of the virtual characters, included in the game environment. The virtual characters either facilitates or complicates the navigation of students, by helping guide students or generating obstacles, respectively.

- The learning path, by providing personalized content and advancing the plot according to player profile.

- The puzzles and tasks needed to be solved, ensuring that they are aligned with the learning objectives and match student current level of proficiency.

The fuzzy control system is composed of three primary components: the fuzzification module, the inference engine, and the defuzzification module. In the fuzzification phase, it takes the input crisp set and transforms it into a fuzzy set using triangular membership functions. Afterwards, the inference mechanism computes fuzzy outputs by applying the Mamdani method to combine the active rules of the IF-THEN rule base, consisted of 81 statements. Finally, the defuzzification module applies the Center of Gravity (COG) technique to translate the resultant fuzzy output into a crisp value based on triangular membership functions. Fig. 1 illustrates the architecture of the fuzzy control system employed.

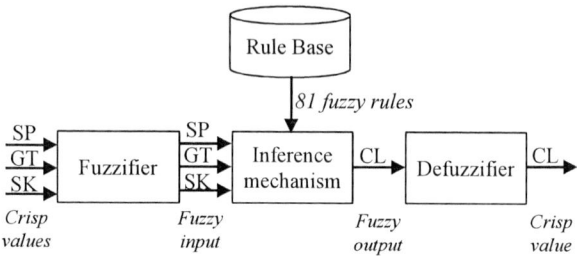

Fig. 1. Architecture of the fuzzy control system

The input set of the fuzzy model includes the following:

- Student progress (SP): It refers to the ratio of the completed puzzles, tasks and missions to the total ones required from the game.

- Gaming time (GT): It is about the ratio of time the student has spent playing to the set timeframe.

- Student knowledge (SK): It concerns the knowledge acquisition measured by the scores achieved in the quizzes and assessments within the game.

The output of the fuzzy model is the following:

- Challenging level (CL): It refers to the degree of difficulty or complexity presented to the player while engaging with the game's educational content.

All the variables are arithmetical values converting into fuzzy ones based on triangular membership functions. Table 1 shows the fuzzy input and output set. For instance, the equations of the membership functions of student progress are the following:

$$
\mu_{LSP}(x) = \begin{cases} 0, & x \leq 0 \\ \dfrac{x}{0.2}, & 0 \leq x \leq 0.2 \\ \dfrac{0.4 - x}{0.2}, & 0.2 \leq x \leq 0.4 \\ 0, & x \geq 0.4 \end{cases}
$$

$$
\mu_{ISP}(x) = \begin{cases} 0, & x \leq 0.3 \\ \dfrac{x - 0.3}{0.2}, & 0.3 \leq x \leq 0.5 \\ \dfrac{0.7 - x}{0.2}, & 0.5 \leq x \leq 0.7 \\ 0, & x \geq 0.7 \end{cases}
$$

$$
\mu_{ASP}(x) = \begin{cases} 0, & x \leq 0.6 \\ \dfrac{x - 0.6}{0.2}, & 0.6 \leq x \leq 0.8 \\ \dfrac{1 - x}{0.2}, & 0.8 \leq x \leq 1 \\ 0, & x \geq 1 \end{cases}
$$

979-8-3503-8368-3/23 $31.00 © 2023 IEEE

Fig. 2 illustrates the scheme of above membership functions. All the other variables are represented correspondingly.

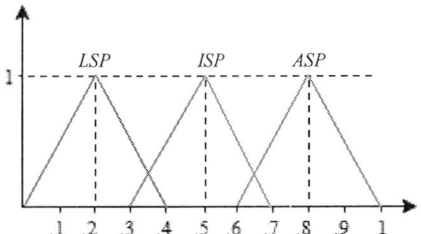

Fig. 2. Scheme of triangular membership functions of Student Progress variable (LSP: Low Student Progress; ISP: Intermediate Student Progress; ASP: Advanced Student Progress)

TABLE I.   FUZZY INPUT AND OUTPUT SET OF DETECTING CHALLEGING LEVEL

Variable	Linguistic Term	Symbol	Interval
Input Set			
Student Progress (SP)	Low	LSP	(0, 0.2, 0.4)
	Intermediate	ISP	(0.3, 0.5, 0.7)
	Advanced	ASP	(0.6, 0.8, 1)
Gaming Time (GT)	Start	SGT	(0, 0.2, 0.4)
	Middle	MGT	(0.3, 0.5, 0.7)
	End	EGT	(0.6, 0.8, 1)
Student Knowledge (SK)	Novice	NSK	(0, 0.2, 0.4)
	Intermediate	ISK	(0.3, 0.5, 0.7)
	Advanced	ASK	(0.6, 0.8, 1)
Output Set			
Challenging Level (CL)	Easy	ECL	(0, 0.2, 0.4)
	Intermediate	ICL	(0.3, 0.5, 0.7)
	Difficult	DCL	(0.6, 0.8, 1)

Regarding the rule base, the formulation of the 81 IF-THEN statements was based on the authors' experience in implementing fuzzy control systems [5, 16, 20], as well as educational games [4, 26]. The rationale behind the construction of these statements is that the more easily and fast a good student solves the game's puzzles/tasks/missions, the more complex the challenges provided to them are. On the other hand, the more a student with low cognitive level finds it difficult and delay to solve the puzzles/tasks/missions, the easier and more supportive the environment becomes. A sample of the defined fuzzy rules is the following:

- **IF** SP = LSP *and* GT = SGT *and* SK = NSK **THEN** CL = ECL

- **IF** SP = LSP *and* GT = SGT *and* SK = ASK **THEN** CL = ICL

- **IF** SP = ISP *and* GT = MGT *and* SK = LSK **THEN** CL = ICL

- **IF** SP = ISP *and* GT = SGT *and* SK = ASK **THEN** CL = DCL

- **IF** SP = ASP *and* GT = EGT *and* SK = ISK **THEN** CL = ICL

- **IF** SP = ASP *and* GT = MGT *and* SK = ISK **THEN** CL = DCL

The scope of the fuzzy rules is to predict the challenging lever properly for adjusting game difficulty dynamically. Therefore, if a student is finding the game too easy, the system can introduce more challenging obstacles and puzzles to keep student experience engaging. Conversely, if the student is struggling, the game simplify tasks and provide additional guidance. To better clarify the functionality of this inference, an example of operation is given. In this case, a student, named Mary, is a very good student (SK = ASK). While she is playing, she solves very fast the puzzles of the game (SP = ISP, GT = SGT). As such, the system provides her more complex challenges to match her needs and abilities (CL = DCL), maintaining her motivation and engagement.

III. SYSTEM EVALUATION AND DISCUSSION

Evaluating educational virtual games is crucial to assess their effectiveness in promoting learning outcomes and engaging students. As such, an evaluation model for assessing the presented system was developed based on the pertinent literature [16]. The model considers four aspects of the game's effectiveness:

- Game engagement and experience, to analyze the game's ability to engage and motivate students.
- Educational effectiveness, to ensure the game meets learning objectives and enhances students' understanding of the subject matter.
- Usability, to ensure the game is user-friendly and effective learning tools.
- User satisfaction, to understand how well the game meets the needs and expectations of learners.

In order to evaluate model's aspects, a questionnaire consists of 12 questions was conducted, namely three question items per aspect. The survey uses a 5-point Likert scale, ranging from "Very Dissatisfying" to "Very Satisfying", to measure respondents' opinion and attitude. Table II shows the questionnaire used for assessing the developed system.

The survey was conducted among 40 university students, namely 22 males and 18 females, having similar characteristics. In particular, all students study at the same class in the same University, being in average 18-22 years old. Prior to using the educational virtual game, students received instructions on its usage and information regarding its features and functionalities. It was noted that students quickly acclimated to the system's usage. Throughout their interaction with the system, they displayed a strong affinity for it and devoted a significant amount of time to using it daily.

TABLE II.   EDUCATIONAL VIRTUAL GAME EVALUATION QUESTIONNAIRE

Aspects	#	Question Item
Game Engagement and Experience	1	How would you rate the overall enjoyment and engagement level of the game?
	2	Did the game motivate you to continue playing and learning?
	3	To which degree the challenging level of the game met your expectations?
Educational Effectiveness	4	Do you believe that this game has helped you improve your knowledge and skills in the subject it covers?
	5	Do you think the game aligns with the educational objectives or learning goals it aims to address?

979-8-3503-8368-3/23 $31.00 © 2023 IEEE   135

	6	Have you noticed any improvements in your problem-solving or critical thinking abilities as a result of playing this game?
Usability	7	How would you rate the ease of use and navigation of the game's interface?
	8	Were the game instructions clear and understandable?
	9	Were the game controls intuitive and responsive?
User Satisfaction	10	How satisfied were you with the game's overall experience?
	11	Were the educational content and challenges in the game relevant to your needs and skill level?
	12	Based on your experience, would you recommend this game to other learners?

The evaluation results are illustrated in Fig. 3 and Fig. 4. The results in Fig. 4 are aggregated into three categories, namely high acceptance emerged from the average percentage of "Satisfying" and "Very Satisfying" answers of the questions belonging to each aspect, average acceptance calculating correspondingly from "Neutral" answers, and low acceptance resulted respectively from "Dissatisfying" and "Very Dissatisfying" answers.

As it is observed, the students expressed their strong acceptance of the system in all evaluation aspects. Regarding the game engagement and experience aspect, the vast majority of the students found the game experience very enjoyable and engaging, with the satisfaction level exceeding 80%. Moreover, it seems that the game motivated them to continue playing and being involved in the learning process (up to 80% the "Satisfying" and "Very Satisfying" answers). More than 80% of the students stated that the challenging level of the game met their expectations.

Concerning the educational effectiveness of the game, the 77.5% of the participants believe that the game helped them improve their knowledge and skills in programming subject. Also, the same percentage found the game consistent with the learning objectives. The 72.5% of the students was satisfied with the extent to which playing the game has helped them improve their problem-solving and critical thinking abilities.

The usability of the system was rated highly, with satisfying level around 80%. The students found the game's interface ease of use, as well as its instructions clear and understandable. Moreover, it seems they did not face technical problems while playing the game, since they found its controls intuitive and responsive.

Regarding user satisfaction, the vast majority of the participants (more than 80%) were immensely satisfied with the overall game experience, having the intention to recommend it to other students. Furthermore, the survey shows that the educational content and the challenging level provided by the game were relevant to players' needs and abilities.

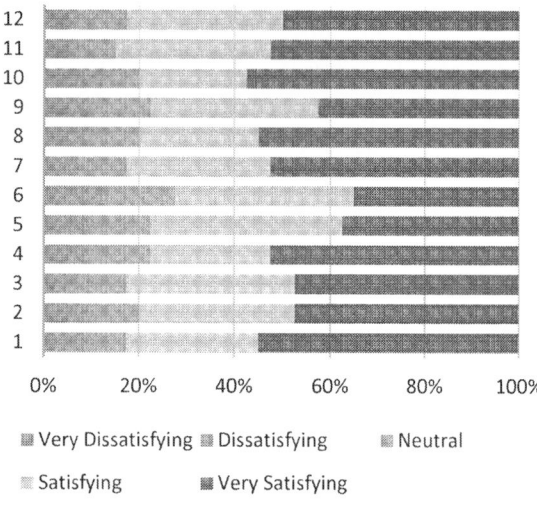

(a)

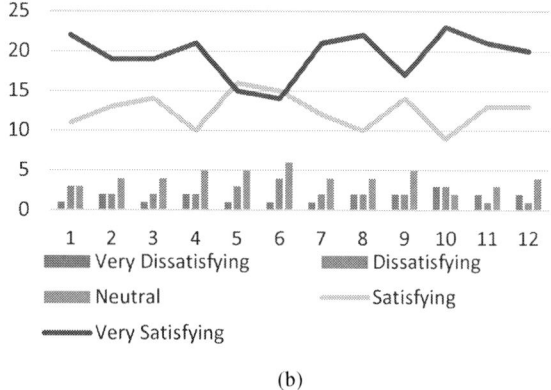

(b)

Fig. 3. Evaluation results.

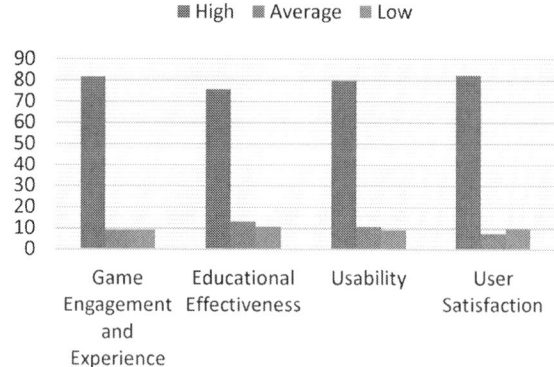

Fig. 4. Aggregated evaluation results.

In order to further assess the educational affordance of the virtual game, a pre-test/post-test evaluation design without a control group was conducted. Students completed a knowledge assessment before the learning process using the game and after the end of the intervention. The test scores ranged from 1 to 100. On average, there was a statistically significant increase in students' performance on the assessment after completing the intervention, having Mean

Difference = 19.28, Standard Deviation = 11.45, t = 7.965, p < 0.01, 95% Confidence Interval [15.31, 23.25]. This improvement can be described as substantial (Cohen's d = 1.58).

To sum up, the positive evaluation results of the educational virtual game are indeed encouraging for adopting this technology in learning process. As it shown, the game has effectively met its educational objectives. Students, after engaging with the game, demonstrated significant improvements in their knowledge and skills. The game's content was not only engaging but also aligned with the curriculum, ensuring its relevance to the educational goals. Furthermore, the interactive elements and clear instructions within the game contributed to a positive learning experience. These positive evaluation results affirm the game's value as an effective educational tool, demonstrating its potential to enhance the learning experience of students.

IV. CONCLUSIONS

In conclusion, this paper has explored the integration of fuzzy logic in order to enhance the learning experience and outcomes of users, learning C++ programming concepts. The system takes as input several characteristics (SP, GT, SK) and applies a set of rules to adjust the complexity and pacing of challenges. The system has been evaluated and the results are very promising in terms of game engagement and experience, educational effectiveness, usability and user satisfaction.

It is in our future plans to further refine the fuzzy logic-based adaptivity mechanism in order to explore whether it enhances its precision and effectiveness. Moreover, we plan to extend the evaluation by incorporating statistical analysis methods such as the student's t-test. This will allow us to gain a more comprehensive understanding of the impact and effectiveness of the presented mechanism in our educational game environment.

ACKNOWLEDGMENT

This research was funded by the European Union and Greece (Partnership Agreement for the Development Framework 2014-2020) under the Regional Operational Programme Ionian Islands 2014-2020, project title: "Indirect costs for project "TRaditional corfU Music PresErvation through digiTal innovation" ", project number: 5030952.

REFERENCES

[1] G. Chen, X. Xie, Z. Yang, R. Deng, K. Huang and C. Wang, "Development of a Virtual Reality Game for Cultural Heritage Education: The Voyage of "Gotheborg"," 2023 9th International Conference on Virtual Reality (ICVR), Xianyang, China, 2023, pp. 531-535, doi: 10.1109/ICVR57957.2023.10169671.

[2] A. Marougkas, C. Troussas, A. Krouska, C. Sgouropoulou, "A Framework for Personalized Fully Immersive Virtual Reality Learning Environments with Gamified Design in Education", Novelties in Intelligent Digital Systems: Proceedings of the 1st International Conference (NIDS 2021), Athens, Greece, IOS Press, vol. 338, pp. 95-104, 2021, doi:10.3233/FAIA210080.

[3] R. Karunakaran and S. S, "Quantifying the New Media Literacy of Rural People: An Online Video Entertainment Perspective," 2022 Interdisciplinary Research in Technology and Management (IRTM), Kolkata, India, 2022, pp. 1-5, doi: 10.1109/IRTM54583.2022.9791650.

[4] A. Marougkas, C. Troussas, A. Krouska, C. Sgouropoulou, "How personalized and effective is immersive virtual reality in education? A systematic literature review for the last decade", Multimedia Tools and Applications, 2023, https://doi.org/10.1007/s11042-023-15986-7

[5] A. Krouska, C. Troussas, C. Sgouropoulou, "Fuzzy Logic for Refining the Evaluation of Learners' Performance in Online Engineering Education", European Journal of Engineering and Technology Research, vol. 4, no. 6, 2019, pp. 50–56, doi: https://doi.org/10.24018/ejeng.2019.4.6.1369.

[6] A. Alam, S. Ullah and N. Ali, "The Effect of Learning-Based Adaptivity on Students' Performance in 3D-Virtual Learning Environments," in IEEE Access, vol. 6, pp. 3400-3407, 2018, doi: 10.1109/ACCESS.2017.2783951.

[7] S. Meacham, V. Pech and D. Nauck, "AdaptiveVLE: An Integrated Framework for Personalized Online Education Using MPS JetBrains Domain-Specific Modeling Environment," in IEEE Access, vol. 8, pp. 184621-184632, 2020, doi: 10.1109/ACCESS.2020.3029888.

[8] M. A. Akour and A. Das, "Developing a Virtual Smart Total Learning Environment for Future Teaching-Learning System," 2020 IEEE International Conference on Teaching, Assessment, and Learning for Engineering (TALE), Takamatsu, Japan, 2020, pp. 576-579, doi: 10.1109/TALE48869.2020.9368373.

[9] H. Singh, K. V. S. Praveena, M. Raja, Nikhilesh, T. Kumar and K. Tongkachok, "Adaptive 3D and VFX Films Virtual Learning Environments," 2022 5th International Conference on Contemporary Computing and Informatics (IC3I), Uttar Pradesh, India, 2022, pp. 1129-1134, doi: 10.1109/IC3I56241.2022.10073177.

[10] A. Prakash, "[DC] Investigating student's learning processes in a Virtual Reality learning environment by analyzing their interaction behavior," 2023 IEEE Conference on Virtual Reality and 3D User Interfaces Abstracts and Workshops (VRW), Shanghai, China, 2023, pp. 995-996, doi: 10.1109/VRW58643.2023.00340.

[11] M. Momenzad, B. Majidi and M. Eshghi, "Deep Summarization of Academic Textbooks for Adaptive Gamified Virtual Learning Environments," 2018 2nd National and 1st International Digital Games Research Conference: Trends, Technologies, and Applications (DGRC), Tehran, Iran, 2018, pp. 88-94, doi: 10.1109/DGRC.2018.8712065.

[12] B. N. Ncube, P. A. Owolawi and T. Mapayi, "Adaptive Virtual Learning System Using Raspberry-Pi," 2020 International Conference on Artificial Intelligence, Big Data, Computing and Data Communication Systems (icABCD), Durban, South Africa, 2020, pp. 1-5, doi: 10.1109/icABCD49160.2020.9183844.

[13] E. Cibuľska and K. Boločko, "Virtual Reality In Education: Structural Design Of An Adaptable Virtual Reality System," 2022 6th International Conference on Computer, Software and Modeling (ICCSM), Rome, Italy, 2022, pp. 76-79, doi: 10.1109/ICCSM57214.2022.00020.

[14] Z. Liu, M. Fan and B. Wang, "Discussion on the Basic Model of Intelligent Technology to Enable the Transformation of Learning Methods," 2022 International Symposium on Educational Technology (ISET), Hong Kong, Hong Kong, 2022, pp. 123-125, doi: 10.1109/ISET55194.2022.00033.

[15] A. Prakash, D. S. Shaikh and R. Rajendran, "Investigating Interaction Behaviors of Learners in VR Learning Environment," 2023 IEEE Conference on Virtual Reality and 3D User Interfaces Abstracts and Workshops (VRW), Shanghai, China, 2023, pp. 931-932, doi: 10.1109/VRW58643.2023.00308.

[16] C. Troussas, A. Krouska, C. Sgouropoulou, "Dynamic Detection of Learning Modalities Using Fuzzy Logic in Students' Interaction Activities", V. Kumar, C. Troussas, Intelligent Tutoring Systems. ITS 2020. Lecture Notes in Computer Science, Springer, Cham, col. 12149, 2020. https://doi.org/10.1007/978-3-030-49663-0_24

[17] K. Kulagin, M. Salikhov and R. Burnashev, "Designing an Educational Intelligent System with Natural Language Processing Based on Fuzzy Logic," 2023 International Russian Smart Industry Conference (SmartIndustryCon), Sochi, Russian Federation, 2023, pp. 690-694, doi: 10.1109/SmartIndustryCon57312.2023.10110734.

[18] A. Bhardwaj, H. Tyagi and C. Dubey, "Evaluation of Student Performance Using Fuzzy Logic," 2021 3rd International Conference on Advances in Computing, Communication Control and Networking (ICAC3N), Greater Noida, India, 2021, pp. 895-897, doi: 10.1109/ICAC3N53548.2021.9725495.

[19] M. Dhokare, S. Teje, S. Jambukar and V. Wangikar, "Evaluation of Academic Performance of Students Using Fuzzy Logic," 2021 International Conference on Advancements in Electrical, Electronics, Communication, Computing and Automation (ICAECA), Coimbatore, India, 2021, pp. 1-5, doi: 10.1109/ICAECA52838.2021.9675557.

[20] C. Papakostas, C. Troussas, A. Krouska, C. Sgouropoulou, "Modeling the Knowledge of Users in an Augmented Reality-Based Learning Environment Using Fuzzy Logic". A. Krouska, C. Troussas, J. Caro (eds), Novel & Intelligent Digital Systems: Proceedings of the 2nd International Conference (NiDS 2022). NiDS 2022. Lecture Notes in Networks and Systems, Springer, Cham, vol. 556, 2023. https://doi.org/10.1007/978-3-031-17601-2_12

[21] M. Gupta, R. Kumar, A. Arora and J. Kaur, "Fuzzy logic-based Student Placement Evaluation and Analysis," 2022 4th International Conference on Advances in Computing, Communication Control and Networking (ICAC3N), Greater Noida, India, 2022, pp. 1503-1507, doi: 10.1109/ICAC3N56670.2022.10074547.

[22] L. Anifah, E. Sulistiyo, M. S. Zuhrie, F. Achmad, Y. S. Nugroho and S. Schulte, "Decision Support System of Student Learning Outcomes Assessment on Digital Electronic Subject using Fuzzy Logic," 2021 Fourth International Conference on Vocational Education and Electrical Engineering (ICVEE), Surabaya, Indonesia, 2021, pp. 1-6, doi: 10.1109/ICVEE54186.2021.9649735.

[23] A. Rachmawati, S. M. Susiki and E. M. Yuniarno, "Implementation Of Fuzzy Logic For Determining The Level Score Historical Character Recognition Puzzle Educational Game," 2022 International Seminar on Intelligent Technology and Its Applications (ISITIA), Surabaya, Indonesia, 2022, pp. 232-237, doi: 10.1109/ISITIA56226.2022.9855377.

[24] C. Troussas, A. Krouska, P. Mylonas and C. Sgouropoulou, "Personalized Learner Assistance Through Dynamic Adaptation of Chatbot Using Fuzzy Logic Knowledge Modeling," 2023 18th International Workshop on Semantic and Social Media Adaptation & Personalization (SMAP)18th International Workshop on Semantic and Social Media Adaptation & Personalization (SMAP 2023), Limassol, Cyprus, 2023, pp. 1-5, doi: 10.1109/SMAP59435.2023.10255169.

[25] M. S. Ivanova, "Fuzzy Set Theory and Fuzzy Logic for Activities Automation in Engineering Education," 2019 IEEE XXVIII International Scientific Conference Electronics (ET), Sozopol, Bulgaria, 2019, pp. 1-4, doi: 10.1109/ET.2019.8878622.

[26] C. Papakostas, C. Troussas, A. Krouska, and C. Sgouropoulou, "Personalization of the Learning Path within an Augmented Reality Spatial Ability Training Application Based on Fuzzy Weights," *Sensors*, vol. 22, no. 18, p. 7059, Sep. 2022, doi: 10.3390/s22187059.

"Technician of Refrigeration, Ventilation, and Air Conditioning Installations": A New Approach of the Modern Curricula in the Mechanical Sector of the 3rd Class of the Vocational School

Konstantinos Korakis
Mechanical Engineer, MSc
2nd Laboratory Center Koropiou
Koropi, Greece
korakisconstantinos@gmail.com

Michael Dossis
Department of Informatics,
University of W. Macedonia,
Kastoria, Greece
mdossis@uowm.gr

Abstract— A modern Curriculum for Technical Education should include teaching methods that are appropriate for the engineering sector. It should be based on teaching methods that are learner-centered and based on collaborative teamwork methods. This paper briefly presents these teaching methods which include: the flipped classroom method, the method with a technical planning process, the inquiry-based learning method, the Problem-Based Learning method, backward planning, and some basic specifications for a modern survey by professional refrigeration technicians for the curriculum of the specialization "Refrigeration, Ventilation and Air Conditioning Technician" of the 3rd class of the Vocational Education and Training. The laboratory lessons of the Specialty operate at the 2nd Laboratory Center of Koropiou.

Keywords— VET, Refrigeration, Mechanical engineering, Curriculum

I. INTRODUCTION

The optimization of the framework of technical education contributes to the upgrading of Vocational Education and Training (VET). In other words, the establishment of appropriate rules regarding students' learning performance and regulatory obligations. This would give weight to the validity of studies and the award of diplomas and qualifications. Vocational high school will cease to be a choice for low-performing students to obtain a degree of no consequence. The vocational pathway should offer graduates certified professional skills from which specific professional rights will derive, as well as opportunities for further study.

The need to modernize vocational education and training to meet both the demands of the market and the needs of learners is changing the approach to the design and development of curricula, which are more focused on improving learning, assessing learners, and strengthening the links between education and the labor market.A modern curriculum includes both the learning outcomes and the content, the teaching and learning methods, the pedagogical and teaching principles, the proposed learning activities, the timetables, and indicative lesson plans.

II. SUGGESTED TEACHING METHODS FOR THE IMPLEMENTATION OF THE VOCATIONAL CURRICULA

In this paragraph, suitable modern teaching methods are briefly presented which can be applied to students at VET's Schools and in a modern curriculum. These methods are in line with the training of Engineering Teachers in the B2 level of Information and Communication Technologies (ICT), training as well as with the introductory training of Newly Appointed Teachers of technical disciplines.

Method of the Inverse Order

The "flipped classroom" is a blended learning model in which students learn by watching video lectures or other educational material at home, while "homework" is done in the classroom with the teacher and students discussing and solving questions [7]. The flipped classroom method is based on reversing the roles of teachers and students in and out of the classroom. Students, after engaging with the course material in their personal space, with the help of ICT applications, return to the classroom and take part in planning and implementing exercises and developing discussions related to the subject matter on which they have prepared at home. In the context of the flipped classroom, teachers transform introductions into homework and transfer the students' work that has so far been related to their homework into the classroom. There is an inversion of tasks in terms of the space and how they are implemented, which fully justifies the name of this method.

Through the flipped classroom process, students take responsibility for their learning process, as they have a major role in the educational process. This technique seems to be gaining increasing acceptance by the educational community and for this reason, it is chosen by teachers who wish to implement educational techniques that are learner-centered [1]. To implement the model, the use of an online education platform is required, and this is where the remote contribution of technology comes in. The reason why this model is chosen is that the Flipped Classroom frees up valuable time for knowledge acquisition through problem-solving, and student interaction with each other, the teacher, and the subject matter.

Technical Design Process

The introduction of technical design as a teaching approach in secondary education is, internationally, relatively recent. Engineering design (engineering design process) is how engineers solve problems and meet people's needs by creating artifacts such as buildings, roads, machines, electronic devices, and production processes. It is a decision-making process, which can lead to different alternatives. The engineer must be able to translate the needs and requirements of users into technical specifications of 'products', considering the relevant legal framework as well as non-expressed needs such as reliability, ergonomics, safety, environmental protection, etc., to evaluate alternatives, taking into account various criteria, both technical and economic, and to safeguard the technical

979-8-3503-8368-3/23 $31.00 © 2023 IEEE

object from possible ways of 'failure'. Important elements involved in technical design are understanding the parameters of the problem, selecting specifications and criteria, creating a model, testing, and evaluation. The teacher and curriculum resources participate in a dynamic and collaborative relationship, interacting with and influencing each other. During lesson planning, teachers work with curricular resources: they interpret them and transform them as they design instruction.

Typical phases of engineering design are:
- Identification of a need or problem,
- Research around the identified need or problem,
- Development of possible solutions,
- Selection of the optimal solution,
- The construction of a prototype,
- Evaluating the solution,
- Communicating the solution, and
- The redesign.

Inquiry-based learning

Dewey's experiential learning (learning through experience) implies the active participation of the learner in personal and authentic experiences, making the process meaningful for students [5]. Inquiry-based learning can be realized through experiential learning to the extent that it involves engagement with the content/material under study, exploration, and collaboration in the effort to create meaning. Vygotsky approached constructivism as learning through experience influenced by society and the facilitator. The meaning constructed from an experience can be both individual and group in nature. Inquiry-based learning relies mainly on students' inquiries, questions, and inquiries rather than on the teacher's presentation of the curriculum. The aim of inquiry-based learning is for the learner to become personally involved in the cognitive process and to learn how to learn autonomously [3].

Some of the key features of inquiry-based teaching are:
- Students are involved in difficult problems or situations that have an open solution to the extent that a range of solutions and answers are acceptable.
- Students have control over the direction of the research/investigation and the methods/approaches used.
- Students build on existing knowledge and identify their own learning needs.
- The different tasks and roles stimulate curiosity in students, which encourages them to keep looking for new data or evidence.
- Students are responsible for analyzing and presenting evidence to defend their solution to the original problem.

Problem-Based Learning (PBL)

It is a student-centered pedagogical approach in which students learn about a topic through the experience of solving problems. It helps build students' confidence and deeper understanding of the concepts they are dealing with and encourages them [4]. A well-designed PBL program provides students with the opportunity to develop skills related to:
- Working in groups.
- Project management and leadership roles.
- Oral and written communication.
- Self-awareness and evaluation of team actions.
- Individual work.

- Critical thinking and analysis.
- Analysis of ideas.
- Self-directed learning.
- Application of course content to real-world examples.
- Research and information literacy.
- Interdisciplinary problem-solving.

Problem-based learning is an active way of learning that offers better retention of knowledge, enhances motivation to learn, and encourages the development of skills needed for the 21st-century labor market.

Starting from the end (backward planning)

"Starting from the end" design is a strategy proposed by Grant Wiggins and Jay McTighe and is a building block of the "Understanding by Design" model, which guides the teacher in planning, evaluating, and organizing teaching. According to the strategy, the teacher begins instructional design by setting clear learning objectives focused on in-depth understanding and then chooses the modes of assessment of learning, the steps, teaching methods, and resources to be used. In this way, he ensures that the content, teaching resources, and assessment methods he chooses are appropriate and focused on achieving the learning objectives [6]. It is embedded in the context of teaching for understanding.

Teachers are required to:
- identify the topics, concepts, and skills that students need to understand.
- set objectives that will help students focus on the most important aspects of these topics, concepts, and skills.
- design meaningful learning projects to achieve these objectives, and,
- develop assessment projects that contribute to broadening understanding [13].

A. Proposed General Curriculum Specifications

The "curriculum" is a broader concept than the Syllabus and the Course Outline because it provides not only for "what" will be taught but also for "when" the material will be taught. In other words, in addition to the content, it must provide for the time of teaching the course, the material and equipment that will be required for either laboratory or theoretical vocational courses. Finally, it is linked to the tasks to be carried out with the corresponding worksheets, how they are to be assessed, and how students are to exercise and work. In general, there are activities related to the organization and planning of a course according to the objectives set for it.

More specifically, as far as modern curricula are concerned, the six categories are:

a) Ideal curriculum which is the expression of the vision for education through programmatic announcements and works prepared by scientific committees, institutions, and governmental bodies.

b) Formal curriculum with the recommendation of the competent body.

c) Interpreted by teachers as a personal theory of pedagogy and education of each teacher, and in this respect, there is often little correlation with the official curriculum.

d) Operational or applied curriculum as that which is put into practice in the classroom.

e) Experienced curriculum related to the learning experiences of students.

f) Mastered curriculum, which is "what students know and can do with what they know" [8].

Another division that is made is the separation of the P.S. into:

(a) science-centered, in which school knowledge is organized into independent subjects in correspondence with the scientific disciplines,

(b) interdisciplinary, which overcomes the deadlocks of science-centered curricula with distinct and independently taught subjects; and,

(c) interdisciplinary, which escapes the logic of individual scientific disciplines and organizes school knowledge around themes, issues, and problems of high generalization, which are of general interest to culture [9].

For a curriculum to be effective, learning objectives must be set. A learning objective is a description of a desired performance that trainees can demonstrate for their training to be considered complete. According to Bloom's taxonomy, the learning objectives in the curriculum of VET schools should be mainly from the intermediate level upwards, i.e., from application to synthesis and assessment. Thus, the curriculum will lead to the acquisition of practical skills [2].

The specifications for each course of the proposed Curriculum should follow the Quality Framework for Curricula of the European vocational training. More specifically, as stated in the Framework, they should aim to (Government Gazette 490B/20-2-2017):

- developing skills in applying knowledge and problem-solving in authentic professional environments,
- in obtaining qualifications following the National Qualifications Framework,
- approaching and assessing the necessary knowledge through a variety of alternative ways to offer learning and development opportunities to all learners, depending on their specific characteristics (interests, aptitudes, experiences, readiness, learning profile, cultural background, special needs and/or disabilities, etc.),
- the cultivation of each student's ability to critically approach and use educational technology with informatics,
- creating conditions that enable each individual to engage in lifelong learning ("learning how to learn", self-directed learning, problem-solving),
- the development of knowledge, skills and competences and problem-solving in the modern professional environment,
- the formation of a responsible professional consciousness and identity,
- the use of both quantitative and qualitative evaluation methods (e.g., student portfolio), which provide feedback on the achievement of the learning objectives and the improvement of the educational practice,
- cultivating values of cooperation and environmental awareness,
- the emphasis on occupational health and safety issues,

- developing innovation and entrepreneurship skills, extroversion, and adaptability,
- the development of the soft skills of the trainees for their application in the modern professional environment, in addition to the knowledge, skills, and competencies necessary for the relevant profession,
- promoting the principles of education for sustainable development,
- facilitating access and retention in the education system for people with special needs and, more generally, for people from vulnerable social groups,
- promoting the transparency of qualifications and mobility within the European vocational training considering that the recognition of qualifications regardless of how they are acquired (through formal, non-formal, or informal learning) helps to enhance permeability and facilitate transitions between different education and training systems,
- the development of positive attitudes towards lifelong learning.

The curriculum to be developed for the engineering sector of the Vocational Lyceum, and therefore the subjects (courses) defined in it, should meet the following requirements to the maximum extent possible [11]. Scientific validity is also necessary because the knowledge presented must be distinguished by accuracy, scientific evidence, and objectivity. In addition, it needs to contain substantial and up-to-date content based on current literature and recent developments in the engineering sciences.

- Completeness-Completeness- Comprehensiveness: the content of the curriculum needs to be rationally organized and complete, without gaps and overlaps in the fundamental fields of knowledge of the areas it is intended to cover. In addition, it must be characterized by logical sequence, linguistic and conceptual coherence, clarity, and comprehensibility.
- Teaching Suitability: The choice of the content of each subject (course) needs to consider the purpose and objectives of each subject (course), the duration of the teaching hours as provided for each module, the characteristics of the target group as well as the context, structure, content and, in general, the philosophy underlying the curriculum. In addition, it is considered appropriate that the material should draw on students' prior experience, focus on authentic professional environments and projects, and have references to real-life labor market conditions.
- Flexibility: The content of the curriculum should provide the flexibility to adapt to the social, cultural, and educational characteristics and specific needs of the intended learners.
- Simple and understandable language: The curriculum must be written in simple and understandable language while maintaining its scientific validity. It is also important to avoid long descriptions.
- Respect for democratic values.

B. Short survey

In that point of our paper, we present a short survey by professional refrigeration professionals for the specialization "Technician of Refrigeration, Ventilation, and Air Conditioning Installations" of the 3rd Class of the Vocational School of Refrigeration and Air Conditioning (Fig 1).

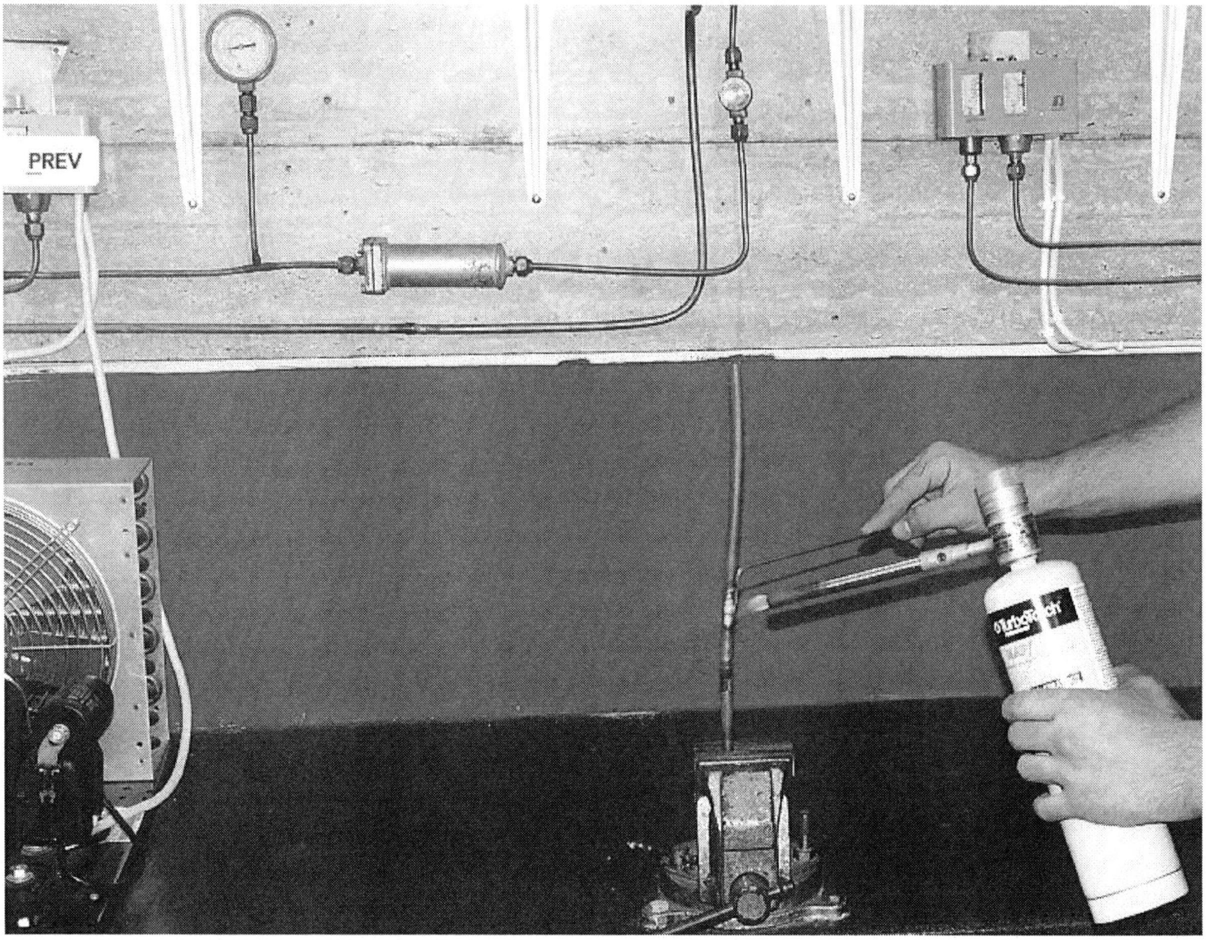

Fig. 1. Laboratory of Refrigeration

The refrigeration technician is a profession that requires a combination of a good theoretical knowledge of sciences, such as mechanics and electrical engineering, and the acquisition of skills and abilities for the practical application of these sciences, as well as skills in the use of tools and machines. It requires concentration, attention, and problem-solving skills because both the installation of air-conditioning and refrigeration equipment and the repair of faulty equipment are often demanding, complex, or even dangerous tasks. The environment (dangerous outdoor areas exposed to the weather conditions, e.g., balconies at high altitude) in which the air conditioning or refrigeration unit may be located and the fact that they contain dangerous, explosive liquids and gases are the main factors contributing to the above difficulties (Figure 2).

The institutional framework is dynamic, has changed several times, and may change in the future because it is directly related to the wider European concern for environmental protection and public health and safety. Finally, the refrigeration technician must have good communication skills to be able to effectively serve the customers they meet and increase sales [12].

The skills considered to be of great importance, for carrying out work to improve energy efficiency such as estimation of cooling loads to control the size of devices (covering real needs, avoiding oversizing), estimation of the appropriate type of air conditioning installation according to the thermal needs of each of the building space, estimation of the mass-volume supply, the speed and dimensions of the air-conditioned air duct network to achieve in practice the efficient operation of the air conditioning system (Fig 3).

Fig. 2. Laboratory of Air Conditioning Installations

979-8-3503-8368-3/23 $31.00 © 2023 IEEE

Fig. 3. Laboratory off Ventilation

These skills relate to assessment activities to determine the most efficient solution, energy-saving activities, application, and installation of machinery, and finally the control of the operation and maintenance of the installations. The workforce in the industry does not have most of these skills and there is no training program to provide them [10].

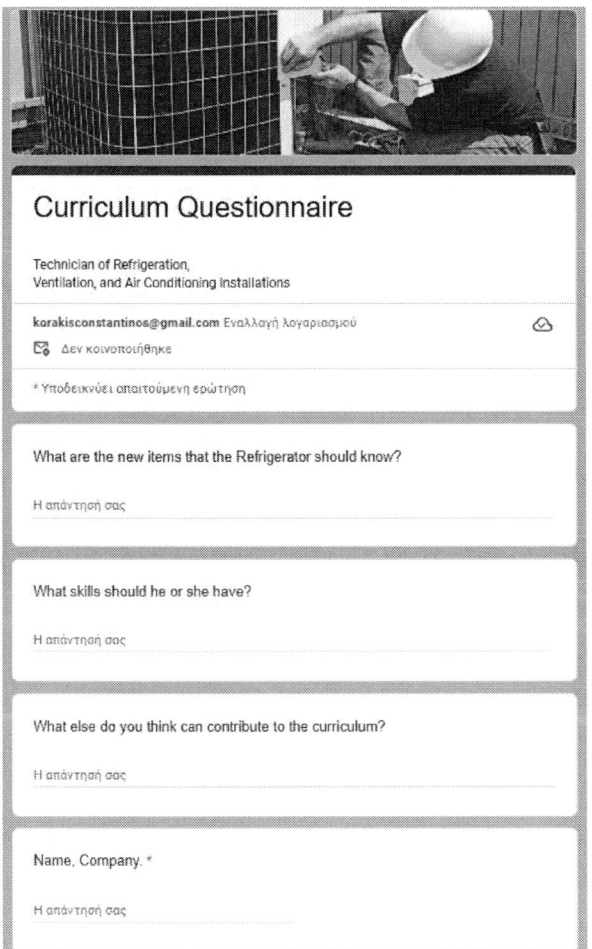

Fig. 4. Google Forms questionnaire

To obtain information from the experts, a questionnaire was created in Google Forms (Figure 4) and published on an online website for refrigeration technicians. Four company representatives and 8 freelancers responded anonymously to the questionnaire. The questionnaire included the following questions:

- What are the new items that the Refrigerator should know?
- What skills should he or she have?
- What else do you think can contribute to the curriculum?

From the answers to the questionnaire, it is clear that the refrigeration technician should acquire knowledge of new refrigerants and the management of refrigerant fluids and should be aware of energy-saving methods. Knowledge of heat pumps, how they work, installation methods, and hydraulic and electrical connections of the respective machines should be included in the curriculum. In the laboratory exercises for refrigeration air conditioning, elements of electrics and electronics should be integrated as a single set and not separately. Maintenance and repair are another area that should be given attention in the refrigeration field, modules should be developed in the curriculum for it. An important aspect is presentations and workshops from refrigeration and air conditioning engineering companies which will help practically to enhance the knowledge acquired by the students.

C. Conclusions

Young refrigeration technicians appear to have multiple and significant gaps in their knowledge, skills, and competencies. The main cause of the deficiencies in fundamental general knowledge and specific professional knowledge is the initial training, which is identified as inadequate, and the way to address this is through the training of refrigeration technicians. The lack of capacity for continuous learning, critical thinking, problem-solving, communication, teamwork, adaptability, responsibility, customer orientation, and goal achievement is due to the lack of motivation of employees and is addressed by introducing financial incentives. Specific professional skills, such as the use of tools and machines, the management of refrigeration equipment, and the ability to assemble, install, and connect the machinery, components, automation, and safety devices of the installation are acquired during the occupation through professional experience.

Research shows that learning through problem-solving also promotes conceptual understanding and the development of thinking strategies. Specifically learning through problem-solving activates students and helps them develop flexible knowledge (i.e., knowledge that can be transferred to other situations), problem-solving skills problems, self-directed learning abilities, collaboration skills, and communication. Problems must be complex and loosely structured to promote flexibility, but they must also be realistic, and they tune into students' experiences to activate their intrinsic motivation.According to the above, a modern curriculum can be a driving force for the change of the school climate of Vocational High School. To inspire teachers with appropriate instruction, space, ways, and means to make the classroom a laboratory of inquiry, communication, action, and expression that leads students to work to achieve: cooperativeness, self-awareness, linguistic awareness, communicative competence, responsibility, tolerance, discipline, sense of justice, democratic sensitivity, fair play, solidarity, aesthetic culture, problem-solving, exploratory learning, digital literacy.

The aim, therefore, is to develop student-centered, functional, and open to the emergence of timeless values, which activate students so that they authentically experience professional skills. Ultimately, to become creative people and conscious citizens of their country, their nation, Europe, and the world, and in particular good professionals.

ACKNOWLEDGMENT

This paper was partially supported by the University of Western Macedonia.

REFERENCES

[1] Baker, J. (2000) The classroom flip: Using web course management tools to become the guide by the side, Selected Papers from the 11th International Conference on College Teaching and Learning, Florida Community College at Jacksonville, pp. 9-17.

[2] Bloom, B. (1956) Taxonomy of educational objectives: The classification of educational goals. New York, NY: Longmans, Green.

[3] Kanellopoulos, J., Papanikolaou, K. & Zalimidis, P. (2017). flipping the classroom to increase students' engagement and interaction in a mechanical engineering course on machine design. International Journal of Engineering Pedagogy (iJEP). 7(4).19-24. retrieved December 22, 2022, from https://www.researchgate.net/publication/321285321_Flipping_The_Classroom_to_Increase_Students'_Engagement_and_Interaction_in_a_Mechanical_Engineering_Course_on_Machine_Design.

[4] Nilson, L. (2010) Teaching at its best: A research-based resource for college instructors (2nd ed.) San Francisco, CA: Jossey-Bass. Unger, C. (1994) What teaching for understanding looks like. Educational Leadership, 51, 8-8. Retrieved 15/12/2022

fromhttps://knilt.arcc.albany.edu/images/b/b9/What_teaching_for_understanding_looks_like.pdf

[5] Vani Veikoso, T. (2010). Teachers' practices, values, and beliefs for successful inquiry-based teaching in the International Baccalaureate Primary Years Programme. Journal of Research in International Education, 9, 40-65. Retrieved December 12, 2022, from https://journals.sagepub.com/doi/10.1177/1475240909356947.

[6] Wiggins G. & McTighe J., (2005), Understanding By Design, PEARSON: Merril Prentice Hall

[7] Gariou, A., Makrodimos, N., Papadakis, S., (2021). Inverted Classroom: A blended learning model for all levels of education. Patras.

[8] Kouloubaritsi, A. (2013). Curriculum and Educational, Curriculum - Theoretical approaches - Current Programmes - Institutional Framework of the Timetable - Educational Material and School Manuals, Training Programme "Training in scientific-pedagogical guidance for the acquisition of a certificate of guidance competence for teachers of Primary and Secondary Education". Athens: National Centre of Public Administration & National Institute of Public Administration and Local Government - Institute of Training.

[9] Matsangouras, H. (2004). Interdisciplinarity in curricula: theory and practice. In.: Athens.

[10] Malamatenios H. (2015), Qualitative study/ Report on the needs of skills diagnosis in the thematic field of "Energy saving in buildings" for three specialties: plumber, electrician, and refrigeration. Athens: FHW GSEBEE.

[11] Mavrikakis, E., Syrigos E., Farantou, E. (2018). Specifications of educational material. Athens: I.E.P. Retrieved 18/12/2022 fromhttp://www.iep.edu.gr/images/IEP/EPISTIMONIKI_YPIRESIA/Epist_Monades/B_Kyklos/Tee/2018/2018-03-26_prodiagrafes_ey_mathiteias_dia_zosis.pdf.

[12] Tsalouhidis, S. (2021). Roadmap for the adaptation of the profession: Refrigeration. Athens: FHW GSEBEE.

[13] Unger, S. H. (1994). Controlling technology: Ethics and the responsible engineer. John Wiley & Sons.

Educational Virtual Worlds for Vocational Education and Training Laboratories

Dimitrios Magetos
NetLab, Department of Informatics,
University of Piraeus,
Piraeus, Greece
dmagetos@unipi.gr

Dimitrios Kotsifakos
NetLab, Department of Informatics,
University of Piraeus,
Piraeus, Greece
kotsifakos@unipi.gr

Christos Douligeris
NetLab, Department of Informatics,
University of Piraeus,
Piraeus, Greece
cdoulig@unipi.gr

Abstract— **The concept of a metaverse, a virtual world that offers immersive experiences, has gained widespread interest in recent years. Designed as a simulation of the physical world, the metaverse covers all areas of human activity and provides a suitable platform for researchers in all disciplines, from health to sport, education, and art. The work presented in this paper is part of the development of a virtual world for the teaching of the "History of Art" course in vocational education and training. Its main goal is to investigate the development process of Open Educational Virtual Worlds (OEVs), and the challenges and issues that arise, starting from the development of the virtual world to its publication and educational use in Vocational Education and Training (VET). This study found that there are online virtual world platforms freely provided for educational purposes as well as free digital repositories of 2D and 3D educational resources. In particular, the use of the spatial.io web environment to develop open educational virtual worlds is suggested. To enrich these worlds with corresponding content, we can use educational resources from digital repositories such as Sketcfab and Photodentro, and the ADDIE educational design model as a methodology for the development of the virtual world.**

Keywords— **metaverse, virtual world, open educational resources, digital repositories.**

I. INTRODUCTION

The COVID-19 pandemic has had a dramatic impact on the way society functions, a shift towards digital technologies and online platforms has been observed. This has led to increased interest in the potential of virtual worlds, also known as metaverse, as educational and communication platforms. Metaverse are online social application that utilizes multiple innovative technologies that aim to provide an immersive experience. Even though it is not a new concept, the readiness of people, social media, and technology is pushing our society into the metaverse era at unprecedented speeds. Metaverse is characterized by interactive, embodied, persistent, real, ubiquitous, interoperable, scalable, immersive, accessible, synthetic, multi-layered, and collaborative modules.

The satisfaction of the learning needs in the era of the 4th industrial revolution requires alternative approaches and innovative educational interventions. Learning is approached in an interdisciplinary way, becoming multidimensional, and developing 21st-century competencies through knowledge, skills, and values [1]. Students require active learning environments, using the capabilities of various technological applications to acquire knowledge. They seek more interesting, fun, motivating, and engaging learning experiences [2]. Educators are looking for modern teaching environments to engage their students and increase motivation to learn.

This paper focuses on the design and development of Open Educational Virtual Worlds (OEVs) oriented towards teaching and learning [3]. The purpose of this paper is to explore virtual worlds as educational tools for Vocational Education and Training (VET) [4]. It also seeks to inform the educational community about the development process of OEVs, and the challenges and issues that arise, starting from the development of specific OEVs to their publication and educational utilization [5]. A partial goal of this is for the teacher to understand the development process of OEVs and to be able to act not only as a consumer of technology but also as a creator of OEVs, as well as to function effectively in educational environments that employ virtual worlds [6]. Thus there is a need to research the effective and efficient utilization of virtual worlds in VET – a topic that has not attracted a wide body of research [7].

Several virtual worlds have been created through lengthy processes with large budgets and with the involvement of many scientists of different disciplines, both in their technological and pedagogical approaches. There is a definite need though to create virtual worlds without excessive development costs and without the need for specialized knowledge to develop them. A flexible, fast, and cheap method of developing virtual worlds by the teacher himself, based on open educational resources, which are freely provided via the internet and are easy to reuse and modify without economic costs is one other goal of this paper.

In the context of the present study, virtual world development environments were investigated, which are supported by cloud computing infrastructures and freely offered as online open educational resources in VET [8]. In particular, the cloud-based (https://www.spatial.io) platform was used in the development of an OEV for the teaching of the History of Art course of the Applied Arts sector of VET [9].

The development of the virtual world was based on the ADDIE instructional design model.

The research questions that the study seeks to answer are:

1. What is an appropriate framework for the development of open educational virtual worlds at no economic cost?

2. Which pedagogical approaches are appropriate for the development of OEVs?

3. What platforms are available in the metaverse that provides free features and hosting for OEVs development?

4. Which digital repositories provide educational resources that can be integrated into OEV?

5. What is the most relevant scenario for the exploitation of the educational virtual world (OEV) under development?

The remainder of the article is structured as follows. Section 2 deals with the necessary details on the required theoretical and technical background. Section 3 presents a

979-8-3503-8368-3/23 $31.00 © 2023 IEEE

literature review on the use of metaverse in various environments. Section 4 describes the virtual world development methodology. Section 5 presents the answers to the corresponding research questions of the study. Section 6 presents a thorough discussion and, finally, section 7 concludes the paper and provides suggestions for future research.

II. THEORETICAL BACKGROUND

The educational use of new Media and Technologies is based on students' experiences and experiential knowledge [10]. The daily life of students is constantly structured employing communication and multimedia. They use various media, both traditional and digital, in all aspects of their lives. The youth of this generation are constantly connected and interacting with technology in every aspect of their lives [11]. This close relationship with media and technology shapes the way students perceive the world around them and affects the pedagogical process [12].

Educators must take this reality into account and integrate new media and technologies into the educational process [13]. The educational process can benefit from the use of digital educational resources, interactive platforms, and technological applications to enhance learning and create educational experiences that reflect the multimedia world of students [14]. In parallel, the teachers should look at proposed pedagogical strategies that will facilitate the teaching and learning process [15].

A. Virtual world, metaverse

Virtual worlds in the metaverse are three-dimensional synthetic computing environments in which multiple users, appearing through their digital incarnations, communicate with each other or with other synthetic entities, explore and interact with the environment, and create new content. Today, the theory and technology of virtual reality are widely applied in various fields, such as education, training, simulation, culture, medicine, entertainment, and collaboration [16].

Virtual reality environments are a suitable tool for safely conducting training and practice in situations that may be difficult in the real world and potentially dangerous for both the trainees and those involved in the training process. The high cost of education can be minimized thanks to Metaverse, where we can create a digital twin of the real world [17].

This can be exploited in appropriately designed educational spaces that support digital classrooms, i.e. virtual distance learning spaces where instructors and students participate through virtual avatars. In addition, for topics involving time-consuming or complex processes (e.g. the functioning of the digestive system, shooting, traffic codes), virtual worlds can create the right conditions for free or controlled experiments, allowing students to try different approaches and observe the results.

B. Repositories, open education resource

Within the framework of open education, open educational resources are emerging as a valuable and innovative resource for the process of education and knowledge at various levels and environments. The application of digital resources, in particular, has a far-reaching and positive effect on the education process, offering a multitude of benefits [18]. The use of open resources is associated with the reduction of costs for educational materials and textbooks, making it possible to use resources more efficiently for other educational needs [19].

At the same time, in the context of open education, digital repositories have been developed that host open educational resources, both two-dimensional and three-dimensional multimedia resources, suitable to be reused, modified, and integrated for free by teachers in the development of virtual worlds [43]. Digital repositories provide appropriate metadata of the available resources as well as the necessary rights for their educational use.

C. Cloud computing platforms

The continuous development of cloud computing technology provides modern platforms for the development of virtual worlds, which are constantly available via the Internet with any device and in any place, at any time, and without financial commitments. Platforms supported by cloud computing infrastructures offer resiliency to a large number of students as well as elasticity of their services according to the usage load.

D. Examples of applications of virtual worlds

The next section presents examples of virtual worlds for educational purposes. River City [20] is a virtual world application created by Harvard University for research purposes where participating students not only actively participate in research but also learn from it. Ubisoft's Discovery Tour [21] is an educational experience that allows users to explore and interact with world history and culture during the periods of Ancient Greece "Fig. 1", Ancient Egypt (49 BC - 43 BC), and the Viking Age (872 AD - 878 AD). The LAVA Virtual World [22] introduces the work of Virtual Archeology in the Acropolis of Ancient Sparta. LAVA is a collaborative archaeological learning environment, developed to address the need for students to be able to engage with realistic archaeological excavation scenarios.

III. LITERATURE REVIEW

Mystakidis [23] defines the metaverse as a multiuser environment that merges physical reality with digital virtuality, enabled by technologies like virtual reality and augmented reality. Weinberger [24] proposes a definition based on a meta-synthesis of existing literature, describing the metaverse as an interconnected web of ubiquitous virtual worlds that overlap with and enhance the physical world, allowing users to connect, interact, and consume user-generated content. Ritterbusch [25] presents two revised definitions for the metaverse, with the first being a simplistic description of a three-dimensional online environment where users interact with each other in virtual spaces, and the second providing a more detailed and comprehensive definition. Almoqbel [26] conducts a systematic literature review and finds that while there are overlaps in the characteristics defined as properties of the metaverse, there is a divergence in the details. In summary, the papers collectively offer various definitions for the metaverse, highlighting its nature as an interconnected virtual world that merges with and enhances the physical world, enabling user interaction, consumption of content, and the presence of avatars.

The following articles collectively suggest that the metaverse platforms have the potential to be used in education. Ortiz [27] highlights that virtual platforms focused on metaverse and virtual reality can enhance teaching and learning processes, allowing for interaction within virtual environments. Rahman [28] emphasizes the transformative impact of metaverse technology on the education sector, including visualizing content, virtual campuses, 3D simulations, and remote quality education. Jeon [29], in a systematic literature review, finds that metaverse platforms are commonly used for communication, collaboration, emotional support, and knowledge formation in educational contexts. However, more research is needed to examine the effects of metaverse use on learning performance and to develop guidelines for incorporating metaverse in educational settings.

Some papers suggest several advantages of using the metaverse in education. Abraham [30] highlights that the metaverse provides a more interactive and immersive learning environment, allowing learners to engage with digital content and collaborate with others globally. Kye [31] discusses the potential of the metaverse in offering new social communication spaces, higher freedom for creation and sharing, and providing new experiences and high immersion through virtualization. The use of the metaverse in educational and training processes to create new learning environments that allow learners to function in a parallel, safe, and personalized reality is gaining ground as discussed in López-Belmonte [32]. Şentürk [33] further supports the benefits of virtual reality technology in education, including increased student performance and interaction between students and educators. In summary, the metaverse in education offers advantages such as interactive and immersive learning experiences, global collaboration opportunities, new social communication spaces, and improved student performance.

Another body of work collectively discusses the disadvantages of using the metaverse in education. Abraham [30] highlights the drawbacks of traditional online meeting software platforms and suggests that the metaverse can overcome these limitations by providing a more interactive and immersive learning environment. Kye [31] categorizes the metaverse into four types and discusses the potential and limitations of its educational applications, including weaker social connections, privacy concerns, and potential maladaptation to the real world. Kaddoura [34] emphasizes the challenges, opportunities, and ethical considerations associated with the rising trend of the metaverse in education, calling for further research in this field.

IV. METHODOLOGY

This study combines theoretical and applied research approaches and was conducted in the following distinct but parallel phases:

A. Exploratory phase of the field for metaverse.

In this phase, we studied related studies that had as their subject the metaverse as well as its application in the educational process. We focused our research on the metaverse technologies used by various researchers, but also on the methodologies used in the development of educational virtual worlds. Our research was done through Google Scholar and the search criteria were metaverse education metaverse or educational virtual worlds development methodologies.

B. Construction phase of the OEV.

For the development of the virtual world, the study was based on the ADDIE [35] (Analysis, Design, Development,

Implementation, Evaluation) methodology which is a structured framework used for the design and development of educational programs. The methodology followed for the development of the virtual worlds based on the ADDIE model is presented below.

The first phase involves three steps: defining the educational objectives, identifying the target audience, and analyzing the environment. The educational objectives include the definition of cognitive goals, skills, and values to be acquired by the students. The target audience is determined by age group, level of education, and specific needs. The environment analysis takes into account technical constraints, available virtual world platforms educational materials, and digital repositories. In phase 2 of the design of the virtual world for education, the design and organization of virtual spaces, objects, and activities are implemented. The educational content is selected from digital repositories, including courses, exercises, and tools. The necessary platforms, applications, and technologies for the creation and operation of the virtual world are selected. In phase 3 the virtual world is created: In this step, the virtual world is implemented based on the design and content selected in the previous phase.

The implementation was made using the capabilities of the web platform spatial.io. Initially, the gathering space for students and teachers was created from the ready-made spaces provided by the environment. Then independent spaces [36] were created that resemble museums. Then 2D and 3D educational resources from various digital repositories were added to these spaces. Various learning paths were also created so that groups of students could explore and interact with the educational resources. Communication and collaboration between students and teachers can be done through the communication and collaboration tools provided by the spatial.io environment.

The virtual space exploitation and evaluation phase (phase 4) will take place in a future application of the virtual world by students and teachers. This phase involves conducting instruction using the virtual world.

V. STUDY RESULTS

This study is the product of the development of a virtual world for teaching the History of Art course in the field of applied "Fig. 1" arts of the VET. Through this development and research activity, answers to the specific research questions were provided. The virtual world shown in Figure 1 is a virtual reality office space.

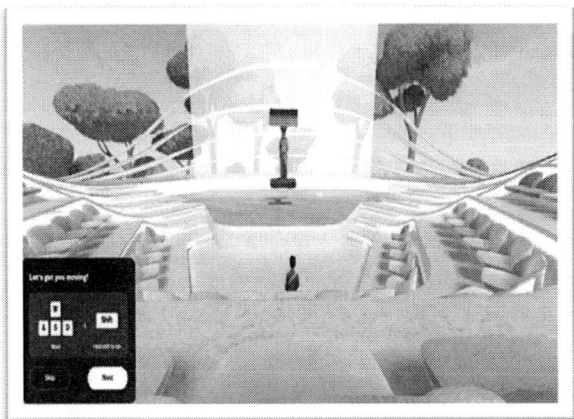

Fig 1. The initial image of the virtual world for the Art History lesson.

This term refers to virtual environments that run on a home computer and, therefore, the interface is largely conventional. The visualization of the virtual world is projected on the computer screen without the use of specialized equipment, user interaction is implemented via keyboard and mouse, and communication is supported by simple headphones and a webcam.

In the following, we present the answers to the questions posed and present a scenario for using the virtual world we developed.

A. Pedagogical models of OEV development

Some pedagogical and teaching approaches can be used individually or in combination to develop OEV. The first pedagogical approach is Experience Design. In this approach, the virtual world is designed around the experiences that students have. Another approach is collaborative design. According to this, students, teachers, and designers are all involved from the beginning in the process of designing the virtual world. Another alternative approach is for the virtual world to use simulation (Simulation-Based Learning). This approach utilizes simulators to create immersive learning experiences where students can develop skills in a safe environment. Another approach is Problem-Based Learning in which virtual scenarios are designed where students have to solve problems or face challenges. A further approach is for the virtual world to include educational games. Games with educational objectives have been shown to provide learning through interactive experiences. Another approach to designing the virtual world is to support authentic learning scenarios. Virtual worlds that correspond to real environments and everyday situations are appropriate so that learners can apply their knowledge to real-life situations. Several of the above approaches were used in this research.

B. Platforms for developing virtual worlds

In this study, several platforms for the development of virtual worlds were investigated which were mostly free and did not require a subscription for their basic functions. The next section presents some of these virtual world platforms. Second Life is one of the first virtual worlds [37] where users can create content, interact, and train within the virtual space. OpenSimulator [38] is an open-source platform that allows the creation and hosting of virtual worlds. The Mozilla Hubs web platform [39] allows users to easily create and share virtual spaces. The Engage platform [40] focuses on education and provides tools for creating virtual learning experiences. We found that there are several environments for developing virtual worlds and we chose the spatial.io web environment because it freely provides the necessary functionality for communication and collaboration between students and teachers, as well as enriching the environment with additional educational content that can be drawn from public repositories.

C. Digital repositories and open educational resources

In a virtual world development environment, the digital content must be readily available and free for educational use, so that the teacher can easily integrate it into the virtual world. This need can be met by open educational content made available online - without time, space, and financial constraints - from digital repositories and digital libraries. Several open-source repositories, (Greek or international),

exist for various scientific fields, which can be easily exploited in the virtual world.

Some of these repositories are presented below. The Open Educational Resources Commons (OER Commons) [41], is a platform that brings together educational resources from various fields, including virtual worlds. The MERLOT Repository [42] is an online catalog that provides free educational resources, including those that can, be used in educational virtual worlds. The Sketchfab [43] repository, provides 3D models "Fig. 2", that can be used in educational virtual worlds. The NASA 3D [44] repository provides free 3D models and data that can be used for educational purposes. TurboSquid [45] is a repository of 3D models that provides a wide range of content, some of which can be used for educational purposes. Photodentro [46] Learning Objects is the Panhellenic Learning Objects Repository for primary and secondary education. Photodentro works as an open tool for everyone, students, teachers, and any interested party.

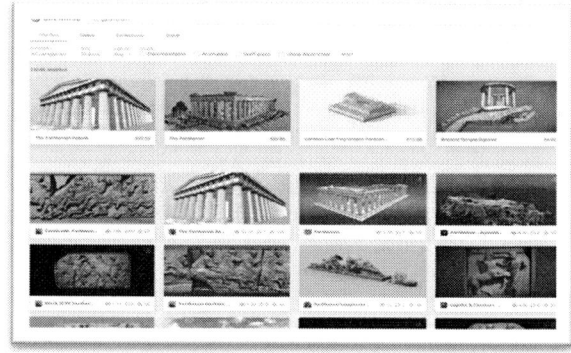

Fig 2. Digital repository with 3D resources.

D. Proposed teaching utilization of VET

The OEV developed is in its first version and in trial mode, which is immediately available via the Internet. For its educational use, the teacher and the students must first understand its functions, use it on a trial basis, and create their virtual character, the avatar, to represent them in the metaverse. The necessary equipment that teachers and students must have is a modern computer with headphones, a mouse, a keyboard, and a webcam with a reliable internet connection. Smart mobile devices can also access the world from the internet. No specialized VR equipment is required since an OEV belongs to the category of virtual worlds represented on a simple computer screen. The OEV can be used either in live computer classrooms or in distance online education since it is supported by cloud computing infrastructure and delivered over the internet. The OEV's environment of "Fig. 3", consists of:

a) the space for gathering, slide presentation, and whole group discussion. This space is offered for informing and guiding the students, for providing information about the use of the world, and for anything useful for the educational process,

b) spaces for interaction and cooperation between subgroups. These spaces are virtual classrooms with digital exhibits with which students interact to understand the relevant cognitive content of the course.

The navigation of the avatars in the space is facilitated by:

a) the creation of corridors with the appropriate placement of objects and the use of pathways, signs, and markers,
b) the use of text floating over objects, colors, and images.

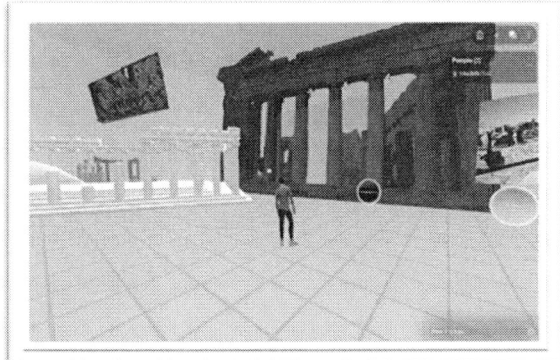

Fig. 3. A view of the virtual world with 3D representations of Ancient Athens.

The exhibits are appropriate to open educational resources where they are either embedded in the virtual world or provided as external links to the corresponding repositories. OEV can support many simultaneously connected users where they coexist in a single space and can perceive each other through their avatars and refer to objects or data in the world they observe together. Their placement can be organized according to the conventions that apply in corresponding physical spaces (e.g., collaboration roundtable, headquarters), roles can be assigned to users which are even reflected in their appearance, and the communication and collaboration model can be organized through offered actions and functions (e.g., polls, upvote, accept/reject, text commenting, logging of minutes).

The pedagogical use of OEV can be achieved through teamwork techniques, such as jigsaw puzzles, in which the members of each team must work together as a whole to achieve a common goal. The teacher suggests that students log on to the virtual meeting place, where he or she describes the objectives and the course of instruction. The students are divided into groups so that each group can explore a different area and research the issues assigned by the teacher. Users (students, teachers) can embody inhabitants of each era, experience their stories and adventures, and interact with the world around them, either as Pericles, Ictinus, Kallikrates, Phidias, etc. The students can also embody the roles of respective groups such as architects, sculptors, and philosophers.

Each group can pass through all the spaces in which there are appropriate educational resources to do the corresponding activities assigned to it. After the groups have visited the selected places in a reasonable period, they return to the original meeting place. At this point, in the context of the plenary session, the groups should announce their achievements and the teacher should give the appropriate feedback. The teacher can motivate students to visit the OEV anytime and from anywhere since it remains open 24/7 and outside school hours. Alternatively, it can enrich the OEV with new content and motivate students to log in to do additional activities.

The above proposal is indicative and can be modified depending on the intended goals and training conditions.

VI. DISCUSSION

Today's students, who, according to the Cambridge dictionary [47], are described as digital natives are fully familiar with digital technology and have rich experience of using it in virtual worlds, from their constant engagement with 3D games, such as Roblox, Minecraft, and Fortnight. The educators should include in their teaching practices the habits and interests of modern students to attract them to learning. The use of these three-dimensional worlds requires special knowledge and skills from teachers both for their development and for their effective educational use. It is therefore necessary to have modern environments for the development of educational virtual worlds as well as available educational resources that can be easily integrated into the virtual worlds being developed. Virtual world environments should provide teachers with the rapid development of OEVs to adopt them in their teaching practice.

It is also important to determine whether virtual worlds are indeed the most appropriate solution for the application, deciding whether the problem we want the application to solve is essentially three-dimensional, whether it evolves, whether active user involvement is required, whether the simultaneous presence of multiple users is mandatory if there is a need to present some three-dimensional structure or space, or if the representation of some realistic environment is required.

We should also recognize that knowledge of technology alone is not sufficient for teachers to implement truly sound teaching and learning practices utilizing virtual worlds. It is necessary to combine knowledge of technology, the content of the cognitive object, and its teaching approach through technology. In this direction, the framework of Technological Pedagogical Content Knowledge (TPACK) [48] has been proposed, which focuses on the connections between the three critical parameters related to the effective integration of technology: content, technology, and pedagogy, where knowledge in matters of pedagogy it is combined with knowledge of the subject and knowledge of technology.

The virtual world was developed for the didactic needs of a module of the Art History course at the VET. The authors developed the virtual world using the functions provided by the platform spatial.io. To enrich it with relevant educational material they mainly used 3D open educational resources available in digital repositories.

The usefulness of these virtual worlds will have to be assessed based on various technological, scientific, and pedagogical criteria. For the development of effective OEVs, the above principles and approaches must be taken seriously.

VII. CONCLUSIONS - FUTURE WORK

The present study focuses on the development of a historical virtual world with the aim of its didactic use in the teaching of art history, the field of applied arts of the VET. The study aimed to describe the framework for the development of an educational virtual world (OEV), both from a technical and pedagogical point of view. Virtual worlds are usually created through time-consuming processes with large budgets and the involvement of many experts in their development. This study proposed a flexible, fast, and inexpensive method of developing virtual worlds by the teacher himself. Our approach was based on open educational resources and free development environments that are freely

979-8-3503-8368-3/23 $31.00 © 2023 IEEE

provided over the Internet and are easy to reuse and modify at no financial cost.

A key limitation of our proposal is the possible lack of relevant open resources that are suitable for the respective OEV. However, the development of free software and open educational resources, by educators, independent creators, and public bodies, increases the possibility of easy finding and utilization of educational resources from the internet. Another limitation is that we used the free version of the Spatial.io environment which provides basic functions, while in its subscription form, there are no restrictions and there are many options for developing the virtual world. The virtual world that we developed, although based on the curricula of VET, could be used in similar courses in general education and higher education if the necessary adaptations are made.

In our future studies, the virtual world should be used educationally and it should be evaluated by students and teachers, in terms of its usefulness, its ease of use, and its educational value. Also, our future pursuit is the development of OEVs for other sectors of VET, such as Health, Mechanical and Agriculture and Environment, Administration and Economy, and Construction Projects. Another future study of ours will investigate security and privacy protection in virtual world environments.

ACKNOWLEDGMENT

This work has been partly supported by COSMOTE under a PEDION24 grant.

REFERENCES

[1] S.Elayyan, "The future of education according to the fourth industrial revolution." Journal of Educational Technology and Online Learning 4.1, 2021: 23-30.

[2] T.Anastasiadis, G. Lampropoulos, and K. Siakas. "Digital game-based learning and serious games in education." International Journal of Advances in Scientific Research and Engineering 4.12, 2018 139-144.

[3] N. Lete, A. Beristain, and A. García-Alonso, "Survey on virtual coaching for older adults." Health Informatics Journal 26.4, 2020.

[4] F. Dahalan, A. Norlidah, and N. Mohd Shahril, "Gamification and game-based learning for vocational education and training: A systematic literature review." Education and Information Technologies, 2023: 1-39.

[5] G.Burnett, H.Catherine, and R. Kay, "Bringing the Metaverse to Higher Education: Engaging University Students in Virtual Worlds." Methodologies and Use Cases on Extended Reality for Training and Education. IGI Global, 2022. 48-72.

[6] S. Shorey, and D. Esperanza, "The use of virtual reality simulation among nursing students and registered nurses: A systematic review." Nurse Education Today 98, 2021: 104662.

[7] J. Jayalath, and E. Vatcharaporn, "Gamification to enhance motivation and engagement in blended eLearning for technical and vocational education and training." Technology, Knowledge and Learning 27.1,2022: 91-118.

[8] J., M. Pilz, "International transfer of vocational education and training: A literature review." Journal of Vocational Education & Training 75.2, 2023: 185-218.

[9] M.Fragkaki, S.Mystakidis, I.Hatzilygeroudis, K.Kovas, Z.Palkova, A.Salah, A.Ewais, Tpack instructional design model in virtual reality for deeper learning in science and higher education: From "apathy" to "empathy". In EDULEARN20 Proceedings (pp. 3286-3292). 2020.

[10] N. Gyimah, "Assessment of Technical and Vocational Education and Training (TVET) on the development of the World's Economy: Perspective of Africa, Asia and Europe." Asia and Europe (February 19, 2020), 2020.

[11] L., E.Liu, and Y. Lam Introducing participatory action research to vocational fashion education: theories, practices, and implications. Journal of Vocational Education & Training, 74(3), 2022, 415-433.

[12] A.Cattaneo, J.Gurtner, and J.Felder, "Digital tools as boundary objects to support connectivity in dual vocational education." Developing Connectivity between Education and Work, 2021: Principles and Practices.

[13] W. Nouwen, N. Clycq, A. Struyf, and D. Donche, The role of work-based learning for student engagement in vocational education and training: an application of the self-system model of motivational development. European Journal of Psychology of Education, 2022, 37(3), 877-900.

[14] D.Assante, M.Gerardo, and L.Placidi, "3D printing in Education: a European perspective." 2020 IEEE Global Engineering Education Conference (EDUCON). IEEE, 2020.

[15] H.Serin, "Virtual reality in education from the perspective of teachers." Amazonia investiga 9.26, 2020: 291-303.

[16] N.Sala, "Virtual reality, augmented reality, and mixed reality in education: A brief overview." Current and prospective applications of virtual reality in higher education, 2021: 48-73.

[17] N.Elmqaddem, "Augmented reality and virtual reality in education. Myth or reality?." International journal of emerging technologies in learning 14.3, 2019.

[18] V. Chrismadinata, N.Jalinus, F.Rizal, S.Sukardi, D.Ramadhani, L.Lubis, "Blended learning as an instructional model in vocational education: a literature review." Universal Journal of Educational Research 8.11B. 2019.

[19] S.Armakolas, P.Robolas, I.Karachalios, A.Karachasani, P. Anastopoulou, L.Gomatos, "Constructing and implementing an OER regarding sustainability issues in vocational education." Educational Journal of the University of Patras UNESCO Chair. 2019.

[20] https://muve.gse.harvard.edu/

[21] https://www.ubisoft.com/en-gb/

[22] http://lava.cs.st-andrews.ac.uk/

[23] S.Mystakidis, Metaverse. Encyclopedia, 2(1), 486-497. 2022.

[24] M. Weinberger, What Is Metaverse?—A Definition Based on Qualitative Meta-Synthesis. Future Internet, 14(11), 310. 2022.

[25] D.Ritterbusch, R.Teichmann, Defining the metaverse: A systematic literature review. IEEE Access. 2023.

[26] Y.Almoqbel, A.Naderi, Y.Wohn, and N.Goyal, The metaverse: a systematic literature review to map scholarly definitions. In Companion Publication of the 2022 Conference on Computer Supported Cooperative Work and Social Computing (pp. 80-84). 2022.

[27] A. Ortiz, m.Rojas, G.Cano, Application of Metaverse and virtual reality in education. Metaverse, 3(2), 13. 2022.

[28] K. Rahman, K.Shitol, S.Islam, T.Iftekhar, S.Pranto, Use of Metaverse Technology in the Education Domain. Journal of Metaverse, 3(1), 79-86. 2023.

[29] J.Jeon, K.Jung, Exploring the educational applicability of Metaverse-based platforms. 2021.

[30] A.Abraham, B.Suseelan, J.Mathew, P.Sabarinath, K.Arun, Study on Metaverse in Education. In 2023 7th International Conference on Computing Methodologies and Communication (ICCMC) (pp. 1570-1573). IEEE. 2023.

[31] B.Kye, N.Han, E.Kim, Y.Park, S. Jo, Educational applications of metaverse: possibilities and limitations. Journal of educational evaluation for health professions, 18. 2021.

[32] J.López-Belmonte, S.Pozo-Sánchez, J.Moreno-Guerrero, G.Lampropoulos, Metaverse in education: A systematic review. 2023.

[33] F.Şentürk, Z.Gürkaş-Aydın, M. Aydin, A Study on Metaverse and Its Applications in Education. El-Cezeri Journal of Science and Engineering. 2022.

[34] S. Kaddoura, S. Al-Husseiny, The rising trend of Metaverse in education: challenges, opportunities, and ethical considerations. PeerJ Computer Science, 9, e1252. 2023.

[35] N. Hess, K. Greer, Designing for engagement: Using the ADDIE model to integrate high-impact practices into an online information literacy course. Communications in information literacy, 10(2), 6. 2016.

[36] https://tinyurl.com/wusxp394

[37] https://secondlife.com/

[38] http://opensimulator.org/

[39] https://hubs.mozilla.com/

[40] https://engagevr.io/

[41] https://oercommons.org/

[42] https://www.merlot.org/merlot/

[43] https://sketchfab.com/

[44] https://nasa3d.arc.nasa.gov/

[45] https://www.turbosquid.com/

[46] http://photodentro.edu.gr/aggregator/

[47] https://tinyurl.com/mtt9hpdd

[48] J.Cowin, Chain of Worlds: Education in the Age of Metaverses. Proceedings of the 26th World Multi-Conference on Systemics, Cybernetics and Informatics. 2022

Area Allocation for Electric Vehicle Coverage Path Planning

1st Nikolaos Baras
Department of Electrical and Computer Engineering
University of Western Macedonia
Kozani 50100 Greece
nbaras@uowm.gr

2nd Antonios Chatzisavvas
Department of Electrical and Computer Engineering
University of Western Macedonia
Kozani 50100 Greece
achatzisavvas@uowm.gr

3rd Dimitris Ziouzios
Department of Electrical and Computer Engineering
University of Western Macedonia
Kozani 50100 Greece
dziouzios@uowm.gr

4th Ioannis Vanidis
Siatista Kozanis 50300
jvanidis@gmail.com

5th Minas Dasygenis
Department of Electrical and Computer Engineering
University of Western Macedonia
Kozani 50100 Greece
mdasyg@ieee.org

Abstract— Coverage path planning (CPP) plays a pivotal role in several application domains, such as agriculture, robotics, and surveillance. At its core, CPP aims to identify a route that ensures complete coverage of a specified area in the least amount of time. This study introduces a novel method for spatial allocation in CPP by leveraging affinity propagation clustering (APC). By employing APC, the proposed technique groups the target area into clusters based on shared attributes. Subsequently, a robot is designated to each of these clusters, and a unique route is charted for each to ensure comprehensive coverage of its assigned region. The primary objective of the proposed method is to enhance cluster distribution among robots, thus minimizing both communication overhead and path length. The efficacy of the approach is evaluated through simulation experiments, where it is benchmarked against other prevailing methods. The results indicate that the proposed methodology surpasses its counterparts and yields sub-areas of higher quality to the robots. Due to its effectiveness, the suggested approach may effectively handle a wide range of CPP area division problems in various environment designs and sizes.

Keywords—path planning, robotics, electric vehicles, affinity propagation

I. INTRODUCTION

In recent decades, profound technological advancements have been observed, significantly altering the global paradigm. These strides, spanning areas like telecommunications, computing, artificial intelligence, and robotics, have transformed human operations and task execution [1], [2]. Tasks once deemed labor-intensive or mundane are now accomplished with heightened speed, accuracy, and efficiency, primarily attributed to automation and the introduction of intelligent systems.

Robotics has been a driving force in this transformative era. Once a staple of science fiction, robots now play pivotal roles across diverse sectors, including manufacturing [3], healthcare [4], agriculture [5], and logistics [6]. They have proved indispensable in boosting productivity and refining service quality. In the realm of disaster management, robots have been crucial for search-and-rescue operations in perilous environments [7], [8]. Furthermore, as technology continues to evolve, the potential applications and capabilities of robots are bound to expand, offering unprecedented solutions to modern challenges.

However, the rapid advancement of robotics has brought out new difficulties that call for creative solutions. One such challenge in the robotic sphere is Coverage Path Planning (CPP) [9]–[11]. Essentially, CPP entails designing a trajectory enabling a robot to traverse its entire operational domain efficiently. CPP is crucial for various applications, like field inspection in agriculture, where a robot is required to navigate an entire field to evaluate crop health.

The intricacies of CPP amplify when transitioning from single to multi-robot systems. Multi-robot CPP necessitates crafting paths for several robots to guarantee thorough and efficient coverage of expansive or intricate areas. A critical concern in multi-robot CPP is the partitioning of the operational space amongst robots, an issue termed the 'area division problem'. Envision a squadron of drones tasked with extensive environmental monitoring or a convoy of autonomous vehicles engaged in a search-and-rescue mission within a calamity-affected zone. Optimal region assignment to individual robots can considerably elevate coverage efficiency and streamline operations, mitigating redundancies and conserving precious time. Addressing the area division challenge is crucial for the effective deployment of multi-robot systems in diverse practical applications, making it an active research domain, beckoning innovative solutions and techniques tailored to contemporary robotic system needs.

In this study, a method for area allocation in CPP utilizing Affinity Propagation Clustering (APC) [12] is introduced. By implementing the APC clustering technique, area points are categorized based on their similarities and differences. Once areas are grouped into clusters with similar features using APC, each cluster is designated to a robot for coverage. The proposed approach seeks to allocate operational regions to robots according to their attributes. Simulation tests are conducted to gauge the efficacy of the proposed method against alternative strategies.

The structure of the paper is as follows: Section 2 of this paper reviews related works found in literature, shedding light on their merits and limitations. Section 3 offers a formal problem definition. Section 4 delves into the proposed algorithm. Section 5 showcases the experimental findings and methodologies adopted to evaluate the algorithm's potency. Section 6 concludes the study.

II. RELATED WORK

In recent decades, various strategies for multi-robot CPP have been proposed. Voronoi diagrams frequently serve as a method for space allocation. Using these diagrams, the desired area is partitioned into distinct regions, with each region then

979-8-3503-8368-3/23 $31.00 © 2023 IEEE

allocated to a robot for coverage. Yet, Voronoi-based techniques [13] exhibit certain limitations, such as a considerable computational overhead and a pronounced sensitivity to initial point selection. Another approach for space allocation is the k-means [14] clustering, which endeavors to group data points into k clusters based on their similarity. Despite its application in CPP for area division, the k-means algorithm struggles with non-convex geometries and irregular data point distributions.

A notable multi-robot CPP algorithm is MSTC* introduced by Tang et al. in [11]. This algorithm's primary aim is to design coverage paths for multiple robots, accounting for tangible constraints such as obstacles and inter-robot communication pathways. The MST method [15], instrumental in segmenting the target environment into smaller sub-regions, underpins this algorithm. Robots are then allocated to these sub-regions based on their capabilities and task requirements. A significant merit of this algorithm is its ability to accommodate physical limitations, essential in real-world scenarios where robots traverse complex, dynamic environments. However, a limitation resides in its reliance on a predetermined robot count to ascertain both the number of robots and sub-regions. Consequently, the algorithm doesn't compute the number of produced sub-regions, leaving its optimality unassured. It's pertinent to mention that while the problem addressed by Tang et al. bears similarities to the problem defined in this study regarding area division and allocation, the specific problem nuances presented in [11] diverge considerably.

III. PROBLEM FORMULATION

Before we present the details of the proposed algorithm, it is important to define the actual problem of area division for multirobot CPP. The objective is to divide a grid-based area of interest into several sub-areas and assign a robot to each sub-area for coverage. The distribution should ensure that each sub-area is completely covered. The issue can be formalized as follows:

Let A represent the N-point grid-based area of interest (Figure 1). Each cell of the grid is either accessible traversable area, or inaccessible (obstacle). The area A must be divided into K spatially connected sub-areas $(S_1, S_2, \ldots, S_K)$, each of which must be assigned to a robot for coverage. The number of available robots is considered to be *infinite*. This allows the APC algorithm to find the optimum number of sub-areas.

To achieve coverage, it's crucial to understand the environment's heterogeneity. The grid's cells, labeled as accessible or inaccessible, present varying challenges for robot navigation. Inaccessible cells, acting as obstacles, can introduce complexities in the path planning, whereas accessible cells offer smoother traversal. Given the infinite availability of robots, the APC algorithm's strength lies in determining the optimal number of sub-areas for coverage, ensuring that the entire accessible region is addressed without redundancy. By precisely partitioning the area into well-defined sub-areas and delegating them to robots, efficient and comprehensive coverage can be achieved, minimizing potential overlaps and ensuring that no section remains unattended.

IV. IMPLEMENTATION

APC clustering emerges as a pivotal technique in the domain of area division for CPP. Contrary to many clustering algorithms, APC possesses the inherent capability to determine the optimal number of clusters without making arbitrary assumptions, offering a significant edge in multi-robot CPP by ascertaining the optimal division of the area into sub-regions for robotic traversal.

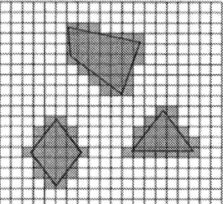

Fig. 1. Graphical representation of an example grid-based environment with 3 obstacles. Due to the nature of approximate cellular decomposition, each cell is either accessible or inaccessible to the robots. Cells containing any portion of an obstacle are deemed inaccessible.

The APC algorithm operates based on a principle of "message passing" between data points. At its essence, there are two types of messages that are exchanged: "responsibility" and "availability." The "responsibility" message sent from a data point i to a candidate exemplar k reflects the accumulated evidence for how well-suited point k is to serve as the exemplar for point i, considering other potential exemplars. Conversely, the "availability" message sent from a candidate exemplar k to a data point i conveys the accumulated evidence for how appropriate it would be for i to select k as its exemplar, considering the support from other points that k should be an exemplar.

The APC algorithm iteratively updates these messages, and over time, evidence accumulates, leading to a decision on the optimal number of clusters and the assignment of data points to these clusters. This iterative process continues until convergence, resulting in a set of "exemplars" that represent the centers of the clusters and the corresponding assignments of data points to these exemplars.

In the context of CPP, the data points correspond to the cells within the grid-based area of interest. Traditional clustering would employ a standard Euclidean distance metric to compute similarities between data points. However, this metric proves inadequate for CPP due to the presence of obstacles, rendering certain paths between points non-traversable. Thus, direct Euclidean distances can be misleading.

To overcome this limitation, a custom distance metric is introduced, which is based on the Breadth-First Search (BFS) algorithm. This BFS-rooted metric effectively gauges the navigational distance between grid cells, factoring in spatial constraints imposed by the grid's design and obstructions present. By using BFS, the algorithm determines the shortest path between two points that avoids obstacles, providing a more accurate representation of the "true" distance that a robot would need to traverse.

Once this BFS-based metric is established, it serves as the foundation for the similarity matrix in the APC algorithm. As APC progresses, it uses this custom similarity matrix to determine the optimal division of the area into sub-regions, effectively addressing the unique challenges posed by CPP. It is worth noting that during the similarity matrix calculation, several more modifications can be made, to enhance the homogeneity of the clusters, based on other characteristics that

979-8-3503-8368-3/23 $31.00 © 2023 IEEE

they may have, such as different elevation levels or floor types.

The APC algorithm (Algorithm 1) delineates a systematic procedure for deriving and presenting the final clusters, given a grid-based area as input. Initially, a similarity matrix for the data points within the study area is constructed. This matrix embodies the pairwise similarity among the data points, with each grid point considered a distinct data point. Subsequent to this, the responsibility matrix is computed. This matrix conveys the suitability of each data point in acting as an exemplar. Its computation involves the previously established similarity matrix and incorporates a damping factor. Following this, the availability matrix is formulated, indicating the viability of each data point slated for exemplar assignment. This matrix is derived using both the damping factor and the responsibility matrix. An iterative process is then initiated, where the availability and responsibility matrices undergo repeated updates until a point of convergence is reached. This iterative refinement utilizes the preceding matrix values and the damping factor to obtain the updated matrices.

Upon convergence, exemplars are identified using the refined availability and responsibility matrices. A data point exhibiting elevated values in both matrices is designated as an exemplar. In the concluding phase, data points are clustered based on their similarity to these identified exemplars. Each resulting cluster corresponds to a specific sub-area, optimally positioned for allocation to a robot for comprehensive coverage.

ALGORITHM 1. THE PROPOSED APC BASED ALGORITHM PSEUDOCODE.

```
Algorithm: Affinity Propagation for CPP

Input:

- Data points (cells within the grid-based area of
interest)

- Maximum iterations

- Convergence criterion

Output:

- Exemplars (cluster centers)

- Cluster assignments for each data point

Begin:

1. Initialize the similarity matrix S using BFS-
based metric

    For each pair of data points i and j:

    S(i, j) = BFS_distance(i, j)

2. Initialize responsibility matrix R and
availability matrix A to zero

3. For iter = 1 to Maximum iterations:

    a. Update Responsibility matrix:

    For each data point i: R(i, k) = S(i, k) -
max(S(i, j) + A(i, j) for all j ≠ k)

    b. Update Availability matrix:

      For each data point i:

        A(i, k) = min(0, R(k, k) + sum(max(0, R(j,
k)) for all j ≠ i, k))

        A(k, k) = sum(max(0, R(j, k)) for all j ≠
k)
```

```
    c. Check convergence:

        If change in matrices R and A is below the
        convergence criterion: Break

4. Determine exemplars:

    For each data point i:

        If R(i, i) + A(i, i) > 0:  i is an exemplar

5. Assign data points to exemplars:

    For each data point i:

        Assign i to the exemplar k that maximizes
        R(i, k) + A(i, k)

End
```

Note that the specific details and implementations of the functions that calculate the similarity and availability matrix are out of the scope of this paper and are not presented.

IV. EXPERIMENTAL RESULTS

Experimental validation serves as a foundational pillar in algorithmic research, offering a lens through which the practical applicability, efficiency, and constraints of an algorithm can be discerned. Particularly for CPP algorithms tailored for EVs, the diverse terrains where these vehicles function become instrumental in appraising the viability of any introduced technique. As the proposed APC based approach departs from traditional methods, it's imperative to rigorously test its mettle against benchmark datasets and the renowned k-means clustering algorithm.

For performance assessment of the proposed methodology, three distinctive environment configurations were utilized: (A) 30x30, (B) 60x60, and (C) 120x120. In each environment, a pseudo-random allocation designated 25% of its cells as obstructions and the remaining 75% as navigable terrain. These environments were fed into each algorithm as binary matrices, with the k-means clustering algorithm selected as the comparative benchmark.

The Davies-Bouldin Index (DBI) [16] was chosen to evaluate clustering performance in this study. The DBI quantifies the average 'similarity' ratio of each cluster with its most similar cluster. This similarity measure is the ratio of the sum of the intra-cluster distances to the inter-cluster distances. Lower values of the DBI indicate better clustering. For this purpose, a modified DBI function was employed that takes into account the 4-neighbor connectivity of cells and their normalized distance, drawing from the principles of Dijkstra's shortest path algorithm.

$$\text{DBI}(i) = \frac{intra-cluster\ distance + inter-cluster\ distance}{inter-cluster\ distance},$$

where:

- DBI(i) represents the Davies-Bouldin Index for cluster i.
- The intra-cluster distance is the average distance within cluster i.
- The inter-cluster distance is the minimum average distance from cluster ii to every other cluster.

In the conducted experiments, the proposed Affinity Propagation clustering algorithm's performance was compared to the traditional k-means clustering algorithm across various environments, using the DBI as the evaluative metric. In most of the tested environments, the proposed algorithm consistently outperformed the k-means method, yielding 6% better results in terms of the DBI.

However, there was an exception in one specific environment, where the proposed algorithm did not demonstrate superior performance. In this setting, the proposed method's efficacy was somewhat diminished. This divergence can be attributed to the unique spatial characteristics of this environment. The presence of expansive unobstructed spaces in this environment might have been more aptly clustered by the k-means algorithm. Such results underline the notion that while an algorithm may exhibit exemplary performance in most scenarios, it may not be universally superior across all settings, underscoring the significance of tailoring solutions based on specific environmental intricacies.

V. CONCLUSIONS

In this research endeavor, an innovative approach to area allocation in coverage path planning was introduced, utilizing the principles of APC. The target area was meticulously segmented into clusters bearing analogous characteristics through the application of APC, subsequently assigning each distinct cluster to a dedicated robot for comprehensive coverage. The primary ambition of the proposed method was to enhance the sub-area allocation to the robots. To validate the merits of the presented approach, a series of simulation experiments were orchestrated, wherein its performance was juxtaposed against alternative clustering methodologies.

Looking ahead, there's a plethora of avenues to expand upon this foundational work. Future endeavors could delve into the real-world applicability of the proposed strategy, assessing its mettle in tangible, on-ground scenarios. Additionally, given the iterative nature of APC, it is possible to incorporate parallelization techniques, thereby vastly accelerating the clustering process. This parallel approach could be particularly beneficial when dealing with vast environments or when real-time decisions are paramount.

ACKNOWLEDGEMENT

This research was carried out as part of the project "Smart Safe Navigation for Electric Bicycles and Skateboards" (project code: KMP6-0292520) under the framework of the Action "Investment Plans of Innovation" of the Operational Program "Central Macedonia 2014 2020", that is co-funded by the European Regional Development Fund and Greece.

REFERENCES

[1] H. Lasi, P. Fettke, H.-G. Kemper, T. Feld, and M. Hoffmann, "Industry 4.0," *Bus Inf Syst Eng*, vol. 6, no. 4, pp. 239–242, Aug. 2014, doi: 10.1007/s12599-014-0334-4.

[2] L. S. Dalenogare, G. B. Benitez, N. F. Ayala, and A. G. Frank, "The expected contribution of Industry 4.0 technologies for industrial performance," *International Journal of Production Economics*, vol. 204, pp. 383–394, Oct. 2018, doi: 10.1016/j.ijpe.2018.08.019.

[3] Z. Pan, J. Polden, N. Larkin, S. Van Duin, and J. Norrish, "Recent progress on programming methods for industrial robots," *Robotics and Computer-Integrated Manufacturing*, vol. 28, no. 2, pp. 87–94, Apr. 2012, doi: 10.1016/j.rcim.2011.08.004.

[4] M. Kyrarini *et al.*, "A Survey of Robots in Healthcare," *Technologies*, vol. 9, no. 1, Art. no. 1, Mar. 2021, doi: 10.3390/technologies9010008.

[5] C. Lytridis *et al.*, "An Overview of Cooperative Robotics in Agriculture," *Agronomy*, vol. 11, no. 9, Art. no. 9, Sep. 2021, doi: 10.3390/agronomy11091818.

[6] N. Baras, A. Chatzisavvas, D. Ziouzios, and M. Dasygenis, "Improving Automatic Warehouse Throughput by Optimizing Task Allocation and Validating the Algorithm in a Developed Simulation Tool," *Automation*, vol. 2, no. 3, Art. no. 3, Sep. 2021, doi: 10.3390/automation2030007.

[7] J. Trevelyan, W. R. Hamel, and S.-C. Kang, "Robotics in Hazardous Applications," in *Springer Handbook of Robotics*, B. Siciliano and O. Khatib, Eds., in Springer Handbooks. , Cham: Springer International Publishing, 2016, pp. 1521–1548. doi: 10.1007/978-3-319-32552-1_58.

[8] G. Angelopoulos, N. Baras, and M. Dasygenis, "Secure Autonomous Cloud Brained Humanoid Robot Assisting Rescuers in Hazardous Environments," *Electronics*, vol. 10, no. 2, Art. no. 2, Jan. 2021, doi: 10.3390/electronics10020124.

[9] A. Zelinsky, *Planning paths of complete coverage of an unstructured environment by a mobile robot*, vol. 13. 1993.

[10] R. Almadhoun, T. Taha, L. Seneviratne, and Y. Zweiri, "A survey on multi-robot coverage path planning for model reconstruction and mapping," *SN Appl. Sci.*, vol. 1, no. 8, p. 847, Jul. 2019, doi: 10.1007/s42452-019-0872-y.

[11] J. Tang, C. Sun, and X. Zhang, "MSTC∗:Multi-robot Coverage Path Planning under Physical Constrain," in *2021 IEEE International Conference on Robotics and Automation (ICRA)*, May 2021, pp. 2518–2524. doi: 10.1109/ICRA48506.2021.9561371.

[12] B. J. Frey and D. Dueck, "Clustering by Passing Messages Between Data Points," *Science*, vol. 315, no. 5814, pp. 972–976, Feb. 2007, doi: 10.1126/science.1136800.

[13] V. G. Nair and K. R. Guruprasad, "GM-VPC: An Algorithm for Multi-robot Coverage of Known Spaces Using Generalized Voronoi Partition," *Robotica*, vol. 38, no. 5, pp. 845–860, May 2020, doi: 10.1017/S0263574719001127.

[14] H. Guo, J. Ma, and Z. Li, "Active Semi-supervised K-Means Clustering Based on Silhouette Coefficient," in *Advances in Intelligent, Interactive Systems and Applications*, F. Xhafa, S. Patnaik, and M. Tavana, Eds., in Advances in Intelligent Systems and Computing. Cham: Springer International Publishing, 2019, pp. 202–209. doi: 10.1007/978-3-030-02804-6_27.

[15] Y. Gabriely and E. Rimon, "Spanning-tree based coverage of continuous areas by a mobile robot," *Annals of Mathematics and Artificial Intelligence*, vol. 31, no. 1, pp. 77–98, Oct. 2001, doi: 10.1023/A:1016610507833.

[16] D. L. Davies and D. W. Bouldin, "A Cluster Separation Measure," *IEEE Transactions on Pattern Analysis and Machine Intelligence*, vol. PAMI-1, no. 2, pp. 224–227, Apr. 1979, doi: 10.1109/TPAMI.1979.4766909.

Modeling Network Traffic and Exploring Distribution Fitting: A Case Study on Spotify

Odysseas Karadimas
Dept. of Informatics and Telecommunications
University of Ioannina
Arta, Greece
pint00156@uoi.gr

Aikaterini Florou
Dept. of Informatics and Telecommunications
University of Ioannina
Arta, Greece
pint00158@uoi.gr

Spiridoula V. Margariti
Dept. of Informatics and Telecommunications
University of Ioannina
Arta, Greece
smargar@uoi.gr

Eleftherios Stergiou
Dept. of Informatics and Telecommunications
University of Ioannina
Arta, Greece
ster@uoi.gr

Chrysostomos D. Stylios
Industrial Systems Institute,
Athena Research Center
Patra, Greece
stylios@isi.gr & stylios@athenarc.gr

Abstract—As Internet traffic grows exponentially due to user demands and economic competition, the characterization of network workload and its resource utilization becomes more complex but also more important. This study aims to gain insights into the network's dynamics and explore various aspects of its behavior and performance. It presents an empirical analysis of network traffic based on a 24-hour dataset obtained at a university campus from internet users on the 'Spotify' platform. The analysis is built on data features, and it reveals several traffic models. More specifically, using the Cullen and Frey graphs, it investigates which are the optimal distributions fitting to data traffic. It focuses on patterns observed in the number of connections per minute and brings out an initial burst of activity followed by a stable level of connectivity. The study also examines the duration of TCP (Transmission Control Protocol) connections and the size of transfer data, indicating that most connections are short-lived and data transfers predominantly involve smaller sizes. These findings contribute to a comprehensive understanding of the network's behavior and can shape network management strategies for optimizing performance and resource allocation.

Index Terms—Network Traffic, distribution fitting, traffic analysis, flow duration, network traffic modeling

I. INTRODUCTION

The Internet is an evolutionary, dynamic, and complex ecosystem that serves a vast collection of data flows that originate from various sources and have different requirements. Concurrently, the demand for various Internet services is continuously growing, forming various traffic patterns that strongly impact network efficiency. This demand highlights the importance of efficient management and allocation of network resources, while design, specialized service provision, and traffic engineering decisions are defined by traffic characteristics [1]. These developments compel continuous network operations without interruption, the provision of sufficient network capacity and capability for network maintenance, and carrying out critical tasks. Thus, a scientific and compre-

hensive understanding of network traffic patterns is vital for ensuring smooth operations and optimal network performance.

A large fraction of internet traffic emanates from popular network applications, such as Instagram, Facebook, and Spotify, hosted on online platforms, as they engage the users who consume their services. In this context, and as technology evolves while application capabilities grow and improve, a model a model is needed to reliably describe the essential characteristics of internet traffic. In this study, we mention Spotify, a popular online platform that provides music and other audio products as a streaming subscription service [2]. Launched in 2008, it targets a dynamic market of millions of users from different regions of the world. The introduction of new products (e.g., podcasts) attracts the attention of users and leads to the generation of considerable internet traffic.

The purpose of this work is to perform an analysis of a one-day empirical dataset obtained from a online platform's internet users, specifically the platform Spotify. The real traffic measurements have been collected at Calgary University [3] and provided to us in our investigation. Using statistical tools, such as R software, probability distributions and Cullen-Frey graph, we examine the data flow characteristics and strive to find a useful model to fit on network traffic. We believe that the use of such tools, can help extract valuable insights for modeling network traffic. Through an in-depth investigation of this data, we expected to gain insights into the network's behavior, explore various aspects of network performance, reveal general traffic patterns through these analyses, and provide a comprehensive view of network behavior and traffic patterns.

Analyzing network data at the link level provides valuable insights into the factors that govern traffic within a network. The findings could guide network managers and operators in making optimal decisions for network operation. More specifically, this knowledge can be useful for determining network management strategies, making decisions on resource

979-8-3503-8368-3/23 $31.00 © 2023 IEEE

allocation, and improving overall network performance.

We use the aforementioned datasets in order to examine whether the traffic patterns are approached by well-known distributions in terms of connection number (workload) and connection duration. We focus on the arrival time of users and analyze how TCP connection flows are distributed over time. Several graphs are created in order to illustrate this pattern and discuss observations and insights. Additionally, the duration of TCP connections is examined, and indicate that the Gamma distribution is more appropriate to represent the distribution of connection duration per minute. Furthermore, the transfer size of data is investigated, by analyzing the number of packets sent and received in each TCP connection. Basic statistics, along with graphical representations, are used to understand the size distribution and its implications.

This work is further structured as follows: Section II establishes the context of this work and introduces the main findings from previous related works. Section III describes the dataset and presents the tools and methodologies used in this work. Section IV discusses the results obtained from the data analyses and, finally, in Section V, the conclusions are provided.

II. BACKGROUND AND RELATED WORK

A. Spotify

Spotify is an online platform that provides music streaming services and is available in the global market with more than 550 million active users [2]. The service is offered for various types of devices, giving on-demand access to millions of music tracks and podcast titles, as well as several thousand audiobooks. It is based on the client-server architecture and uses the TCP protocol at the transport layer [4]. In order to eliminate the problem of delay due to TCP acknowledgements, it works with "pre-fetching" data from the P2P (peer-to-peer) network. Spotify has attracted the attention of researchers, who are investigating issues related to its traffic and effectiveness.

In a recent study [5], the quality of experience (QoE) of Spotify's audio streaming and app browsing under various network conditions was explored and its performance in terms of key performance indicators (KPIs) for different network conditions was examined. The study provides valuable insights into the network dynamics and traffic patterns of the Spotify platform. Its findings can be used by ISPs (Internet Service Provider) to develop new techniques to improve their traffic and network management.

Setty et al. [6] provide a case study of how Spotify uses the pub/sub paradigm to enhance its users' music experience. It includes a detailed analysis of the pub/sub traffic and workload of the platform and characterizes the workload of the system in terms of event publication rates, topic popularity, subscription sizes, subscription cardinality, and temporal subscription/unsubscription patterns. Additionally, it analyzes the pub/sub traffic to derive trends and patterns, which can be used to inform further research on network dynamics and traffic patterns in the context of music streaming platforms like Spotify.

B. Literature review

The importance of network traffic analysis and modeling has been highlighted by many researchers, as it is considered essential for the design and development of a reliable and efficient network [7]. In modern networks, users' interaction with online platforms shapes access patterns depending on traffic intensity and user behavior. Several researchers based their conclusions on empirical traffic in an attempt to conceptualize network traffic and understand performance's characteristics.

Empirical measurements show changes in network traffic as a result of user behavior. Researchers agree that a critical factor in network performance is good traffic estimation [8]. Such estimation can reveal bottlenecks in the network and help improve its performance by eliminating them. The majority of these studies focus on the analysis and modeling of real network traffic using datasets that the authors themselves collect, mainly from the local networks of universities. Campus networks can be seen as mini-Internets as they emerge from the consolidation of many local networks. The university network is a "rich and fertile" [9] field for further research and study and includes a wide range of internet traffic examples, such as traffic from instant messaging, email, social networks, or other network applications [5], [10], [11].

Unrelated to Spotify, there are studies that provide a framework for analyzing network traffic and could potentially shed light into the analysis of traffic patterns on the Spotify platform. Particularly, [10] provides a methodology for collecting and analyzing flow traces from a network, as well as constructing flow models that accurately describe the traffic. The authors emphasize the importance of accurate flow models in assessing the efficiency of flow-based networking mechanisms and note the lack of credible models or data in existing literature. In addition, [9] provides valuable insights into the traffic characterization of Instagram, a popular network application, despite the challenges of end-to-end encryption, NAT, DHCP, and high traffic volume. Some techniques used in this paper include measurement infrastructure and software tools for data collection and analysis.

Moore et al. [12] provide a comparative analysis of the network performance of Spotify and YouTube Music. The study characterizes the network performance of audio streaming traffic for both services at a micro and macro scale, capturing specific packet-level data traffic on the campus network. The paper also discusses the differences in network infrastructure, protocol, usage, and popularity between the two services. In a similar work, Keshvadi et al. [11] concern the traffic characterization of instant messaging apps on a large campus edge network. While the focus of this study is on instant messaging apps, it provides insights into network traffic measurement and analysis in general.

All the above-mentioned studies present exciting results, but almost none of them propose distribution fitting models related to workload (the number of connections) or the duration of each connection. We use Cullen and Frey graphs [13] to identify the best-fit distribution on Spotify traffic datasets for these specific features.

III. MATERIALS AND METHODS

A. Dataset Description

In our study, we used the Spotify traffic measurements that have been collected by a research group at the University of Calgary and shared with us [3]. The dataset comprises 24-hour network traffic data from users of the campus network who consume Spotify platform services. The dataset consists of tuples of key elements like timestamps, IP addresses, transport-layer protocol, connection state, and more. Table I shows the basic statistics of the collected data. The dataset underwent preprocessing to address missing values and ensure consistency.

TABLE I
SUMMARIZED DATA OF SPOTIFY TRAFFIC

Data Description	Protocol		
	ICMP	*TCP*	*UDP*
Number of flows	614	935 066	59
Number of bytes-sent	2 544 873	40 507 073 495	70 310
Number of bytes-received	69 552	110 692 714 638	141 934

B. Methodology

Data collection was based on timestamps. For better processing and visualization, the data is analyzed by time unit defined in 1, 15, or 60-minute time bins. More specifically, the data is aggregated or averaged and placed into bins according to their timestamps to reduce the volume of data processed in the next steps. The data set is analyzed based on specific metrics, such as the number of connections in the time unit, the duration of each connection, the number of packets sent or received, and the state of the connection. Initially, the data are cleaned, and separate records are merged while the data binning is performed at the process phase. In the data analysis process, TCP connections were examined for temporal patterns. More specifically, the 24-hour dataset was divided into 15-minute segments, and the number of connections was observed. In addition, connection duration and bytes sent and received were subject to statistical analysis, offering insights into network behavior.

Then follows the fitting process, namely the finding of a probability distribution that best fits the data set of Spotify traffic. The investigation of this problem is based on the skewness-kurtosis or Cullen and Frey Graph [13], which potentially indicates the proper distribution that models the gathered traffic depending on some aspect. The aim of Cullen and Frey Graph is to provide a suitable distribution for data series among predefined distributions (normal, exponential, gamma, and others). The best distribution fitting was inferred from skewness and kurtosis level. To determine the goodness-of-fit of probability distribution, classical graphical tools (density plot, Cumulative Distribution Function (CDF) plot, Q-Q plot, and P-P plot) are used.

C. Tools and Software

The data analysis process involved several key steps. Initially, MS Excel was used for dataset curation and formatting, addressing missing values, and ensuring consistency. Subsequently, R language and the RStudio IDE were employed for statistical analysis, distribution approximation and fitness with the "fitdistrplus" package, and statistical metrics calculation via the "statistics" library. Finally, the RStudio IDE's plotting device was utilized for creating high-quality visualizations. These methods contributed to a comprehensive dataset analysis, offering valuable insights into network behavior and traffic patterns. By utilizing these tools, the dataset was effectively curated, and a comprehensive data analysis was performed. These techniques collectively contribute to a robust analysis of the dataset, providing valuable insights into network behavior within the study's context.

IV. TRAFFIC ANALYSIS

This section provides a comprehensive overview of the observations, measurements, and statistical analyses performed on the network dataset obtained from Spotify. Through the application of various analytical techniques, the chapter delves into the distribution properties, and traffic behavior within the network. The results offer valuable insights into the behavior and dynamics of the network, enabling a deeper understanding of its performance and optimization strategies. By presenting the key findings, trends, and statistical summaries, this section serves as a comprehensive account of the analysis outcomes, laying the foundation for the subsequent discussions and interpretations in the following sections.

A. TCP connection flows

Regarding TCP connections, the following plots were created, which represent the number of connections per 15-min period.

Fig. 1 illustrates the aggregated traffic volume in the number of all connections at the transport layer per quarter interval throughout the day. The number of connections ranges from

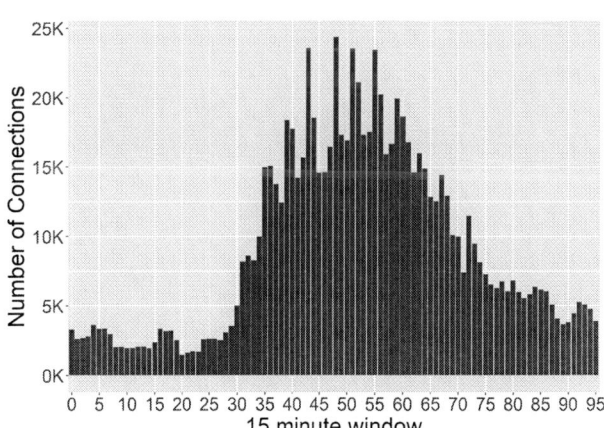

Fig. 1. Daily number of connections per 15 minutes of Spotify observed traffic.

979-8-3503-8368-3/23 $31.00 © 2023 IEEE

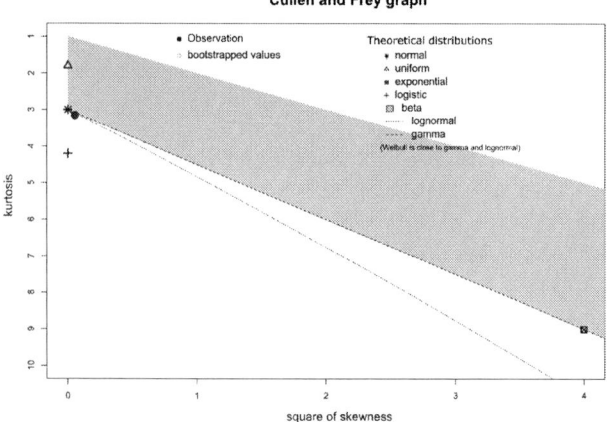

Fig. 2. The Cullen Frey plot for the produced workload from daily connections to Spotify.

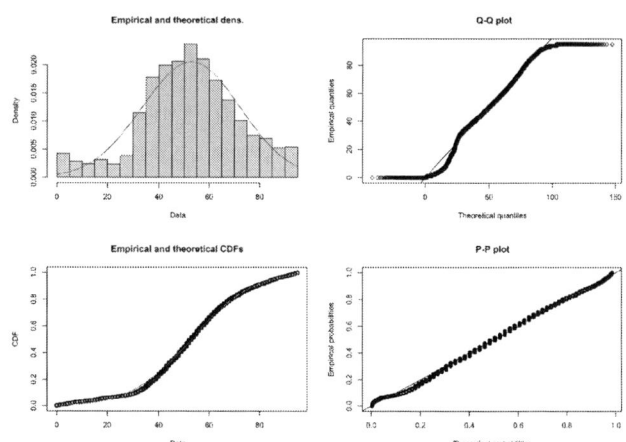

Fig. 3. Assessing the fitting to normal distribution.

2,000 to 24,000. Notably, there is a significant increase in connections during the midday hours, with a decrease in the early morning and late evening. The distribution of connections per quarter appears to resemble a normal distribution. Fig. 2 shows a Cullen-Frey plot that was constructed to explore the distribution of connection counts. This plot suggests a close approximation to a normal distribution, providing strong evidence that the connection counts follow a normal distribution. The normal distribution function is defined as:

$$F(x) = \frac{1}{\sigma\sqrt{2\pi}} e^{-(x-\mu)^2/2\sigma^2}$$

where $\mu = 52.82$ and $\sigma = 19.41$

Fig. 3 compares the empirical density of the variable with the theoretical density of the normal distribution. The satisfactory fit observed in histograms, quantile-quantile (Q-Q) plots, cumulative density function (CDF) plots, and probability plots indicates that the data aligns well with a normal distribution.

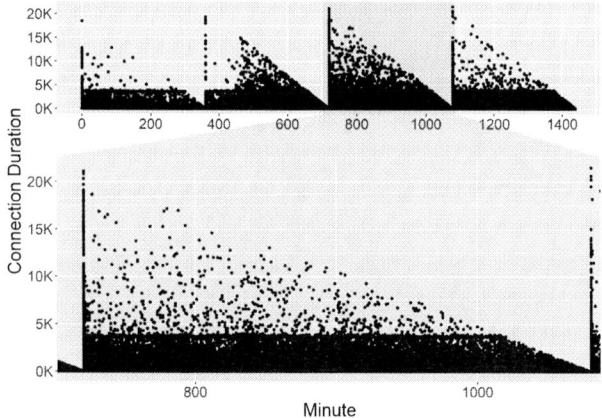

Fig. 4. User connection duration per minute

B. TCP connections duration

In Fig. 4, we focus on the duration of connections per minute over a 6-hour window. By modeling this recurring pattern, we aim to gain a comprehensive understanding of the distribution of connection duration per minute. Fig. 5 begins the modeling process by constructing a Cullen-Frey plot, which suggests that the variable approximates a gamma distribution. However, further exploration is required to determine the exact fit.

The Gamma probability density function is defined as:

$$f(x_i) = \frac{1}{\beta^\alpha \Gamma(\alpha)} x_i^{(\alpha-1)} e^{-(\frac{x_i}{\beta})}$$

where x_i is the Gamma distributed variable, $\alpha = 0.1025$ and $\beta = 0.0005$.

Fig. 6 illustrates the theoretical and empirical probability density function (Pdf), cumulative distribution function (Cdf), and provides the gamma distribution's Q-Q and P-P plots relative to connection duration. It evaluates the quality of the fit, indicating that while the empirical and theoretical cumulative density functions are reasonably close, quantile-quantile and probability plots suggest deviations from a perfect gamma distribution.

C. Inbound and Outbound traffic distribution

Fig. 7 displays the average incoming and outgoing packets per hour over the 24-hour period. Notably, incoming packets exceed outgoing packets, particularly during peak midday hours when the highest number of connections occur. The results of the Fig. 7 lead us to two conclusions: a) the difference between sent and received packets is due to transmission errors and is an inherent property of a network environment; b) as the number of connections increases (or decreases) (Fig. 1), the network cannot handle the load and packets are dropped due to congestion. In addition, the TCP protocol, which performs algorithms to control congestion, reduces the

979-8-3503-8368-3/23 $31.00 © 2023 IEEE

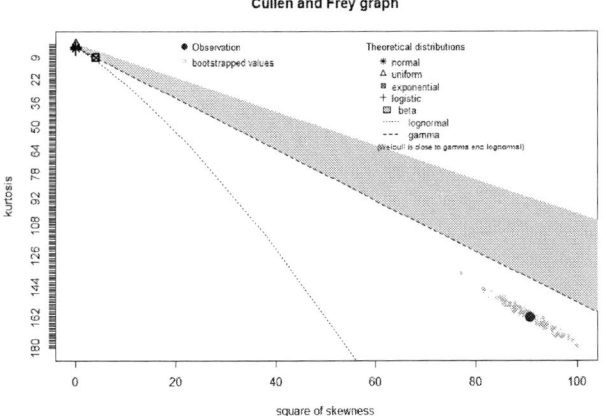

Fig. 5. The Cullen and Frey plot of estimated distributions for connection duration.

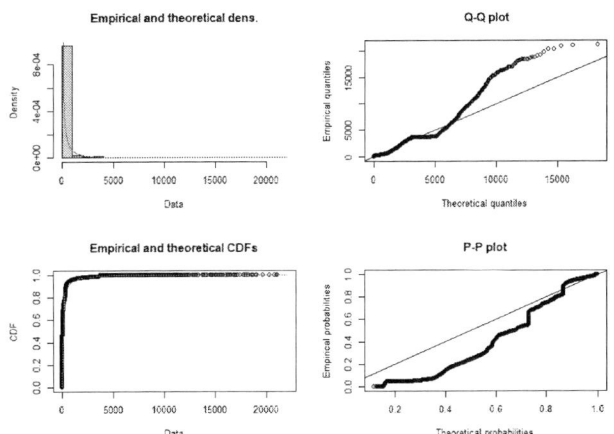

Fig. 6. Theoretical and empirical Pdf and Cdf with Q-Q plot and P-P plot for gamma distribution for connection duration.

number of packets sent. Thus, it is observed a decrease in packets during the midday hours when the workload increases.

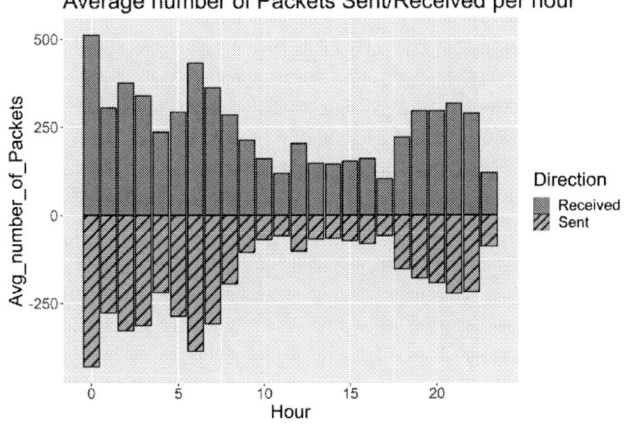

Fig. 7. Traffic patterns of Inbound and Outbound traffic (average # of packets per hour).

D. Flow connections state

The connection state is a field of the dataset that describes the state of the connection at the transport layer using 13 different indicators. For example, if a communication flow is "normally established and terminated" [14] labeled by the SF indicator. Table II lists the number of flow connections per state, while Fig. 8 shows the frequency of occurrence of each state aggregated per 15 minutes.

Most connections indicated by SF are successful (33,3%), followed by connections indicated by SHR, which means the sequence SYN-ACK-FIN does not include the expected SYN for the correct termination of the connection (20,6%). There is also a small number of connection requests (542) that are rejected (REJ). From the graphical representation of the measurements, it can be seen that in almost all cases, there is a peak hour volume during peak hours. Variations in

connection counts across different connection states are also observed, with some states having over 5,000 connections per 15 minutes, while others have fewer than 100. The "REJ" state is excluded from the comparative histogram due to its limited data.

In short, these figures collectively provide valuable insights into the behavior and distribution of connection data, with a strong indication that it follows a normal distribution in several cases. However, further investigation is required as to their distribution and impact on network performance. These findings contribute to a comprehensive understanding of the network connections under analysis.

V. CONCLUSION

In conclusion, this work analyzes Spotify traffic from a campus network based on a one-day data set. Distinct patterns were observed in the number of connections per minute, with an initial burst of activity followed by a stable level of connectivity. The data was classified into in normal-distribution or in Gamma-distribution in terms of the number or duration of connections, respectively. Both the Cullen & Frey plot, as well as the CDF, Q-Q plot, and P-P plot support the hypothesis that the observed values follow these distributions. This hypothesis is not confirmed by other tests (e.g., Kolmogorov-Smirnov).

These findings encourage further research. We plan to extend this study by investigating more candidate distributions for more accurate fitting and analyzing their reliability using various goodness-of-fit tests.

ACKNOWLEDGMENT

This research has been financed by the European Union: Next Generation EU through the Program Greece 2.0 National Recovery and Resilience Plan, under the call RESEARCH – CREATE – INNOVATE, project name "iCREW: Intelligent small craft simulator for advanced crew training using Virtual Reality techniques" (project code:TAEDK-06195). Also, we would like to thank Dr. Carey Williamson, who provided

979-8-3503-8368-3/23 $31.00 © 2023 IEEE

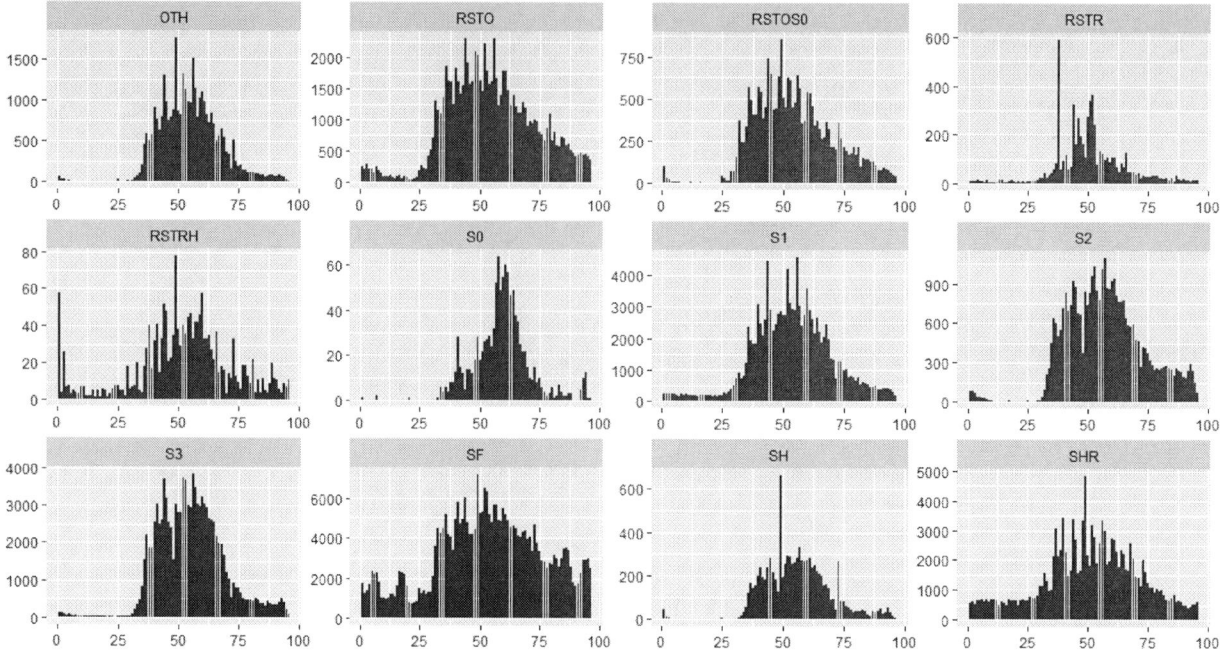

Fig. 8. Detail statistics for the flow connection state.

TABLE II
SUMMARY OF STATISTICS FOR THE FLOW CONNECTION STATE[1].

Connection State	OTH	REJ	RSTO	RSTOS0	RSTR	RSTRH	S0	S1	S2	S3	SF	SH	SHR
Number of Connections	34941	542	84087	22377	6112	1742	1121	121221	34922	109018	311184	9508	132114

[1] **SF**: connection successful established and completed **REJ**: Connection rejected; **S0**: no response to initial SYN; **S1**: connection established, nothing more seen; **S2**: connection established, attempt to closed only by initiator side; **S3**: connection established, responder has closed their side; **RSTO**: Connection established, initiator reset; **RSTOS0**: initiator sent a SYN and then reset, no response (SYN-ACK); **RSTR**: established, responder reset; **RSTRH**: Responder sent a SYN-ACK and RST, no SYN from the initiator; **SH**: connection closed by initiator, no SYN-ACK by resp; **SHR**: connection closed by responder, no SYN by initiator; **OTH**: No SYN seen, not terminated, midstream traffic.

us with the datasets, as without them it was impossible to complete this work.

REFERENCES

[1] P. Tune, M. Roughan, H. Haddadi, and O. Bonaventure, "Internet traffic matrices: A primer," *Recent Advances in Networking*, vol. 1, pp. 1–56, 2013.

[2] "Spotify — About Spotify — newsroom.spotify.com," https://newsroom.spotify.com/company-info/, [Accessed 13-07-2023].

[3] "CPSC Home — science.ucalgary.ca," https://science.ucalgary.ca/computer-science, [Accessed 0-10-2023].

[4] G. Kreitz and F. Niemela, "Spotify–large scale, low latency, p2p music-on-demand streaming," in *2010 IEEE Tenth International Conference on Peer-to-Peer Computing (P2P)*. IEEE, 2010, pp. 1–10.

[5] A. Schwind, L. Haberzettl, F. Wamser, and T. Hoßfeld, "Qoe analysis of spotify audio streaming and app browsing," in *Proceedings of the 4th internet-QoE workshop on QoE-based analysis and management of data communication networks*, 2019, pp. 25–30.

[6] V. Setty, G. Kreitz, R. Vitenberg, M. Van Steen, G. Urdaneta, and S. Gimåker, "The hidden pub/sub of spotify: (industry article)," in *Proceedings of the 7th ACM international conference on Distributed event-based systems*, 2013, pp. 231–240.

[7] O. Markelov, V. N. Duc, and M. Bogachev, "Statistical modeling of the internet traffic dynamics: To which extent do we need long-term correlations?" *Physica A: Statistical Mechanics and its Applications*, vol. 485, pp. 48–60, 2017.

[8] I. Antoniou, V. V. Ivanov, V. V. Ivanov, and P. Zrelov, "On the log-normal distribution of network traffic," *Physica D: Nonlinear Phenomena*, vol. 167, no. 1-2, pp. 72–85, 2002.

[9] S. B. Klenow, C. Williamson, M. Arlitt, and S. Keshvadi, "Campus-level instagram traffic: a case study," in *2019 IEEE 27th International Symposium on Modeling, Analysis, and Simulation of Computer and Telecommunication Systems (MASCOTS)*. IEEE, 2019, pp. 228–234.

[10] P. Jurkiewicz, G. Rzym, and P. Boryło, "Flow length and size distributions in campus internet traffic," *Computer Communications*, vol. 167, pp. 15–30, 2021.

[11] S. Keshvadi, M. Karamollahi, and C. Williamson, "Traffic characterization of instant messaging apps: A campus-level view," in *2020 IEEE 45th Conference on Local Computer Networks (LCN)*. IEEE, 2020, pp. 225–232.

[12] L. Moore, "Audio streaming application performance: A comparative study of spotify and youtube music," https://pages.cpsc.ucalgary.ca/ carey/CPSC641/slides/measurement/Lachlan-CPSC503-Paper.pd, 2021, accessed: 2023-09-06.

[13] A. C. Cullen and H. C. Frey, *Probabilistic techniques in exposure assessment: a handbook for dealing with variability and uncertainty in models and inputs*. Springer Science & Business, 1999.

[14] B. A. Alahmadi, E. Mariconti, R. Spolaor, G. Stringhini, and I. Martinovic, "Botection: Bot detection by building markov chain models of bots network behavior," in *Proceedings of the 15th ACM Asia Conference on Computer and Communications Security*, 2020, pp. 652–664.

979-8-3503-8368-3/23 $31.00 © 2023 IEEE

Vehicle Density in mmWave 5G V2X Highway Communication Systems: A Channel Coding Approach

Dimitrios Chatzoulis
Department of Digital Systems
University of Thessaly
Larissa, Greece
dchatzoulis@uth.gr

Costas Chaikalis
Department of Digital Systems
University of Thessaly
Larissa, Greece
kchaikalis@uth.gr

Dimitrios Kosmanos
Department of Digital Systems
University of Thessaly
Larissa, Greece
dikosman@uth.gr

Apostolos Xenakis
Department of Digital Systems
University of Thessaly
Larissa, Greece
axenakis@uth.gr

Kostas Anagnostou
Department of Informatics &
Telecommunications
University of Thessaly
Lamia, Greece
anagko@uth.gr

Abstract— **Vehicle-to-everything (V2X) is an emerging communication technology that enables data communication between vehicles and inevitably has become part of the 5th-Generation New Radio (5G-NR) and the millimeter wave 5G-NR (mmWave 5G-NR) technologies. The V2X use cases and their quality of service (QoS) parameters were defined by the 3rd-Generation Partnership Project (3GPP). These parameters, namely reliability, end-to-end latency, data rate and range, are highly dependent on channel coding, which is a method of improving communication quality. Apart from these parameters the 5G Automotive Association (5GAA) introduced vehicle density as another crucial parameter that is significantly affected by the 3GPP QoS parameters and the channel coding scheme, affecting the overall communication quality. In this paper, we investigate the vehicle density constraints in a highway 5G V2X communication environment and we measure the impact of use of 4th-Generation Long-Term Evolution (4G-LTE) turbo codes, 5G-NR polar codes and low-density parity-check codes (LDPC) on vehicle density for all V2X use cases.**

Keywords—V2X, 3GPP, 5G, mmWave, channel coding, vehicle density

I. INTRODUCTION

Channel coding is crucial in wireless communications since it greatly affects the reliability of a communication system [1]. The first near-optimal class of coding techniques was Turbo codes, [2, 3], and it consisted of two algorithms, the maximum a posteriori probability algorithm (MAP or BCJR) [4], and the Soft-Output Viterbi Algorithm (SOVA) [5]. Being complex the MAP algorithm was replaced by less complex implementations such as the Max-Log-MAP (maxlogBCJR) [6, 7] and Log-MAP (logBCJR) [8] algorithm and became part of 4G-LTE [9]. In 5G-NR LDPC [10] and Polar coding [11] were introduced as less complex coding schemes, especially for large size data frames.

V2X communications are an emerging field of wireless communications and have attracted much of the research community and the telecommunications and automotive industry. 3GPP embed the V2X services in the 5G NR, by introducing four advanced use cases, specifically vehicle platooning, extended sensors, advanced driving, and remote driving [12]. In [13] 3GPP listed the payload, data rate, maximum end-to-end latency, reliability and required

communication range as the key requirements for the aforementioned use cases and defined their typical values. Furthermore, 5GAA [14] extended these 3GPP requirements, and it introduced vehicle velocity and vehicle density as important QoS requirements.

The authors in [15] proved that turbo codes can be used for enhancing safety-based 5G V2X services for short frame lengths and flat Rayleigh fading channel. The authors in [16] and [17] examined the QoS efficiency and the bit error rate (BER) performance of turbo codes in a 3GPP 5G NR V2X geometry-based V2X channel and small length frames. In [18] the authors made extensive research on BER and frame error rate (FER) performance of turbo, LDPC and Polar codes, implementing a stochastic V2X propagation model, based on 3GPP 5G V2X specifications and they determined that turbo schemes are more efficient for small frame 5G V2X services in terms of BER and FER performance. In [19] the authors studied the effect of coding schemes on the 3GPP and 5GAA QoS parameters for a 3GPP 5G V2X stochastic propagation environment and for small data frame communications.

Based on the above discussion, we conclude that there has been a lot of study of QoS parameters in 5G communication systems, but not in mmWave 5G at an adequate level. Therefore, in this paper, we simulate an mmWave 5G V2X Release 16 highway propagation environment and we examine the vehicle density of the propagation environment by using different coding schemes. In this study, we chose to investigate 128-bit frames, as a representative for small-frame V2X communications.

II. 3GPP 5G V2X QoS SCENARIOS

3GPP specified six different QoS scenarios [13] for 5G and mmWave 5G V2X communication systems, namely cooperative awareness, cooperative sensing, cooperative maneuver, vulnerable road user, traffic efficiency and tele-operated driving, which are shown at TABLE I. In this table the communication range is considered as short for distances smaller than 200 m, short to medium for distances between 200 m and 500 m, and long for distances larger than 500 m. Thus, we have selected 200 m for Q2 and Q4 scenarios, 375 m for Q1 and Q3 scenarios and 500 m for Q5 and Q6

979-8-3503-8368-3/23 $31.00 © 2023 IEEE

scenarios as maximum typical range values for our simulations.

TABLE I. 3GPP 5G AND MMWAVE 5G QoS SCENARIOS.

QoS Scenario	Reliability (%)	End-to-End Latency (ms)	Data Rate per Vehicle (Kbps)	Range
Q1 - Cooperative awareness	90-95	100 to 1000	5 to 96	Short to medium
Q2 - Cooperative sensing	>95	3 to 1000	5 to 25000	Short
Q3 - Cooperative maneuver	>99	<3 to 100	10 to 5000	Short to medium
Q4 - Vulnerable road user	95	100 to 1000	5 to 10	Short
Q5 - Traffic efficiency	<90	>1000	10 to 2000	Long
Q6 - Tele-operated driving	>99	5 to 20	>25,000	Long

III. SYSTEM SIMULATION MODEL

The V2X system simulation model is presented in Fig. 1. 128-bit data-frames are encoded by BCJR, logBCJR, maxlogBCJR, SOVA, Polar and LDPC coding schemes respectively. The coding rate for all the coding schemes is 1/3. The coded data frames are modulated by a Binary Phase-Shift Keying (BPSK) modulator and are transmitted through a stochastic mmWave 5G 3GPP V2X channel with additive white gaussian noise (AWGN). This channel is implemented assuming a highway traffic scenario, where all vehicles are moving at a constant speed $v = 100\ Km/h$. The implementation details of the channel are fully explained in [18] and [19]. For this traffic model we assume two different channel states, namely the line of sight (LOS) state and the non-line of sight with vehicle blockage (NLOSv) [20]. The carrier frequency is equal to 63GHz [20]. The received data in the receiver are demodulated by a BPSK demodulator and finally they are decoded by the corresponding decoder. Turbo and LDPC decoding schemes are implemented by using 4 and 12 decoding iterations, according to [19].

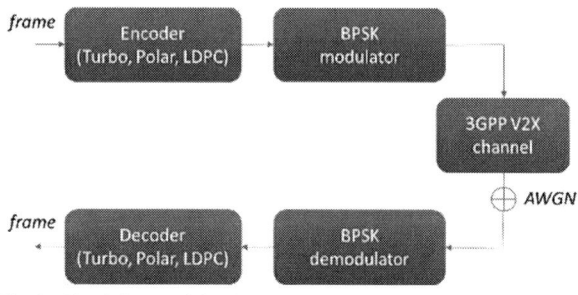

Fig. 1. Simulation model using 3GPP V2X stochastic channel model, additive white gaussian noise and BPSK modulation [19].

For each simulation scenario, we transmit 100,000 frames from the transmitter to the receiver through the V2X channel. Finally, MATLAB was used for the implementation of the simulation model and the generation of the results using floating-point precision.

IV. VEHICLE DENSITY ESTIMATION

In our simulation model we assume the worst-case scenario, where each vehicle receives and sends data frame from and to all other vehicles around it. In such scenario the acceptable vehicle density D i.e., the number of vehicles per Km, is equal to the number n_{rel} of the data frames a vehicle can reliably receive.

A. Road Topology

In order to estimate vehicle density, we use the road configuration topology defined by 3GPP for V2X highway traffic [20] (p. 36), i.e., we assume an 1 Km road that has $l = 6$ lanes in total, 3 in each direction. This topology is presented in Fig. 2.

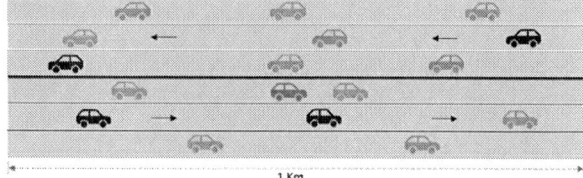

Fig. 2. 3GPP for V2X highway traffic topology.

In this topology we assume that there are only Type 2 vehicles [20] i.e., the vehicle length v_ℓ is 5 m. Also, the distance d_v between the rear bumper of a vehicle and the front bumper of the following vehicle is formulated as $max\{2, \chi\}$, where χ is an exponential random variable $\chi \sim Exp(2v)$ [20]. Considering the topology of Fig. 2, our goal is to estimate the mean number n_{rel} for a vehicle in the middle of this road, i.e., for the grey vehicle.

B. Number of Simultaneous Reliable Communications

The mean number n_{rel} of the data frames a vehicle can reliably receive and process simultaneously is computed by the equation:

$$E\{n_{rel}\} = \frac{t_{E2E,max}}{E\{t_{E2E}\}} \quad (1)$$

where $t_{E2E,max}$ is the maximum end-to-end latency constraint given by the communication scenario of TABLE I and t_{E2E} is the actual end-to-end latency [19], which is computed by the equations of TABLE II.

TABLE II. END-TO-END LATENCY

Coding Scheme	End-to-End latency
SOVA	$t_{E2E} = \dfrac{d_v}{c} + \dfrac{K}{r_{ef}} \times I_{Turbo}$
BCJR	$t_{E2E} = \dfrac{d_v}{c} + 3 \times \dfrac{K}{r_{ef}} \times I_{Turbo}$
logBCJR	$t_{E2E} = \dfrac{d_v}{c} + 2.8 \times \dfrac{K}{r_{ef}} \times I_{Turbo}$
maxlogBCJR	$t_{E2E} = \dfrac{d_v}{c} + 1.8 \times \dfrac{K}{r_{ef}} \times I_{Turbo}$
LDPC	$t_{E2E} = \dfrac{d_v}{c} + 0.57 \times \dfrac{K}{r_{ef}} \times I_{LDPC}$

Coding Scheme	End-to-End latency
Polar	$t_{E2E} = \dfrac{d_v}{c} + 0.47 \times \log_2 N \times \dfrac{K}{r_{ef}}$

In the above equations d_v is the distance between the two vehicles, c is the speed of light, K is the frame size, N is the size of the coded frame, I_{Turbo} is the number of iterations for SOVA and the BCJR-based algorithms (4 iterations in our simulations) and I_{LDPC} is the number of iterations for the LDPC decoder (12 iterations in our simulations). Finally, r_{ef} is the effective data rate, i.e., the data rate that can achieve the desired reliability for each QoS scenario as defined by TABLE I.

C. Effective Data Rate

In order to have a measure for the mean r_{ef} we relate to the mmWave 5G 3GPP V2X channel and its large-scale fading characteristics. In an ideal receiver, the power P_R is given from the formula $P_R = E_b r_b$, where E_b is the energy per bit at the receiver and r_b is the data rate. To be able to calculate P_R and at the same time relate it to the system reliability, we reveal the energy per bit to noise power spectral density ratio $\gamma_{b,in} = E_b/N_0$ in the aforementioned equation, so:

$$P_R = r_b N_0 \gamma_{b,in} \qquad (2)$$

In real communication systems, there is always a difference between the actual and the ideal receiver, which is modeled as an extra factor of degradation of the $\gamma_{b,in}$ at the receiver, called noise figure F. So, the actual energy per bit to noise power spectral density ratio γ_b is modeled by the equation $\gamma_{b,in} = \gamma_b F$. Taking these values into account and presenting (2) in dBm, we conclude that the transmitted power P_T in dBm is:

$$\begin{aligned} P_T^{dBm} &= PL^{dB} + P_R^{dBm} \\ &= PL^{dB} + 10\log_{10} r_b + N_0^{dBm} + F^{dB} + \gamma_b^{dB} \end{aligned} \qquad (3)$$

where PL^{dB} is the path loss in dB of the propagation model. The average path loss PL^{dB} in an mmWave V2X highway communication system [19], [20] is computed by the equation:

$$\begin{aligned} E\left\{PL^{dB}\right\} = {}&32.4 + 20\log_{10} E\{d\} + 20\log_{10} f_c \\ &+ 0.0105 E\{d\} + PL_v \end{aligned} \qquad (4)$$

where $PL_v = 0$ dB if there is LOS channel state and $PL_v = 4.67$ dB if there is NLOS channel state [19]. Reference [20] defines that in mmWave 5G V2X highway communication systems, the noise figure F^{dB} is set to 13 dB, the transmitted power P_T^{dBm} is set to 21 dBm and $f_c = 63$ GHz. Additionally, the noise power spectral density N_0^{dBm} for thermal white noise is -174 dBm. So, by using (3) and (4), the mean data rate is computed as follows:

$$E\{r_b\} = 10^{\left(113.61 - E\left\{\gamma_b^{dB}\right\} - 20\log_{10} E\{d\} - 0.0105 E\{d\} - PL_v\right)/10} \qquad (5)$$

Taking into account the fact that each V2X communication scenario has a maximum acceptable data rate r_{max}, defined by TABLE I, we conclude that the mean effective data rate r_{ef} is:

$$E\{r_{ef}\} = \min\{r_{max}, E\{r_b\}\} \qquad (6)$$

D. Upper bound Density Constraints

Vehicle density is practically limited by the road configuration parameters. For this purpose, we use the 3GPP V2X road configuration parameters for highway traffic [20] (p. 36), i.e., we use the topology of Fig. 2. Also, we assume that there are only Type 2 vehicles [20] i.e., the vehicle length v_ℓ is 5 m and that the distance d_v between the rear bumper of a vehicle and the front bumper of the following vehicle is the minimum possible, i.e., 2 m, according to [20] (p. 10). Taking these parameters into consideration, we determine the maximum density D_{max} as:

$$D_{max} = \left\lfloor \frac{1000}{v_\ell + d_v} \right\rfloor l = 852 \text{ vehicles/Km} \qquad (7)$$

This value is an upper bound for the number of vehicles that can communicate simultaneously. So, by using (1) and (7) the mean vehicle density is computed by the equation:

$$E\{D\} = \min\{D_{max}, E\{n_{rel}\}\} \qquad (8)$$

E. Vehilcle Density Estimation Method

Concluding the steps to find the density are:

- Choose the channel state (LOS, NLOSv) and compute PL_v.

- Choose the QoS scenario from TABLE I.

- Choose the coding scheme (BCJR, logBCJR, maxlogBCJR, SOVA, LDPC, Polar).

- Simulate 100,000 data transmissions and calculate the mean FER as a function of the γ_b^{dB}. Then, find the $\mathbb{E}\{\gamma_b^{dB}\}$ that corresponds to the minimum reliability requirements (FER) of the QoS scenario for the specific coding scheme.

- Select random number of vehicles and place them in random distances in a 6-lane road and in a 1 Km range. Repeat it 100 times. Then, calculate the mean distance $\mathbb{E}\{d\}$ of the vehicles.

- Calculate the mean effective data rate $\mathbb{E}\{r_{ef}\}$ for the specific QoS scenario from (6), using (5) and TABLE I for the specified $\mathbb{E}\{\gamma_b^{dB}\}$ and $\mathbb{E}\{d\}$.

- Compute $\mathbb{E}\{t_{E2E}\}$ from TABLE II for the specified coding scheme.

- Calculate $\mathbb{E}\{D\}$ from (8), using (7) and (1).

SIMULATION RESULTS

In TABLE III we show the mean vehicle density for each QoS scenario of TABLE I and for each decoding algorithm.

TABLE III.   MEAN VEHICLE DENSITY RESULTS

QoS Scen.	Channel State	BCJR	logBCJR	maxlogBCJR	SOVA	LDPC	Polar
Q1	LOS	62	66	104	187	109	185
Q1	NLOSv	62	66	104	187	109	185
Q2	LOS	852	852	852	852	852	852
Q2	NLOSv	294	316	491	852	517	852
Q3	LOS	19	21	32	59	34	58
Q3	NLOSv	5	5	8	15	9	15
Q4	LOS	6	6	10	19	11	19
Q4	NLOSv	6	6	10	19	11	19
Q5	LOS	96	103	160	289	169	286
Q5	NLOSv	26	28	44	80	46	79
Q6	LOS	1	1	2	4	2	4
Q6	NLOSv	0	0	0	1	0	1

The tables above reveal the superiority of polar coding in vehicle density in all scenarios for both LOS and NLOS channel states. The main reason for that is the low $\mathbb{E}\{t_{E2E}\}$ due to the the non-iterative nature and the low complexity of the Polar decoding algorithm. Scenarios Q1 and Q4 have the same density in both LOS and NLOSv channel states due to the fact that they achieve the maximum possible data rate ($r_{ef} = r_{max}$) in both cases. Also, in scenarios Q2, Q3, Q5 and Q6 the LOS channel state have superior density performance in contrast to NLOSv channel state. In the second case reliable communication can be achieved only using data rate smaller than the maximum data rate defined of TABLE I, resulting in a significant increase in the decoding delay and consequently in end-to-end latency and vehicle density. It is obvious that in this channel state these scenarios have at least 75% lower density performance in average. Also, Q2 scenario has the best density performance mainly because it uses very high data rates causing small latency. Therefore, in this scenario almost all performances are limited by the theoretical maximum density D_{max}. Additionally, scenarios Q3 and Q4 have low density performance highlighting the effect of the high reliability and low transmission rate respectively on the vehicle density. Finally, Q6 scenario has the worst performance because it has small maximum latency and very high reliability. In this scenario all coding schemes, except SOVA and Polar, fail to achieve reliable communication with more than one vehicle, making obvious the need of MIMO communication systems.

CONCLUSIONS

In this work we investigate the mean vehicle density of a highway mmWave 5G-NR V2X propagation environment. For this purpose, we simulate a stochastic mmWave 5G V2X channel model that is based on the 3GPP V2X specifications and we examine the performance, in terms of vehicle density, of BCJR, logBCJR, maxlogBCJR, SOVA, LDPC and Polar coding algorithms. Our goal is to examine the effectiveness of the Turbo-based algorithms in relation to LDPC and Polar coding and investigate their suitability to an mmWave 5G V2X communication system.

Our analysis shows that Polar is the best coding scheme, mainly because of his non-iterative implementation and his very good performance in case of medium to good communication reliability [19]. Also, LDPC has inferior density performance but can be used efficiently in medium density propagation environments. Furthermore, BCJR-based algorithms can serve enough vehicles simultaneously and respond efficiently to medium-density highway traffic environments, and maxlogBCJR in particular has almost identical performance to LDPC, making it a reliable coding candidate for mmWave 5G V2X communications with small data frames and medium density use cases.

For future work, we plan to investigate the performance of the 3GPP and 5GAA QoS parameters in a dynamic propagation model, where the channel state and traffic model change dynamically as a function of time, revealing the potential pros and cons of the decoding algorithms.

REFERENCES

[1] K. Tsoukatos and D. Chatzoulis, *Wireless Communications*. Athens, Greece: Kallipos, Open Academic Editions, 2023. doi: https://doi.org/10.57713/KALLIPOS-202.

[2] C. Berrou and A. Glavieux, "Near optimum error correcting coding and decoding: turbo-codes," *IEEE Transactions on Communications*, vol. 44, Art. no. 10, 1996, doi: https://doi.org/10.1109/26.539767.

[3] C. Berrou, A. Glavieux, and P. Thitimajshima, "Near Shannon limit error-correcting coding and decoding: Turbo-codes. 1," *Proceedings of ICC '93 - IEEE International Conference on Communications*, 1993, doi: https://doi.org/10.1109/icc.1993.397441.

[4] L. Bahl, J. Cocke, F. Jelinek, and J. Raviv, "Optimal decoding of linear codes for minimizing symbol error rate (Corresp.)," *IEEE Transactions on Information Theory*, vol. 20, Art. no. 2, 1974, doi: https://doi.org/10.1109/tit.1974.1055186.

[5] J. Hagenauer and P. Hoeher, "A Viterbi algorithm with soft-decision outputs and its applications," *IEEE Global Telecommunications Conference, 1989, and Exhibition. 'Communications Technology for the 1990s and Beyond*, 1989, doi: https://doi.org/10.1109/glocom.1989.64230.

[6] W. Koch and A. Baier, "Optimum and sub-optimum detection of coded data disturbed by time-varying intersymbol interference (applicable to digital mobile radio receivers)," *[Proceedings] GLOBECOM '90: IEEE Global Telecommunications Conference and Exhibition*, 1990, doi: https://doi.org/10.1109/glocom.1990.116774.

[7] J. Erfanian, S. Pasupathy, and G. Gulak, "Reduced complexity symbol detectors with parallel structure for ISI channels," *IEEE Transactions on Communications*, vol. 42, Art. no. 2/3/4, 1994, doi: https://doi.org/10.1109/tcomm.1994.582868.

[8] P. Robertson, E. Villebrun, and P. Hoeher, "A comparison of optimal and sub-optimal MAP decoding algorithms operating in the log domain," *Proceedings IEEE International Conference on Communications ICC '95*, vol. 2, pp. 1009–1013, 1995, doi: https://doi.org/10.1109/icc.1995.524253.

[9] 3GPP, "3GPP TS 36.212 V14.16.0 Release 14, LTE; evolved universal terrestrial radio access (E- UTRA); multiplexing and channel coding," 3rd Generation Partnership Project, 2021.

[10] R. Gallager, "Low-density parity-check codes," *IEEE Transactions on Information Theory*, vol. 8, Art. no. 1, 1962, doi: https://doi.org/10.1109/tit.1962.1057683.

[11] E. Arikan, "Channel polarization: A method for constructing capacity-achieving codes for symmetric binary-input memoryless channels," *IEEE Transactions on Information Theory*, vol. 55, Art. no. 7, 2009, doi: https://doi.org/10.1109/tit.2009.2021379.

[12] M. H. Castaneda *et al.*, "A tutorial on 5G NR V2X communications," *IEEE Communications Surveys & Tutorials*, vol. 23, Art. no. 3, 2021, doi: https://doi.org/10.1109/comst.2021.3057017.

[13] 3GPP, "3GPP TS 22.186 V17.0.0 release 17 service requirements for enhanced V2X scenarios (v16.2.0, release 16)," 3rd Generation Partnership Project, 2019.

[14] 5GAA *White Paper C-V2X Use Cases: Methodology, Examples and Service Level Requirements*; 5GAA: München, Germany, 2019;

[15] C. Chaikalis, D. Kosmanos, and N. S. Samaras, "Utilizing turbo codes for secure 5G V2X," *2020 IEEE Microwave Theory and Techniques in Wireless Communications (MTTW)*, 2020, doi: https://doi.org/10.1109/mttw51045.2020.9245035.

[16] D. Kosmanos, C. Chaikalis, I. K. Savvas, K. E. Anagnostou, and D. Bargiotas, "Investigating 5G V2X QoS using turbo codes," *2021 IEEE Microwave Theory and Techniques in Wireless Communications (MTTW)*, 2021, doi: https://doi.org/10.1109/mttw53539.2021.9607131.

[17] D. Kosmanos, C. Chaikalis, and I. K. Savvas, "3GPP 5G V2X scenarios: Performance of QoS parameters using turbo codes,"

Telecom, vol. 3, Art. no. 1, 2022, doi: https://doi.org/10.3390/telecom3010012.

[18] D. Chatzoulis, C. Chaikalis, D. Kosmanos, K. E. Anagnostou, and G. T. Karetsos, "5G V2X performance comparison for different channel coding schemes and propagation models," *Sensors (Basel)*, vol. 23, Art. no. 5, 2023, doi: https://doi.org/10.3390/s23052436.

[19] D. Chatzoulis, C. Chaikalis, D. Kosmanos, K. E. Anagnostou, and A. Xenakis, "3GPP 5G V2X Error Correction Coding for Various Propagation Environments: a QoS Approach," *Electronics*, vol. 12, no. 13, pp. 2898–2898, Jul. 2023, doi: https://doi.org/10.3390/electronics12132898.

[20] 3rd Generation Partnership Project *3GPP TR 37.885 V15.3.0 Release 15, Study on Evaluation Methodology of New Vehicle-To-Everything (V2X) Use Cases for LTE and NR*; 2019;

An Early Warning Opportunistic Interference Method in Tactical Voice and Data Communications

Nikiforos Kontopoulos
8th Laboratory Center
Directorate of Secondary Education of
D´ Department of Athens, Greece
nkontopoul@sch.gr

Dimitrios Kotsifakos
NetLab, Department of Informatics,
University of Piraeus,
Piraeus, Greece
kotsifakos@unipi.gr

Christos Douligeris
NetLab, Department of Informatics,
University of Piraeus,
Piraeus, Greece
cdoulig@unipi.gr

Abstract— Sporadic E and tropospheric ducting in the Very High Frequency (VHF) and the lower Ultra High-Frequency (UHF) bands are critical for military and emergency response operations nevertheless their reliability is often compromised by various phenomena, leading to degraded communication quality, and reduced operational effectiveness. This article presents an early warning method for detecting and predicting opportunistic interference due to irregular propagation phenomena. By providing an early warning of opportunistic interference, our method can enable tactical communication users to plan accordingly and take appropriate measures to mitigate the impact of these events in their operations. The proposed approach uses amateur radio unknown distance cluster servers as a source of data for detecting and predicting opportunistic interference in the VHF and lower UHF bands. This approach offers a cost-effective and widely available alternative to traditional methods of monitoring and predicting these phenomena.

Keywords— Communications, Propagation, Prediction, Tactical, Interference, VHF, UHF, DX Cluster

I. INTRODUCTION

The warning opportunistic interference methods in tactical voice and data communications over-the-line-of-sight propagation refers to the ability of VHF (Very High Frequency) and UHF (Ultra High Frequency) radio waves to travel beyond the expected radio horizon, which is typically limited to line-of-sight communication. Tropospheric ducting and sporadic E propagation are phenomena that significantly increase the range of VHF and UHF radio signals, often leading to unexpected long-distance contacts and, consequently, potential interference. Specifically, we posit that observing frequency peaks during occurrences of tropospheric ducting or sporadic E (Es) propagation can provide critical insights into potential interference scenarios. Serious problems could arise

i) from interference between topologically isolated users of the same frequency that are beyond nominal propagation boundaries that broadcast in the same timeslots and,

ii) from threat actors or friends of the enemy(foe) that may monitor large portions of usual military frequencies for such propagation irregularities.

These phenomena can cause signals to travel much further than expected, resulting in interference with other communications systems. This can be a major problem in military applications, where reliable communication is essential. It should also be noted that during these events an extension of the operational range of voice and data communications outside of their expected service area could expose them to unwanted open-source intelligence (OSINT) agents or Signals intelligence (SIGINT). Open-source intelligence (OSINT) on voice and data VHF and UHF tactical communications is collecting and analyzing publicly available information about these communications systems. OSINT can be used to collect SIGINT, which is intelligence gathered from intercepted communications. SIGINT can be used to track the enemy's movements, identify their assets, and learn about their plans. This information can be used to gain insights into the capabilities, weaknesses, and vulnerabilities of these systems. For such incidents, no solution has been proposed in the scientific articles or the relative literature. Our study proposes a novel approach to predict interference in tactical voice and data communications by monitoring over-the-line-of-sight or over-the-horizon propagation during instances of enhanced VHF/UHF propagation. This information can be derived potentially from the publicly and free available DX cluster network. DX is an abbreviation for long-distance communications and a DX cluster is a kind of web-based nearly real-time service used by licensed radio amateurs [19].

A DX clusters, real-time networks where amateur radio operators report long-distance contacts, we can identify when these propagation events are occurring. Coupled with monitoring tools like the Weak Signal Propagation Reporter (WSPR) on these frequency bands, we can effectively track specific frequency range elevations. This proposed methodology offers a proactive approach to managing and mitigating interference, providing tactical voice and data communication systems with a valuable tool for maintaining optimal performance. We aim to address the issue of opportunistic interference in VHF and lower UHF bands by proposing a method for early warning. Our secondary goal is to demonstrate the feasibility of using data from amateur radio cluster servers to extract information on tropospheric Es and Tropospheric ducting phenomena and predict their occurrence. The average theoretical distance of VHF/UHF signals is difficult to determine as it can vary greatly depending on several factors such as the height of the antenna, the terrain, the weather conditions, and other factors. However, as a rough estimate, VHF signals can travel up to around 100-150 kilometers in ideal conditions with line-of-sight propagation.

Beyond the introduction, in Section II, we present the main purpose of this research, and in Section III we focus on the basic implementation ideas. Section IV presents the two basic scientific areas of the article (Theoretical Panorama), the Sporadic Es and Multi-hop and Multimode propagation events, two of the crucial areas of our proposal. In Section V we suggest the future works of our research.

II. THE PREDICTION APPROACH

To face those kinds of problems, we have organized and proposed an indirect prediction procedure. Our proposal is based on the principle that often the phenomena of sporadic Es and tropospheric propagation before manifesting to the military frequencies are noticed by radio amateurs, and they sometimes report them on web services known as dx cluster.

979-8-3503-8368-3/23 $31.00 © 2023 IEEE

🖊 Spotter	📶 Freq.	📡 DX	⏱ Time	📍 Info	🗺 Country
DO3GE	144175.0	DL6BF	18:17 12 Feb	sm7koj in qso cluster king	Fed. Rep. of Germany
DO3GE	144175.0	EI9KP	18:16 12 Feb	-9 0.3 2352 sm7koj EI9KP IO5	Ireland
LA9AKA	144174.0	DJ4JB	18:15 12 Feb	FT8 JP20mm -> JO51	Fed. Rep. of Germany
EA1HRR	144174.0	G0JCC	18:13 12 Feb	IN83JJ<TR>IO82MA FT8 Andrew 73	England
G0JCC	144174.0	EA1UQ	18:13 12 Feb	IO82MA<TR>IN83FL B-15 tnx	Spain
G0JCC	144174.0	EA1HRR	18:12 12 Feb	IO82MA<TR>IN83JJ B-15 tnx	Spain
DO3GE	144175.0	DL6BF	18:11 12 Feb	sm7koj cluster king	Fed. Rep. of Germany
DL6BF	144174.0	EI9KP	18:09 12 Feb	IO54,tnx ft8,1102 km	Ireland
GI4SNA	144174.0	DL8LAQ	18:08 12 Feb	FT8 -19dB from JO43 1712Hz	Fed. Rep. of Germany
DL8LAQ	144175.6	G4ITR	18:08 12 Feb	JO43XU<TR>IO95 20W	England

Fig. 1: Typical output of a free web-accessed DX Cluster

The critical point of the proposal lies in the fact that the VHF and the Lower UHF band propagation cover a wide range of values and often "shift" the frequency range upwards or downwards. The scientific stake of the article lies precisely in the fact that if we detect propagation in these bands, then we could also expect propagation in the frequency ranges used by military transmissions. The management of the solution we propose starts from this very assumption: since we perceive and detect the warnings sent by radio amateurs, we could expect possible band "openings" for propagation on the military frequencies neighboring the ones we detected.

The hardest point should such a propagation event occur, is to identify it. Unless some sort of disturbance or reception of any other signal that has an origin above the horizon is noticed, no other incident of these events presents audible evidence or another form of manifestation. If by any chance the radio is using analog circuitry and the frequency is used by other parties on the other point of the propagation aperture, the user may notice channel or frequency occupancy. But most of the operation time of the radio is spent in reception mode. A threat actor could be monitoring certain parts of the radio spectrum in VHF/UHF bands waiting for noticeable carriers to intercept friendly communications. In military and tactical usage, the real-time propagation predictions derived from DX clusters can be used to optimize communication strategies and to improve the effectiveness of radio communications in various environments [18]. The possibility of over-the-horizon propagation due to sporadic E (Es) and tropospheric ducting [11] can be derived from DX clusters up to UHF frequencies using real-time propagation prediction tools that use the data from the DX clusters. These tools are used in amateur radio usage to track the occurrence of Es and tropospheric ducting and to predict the maximum usable frequencies for these forms of propagation, and in military and tactical usage to optimize communication strategies and to improve the effectiveness of radio communications in various environments (Fig. 1). The server software, known as "cluster software," is responsible for collecting and distributing information among the connected stations.

The client software, known as "node software," is responsible for displaying the information on the user's computer and for allowing the user to enter and transmit information to the cluster.

III. IMPLEMENTATION ISSUES

By utilizing DX clusters, real-time networks where amateur radio operators report long-distance contacts, we can identify when these propagation events are occurring. We can effectively track MUF elevations with monitoring tools like the Weak Signal Propagation Reporter (WSPR) on these frequency bands. This proposed methodology offers a proactive approach to managing and mitigating interference, providing tactical voice and data communication systems with a valuable tool for maintaining optimal performance.

A grid or QRA (Quick Response Area) locator is a system used to identify a specific location on the Earth's surface with a unique identifier. It is commonly used in amateur radio and other forms of wireless communication to specify the location of a station or signal source. The QRA locator system is based on a grid system that divides the Earth's surface into squares of 1-degree latitude by 2-degree longitude. Each square is identified by a two-letter combination indicating the latitude band, followed by two digits indicating the longitude band. The squares are then divided into smaller squares of ten (10) minutes of latitude by twenty (20) minutes of longitude and identified by two letters indicating the position within the larger square. For example, a QRA locator of "IO91" would indicate a location within the square bounded by latitudes 51 degrees and 52 degrees North, and longitudes 0 degrees and 2 degrees West. The additional two letters would determine the precise location within this square. It's worth mentioning that different systems may use different grid sizes and letter/number combinations but the idea is the same, to have a unique identifier to indicate a location on the earth's surface. The information displayed on a DX cluster typically includes Callsign (the callsign of the station that made the contact or spotted the propagation), Frequency (the frequency on which the contact or propagation was made, Mode (the mode of operation used for the contact or propagation Time (the date

979-8-3503-8368-3/23 $31.00 © 2023 IEEE

and time of the contact or propagation, DX station (the callsign of the DX station that was contacted or spotted, Report (the signal report given for the contact or propagation, QRA locator (the QRA locator of the station that made the contact or spotted the propagation, and finally Comment (additional information provided by the station that made the contact or spotted the propagation, such as the name of the operator, the equipment used, or the conditions of the contact). Some DX clusters may also include information such as the band, the power, the antenna, and the distance of the contact made. It's important to note that the format and information displayed on a DX cluster can vary depending on the specific software and settings used, but the above elements are typically included in most DX clusters. In our proposed method, we are going to harvest the occurrences of frequencies in the bands of VHF (144 MHz and 220MHz) or UHF (430-440MHz) providing that a reference concerning either Sporadic Es or TR (Tropospheric ducting) should occur. The primary military-only bands in the VHF high band are 138.00-144.00 and 148.00-150.775 but not on a primary basis and depending on individual country legislation. UHF segments are located between 240-400 MHz (Fig. 2).

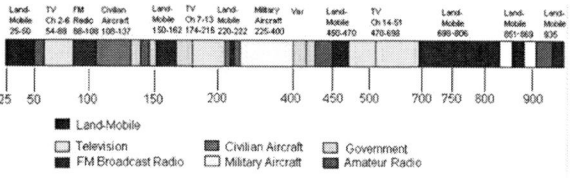

Fig. 2. Frequency allocation spectrum in MHz by ITU concerning main spectrum users and services. Government usage might be simultaneously shared with the military

These bands are the closest neighbors to the frequency bands used by the military either for land communications land-to-air or land-to-satellite.

IV. THEORETICAL PANORAMA

In this theoretical panorama, we introduce Sporadic Es (4.1), Multi-hop, and Multimode propagation events (4.2).

4.1 Sporadic Es

The range covered by Es propagation is affected by the altitude of the Es clouds and the frequency of the radio waves being used. Es clouds located at higher altitudes and formed at frequencies above the critical frequency of the E layer will cover a greater range than those formed at lower altitudes and lower frequencies. The probability of occurrence of sporadic E (Es) propagation in the VHF frequency range can vary depending on the time of day and the season of the year. In general, Es propagation is most commonly observed during the summer months, specifically during the daytime. The probability of Es propagation also varies depending on the season of the year, with a higher likelihood of occurrence during the summer months. According to a study by D. B. Lehnert published in the IEEE Transactions on Antennas and Propagation, Es propagation is most common during June, July, and August [12]. It is worth noting that the probability of Es propagation can also be affected by solar activity, with a higher likelihood of occurrence during periods of increased solar activity. The distance that the radio waves have to travel can also affect the critical frequency of the Es clouds.

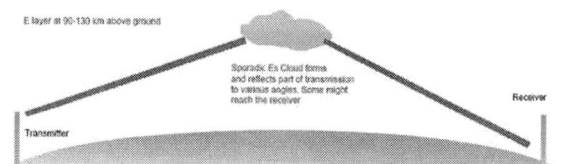

Fig. 3. Sporadic Es clouds forming in the E layer and causing the path of radio signals to be refracted depending on the intensity of ionization and the frequencies of the signals

The critical frequency tends to be lower for radio waves that have to travel longer distances, as the ionized region may not be as dense or tall. Another important aspect to consider is the lifetime of these clouds, which ranges from a few minutes to a few hours, and this means that they tend to cover short distances, generally less than 2,000 km. The critical frequency of Es clouds is affected by many factors, and it can be challenging to predict its behavior. However, understanding the variation of the critical frequency over time and distance can provide insight into the properties of Es clouds and their impact on VHF and UHF propagation. This means that the radio waves travel in a straight line and can be received by a receiver as long as there is no physical obstruction between the transmitter and the receiver. This type of propagation is often used for point-to-point communications, such as in mobile radios or television broadcasting [1].

In the UHF frequency range (300 MHz to 3GHz), radio waves can typically travel shorter distances. Still, they can penetrate through some obstacles, such as walls and buildings (Fig. 3). This type of propagation is often used for point-to-multipoint communications, such as in mobile phone networks, or wireless local area networks (WLANs). Irregularities of propagation, such as sporadic E (Es), can impact the security of military tactical communications [14]. High MUF propagation can allow for unexpected long-distance communications, which can be intercepted by unauthorized parties, and thus pose a security risk. Propagation anomalies can disrupt various types of voice communications for the military [7], such as ground-to-aircraft, inter-unit ground communications, ground units to satellites, and more. These anomalies can cause a security threat vector by disrupting or jamming the communication signals. In the case of analog communications, the danger caused by propagation anomalies is primarily disruption of the signal. E-layer sporadic E (Es) propagation is a type of propagation that occurs in the VHF frequency range (typically between 50-144 MHz and rarely on the 220 MHz bands) and is characterized by the temporary enhancement of the E-layer of the ionosphere [10]. This enhancement allows for long-distance communications over distances that would not normally be possible under normal ionospheric conditions. In general, Es propagation is most commonly observed in the VHF frequency range, specifically between 50-144 MHz, however, it can occur at higher frequencies as well [8].

According to the scientific literature [20], Es propagation has been observed at frequencies as high as 150 MHz. The maximum frequency for Es propagation is affected by the maximum usable frequency (MUF) of the ionosphere. The MUF is the highest frequency that can be refracted by a layer of the ionosphere. As the frequency increases, the MUF decreases and the ionosphere becomes less effective as a reflector, so the maximum frequency for Es propagation decreases as well.

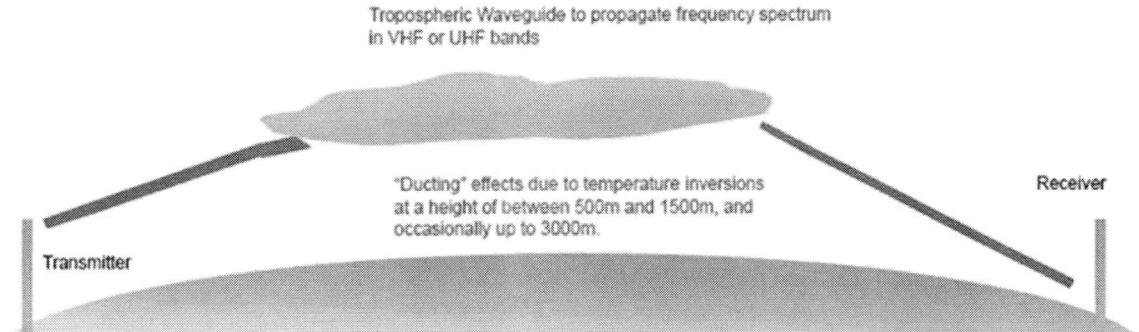

Fig. 4. Tropospheric causes - formation of waveguides - transmitted signals.

Additionally, as the frequency increases, the size of the Es clouds decreases, and the probability of Es propagation decreases. The critical frequency of the F layer is around 50-60 MHz, above which other forms of propagation such as F2 propagation become more likely. The maximum frequency at which sporadic E (Es) propagation can be observed is dependent on the specific ionospheric conditions and can vary in a wide range, with most common occurrences in the VHF frequency range. However, it has been observed at frequencies as high as 432 MHz (70cm) and even higher, as reported in [15].

The maximum frequency for Es propagation is affected by the maximum usable frequency (MUF) of the ionosphere and the probability of Es propagation decreases as frequency increases. The range of frequencies that are hopping over the E layer in VHF and UHF frequencies should such an incident occur, can vary depending on the specific conditions of the ionosphere at a given time. In general, Es propagation allows for long-distance communications over distances that would not normally be possible under normal ionospheric conditions. According to scientific literature, Es propagation has been observed to cover ranges of up to several thousand kilometers [9].

4.2 Tropospheric Ducting

The frequency at which tropospheric ducting vs. frequency ranges occurs may follow a Gaussian distribution, as it is influenced by a variety of factors such as temperature, humidity, and atmospheric pressure which can fluctuate randomly. The ducting occurs when the lower troposphere becomes stratified into two other stable layers and results in bending the propagated signal within a form of waveguide (Fig. 4). A warm dry layer over a cool moist layer with warm and cool being relative terms. This setup is also called a tropospheric inversion. This hypothesis is yet to be confirmed [4]. Many types of research for multi-hop and multimode propagation events discuss how tropospheric ducting can enable multi-hop propagation at VHF and UHF bands [3]. Nevertheless, a threat actor could exploit such events by deriving parts or the entire communication sequence, providing that there are going to be sufficient reception endpoints distributed around the globe.

It is suggested that such endpoints could employ the usage of software-defined radios (also referred to as SDR) as listening stations that monitor and potentially record large portions of the radio spectrum currently used by the military cognitive radio [2]. Multimode propagation alludes to the event of observing different modes of transmission to communicate over long distances. The VHF bands are located between 30 MHz and 300 MHz and are commonly used for voice and digital communications, as well as weak signal and satellite operations. The propagation conditions in VHF bands are different from lower and higher frequency bands, as the radio waves are affected by the ionosphere and the terrain, which can cause reflections, refractions, and scattering. This allows for communications over much longer distances than would be possible using line-of-sight propagation alone.

V. FUTURE WORKS

Several areas of future work could be done with the prediction of propagation irregularities due to Es or TR propagation for military usage. Researchers could verify our hypothesis with real-world data, mined from any dx cluster during May and July. There is the highest probability of irregular propagation occurrences during these months. The accuracy of the predictions could be aided by the integration of data from sources like ionosondes used by weather research facilities, services providing Automatic Packet Reporting System data from sites like [21] tropospheric (TR) prediction maps [22], NOAA space weather reports. Another useful tool should be the WSPR (Weak Signal Propagation Reporter) beacon. The page for monitoring WSPR beacons is WSPR Live, which allows you to analyze real-time WSPR spot data. The database on this site contains all spots ever reported to wsprnet.org and is publicly accessible [23]. A separate permission might be needed for each case. We could use ML algorithms to uncover patterns within the DX data. These patterns may include abrupt changes in the quantity and spatial distribution of spot reports, variations in signal strength, or fluctuations in propagation paths. Also, the prediction model could be adapted for use in different regions, such as mountainous terrain or urban environments, where the characteristics of the propagation irregularities may be different i.e., airplane scattering like in Fig. 5.

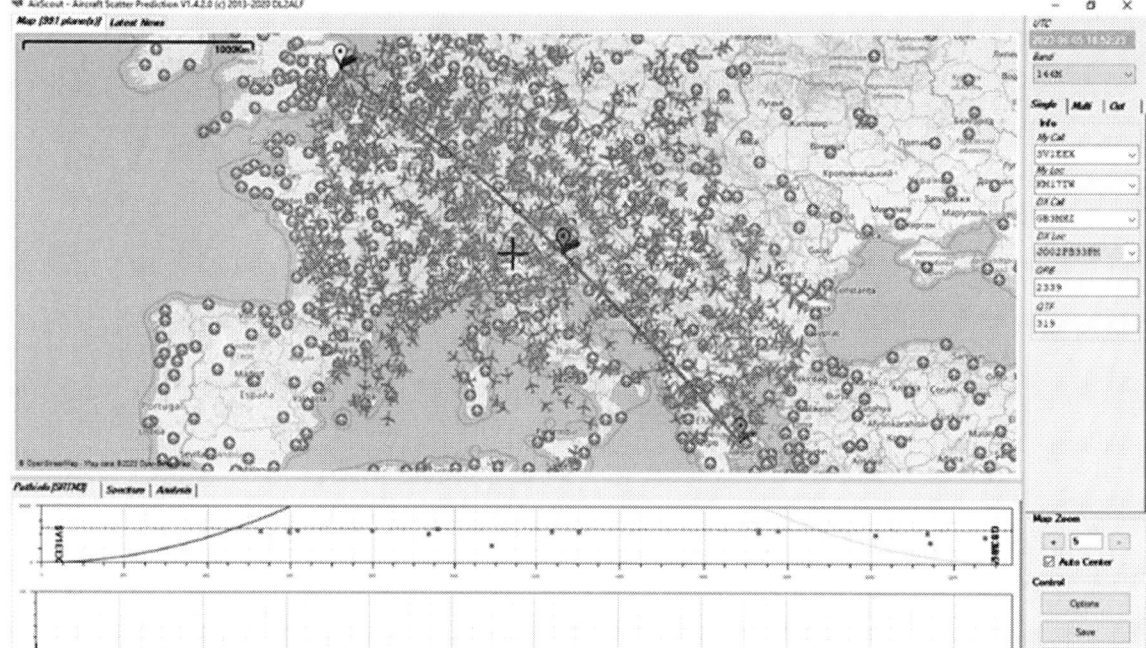

Fig. 5. Civil airplane scatter in Windows 10

A real-time prediction system could be developed that could allow the rapid identification and mitigation of such propagation irregularities, by eliminating the intermediate latencies due to dx cluster propagation methods. Recognizing these patterns can be crucial for detecting potential interference events. A proper ML algorithm like Recurrent Neural Networks (RNNs) could be particularly well-suited for processing sequential data, such as time series information in DX data, enabling the detection of temporal patterns and propagation path changes. The minimum data attributes needed to be considered to train such a model would be:

- Frequency: The frequency of the spot.

- Signal Strength: The strength of the received signal.

- Propagation Paths and affected areas: The paths of propagation as well as the affected areas centered on ham radio operators' locations.

- Time of the Spot: Timestamp of the spot report.

- Solar Activity: Solar events impacting the ionosphere.

- Geomagnetic Activity: Geomagnetic fluctuations and their effects on radio propagation.

- Space Weather: Solar flares and coronal mass ejections that can disrupt radio communications

- Closest usable military frequency that could be affected

The prediction system could be integrated with other tactical systems, such as communication systems and navigation systems. If a frequency range within the usable tactical range is potentially under threat, it should be omitted from the available frequencies for contacts if the phenomenon is expected to last.

The same thing could be considered for frequencies used in Frequency Hopping modes (FHSS), where the ranges under risk should not be considered for communications during such events. Further research could be done to improve the understanding of the underlying physics of propagation irregularities, which would lead to more accurate predictions and a better understanding of the potential impacts of these irregularities. The development of countermeasures for hostile OSINT: Based on the prediction and understanding of the underlying physics, countermeasures can be developed to mitigate the effect of propagation irregularities by hosting fake spot attacks for "Friends of the Enemy" (FOEs).

ACKNOWLEDGMENT

This work has been partly supported by COSMOTE under a PEDION24 grant. (University of Piraeus Research Center).

REFERENCES

[1] A. O. Akanni & K. Odepian, "Comparative Analysis of Propagation Pathloss and Channel Power of VHF and UHF Wireless Signals in Urban Environment", Int. J. Res. Innov. Appl. Sci, 2019.

[2] R. Akeela & B. Dezfouli, "Software-defined Radios: Architecture, state-of-the-art, and challenges", Computer Communications, 128, 106-125, 2018.

J. Aranda, M. Schölzel, D. Mendez & H. Carrillo, "Multimodal wireless sensor networks based on wake-up radio receivers: An analytical model for energy consumption", Revista Facultad de Ingeniería Universidad de Antioquia, (91), 113-124, 2019.

[3] J. Aranda, D. Mendez, H. Carrillo & M. Schölzel, "A framework for multimodal wireless sensor networks", Ad Hoc Networks, 106, 102201, 2020.

[4] M. Calvo-Fullana, F. T. Dagefu, B. M. Sadler & A. Ribeiro, "A Multi-mode autonomous communication systems", in 2019 53rd Asilomar Conference on Signals, Systems, and Computers, pp. 1005-1009, IEEE, 2019, November

[5] T. Jawhly & R. Tiwari, "R. C. Simple VHF and UHF Propagation Loss Model", IEEE Communications Letters, 2023.

[6] F. Liu, J. Pan, X. Zhou & G. Y. Li, "Atmospheric ducting effect in wireless communications: Challenges and opportunities", Journal of Communications and Information Networks, 6(2), 101-109, 2021.

[7] R. Matsushima, K. Hosokawa, J. Sakai, Y. Otsuka, M. K. Ejiri, M. Nishioka & T. Tsugawa, "Propagation characteristics of sporadic E and medium-scale traveling ionospheric disturbances (MSTIDs): statistics using HF Doppler and GPS-TEC data in Japa", Earth, Planets and Space, 74(1), 60, 2022.

[8] S. C. Mora-Partiarroyo, M. Krause, A. Basu, R. Beck, T. Wiegert, J. Irwin & J. English, "CHANG-ES-XIV. Cosmic-ray propagation and magnetic field strengths in the radio halo of NGC 4631", Astronomy & Astrophysics, 632, A10, 2019.

[9] C. Mulas, T. Kalkan, F. von Meyenn, H. G. Leitch, J. Nichols & A. Smith, "Defined conditions for propagation and manipulation of mouse embryonic stem cells", Development, 146(6), dev173146, 2019.

[10] S. Narayanan, A. von Engeln, O. Osechas & N. U. de Haag, "Real-time Capable Compensation of Tropospheric Ducting for Terrestrial Navigation Integrity", in Proceedings of the 2022 International Technical Meeting of The Institute of Navigation, pp. 1190-1201, 2022, January.

[11] K. N. Ngan, C. W. Yap & K. T. Tan, "Video coding for wireless communication systems, CRC Press, 2018.

[12] V. Popov, "Cross-polarization effect of radio waves propagation by forest vegetation in wireless communication systems on transport", Procedia Computer Science, 149, 195-201, 2019.

[13] J. Sánchez-García, J. M. García-Campos, M. Arzamendia, D. G. Reina, S. L. Toral & D. Gregor, "A survey on unmanned aerial and aquatic vehicle multi-hop networks: Wireless communications, evaluation tools, and applications", Computer Communications, 119, 43-65, 2018.

[14] J. J. Sojka & R. A. Helliwell, "Sporadic E at 432 MHz", Journal of Geophysical Research, 82(33), 4891-4896, 1977.

[15] J .R. Wait, "Tropospheric ducting of VHF and UHF signals", Journal of Applied Meteorology, 9(4), 654-660, 1970.

[16] K. Whitman, R. Egeland, I. G. Richardson, C. Allison, P. Quinn, J. Barzilla, & P. Hosseinzadeh, "Review of solar energetic particle models", Advances in Space Research, 2022.

[17] Y. Zhu & C. Deng, "Millimeter-wave dual-polarized multibeam end-fire antenna array with a small ground clearance", IEEE Transactions on Antennas and Propagation, 70(1), 756-761, 2021.

[18] H. T. Friis, "A Note on a Simple Transmission Formula", IRE Proc. 34 (5): 254–256, May 1946.

[19] C. Deacon, C. Mitchell, and R. Watson. "Consolidated Amateur Radio Signal Reports as Indicators of Intense Sporadic E Layers", Atmosphere 13.6 (2022): 906.

[20] C. Deacon, B. Witvliet, C. Mitchell. "Rapid and Accurate Measurement of Polarization and Fading of Weak VHF Signals Obliquely Reflected From Sporadic-E Layers", 2020.

[21] http://www.aprs.fi

[22] https://www.dxinfocentre.com/tropo.html

[23] https://wspr.live

Placing Multi-component Applications in the Multi-access Edge Computing

Asterios Papamichail
Department of Informatics
Ionian University
Corfu, Greece
aspapa@ionio.gr

Athanasios Tsipis
Dept. of Digital Media & Communication
Ionian University
Kefalonia, Greece
atsipis@ionio.gr

Konstantinos Oikonomou
Department of Informatics
Ionian University
Corfu, Greece
okon@ionio.gr

Abstract—Multi-access Edge Computing resolves the problem of the transportation of enormous amounts of data via the Internet by localizing the computation and the storage of data near the user. The offloading of applications to the MEC Servers must be done in a way that minimizes the latency while considering their limited resources. Additionally, there is a trend for applications to be separated into microservices that run on different MEC Servers. This adds complexity to the placement of applications on the MEC Servers because it has to take into consideration the dependencies of its microservices. This problem is defined in the literature as the multi-component application placement problem. This paper proposes an algorithm to tackle this problem efficiently, while the dependencies of services are linear, i.e., a chain of microservices. The algorithm starts with the allocation of the 1-median of the backhaul network. Then, the proposed algorithm extends outwards and moves around until the score function stops improving. In order to evaluate the proposed algorithm a Mixed-integer Quadratic Program is formulated. The results showed that the proposed algorithm achieved great performance in various cases.

Index Terms—Multi-access Edge Computing, multi-component application placement problem, 1-median, microservices

I. INTRODUCTION

In recent years there has been a new era of augmented [1] or virtual [2] reality, artificial intelligence [3], and cloud gaming [4], that consume an enormous amount of mobile data, which is said to reach 165EB of transferred data per month by the year 2025 [5]. In this regard, the classic paradigm of the cloud-only infrastructure cannot scale cost efficiently to address the increasing demand of the applications along with their tight constraints on the quality of service, because the generated data have to be transferred over the Internet at long distances.

Multi-access Edge Computing [6] (MEC) resolves this problem by localizing the computation and the storage of data near the user. This is done by colocating MEC Servers with base stations taking advantage of the existing cellular infrastructure [7]. This adds the benefit that the Internet is not cluttered with the applications' data that use the MEC architecture, since all computation takes place near the end users.

Still, the offloading of applications to the MEC Servers must be done in a way that minimizes latency while taking into consideration their limited computation resources. Addition-

ally, in recent years there has been a trend towards separating applications into microservices that run on different MEC Servers for efficient development [8].

These microservices have dependencies on each other, defining the application as a graph wherein the graph nodes represent the computation of data on the application path while the edges represent the data transfer and communication between the application's microservices. Thus, the offloading of applications must take into consideration the dependencies between the services[1]. This leads to a placement problem that is defined in the literature as the *multi-component application placement problem* (MCAPP) [9].

There are numerous studies that try to solve the MCAPP, e.g., [8]. The general approach followed in these studies is to place the microservices by mapping their computation requirements to the available servers. The question that naturally arises from this approach is to decide on which MEC Servers to ultimately place each microservice, given that the former are typically characterized by limited resources and the fact that there is a growing need for delay guarantees. This is difficult to answer as the MCAPP for edge environments has been proven to be an NP-Hard problem [10], becoming complex to scale as the number of microservices increases [11], and, thus, it necessitates smart placement solutions that can produce feasible offloading solutions in tractable time.

Inspired by these observations, this paper proposes a new algorithm to tackle the MCAPP efficiently. It focuses on the multi-component application placement problem while the dependencies of services are linear, i.e., the application has a chain of services. First, the paper presents the system model containing the descriptions of the backhaul network and the applications. Additionally, it analyzes the different times that are needed for a request of a user to be served. Then, the MCAPP is formulated as a Mixed-integer Quadratic Program in order to use the provided solution as a benchmark for the proposed algorithm. Finally, the proposed algorithm is evaluated on two different network topologies and on two parameter types. The metrics that are used for evaluation are the approximation ratio and the number of iterations.

---
[1]The terms microservice and service are used interchangeably throughout the paper.

The proposed algorithm starts by selecting the 1-median [12] of the backhaul network for the allocation of MEC Servers. Next, the algorithm expands outwards until the number of allocated nodes is the same as the number of services of the application. Then, neighboring nodes of the two "edges" of the allocation are considered for inclusion in the allocation. After the allocation of MEC Servers, the proposed algorithm selects the direction of placement.

The results revealed that the proposed algorithm, on most of the evaluation setups, achieved a solution near the optimal one. The algorithm could not find a solution relatively near the optimal when it was evaluated on a sparsely connected Random Geometric Graph. Additionally, there is not any substantial difference in performance for the different parameter types.

The rest of the paper is organized as follows. In Section II the system model is presented, including the definition of the MCAPP. Section III provides the definition of the problem in a Mixed-integer Quadratic Program. Section IV presents the proposed algorithm that solves the MCAPP. Section V presents the evaluation process along with the results. Finally, the conclusions are drawn in Section VI.

II. SYSTEM

Fig. 1 depicts the architecture that this paper is based on. At the top layer, the cloud infrastructure is located where the backhaul network is connected via the internet where there are long delays. The backhaul network is comprised of multiple base stations (BS) that are interconnected with each other. The User Equipment (UE) is connected to the BSs, and it consists of various devices with different computational resources. Along with the UE, there is a MEC Server connected to each BS. Not all the MEC Servers have the same resources, i.e. processing rate and RAM.

The red arrows of Fig. 1 depict a request and the corresponding response of a UE to an application that is placed at the MEC Server of another BS that the UE is connected to. The lifetime of a request is as follows, first the UE sends its request to the BS. The BS forwards the request to a BS that the service is placed at its MEC Server. When the request arrives at the BS, it is forwarded to the MEC Server for processing. Finally, when the processing of the request is finished, the response is sent back to the UE that sent the request.

A. The Backhaul Network

Let the backhaul network be defined as a graph $\mathbf{B} = (\mathbf{BS}, \mathcal{L})$, where $\mathbf{BS}$ is the set of the BSs and $\mathcal{L}$ is the set of links between the BSs. Let $\mathbf{U}$, be the set of UE that uses the MEC architecture and generates traffic for the applications. Let λ_u^a, be the percentage of traffic that is generated for the application a from the UE u. Let $m^a(u)$, be the size of the messages that are sent by the UE u to the application a. Let $\mathbf{C}$, be the set of MEC servers that are attached to the BSs. Let r_c, be the processing rate of the MEC server c in bytes per millisecond.

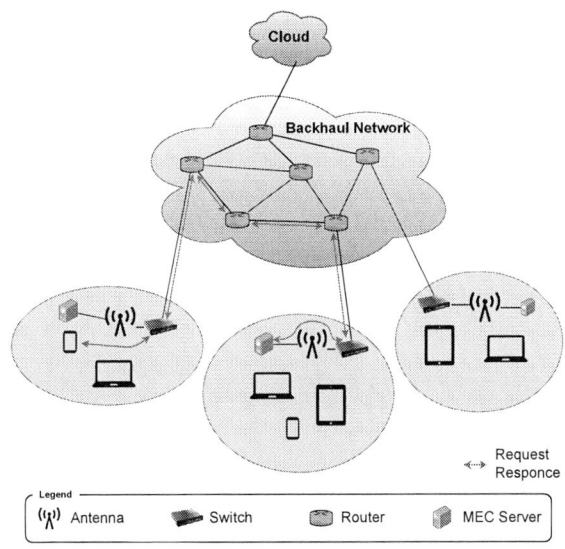

Fig. 1. The architecture of the MEC environment. The red lines depict the path that is taken by a request of UE and its response.

B. Applications

The applications that are considered in this study are comprised of multiple services. More specifically, an application a is defined as a Directed Acyclic Graph (DAG) $a = (\mathbf{S}^a, \mathbf{E}^a)$, where $\mathbf{S}^a$ depicts the set of microservices (computation of data) that need placement and they are indexed by $i = 0, 1, 2, \ldots, |\mathbf{S}^a| - 1$, where $|\cdot|$ denotes the number of elements in the corresponding set. The edges of the DAG $\mathbf{E}^a$ depict the transportation of data between the services and for each edge there is the percentage of the generated data $w_{i \to j}^a \forall e_{i \to j} \in \mathbf{E}^a$ that the service S_i^a sends to the service S_j^a. Here, it is noteworthy that an application has only one input, which is the service S_0^a, and only one output, which is the service $S_{|S^a|-1}^a$. Also, the applications that are considered in this study follow a path topology. An example of an application is shown at the top part of Fig. 2.

Each service S_i^a of an application a is characterized by the following parameters. Let $i_i^a(u)$, be the amount of data that the service S_i^a takes as an input from a request of the UE u, and is given by:

$$i_i^a(u) = \begin{cases} m^a(u), & \text{if } i = 0, \\ m_{i-1 \to i}^a(u), & otherwise \end{cases},$$

where $m_{i \to j}^a(u)$ is the amount of data that is transferred from the service S_i^a to the service S_j^a when a request from UE u is processed and, can be computed as $m_{i \to j}^a(u) = g_i^a(u)w_{i \to j}^a$.

Let d_i^a be the data inflation factor that depicts how much bigger or smaller the size of data has become after the processing on the service S_i^a relative to incoming data. Thus, the size of data that the service S_i^a generates by a request of a UE u is given by: $g_i^a(u) = d_i^a i_i^a(u)$. The computation time

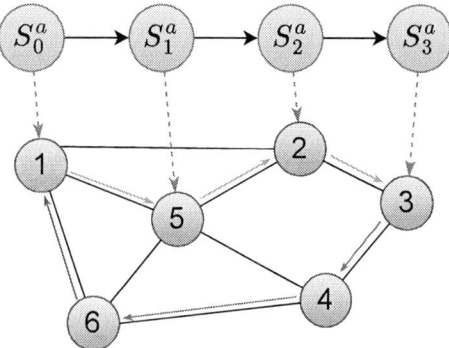

Fig. 2. An assignment of an application that consists of four services to an abstract representation of a backhaul network. The red arrow depicts a request from a UE that is connected to the BS with $id = 6$. The transportation of data between the services is depicted by the green arrows. The orange arrows show the route of the response to the UE that is connected to the BS with $id = 6$.

of a request of a UE u to be processed by the service S_i^a is given by (1)

$$C_i^a(u) = \frac{\mathcal{C}_i^a i_i^a(u)}{r_{p_i^a}}, \quad (1)$$

where p_i^a is the placement of the service S_i^a to a MEC Server $c \in \mathbf{C}$, and $\mathcal{C}_i^a$ denotes the complexity factor of the service S_i^a which indicates how much computationally intensive the service S_i^a is.

C. The Service Time of a Request

Fig. 2 presents the placement (dashed light blue arrows) of an application of four services to a backhaul network and the life of a request of a UE to the placed application. In this figure, the UE and the MEC Servers are omitted for simplicity. In this example, the UE that is connected to the BS with an ID equal to 6 sends a request (red arrow) to the placed application. When the request of the UE arrives at the MEC Server that the service S_0^a is placed on, the request is processed and then the generated data are forwarded to the MEC Server that the next service is placed on (green arrows). These steps are repeated until the data are processed by the last service of the application, where the response to the request is sent back to the UE that sent the request as shown by the orange arrows. Thus, the time that a request of a UE u to be served by an application a can be separated into three different times: i) the ingress time, ii) the computation time, and iii) the egress time.

$$\mathcal{T}_{ingress} = \sum_{u \in \mathbf{U}} \lambda_u^a l_{M(u),M(p_i^a)}, \quad (2)$$

where $M : \cdot \to \mathbf{BS}$ is a function that takes as an input a UE or a server and returns the BS that is connected to it, and $l_{i,j}$ is the minimum latency between the BS i and j.

1) The computation time of a request of a UE u: is the time that is needed for the processing of the request by the services of the application and the transportation of data between its services. Equation (3) expresses the computation time of a request of a UE u to an application a.

$$\mathcal{T}_{comp}(u) = \sum_{\forall e_{i \to j} \in \mathbf{E}^a} \left(C_i^a(u) + l_{M(p_i^a),M(p_j^a)} \right) + C_{|\mathbf{S}^a|-1}^a(u) \quad (3)$$

2) The Egress Time: is the time that is needed for the response to be sent back to the UE that sent the request. Equation (4) formulates the egress time, which is similar to the ingress time with the difference that now the minimum latency is measured from the last service of the application $S_{|\mathbf{S}^a|-1}^a$ back to UE that sent the request.

$$\mathcal{T}_{egress} = \sum_{u \in \mathbf{U}} \lambda_u^a l_{M(p_{|\mathbf{S}^a|-1}^a),M(u)} \quad (4)$$

Equation (5) shows the objective function of the problem of placing multi-component applications in the Multi-access Edge Computing Environment.

$$min \; \mathcal{T}_{ingress} + \sum_{\forall u \in \mathbf{UE}} \lambda_u^a \mathcal{T}_{comp}(u) + \mathcal{T}_{egress} \quad (5)$$

III. PROBLEM FORMULATION

This section formulates the service placement problem to a Mixed-integer Quadratic Program. In order to simplify the formulation of the service placement problem, the mapping function is substituted by the index of the node for both the UE and the MEC Servers, e.g. r_i is the processing rate of the MEC Server that is connected to the BS $i \; \forall \; i = \{0, 1, 2, \ldots, |\mathbf{BS}|-1\}$, and λ_i^a is the percentage of the generated traffic by the UE that is connected to the BS $i \; \forall \; i = \{0, 1, 2, \ldots, |\mathbf{BS}|-1\}$, similarly for the other parameters. The variable of the problem is an *assignment* binary matrix A that is indexed by the services of the application and the nodes of the network, and an $A_{i,j}$ element of the matrix has the value 1 if service S_i^a is placed on the MEC Server of the BS j, the value of 0 otherwise.

Equation (6) expresses the mean ingress time of a request given that the first service of application is placed at node i. The first part of the right-hand-side of Equation (7) expresses the computation time of a request by a UE given that the service S_i^a is assigned at the MEC Server of BS j. The second part of the right-hand-side of Equation (7) computes the latency of the transportation of data between the services given that the service S_i^a is assigned at the MEC Server of the BS s and the service S_j^a is assigned at the MEC Server of the BS d. Equation (8) expresses the mean egress time of a request given that the last service of the application is placed at node i.

$$\mathcal{T}_{ingress}' = \sum_{i=0}^{|BS|-1} A_{0,i} \sum_{u=0}^{|\mathbf{BS}|-1} \lambda_u l_{i,u} \quad (6)$$

$$\mathcal{T}'_{comp}(u) = \sum_{i=0}^{|\mathbf{S}^a|-1} \sum_{j=0}^{|\mathbf{BS}|-1} \left(A_{i,j} \frac{\mathcal{C}_i^a i_i^a(u)}{r_j} \right) +$$

$$\sum_{\forall e_{i \to j} \in \mathbf{E}^a} \sum_{s=0}^{|\mathbf{BS}|-1} \sum_{d=0}^{|\mathbf{BS}|-1} A_{i,s} A_{j,d} l_{s,d} \quad (7)$$

$$\mathcal{T}'_{egress} = \sum_{i=0}^{|BS|-1} A_{|\mathbf{S}^a|-1,i} \sum_{u=0}^{|BS|-1} \lambda_u l_{i,u} \quad (8)$$

Taking into consideration the above, the formulation of the problem will be as follows:

$$\text{minimize} \quad \mathcal{T}'_{ingress} + \sum_{u=0}^{|\mathbf{BS}|-1} \lambda_u^a \mathcal{T}'_{comp}(u) + \mathcal{T}'_{egress}$$

$$\text{subject to} \quad \begin{array}{l} \sum_{i=0}^{|\mathbf{S}^a|-1} A_{i,j} \leq 1, \forall j = 0,1,\dots,|\mathbf{BS}|-1 \\ \sum_{j=0}^{|\mathbf{BS}|-1} A_{i,j} = 1, \forall i = 0,1,\dots,|\mathbf{BS}|-1 \end{array}$$
$$(9)$$

The first constraint denotes that each server can facilitate at most one service and the second constraint denotes that all services are placed and are placed only once.

IV. PROPOSED ALGORITHM

This section presents the algorithm for placing multi-component applications on the MEC Servers. This algorithm is motivated by the fact that the objective function (5) incorporates the $\mathcal{T}_{ingress}$ and the $\mathcal{T}_{egress}$ that seem to compute the 1-median of the backhaul network with the distance being the minimum latency $l_{i,j}$ between two BSs and the demand being the percentage of the generated traffic λ_u^a from a UE u.

The proposed algorithm selects for the allocation of the multi-component application, the 1-median of the backhaul network. The algorithm expands outwards from the 1-median, by allocating the node that has the maximum score, until the number of allocated nodes is the same as $|\mathbf{S}^a|$. Then, different nodes are iteratively considered for inclusion in the allocation, selecting the ones that maximize the scoring function of allocation. Finally, the algorithm selects the proper direction of the placement of the application's services.

Algorithm 1 depicts the placement of a multi-component application to the backhaul network. The proposed algorithm starts from the 1-median of the backhaul network (line 1). The 1-median of the backhaul network $\mu(\mathbf{B}, \lambda^a)$ can be found by minimizing (11). In order for the algorithm to expand from the 1-median, it computes the $score_i$ (12) of all nodes of the 2-hop neighborhood and allocates the node with the maximum one (lines 2 – 6). Then, the algorithm computes the score of the candidate allocations of the application's services and selects the allocation that has the best allocation score

$$\frac{\sum_{\forall i \in \text{al}} r_i}{\sum_{\forall e_{i \to j} \in \mathbf{E}^a} l_{\text{al}[i],\text{al}[j]}} + \frac{1}{l_{\text{al}[0],\mu(\mathbf{B},\lambda^a)} + l_{\text{al}[|\mathbf{S}^a|-1],\mu(\mathbf{B},\lambda^a)}},$$
$$(10)$$

where al is a candidate allocation. The candidate allocations consider the nodes that are in the 2-hop neighborhood of

the two external nodes of allocation, given the fact that the allocation is path. This process is repeated until the score is not improved with respect to the previous score (lines 8 – 23). Finally, the algorithm selects the direction of the placement of the services by computing a directional score $\sum_{i=0}^{|\mathbf{S}^a|-1} \frac{\hat{\mathcal{C}}_i^a}{\hat{r}_{\text{placement}[i]}}$, where the hats denote normalized values of the parameters, for each placement direction, i.e., normal and reversed (line 24).

$$\sum_{u=0}^{|\mathbf{UE}|-1} \sum_{i=0}^{|\mathbf{BS}|-1} \lambda_u^a l_{u,i} \quad (11)$$

$$score_i = r_i + \sum_{\forall j \in n(i)} \frac{r_j}{l_{i,j}} \quad (12)$$

Algorithm 1

Input: $\mathbf{S}^a$, $\mathbf{B}$, $nodeScores$, $\mu(\mathbf{B}, \lambda^a)$
Output: placement // *contains the indices of BSs that the services are placed on their MEC Servers*
1: cAllocation $\leftarrow$ [$\mu(\mathbf{B}, \lambda^a)$]
2: **while** |cAllocation| $<$ |$\mathbf{S}^a$| **do**
3:     candidates $\leftarrow$ getTwoHopCandidates(cAllocation)
4:     node $\leftarrow$ getNodeWithBestScore(candidates)
5:     cAllocation $\leftarrow$ getPlacement(cAllocation, node)
6: **end while**
7: cAllocationScore $\leftarrow$ getPlacementScore(cAllocation)
8: **while** True **do**
9:     pAllocation $\leftarrow$ cAllocation
10:     pAllocationScore $\leftarrow$ cAllocationScore
11:     candidates $\leftarrow$ getTwoHopCandidates(pAllocation, $\mathbf{B}$)
12:     candidateAScores $\leftarrow$ Map(candAllocation, candAScore)
13:     **for all** c $\in$ candidates **do**
14:         candAllocation $\leftarrow$ getAllocation(pAllocation, c)
15:         candAScore $\leftarrow$ getAllocationScore(candAllocation)
16:         candidateAScores[candAllocation] $\leftarrow$ candAScore
17:     **end for**
18:     cAllocation $\leftarrow$ getKeyWithMaxValue(candAScores)
19:     cAllocationScore $\leftarrow$ candAScores[cAllocation]
20:     **if** cAllocationScore $<$ pAllocationScore **then**
21:         break
22:     **end if**
23: **end while**
24: **return** getDirectedPlacementWithBestScore(cAllocation)

V. PERFORMANCE EVALUATION

This section presents the evaluation process of the proposed algorithm on two different graph topologies and depicts the results. For the evaluation process, the Python programming language is used. More specifically, the evaluation process is as follows. First, each network is generated by the NetworkX [13] which is a Python package that is focused on studying the structure of complex networks. Then the proposed algorithm

TABLE I
THE RANGES OF PARAMETERS

parameter	normal	diverse
link delay	$[10\ 20]ms$	$[5\ 90]ms$
r_i	$[500\ 1000]bytes/ms$	$[200\ 1000]bytes/ms$
request rate	$[10\ 20]$	$[5\ 50]$
$m^a(u)$	$[1500\ 4500]$	$[500\ 4500]$
$w^a_{i \to j}$	$[0.5\ 1.5]$	–
c^a_i	$[5\ 20]$	–
d^a_i	$[0.5\ 1.5]$	–

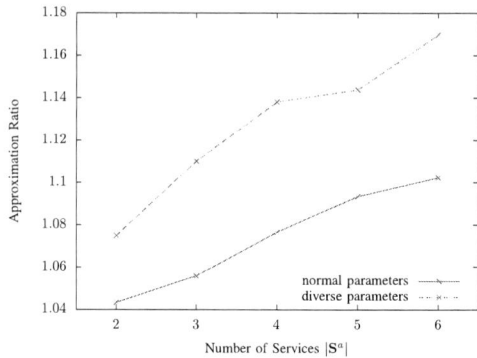

Fig. 3. The approximation ratio of the proposed algorithm on the Grid topology. The solid line depicts the approximation ratio of the proposed algorithm while the parameters are normal. The dashed line depicts the approximation ratio on diverse parameters.

is executed in order to find the placement of services. After the execution of the proposed algorithm, the optimal placement is found using the docplex python package, which is the provided Python API of IBM ILOG CPLEX Optimization Studio 22.11 [14], given the MIQP (9). Finally, the approximation ratio is computed by dividing the cost of the algorithm's placement by the cost of the optimal solution found by the optimizer. This process is repeated 100 times for each parameter setup and then the results are averaged.

The proposed algorithm is evaluated on five different application sizes $|\mathbf{S}^a| = \{2, 3, 4, 5, 6\}$. Additionally, the two topologies that the algorithm is evaluated on are the grid topology and the Random Geometric Graph. For the grid topology, a grid of 100 nodes is considered with dimension 10×10. The link delay of each link is given randomly and it is in the ranges that are depicted in Table I. For the random geometric graph, two different cases of node connectivity are evaluated i.e., sparse ($radious = 0.14$) and dense ($radious = 0.40$) while the link delays are proportional to the Euclidian distance between the nodes.

The algorithm is evaluated on two different types of random parameters namely, normal and diverse. Table I depicts the ranges of the parameters, where there is a dash in a cell it means that the same range is used from the normal parameter type. It is noteworthy here that for each iteration of a parameter setup, all the parameters take random values inside the given ranges of the same table.

A. Grid Topology

Fig. 3 depicts the mean approximation ratio of the proposed algorithm. The solid line presents the mean approximation ratio of the proposed algorithm while the parameters are normal. On the other side, the dashed line depicts the approximation ratio when the parameters are diverse. As expected, the approximation ratio is inversely proportional to the number of services that need placement. This is attributed to the fact that the algorithm has a larger search space as the number of services increases. Also, it is noteworthy here that the proposed algorithm achieves low approximation ratios for both types of parameters. Additionally, the algorithm has a bigger approximation ratio for the case of diverse parameters than the case of the normal ones and this difference starts from 0.02 for $|\mathbf{S}^a| = 2$ and ends close to 0.07 for $|\mathbf{S}^a| = 6$. This shows that the proposed algorithm works better when there is homogeneity in the parameters of the network.

Fig. 4 presents the mean number of iterations that are needed by the proposed algorithm in order to find the allocation of MEC Servers. The solid line depicts the mean number of iterations needed by the proposed algorithm for the case of normal parameters and the dashed line presents the number of iterations needed for the case of diverse parameters. It is shown that the number of iterations is proportional to the number of services that need placement and there is not any notable difference in the number of iterations for the two different types of parameters.

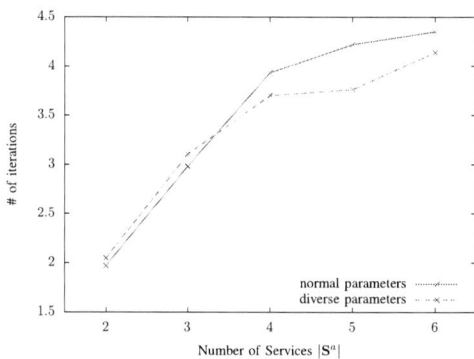

Fig. 4. The number of iterations that were needed by the proposed algorithm to find the placement. The solid line presents the number of iterations on normal parameters, while the dashed line presents the number of iterations on diverse parameters.

B. Random Geometric Graphs

Fig. 5 presents the mean approximation ratio of the proposed algorithm for the two different connectivities of the RGG. The orange lines of the figure depict the approximation ratio when the RGG is densely connected, while the blue ones depict the approximation ratio when the RGG is sparsely connected. Similar to the previous set of results, the solid lines depict the normal parameters while the dashed ones the diverse parameters.

979-8-3503-8368-3/23 $31.00 © 2023 IEEE

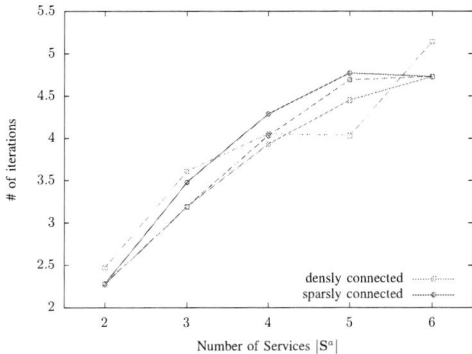

Fig. 6. The mean number of iterations that the proposed algorithm needed in order to find the allocation. The orange lines show the approximation ratio for the case of the densely connected graph and the blue ones depict the sparsely connected graph. The solid lines identify the case of normal parameters and the dashed ones present the case of diverse parameters.

In general, the approximation ratio is proportional to the number of services that need placement. For both the connectivity cases, it is noticeable that for most numbers of services, the approximation ratio of the case of the diverse parameters has greater values than the case of the normal parameters. This behavior is also presented to the grid topology as shown in Fig. 3. Additionally, a noticeable gap exists between the approximation ratio of the proposed algorithm of the two different connectivities of RGG. This is attributed to the fact that in the case of a dense network, there is a noticeably larger search space (more nodes to consider for each iteration) for the algorithm to consider.

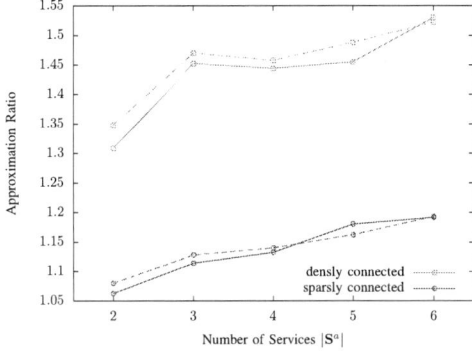

Fig. 5. The mean approximation ratio of the proposed algorithm on RGGs for the two different cases of connectivity. The orange lines show the approximation ratio for the case of the densely connected graph and the blue ones depict the sparsely connected graph. The solid lines identify the case of normal parameters and the dashed ones present the case of diverse parameters.

Fig. 6 depicts the mean number of iterations that the proposed algorithm needs in order to allocate the MEC Servers. The orange lines depict the results of the case of densely connected RGG and the blue ones are the case of sparsely connected RGG. The type of line represents the type of parameters that are used. Similar to the grid topology, the mean number of iterations seems to be proportional to the number of services that need placement and there is not any noteworthy

difference between the two connectivities and between the parameter types as well.

VI. CONCLUSIONS

In this paper, an algorithm is proposed for tackling the MCAPP. The algorithm starts from the 1-median of the backhaul network and then expands and moves around until there is not any improvement in the score function. The proposed algorithm is evaluated against the optimal solution found by the MIQP formulation of the MCAPP. The results showed that the proposed algorithm can find a solution near the optimal within a few number of iterations.

Future work includes the extension of the algorithm and the optimizer to handle multi-component applications with complex dependencies. Additionally, the proposed algorithm will be mathematically analyzed. Further, a variant of the proposed algorithm will be developed to tackle the problem of MCAPP in a dynamic environment.

REFERENCES

[1] T. Braud, F. H. Bijarbooneh, D. Chatzopoulos, and P. Hui, "Future networking challenges: The case of mobile augmented reality," in *2017 IEEE 37th International Conference on Distributed Computing Systems (ICDCS)*. IEEE, 2017, pp. 1796–1807.

[2] A. Tsipis, V. Komianos, K. Oikonomou, and I. Stavrakakis, "Towards fairness and qoe-based edge allocation for multiplayer virtual reality applications in edge computing," 2022.

[3] Y. Lu, "Artificial intelligence: a survey on evolution, models, applications and future trends," *Journal of Management Analytics*, vol. 6, no. 1, pp. 1–29, 2019.

[4] R. Shea, J. Liu, E. C.-H. Ngai, and Y. Cui, "Cloud gaming: architecture and performance," *IEEE network*, vol. 27, no. 4, pp. 16–21, 2013.

[5] 2020. [Online]. Available: https://www.ericsson.com/assets/local/reports-papers/mobility-report/documents/2020/june2020-ericsson-mobility-report.pdf

[6] A. Filali, A. Abouaomar, S. Cherkaoui, A. Kobbane, and M. Guizani, "Multi-access edge computing: A survey," *IEEE Access*, vol. 8, pp. 197 017–197 046, 2020.

[7] A. Tsipis and K. Oikonomou, "Joint optimization of social interactivity and server provisioning for interactive games in edge computing," *Computer Networks*, vol. 212, p. 109028, 2022. [Online]. Available: https://www.sciencedirect.com/science/article/pii/S1389128622001840

[8] T. Bahreini and D. Grosu, "Efficient placement of multi-component applications in edge computing systems," in *Proceedings of the Second ACM/IEEE Symposium on Edge Computing*, 2017, pp. 1–11.

[9] H. T. Malazi, S. R. Chaudhry, A. Kazmi, A. Palade, C. Cabrera, G. White, and S. Clarke, "Dynamic service placement in multi-access edge computing: A systematic literature review," *IEEE Access*, vol. 10, pp. 32 639–32 688, 2022.

[10] P. Han, Y. Liu, and L. Guo, "Interference-aware online multicomponent service placement in edge cloud networks and its ai application," *IEEE Internet of Things Journal*, vol. 8, no. 13, pp. 10 557–10 572, 2021.

[11] A. Papamichail, A. Tsipis, G. Tsoumanis, and K. Oikonomou, "Study of a proposed spectral-based approach for facility location in tree topologies," in *2022 Global Information Infrastructure and Networking Symposium (GIIS)*, 2022, pp. 73–77.

[12] M. S. Daskin and K. L. Maass, "The p-median problem," in *Location science*, 2015, pp. 21–45.

[13] A. A. Hagberg, D. A. Schult, and P. J. Swart, "Exploring network structure, dynamics, and function using networkx," in *Proceedings of the 7th Python in Science Conference*, G. Varoquaux, T. Vaught, and J. Millman, Eds., Pasadena, CA USA, 2008, pp. 11 – 15.

[14] S. Nickel, C. Steinhardt, H. Schlenker, W. Burkart, M. Reuter-Oppermann, S. Nickel, C. Steinhardt, H. Schlenker, W. Burkart, and M. Reuter-Oppermann, "Ibm ilog cplex optimization studio," *Angewandte Optimierung mit IBM ILOG CPLEX Optimization Studio: Modellierung von Planungs-und Entscheidungsproblemen des Operations Research mit OPL*, pp. 9–23, 2021.

Design and Evaluation of a Peripheral for Integrity Checking to Improve RAS in RISC-V Architectures

Daniele Rossi, Nicasio Canino, Stefano Di Matteo, Sergio Saponara, Vasileios Tenentes

Department of Information Engineering, University of Pisa, 56122 Pisa, Italy

{daniele.rossi1, sergio.saponara, vasileios.tenentes}@unipi.it; {nicasio.canino, stefano.dimatteo}@phd.unipi.it

Abstract—**This paper presents a peripheral to check for integrity against errors affecting memories in RISC-V architectures. A HW-SW Interface for Error Logging and Reporting to improve Reliability, Availability, and Serviceability (RAS) is proposed. It defines the facilities developed to log details on detected errors into a set of registers, which are then provided to the system software. The developed architecture has been first synthesized on TSMC** $7nm$ **Standard-Cell technology and then implemented on an FPGA-based test board featuring the RISC-V CV32E40P core. The proposed peripheral provides a significant degree of flexibility and configurability to effectively satisfy the needs of different application scenarios by selectively incorporating or removing specific features.**

Index Terms—**Error Logging, Error Reporting, FPGA, HW-SW Interface, Reliability-Availability-Serviceability, RISC-V.**

I. Introduction

Reliability, Availability, and Serviceability (RAS) describe the ability of a computing system to prevent or recover from failures [1]. Particularly, reliability is the probability that the system produces correct outputs, availability is a measure of the ability of the system to perform the intended service at any time, while serviceability represents the capability of the system to provide information about a failure that has occurred.

The RAS of a system can be improved by adopting error-checking hardware and information redundancy [1], thus allowing the detection and possible correction of possible errors. If the hardware can autonomously correct errors, system operation continues without any noticeable loss in performance.

Different software-based solutions can be embraced to increase the overall system RAS through *fault prediction* and *error recovery* techniques [1]–[7]. Additionally, *error containment* can also be implemented to limit the propagation of erroneous data. This enhances system availability by limiting the effects of errors to a subset of software or hardware resources. However, all software approaches need some hardware that monitors the selected modules and signals the occurrence of an error in hardware or, in general, a specific event (e.g., from a performance monitor of a specific module).

Concurrently, there are several hardware-based solutions to *monitor* and *correct errors* [4], [8]–[10]. For example, one of the most frequent types of error that affects memories is *soft error*, [11], usually caused by the interaction of high-energy particles, such as cosmic rays, with electronic components in computer memory. Indeed, the advancement of technology, increasing miniaturization, higher operating frequencies, and lower voltage levels of electronic components make them more susceptible to soft errors.

All these approaches have a detrimental impact on the availability and performance of the system because the software needs to handle every error event, even if it does not lead to a failure, e.g. erroneous data corrected by an Error Correcting Code (ECC). Synergistically with the previously described approaches, a HW-SW Interface to log error information can be implemented. It will call system software intervention only when necessary, to restore correct operation or contain the error effect. This way, it will allow a significant increase in the system's availability and serviceability, affecting its performance as little as possible [5].

The HW-SW Interface acts as a monitoring device for error messages from error-checking hardware. When an error is detected, it stores relevant error-related information such as severity and source of the error, address location (if available), and timestamp, in a set of ad hoc registers (*Error Record*). A group of error records is referred to as a *Bank*. Additionally, it can also alert software about potential failures, which could potentially result in the loss of ongoing tasks and data integrity.

Some current high-performance processor architectures, implement such an error logging and reporting peripheral, including various generations of Intel Xeon [12] and AMD EPYC [13] (Machine Check Architecture, MCA), and several Arm platforms such as the v8-A architecture [14] (RAS Extension). Instead, so far there has been no development of such a RAS feature within the RISC-V community or literature. Some preliminary results have been presented in [15].

To fill this gap, in the paper, we present the architecture of a HW-SW Interface for integrity check of RISC-V-based systems, hereafter referred to as *RAS Peripheral*. It features an error record interface to store relevant error-related information. This work has been carried out within the European Processor Initiative (EPI SGA2). The RAS Peripheral has been developed by using System Verilog, synthesized with a 7 nm standard cell library for area occupation evaluation, and finally implemented in an FPGA-based test platform for verification and validation.

The rest of the paper is organized as follows: In Section II, a system-level description of the designed architecture will be provided, also discussing the purpose and functionality of its building blocks. In Section III, the results of the syntheses performed on Standard-Cell will be discussed. Then, the testing environment in which the developed architecture has

---

979-8-3503-8368-3/23 $31.00 © 2023 IEEE

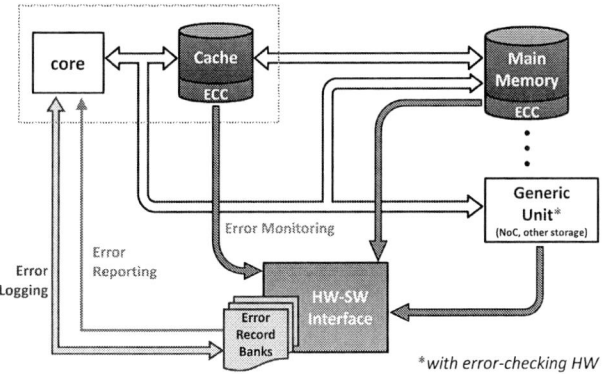

Fig. 1. Placement of the HW-SW Interface in a generic computing system that includes units monitored by error-checking HW.

TABLE I
ERROR CODE IMPLEMENTED, CHARACTERIZED BY DIFFERENT FORMATS IN WHICH THE FIELDS HAVE SPECIFIC MEANINGS.

error_code[7:0]	Type of Source	Description
0b 000 01 T LL	TLB	T = Transaction Content LL = Cache Level
0b 001 RR T LL	Memory	Errors in the cache hierarchy RR = Transaction Type T = Transaction Content LL = Cache Level
0b 010 RR A LL	Interconnects	General bus errors RR = Transaction Type A = Memory or I/O LL = *unused*

been validated will be described. In the end, Section V will summarize the work done and all the major results obtained.

II. PROPOSED HW-SW INTERFACE FOR RAS IMPROVEMENT

The system-level architecture of the proposed HW-SW Interface, called *RAS Peripheral*, is depicted in Figure 1. It gathers information about hardware errors detected by error-checking hardware through specific input ports (error monitoring). After storing the error records, the CPU will access them by a standard memory-mapped interface (e.g., AXI [16]), providing the error logging feature. An interrupt interface is implemented to provide the error reporting capability. Moreover, system software can configure at run-time which events generate an interrupt request. This capability significantly increases system availability and overall performance because the non-relevant conditions are filtered out, thus providing high flexibility and efficiency in the use of error records.

The developed RAS Peripheral has been designed according to the following error classification criteria, according to error severity for the proper system operation.

- *Corrected Error* (CE), if the detected error has also been corrected by specific hardware. As an example, such hardware can implement an ECC protection that can detect and correct errors in protected data words [17], [18].
- *Uncorrected Deferred Error* (UDE), if the detected error can not be corrected but has no immediate impact on the operation of the system. In this case, the operation can continue, and dealing with the error is eventually deferred to a later point in time when the corrupted data is consumed. For instance, if an uncorrected error is detected in a memory-to-memory transfer, it will be classified as deferred since no unit is going to immediately consume the wrong data. However, such erroneous data is tagged as *poisoned* and monitored throughout its lifetime. Since the system can track the poisoned data, when it is about to be consumed, altering the state of the system, it will be escalated to *Uncorrected Urgent Error*;

- *Uncorrected Urgent Error* (UUE), if the detected error can not be corrected and requires immediate action from system software. For example, erroneous data that is going to be consumed by the processor, thus altering the correct state of the system.

Upon the occurrence of an error, it needs to be classified according to its detectability. Indeed, undetected errors cannot be handled by the developed RAS Peripheral because of the very nature of this type of error. Therefore, they are not further analyzed, even though they may cause a system failure.

A high-level block illustration of the architecture of the developed RAS Peripheral is depicted in Figure 2.

The *Error Mux* receives the error control signals generated by the error-checking circuitry of the monitored blocks and generates an error message containing the information for the identification of the monitored HW block generating the error message (error code) and the error severity (CE, UDE, UUE). We propose the error code as reported in Table I. Each row of the table represents a different error code format, depending on the source of the error which can be, for example, TLB (Translation Lookaside Buffer), Memory, or Interconnects.

The *Error Synchronization Interface* synchronizes and combines the information generated by the Error Mux (error_message in Figure 2) with the address associated with that specific error. The *Address Synchronization Buffer*, within this block, delays the input signals by the required clock cycles. Also, a circular *First-In-First-Out (FIFO) Buffer* stores the error message paired with the synchronized address of the erroneous location and sends it to the Main Controller.

Once an error is detected, it should be stored in one of the error records within a Bank, whose number is limited. The *Main Controller* selects the error record where to locate the newly detected error. For efficiency purposes, it adopts a fully associative cache-like approach, selecting the first free error record it finds. When there are no error records available, an overwriting algorithm must be activated, so that the Main Controller makes decisions to resolve overwriting issues based on the state of the records, as reported in Table II.

More in detail, the Main Controller first retrieves the information on the detected error from the FIFO Buffer of the first stage of the RAS Peripheral and the timestamp (if implemented). Then, it explores the current state of each record. For example, whether the record contains valid information

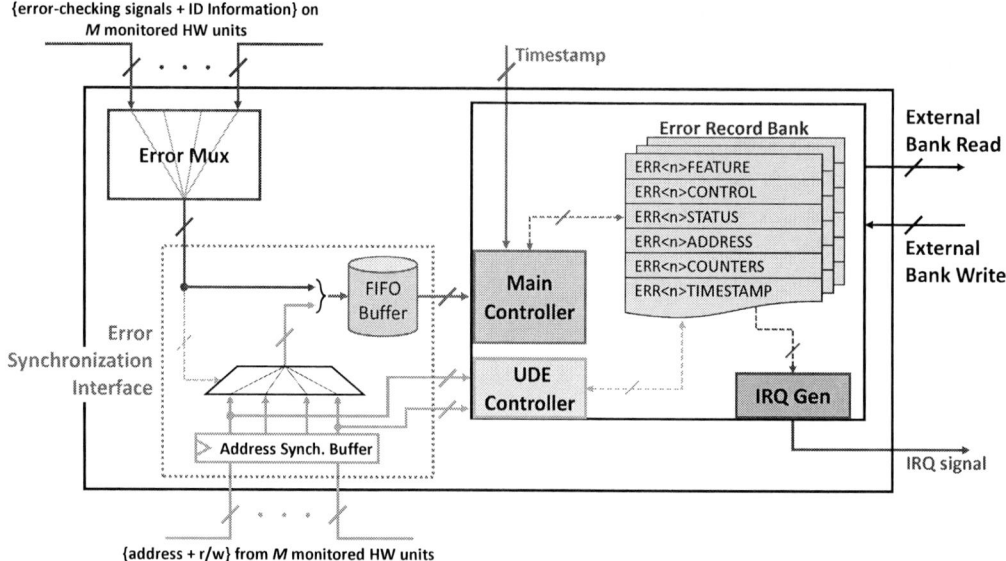

Fig. 2. Architecture of the developed RAS Peripheral.

TABLE II
ACTION TAKEN BY THE MAIN CONTROLLER IN THE SELECTED RECORD,
DEPENDING ON THE SEVERITY OF THE NEW ERROR AND THE STORED ONE
IN SAID RECORD.

	new CE	new UDE	new UUE
-free record-	Write	Write	Write
stored CE	Count/Discard	Overwrite	Overwrite
stored UDE	Discard	Discard	Overwrite
stored UUE	Discard	Discard	Discard

and the severity of the stored errors. Finally, it determines which record to select and how to act (see Table II), based on the state of the records and the new error.

If one or more free error records are available, any new error-related information will be written to the lowest indexed free record within the bank. However, if there are no free error records and the new error is classified as UUE or UDE, it will overwrite a stored error with lower severity. If all stored error records have higher or equal severity, the new error will be discarded according to the policy prioritizing older errors. If a new CE occurs and all the records contain higher severity errors (i.e., UDE or UUE), it will be discarded. When the proper record has been identified, the error information is logged.

The *UDE Controller* operates in parallel with the Main Controller to monitor *poisoned data*, which refers to detected erroneous data whose error information is stored in the records of a bank and classified as UDE. Additionally, the UDE Controller needs information on the nature of the transactions related to the monitored units, read or write transactions, to determine whether the poisoned data are being accessed in read or write mode.

More in detail, the UDE Controller searches for logged

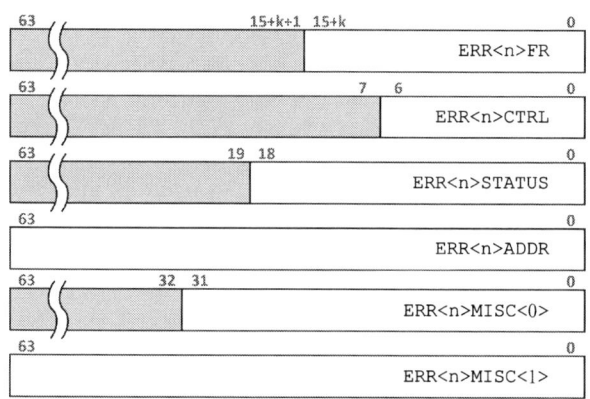

Fig. 3. Registers composing the nth error record. For each register, the maximum number of bits that might be implemented is delimited.

errors with a UDE severity among all the records owned by the RAS Peripheral and, simultaneously, monitors each read or write transaction related to the monitored units containing poisoned data. Therefore, whenever poisoned data is going to be consumed or overwritten by new data, the monitor will determine how to update the related error record. Lastly, the UDE Controller invalidates or upgrades the specific record in the bank, according to the decision process described above.

The developed RAS Peripheral provides the error logging feature through its *Error Records*, where an error record consists of a set of 64-bit registers used to record all necessary error-related information. Figure 3 depicts the set of registers within the generic n^{th} *Error Record* and the maximum number of available bits within each register (grey fields are not used).

Feature Register (ERR<n>FR) is a Read-Only register that defines the main characteristics of the RAS Peripheral and is configured at implementation stage. As an example, it specifies the type of errors that are logged (CE, UDE, UUE), whether the counter for corrected errors is implemented or not, if the timestamp is stored, kind of errors generating an interrupt.

Control Register (ERR<n>CTRL) is a Read-Write register that stores a set of enable signals for the different features (for instance, if the error reporting is enabled, for which cases the interrupt is enabled, etc). It can be configured by the user at runtime.

Status Register (ERR<n>STATUS) is a Read-Write register that stores all the major information about the error logged in the nth record. It contains the error message along with some valid bits related to the address and miscellaneous registers, if a corrected error counter overflow is implemented, or whether the error record has been overwritten.

Address Register (ERR<n>ADDR) is a Read-Only register storing the address associated with the location of the erroneous data.

Two *Miscellaneous Registers* (ERR<n>MISC<0>, Read-Write register, and ERR<n>MISC<1>, Read-Only register) which store additional information about the logged error. Its timestamp is also stored in one of these registers.

In a RAS Peripheral, the Feature and Control Registers are only implemented in the first record of a bank. So, from the second record onward, they will be hardwired to zero.

Finally, The IRQ Generator implements the Error Reporting feature through the generation of an interrupt signal connected to an Interrupt Controller, according to the features of the RAS Peripheral chosen at implementation time (contained in the Feature Register).

III. STANDARD CEALL SYNTHESIS RESULTS

The proposed RAS Peripheral has been equipped with an AXI4 Slave memory-mapped interface to access the error record registers. The overall system has been designed in HDL SystemVerilog language and synthesized on a $7nm$ TSMC Standard-Cell technology. Additionally, it has been implemented and prototyped to verify and validate its behavior using an FPGA board Xilinx Zynq ZCU104.

For both synthesis and implementation, the RAS Peripheral was configured with the following error logging and reporting features:

- Two monitored units ($N_{HW} = 2$);
- Two error records implemented ($N_{REC} = 2$);
- Each record implements both CE Counters of 8-bit ($b_{CEC} = 16$);
- Each record includes the timestamp information, considering a 32-bit value ($b_{TS} = 32$);
- AXI4 address bus of 32-bit ($b_{ADDR} = 32$);
- FIFO Buffer with 2 entries.

Table III reports the synthesis results for the RAS peripheral in terms of maximum operating frequency, total estimated power (leakage + switch), and resource utilization of the composing sub-modules (respecting the hierarchy of the synthesized design). The maximum operating frequency has been obtained by performing multiple syntheses with a

TABLE III
MAXIMUM OPERATING FREQUENCY, TOTAL ESTIMATED POWER, AND RESOURCE UTILIZATION FOR *7nm* STANDARD-CELL TECHNOLOGY.

Hierarchical Cell	Area [μm^2]			Gate Equivalent [GE]		
	Comb.	Seq.	Tot.	Comb.	Seq.	Tot.
AXI4 Slave Interface	64,63	64,65	129,27	841,50	841,75	1683,25
RAS Peripheral	177,50	163,18	338,27	2311,25	2124,75	4404,50
Error Mux	3,51	5,24	8,76	45,75	68,25	114,00
Error Synch. Intf.	30,74	66,60	97,34	400,25	867,25	1267,50
FIFO buffer	30,62	39,19	69,81	398,75	510,25	909,00
Bank	61,15	84,79	145,94	796,25	1104,00	1900,25
Controllers	71,96	3,72	75,69	937,00	48,50	985,50
IRQ Generator	1,86	0,40	2,27	24,25	5,25	29,50
Total Design	~ 468.02			~ 6090		
Maximum Frequency	$5.50\ GHz$					
Power (leak + switch)	$8.23\ mW$					

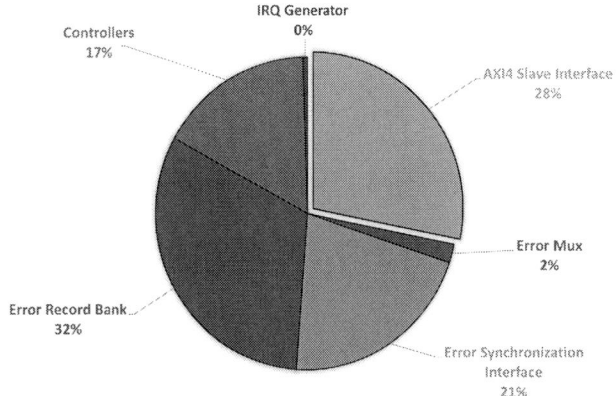

Fig. 4. Relative resource utilization of the RAS Peripheral from the synthesis results on the 7nm Standard-Cell technology.

frequency sweep of the clock signal, reaching $5.50\ GHz$ for the considered $7nm$ technology.

The value obtained for the total power consumption has been estimated by the synthesis tool. By looking at the resource utilization results, expressed both in Area (μm^2) and Gate Equivalents (GE), they have been distinguished between *Combinational* (*Comb.* in Table III) and *Sequential* logic (*Seq.* in Table III). Moreover, the *Total* (*Tot.* in Table III) resource utilization of each sub-module is provided.

Figure 4 shows a pie chart with the relative resource utilization of each sub-module. Even though the resource overhead introduced by the two controllers cannot be reduced, the one due to the error records and to the FIFO Buffer can be adjusted as required by the specific implementation.

IV. FPGA TEST PLATFORM IMPLEMENTATION

The proposed RAS Peripheral has been implemented and prototyped on FPGA to validate its functionality using the board Xilinx Zynq ZCU104. The structure of the developed FPGA test platform is depicted in Figure 5. All blocks communicate via memory-mapped accesses using the AXI4 communication protocol.

The *RISC-V CV32E40P core*, as detailed in [19], is a 4-stage, in-order 32-bit RISC-V processor that implements the RV32IMFC Instruction Set Architecture (ISA) [20]. This core is equipped with two master AXI4 interfaces, one for data and one for instructions, providing access to all the available resources within the test platform. Its interrupt controller [21], has been employed to receive the interrupts generated by the *IRQ Generator* of the RAS Peripheral (depicted as red blocks in Figure 5).

Our *RAS Peripheral* has been configured to own two records with all error logging features implemented and to have full error reporting reconfigurability, by writing into the `ERR<0>CTRL` register. Additionally, we have integrated an Error Injection register which the processor can access during runtime to validate the architecture by injecting errors into the *2° and 3° BRAMs*.

Main Block RAM (BRAM) has been provided as Xilinx IP (Intellectual Property), and it has been implemented as the primary memory of the prototyped system, serving as the instruction and data memory. Its capacity is fixed at $128\ kB$.

Concurrently, the *2° and 3° BRAMs*, also provided as Xilinx IP, are two small auxiliary BRAMs (one of $8\ kB$ and the other of $4\ kB$), configured to implement a *SEC-DED* type ECC code [22]. Also, this IP incorporates functionality for injecting single and double errors at run-time on the protected words stored in these memories (one word is 32-bit). We can activate this feature by writing within the *Error Injection* register, which is incorporated in the RAS Peripheral.

Then, the *Central DMA (CDMA)*, provided as Xilinx IP, and has two AXI4 interfaces, a master and a slave. It can transfer large groups of bytes between memory-mapped devices and is commonly used for simulating memory-to-memory transfers of large data blocks, emulating data transfers between adjacent cache layers.

Finally, the *UART* and *JTAG* blocks are provided as Xilinx IP to, respectively, display the debug information through the terminal and to load the program binary into the Main BRAM at run time.

Since the test platform is equipped with a 32-bit processor, the data bus width is 32-bit. This implies that each 64-bit register of an error record will be accessed as two separate 32-bit registers. Moreover, it is worth noting that the developed RAS Peripheral can be adopted also with 64-bit RISC-V processors with minimal modifications.

The RAS Peripheral monitors two units with error-checking HW, 2° and 3° BRAMs. The CDMA has been used to induce memory-to-memory data transfers between the two ECC-protected BRAMs (2° and 3° BRAMs), as it happens with two adjacent cache levels (e.g., L1 and L2 caches). This configuration allows us to explore all the considered error severities:

- The injection of single-bit errors and their correction through the ECC SEC allows the triggering of errors with CE severity.
- The injection of double-bit errors and their detection through the ECC DED in memory-to-processor transactions triggers errors with UUE severity; they are classified as UUE since data will be consumed immediately by the processor, potentially inducing a system failure.

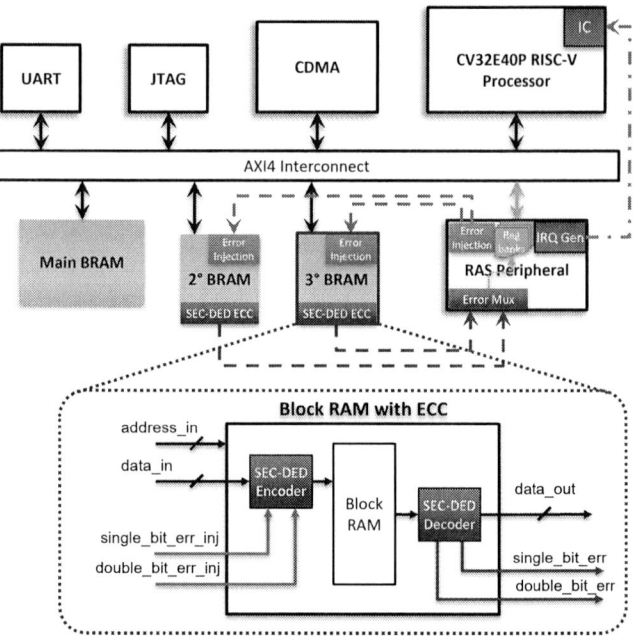

Fig. 5. Architecture of the test SoC implemented. The colored dashed lines represent the main interconnections, independent of the AXI4 interface, to and from the RAS Peripheral.

- The injection of double-bit errors and their detection through the ECC DED in memory-to-memory transactions triggers errors with UDE severity; they are classified as UDE errors since data may not be consumed immediately. The erroneous data will be tagged as poisoned.

To test the functionality of the features of the proposed RAS Peripheral, we wrote several software routines in C language that run on the RISC-V processor and stimulate error conditions, according to the severity and procedure listed above. First of all, we tested the generation of errors with severity CE, UDE, and UUE and checked the correctness of the related error information stored inside the error record registers. Afterwards, we tested several corner cases:

- Overwrite of a stored error with a higher severity one (eg., CE overwritten by new UDEs of UUEs);
- Discard of a new CE, UDE, or UUE due to higher severity errors already stored (or older ones with the same severity);
- CE counting and overflow of the CE Counters (CECs);
- Escalation of a stored error record from UDE to UUE severity (consumption of poisoned data);
- Invalidation of a stored error record of UDE severity (overwrite of the poisoned data);
- Controllability of interrupt generation via the `ERR<0>CTRL` register.

V. CONCLUSION

This work presented the design and implementation of a HW-SW Interface for Error Logging and Reporting, called RAS Peripheral, to improve RAS in both 32- and 64-bit RISC-V architectures. Overall, we have developed the RAS

Peripheral to be flexible and configurable, allowing designers to implement only the desired features, and also optimizing the hardware overhead introduced. Although only ECC-protected memories were considered as monitored HW units, the proposed architecture also allows for other types of units to be monitored, including TLBs and interconnections. The RAS Peripheral has been synthesized on TSMC $7nm$ Standard-Cell technology. The synthesis results have shown that the total complexity of the module heavily depends on parameters such as the number of records implemented and the amount of information that they store. Then, we developed a test SoC, featuring the RAS Peripheral and the RISC-V CV32E40P core, that has been implemented on a ZCU104 board. This allows us to validate the proposed RAS Peripheral system.

ACKNOWLEDGEMENT

This research was partially funded by the European Union's Horizon 2020 research and innovation program "European Processor Initiative" (EPI) under grant agreement No. 101036168 (EPI SGA2). This work was partially supported by Dipartimento di Eccellenza CrossLab and FoReLab projects, Italian Ministry of University and Research (MUR), Department of Information Engineering, Unversity of Pisa.

REFERENCES

[1] R. Canal, C. Hernández, R. Tornero, et al., "Predictive Reliability and Fault Management in Exascale Systems: State of the Art and Perspectives," *ACM Computing Surveys*, vol. 53, pp. 1–32, Oct. 2020.

[2] J. Brandt, F. Chen, V. De Sapio, et al., "Quantifying Effectiveness of Failure Prediction and Response in HPC systems: Methodology and example," in *2010 International Conference on Dependable Systems and Networks Workshops (DSN-W)*, 2010, pp. 2–7.

[3] F. Cappello, H. Casanova, and Y. Robert, "Checkpointing vs. Migration for Post-Petascale Supercomputers," in *2010 39th International Conference on Parallel Processing*, 2010, pp. 168–177.

[4] O. Khan and S. Kundu, "Hardware/Software Codesign Architecture for Online Testing in Chip Multiprocessors," *IEEE Transactions on Dependable and Secure Computing*, vol. 8, no. 5, pp. 714–727, 2011.

[5] A. Chatzidimitriou, G. Papadimitriou, and D. Gizopoulos, "Healthlog Monitor: Errors, Symptoms and Reactions Consolidated," *IEEE Transactions on Device and Materials Reliability*, vol. 19, no. 1, pp. 46–54, 2019.

[6] Z. Zheng, L. Strigini, N. Antunes, and K. Trivedi, "Editorial: Software reliability and dependability engineering," *IEEE Transactions on Dependable and Secure Computing*, vol. 20, no. 4, pp. 2674–2676, 2023.

[7] I. Kaitovic and M. Malek, "Impact of failure prediction on availability: Modeling and comparative analysis of predictive and reactive methods," *IEEE Transactions on Dependable and Secure Computing*, vol. 17, no. 3, pp. 493–505, 2020.

[8] J. K. Park, D. Kim, and J. T. Kim, "Efficient error-resilient bus coding method using bit-basis orthogonal integrative multiplexing," *IEEE Transactions on Emerging Topics in Computing*, vol. 10, no. 2, pp. 1178–1191, 2022.

[9] J. Li, P. Reviriego, L. Xiao, and H. Wu, "Protecting memories against soft errors: The case for customizable error correction codes," *IEEE Transactions on Emerging Topics in Computing*, vol. 9, no. 2, pp. 651–663, 2021.

[10] Y. Sazeides, A. Bramnik, R. Gabor, and R. Canal, "A real-time error detection (rtd) architecture and its use for reliability and post-silicon validation for f/f based memory arrays," *IEEE Transactions on Emerging Topics in Computing*, vol. 10, no. 2, pp. 524–536, 2022.

[11] V. Sridharan, N. DeBardeleben, S. Blanchard, et al., "Memory Errors in Modern Systems: The Good, The Bad, and The Ugly," in *Proceedings of the Twentieth International Conference on Architectural Support for Programming Languages and Operating Systems*, New York, NY, USA: Association for Computing Machinery, 2015, pp. 297–310, ISBN: 9781450328357. [Online]. Available: https://doi.org/10.1145/2694344.2694348.

[12] *Intel Xeon Processor E7 Family: Reliability, Availability, and Serviceability - White paper*, 2011.

[13] M. Insignths and Strategy, *AMD EPYC Brings New RAS Capability - White paper*, 2017.

[14] *Arm Reliability, Availability, and Serviceability (RAS) Specification Armv8, for the Armv8-A architecture profile - White paper*, 2022.

[15] D. Rossi, N. Canino, S. Di Matteo, and S. Saponara, *Hw-sw interface for ras in risc-v architectures*. [Online]. Available: https://riscv-europe.org/media/proceedings/posters/2023-06-06-Daniele-ROSSI-abstract.pdf.

[16] *AXI (Advanced eXtensible Interface) memory-mapped interface*. [Online]. Available: https://developer.arm.com/documentation/ihi0022/e/AMBA-AXI3-and-AXI4-Protocol-Specification.

[17] R. W. Hamming, "Error Detecting and Error Correcting Codes," *The Bell System Technical Journal*, vol. 29, no. 2, pp. 147–160, 1950.

[18] M. Y. Hsiao, "A Class of Optimal Minimum Odd-Weight-Column SEC-DED Codes," *IBM J. Res Develop*, vol. 14, no. 4, Jul. 1970.

[19] OpenHW-Group, *Cv23e40p risc-v processor core*. [Online]. Available: https://github.com/openhwgroup/cv32e40p.

[20] *RISC-V International*. [Online]. Available: https://riscv.org/.

[21] Exceptions and I. D. for the CV32E40P core. [Online]. Available: https://docs.openhwgroup.org/projects/cv32e40p-user-manual/en/latest/exceptions_interrupts.html.

[22] Xilinx, *Block Memory Generator v8.3 Product Guide*. [Online]. Available: https://docs.xilinx.com/v/u/8.3-English/pg058-blk-mem-gen.

Compressing time series towards lightweight integrity commitments

Angeliki Katsika[*], Konstantinos Papageorgiou[*], Alexandros Fakis[†], Athanasios kakarountas[*],
Fotis Andritsopoulos[‡], Vassilis Plagianakos[*], Georgios Spathoulas[*][§]

[*]Department of Computer Science and Biomedical Informatics
University of Thessaly, Lamia, Greece
{akatsika, kopapageorgiou, kakarountas, vpp, gspathoulas}@uth.gr
[†]Department of Information and Communication Systems Engineering
University of the Aegean, Samos, Greece
alfa@aegean.gr
[‡]iTrack, Athens, Greece
fandrit@itrack.gr
[§]Department of Information Security and Communication Technology
Norwegian University of Science and Technology (NTNU), Gjøvik, Norway
georgios.spathoulas@ntnu.no

Abstract—In the current era of big data, the generation and preservation of time series data has seen widespread adoption in various sectors such as trade, healthcare, and industry. At the same time, data have become a prized asset, as we are witnessing an unprecedented explosion in the volume of data generated and collected. This includes data from various sources, such as sensors, social media, and transactions, offering a competitive edge to organizations that harness the power of data, leading them to informed decisions and rapid responses to changing conditions. However, it is essential to responsibly manage data, addressing privacy and security concerns to ensure their integrity and availability, while efficiently managing the cost of storage and operations. To address these challenges, there is a compelling demand for methods that compress these time series while ensuring that the compressed versions faithfully represent the original data and remain secure against tampering or unauthorized alterations. This paper introduces a novel approach to condensing time-series data into lightweight integrity commitments, leveraging polynomial vector commitments as a key component. These integrity commitments are integrated with blockchain technology, serving as a cornerstone in guaranteeing data integrity and authenticity, particularly in supply chain scenarios where trust and transparency are of the utmost importance.

Index Terms—time series, compression, commitments, validity proof, integrity

I. INTRODUCTION

Integrity in time-series data is of paramount importance for various reasons. Time-series data often serve as the foundation for critical decision-making processes in fields such as finance, economics, and environmental monitoring. Maintaining data integrity ensures that the information collected over time is accurate, consistent, and reliable. Without integrity, errors, omissions, or inconsistencies can lead to faulty analyses and

This research has been co-financed by the European Union and Greek national funds through the Operational Program Competitiveness, Entrepreneurship and Innovation, under the call RESEARCH – CREATE – INNOVATE (Project ARTEMIS. Project code:T2EDK-02836)

misguided conclusions. Furthermore, the credibility of research, forecasts, and models is heavily based on the integrity of the underlying time-series data. To maintain the trust and validity of these data-driven applications, it is essential to uphold the highest standards of integrity, which includes data validation, error handling, and proper documentation to ensure the accuracy and reliability of time series datasets.

Although multiple approaches have already attempted to generate commitments for time series, those commitments usually are related to the time series as a whole. There exist several efficient methods to achieve this, but the main assumption is that the prover has access to all the data points in the series. In the current technological landscape, where data relate to multiple actors and granularity in access control is of high importance, it is common for users to have access to only parts of the time series. In a previous work of ours [1] we proposed the use of vector commitments for the construction of a flexible system that enables a prover to commit to a time series and to be able to provide a subset of the values of the time series along with a proof related to that time subset of values. The verifier can verify the validity of the values without having access to the entire time series.

In the present work, we extend the already proposed approach to make it more time efficient. The main contributions are as follows:

- Compression of a time series
- Generation of a commitment for the compressed time series to be used to support integrity verification of the initial one
- Testing of the proposed approach with respect to complexity and accuracy

The rest of the paper is summarized as follows; Section II presents related work, Section III discusses required background concepts, Section IV presents our methodology,

979-8-3503-8368-3/23 $31.00 © 2023 IEEE

Section V discusses conducted experiments and presents the corresponding results, while Section VI draws conclusions on the proposed method.

II. RELATED WORK

Various techniques have been introduced to create condensed representations of time-series data, resulting in smaller representations or abstractions of time series data compared to the original. In the dynamic domain of IoT applications, efficient compression of time series data is essential to address the limitations associated with IoT devices, such as restricted storage and processing capacities and the constraints of network resources. Numerous research works have explored this area, enhancing the processing and analysis of data, especially in the domain of energy management, where real-time data concerning energy consumption are of great importance. In [2]the authors introduce a compression technique based on piecewise regression, accompanied by two performance evaluation methods. The proposed technique is applied to real-world energy datasets, demonstrating its effectiveness in different scenarios and highlighting the feasibility of implementing this compression technique within contemporary database management systems. In Wen et al.'s study [3], the authors examine compression techniques to manage large volumes of data generated by smart meters in the context of smart grids. They focus on electrical measurements presented in time series format and explore various compression methods tailored to this data type. Similarly, in the emerging Smart Manufacturing domain, Gómez-Brandón et al. [4] present Direct Access Compression of time-series (DACTS), a new lossless compression method designed for time series from industrial environments that allows decompressing portions of the initial dataset. In [5] the authors introduce a novel time series compression algorithm that is specifically suitable for low-power sensing devices and efficient for data storage and retrieval on servers. Extensive experiments demonstrate the algorithm's performance across various domains and types of time series data, for a broad range of applications such as sensor data management, server-based data storage, and query operations. In [6] authors proposed a proficient compression methodology for both univariate and multivariate time series, which integrates the lifting implementation of the discrete wavelet transform with an error-bound compression technique, specifically referred to as Squeeze (SZ), in order to achieve an optimal balance between data compression and data fidelity. Similarly, in the domain of genetic data management where vast DNA sequences arise challenges for storage reduction and efficiency, the authors of [7], introduced the Blockchain Applied FASTQ and FASTA Lossless Compression (BAQALC) algorithm, aiming to efficiently transmit and store massive DNA sequence data generated through Next Generation Sequencing. Although this research focuses on sequence data rather than time series data, it serves as a notable example of efficient data compression, storage, and transmission. To the best of our knowledge, the present work is the first attempt to merge commitment schemes with reduced and compressed time-series data. Our work, which presents a cooperative fusion of methodologies, ensures data integrity and reliability while optimizing storage efficiency, but also opens new avenues for secure and efficient data management across diverse fields of application.

III. BACKGROUND

A. Time Series Compression

Time series compression techniques aim to reduce the data's size while retaining its essential information and can be categorized in two primary approaches:

- Lossless Compression reduces the size of time series data while guaranteeing the precise reconstruction of the original data from the compressed version. These techniques are commonly employed in applications where the integrity of data must be maintained, such as critical measurements and financial or medical records. Some of the widely used lossless methods are Delta Encoding [8], [9] , Run-Length Encoding (RLE) [10], [11] and Dictionary-Based Methods [12], [13].

- Lossy compression achieves higher compression ratios by allowing some loss of data fidelity when this is acceptable within the context of the application's goals, as the original data cannot be perfectly reconstructed. Some common lossy compression methods [3] used for time series data are the following: Wavelet Transform (WT) [14], [15], Symbolic Aggregation Approximation (SAX) [16], [17], Piecewise Polynomial Approximation (PPA) [2], [9], Linear Regression-Based Dimension Reduction [18], and Sparse Coding (SC) [19], [20].

The Run-Length Encoding (RLE) [10], [11] algorithm falls under the category of lossless data compression techniques. RLE is a straightforward method used to compress data, and its principle is based on identifying runs of data, which are sequences where the same value occurs consecutively, and representing them more efficiently. Instead of storing each individual element, RLE stores the data value and a count of how many times it appears in the data series. This approach significantly reduces the amount of data needed to represent the original run, making it an effective method for lossless data compression.

B. KZG Vector Commitments (for IoT data streams)

The KZG Polynomial Commitment Scheme represents a fundamental innovation in the domain of cryptographic commitments. It provides a robust framework for binding a set of elements to a specific polynomial, thereby facilitating a compelling verification process. This commitment scheme, as introduced by Kate, Zaverucha, and Goldberg in their seminal work [21], is distinguished by its efficiency in enabling the commitment of polynomials within a bilinear pairing group, while maintaining commitments of fixed and constant size. Using this feature of fixed-sized commitments and proofs, regardless of the number of measurements, we envisaged a scenario involving an untampered IoT device producing valid data streams [1]. The protocol, described in the following operations, served to ensure the integrity of the data, protecting against both deliberate and accidental mishandling of information. The operations outlined below, based on the KZG polynomial commitment scheme [21], are implemented to create and validate commitments and proofs.

- **Setup:** Initialize the system components with the public parameters for the KZG algorithm. These public parameters denoted pp can be generated via a trusted ceremony [22]. It is crucial to ensure that the count of public parameters is equal to or exceeds the maximum number of values that the device commits to ($d \geq n$).
- **Vector Commitment generation:** As the device generates a set S of n values, each element within $S = (e_0, e_1, ..., e_{n-1})$ is associated with a different index $(k_0, k_1, ..., k_{n-1})$. Using interpolation methodologies, a polynomial P_c, that traverses through all the data points $(k_0, e_0), (k_1, e_1), ..., (k_{n-1}, e_{n-1})$ is produced. Subsequently, a KZG commitment $comm$ is generated for this polynomial, utilizing the public parameters of the system. The device also stores the polynomial coefficients c and the commitment $comm$ to storage that is accessible exclusively to the prover p.
- **Proof Request:** The data owner (prover) receives a request from a user (verifier) to verify the integrity of a specified subset of data, let us assume a subset of X with the indexes $ind = \{i_0, i_1, ..., i_{k-1}\}$ where $i_j \in [0, n-1]$.
- **Proof Generation:** The prover retrieves the polynomial coefficients, denoted as c and the commitment $comm$ from storage. Subsequently, the prover evaluates the polynomial at the requested indices $values = v_0, v_1, ..., v_{k-1}$. Leveraging these coefficients, indices, and their corresponding values, the prover proceeds to generate a KZG batch proof, $proof$.
- **Proof Validation:** The prover sends the values and proof to the verifier, who validates the proof using KZG proof validation operation that returns a value res, which can be true or false.

Using this vector commitment scheme as already explored in our previous work [1], an IoT device can securely commit to a collection of data measurements, which are then stored on a public blockchain with the aid of a smart contract. Armed with this commitment and access to the data, the prover can provide the verifier with a subset of measurements and proof of constant size, thus verifying the integrity of the data.

IV. METHODOLOGY

In the present Section, we present the steps of the proposed methodology in detail. The main goal is to provide the means for creating lightweight integrity commitments for long-time series, which tend to have identical values for more than one consecutive index. The proposed methodology is an extension of our previous work [1].

Let us assume that an actor (prover) produces a time series. The methodology allows the creation of a commitment that is made publicly available at the time of the time-series generation. At a later stage, another actor (verifier) requests access to a subset of the time series, and the methodology presented allows the prover to produce a proof that will enable the verifier to validate that the subset of the time series that has been made available to him has not been tampered with in any of the intermediate steps (storage, transmission, etc.).

A. Commitment generation

The prover (denoted as P) produces a time series X of length n:

$$X = \{x_1, x_2, ..., x_n\} \tag{1}$$

Smoothing: The first step relates to smoothing the time series. A stronger effect of having consequent identical values makes the commitment generation process much lighter. Thus, smoothing out tiny peaks in the time series greatly reduces complexity in the commitment generation with a bearable loss of accuracy with regard to the originally produced time series.

The time series X is smoothed by using a moving average window of width w and a new time series Y is produced.

$$Y[i] = \begin{cases} \frac{\sum_{i-w}^{i} X[i]}{w}, i \geq w \\ \frac{\sum_{0}^{i} X[i]}{i}, i < w \end{cases} \tag{2}$$

As expected, this step introduces a loss of accuracy which is strongly coupled with the size of the data. The average accuracy loss is defined as L.

$$L = \frac{\sum |X_i - Y_i|}{n} \tag{3}$$

It should be noted that if $w = 1$ then $L = 0$.

In order to limit the loss of accuracy, we add a correction step that cancels this substitution if the averaged values exceed a specific threshold of loss $loss_{thr}$.

$$Y[i] = \begin{cases} Y[i], |\frac{Y[i] - X[i]}{X[i]}| \leq loss_{thr} \\ X[i], otherwise \end{cases} \tag{4}$$

Compression: The next step is compression of the time series Y using Run-Length Encoding (RLE). Time series that tend to have repeated values for long consecutive indices subsets can be represented by much shorter time series. This alternative representation can greatly benefit the next step of commitment generation. Instead of storing each value of the time series, we opt for storing only those values at indices where the value changes, along with those indices. So we end up with two shorter time series to represent the initial longer one. The operation of the RLE compression step is described in Algorithm 2.

The compression step keeps pairs of indices and values for indices in which the values differ from the value of the previous index. On top of that, the algorithm retains the first and the last pairs of index and value, to enable the reconstruction of the original time-series. If the original time series is $Y = [3, 3, 3, 3, 5, 5, 5, 5, 5, 5, 7, 7, 7]$, then the RLE algorithm will produce $I = [1, 5, 11, 13]$ and $Z = [3, 5, 7, 7]$.

979-8-3503-8368-3/23 $31.00 © 2023 IEEE

Algorithm 1 Operation of RLE compression

1: **procedure** RLE(Y)
2:     Vector Y contains averaged time series values
3:     Empty vectors I, Z initialised
4:     $i \leftarrow 1$
5:     $j \leftarrow 1$
6:     **while** $i \leq n$ **do**
7:         **if** $i = 0 \lor i = n$ **then**
8:             $I[j] \leftarrow i$
9:             Z[j] $\leftarrow Y[i]$
10:        **else if** $Y[i] \neq Y[i-1]$ **then**
11:            $I[j] \leftarrow i$
12:            $Z[j] \leftarrow Y[i]$
13:        **end if**
14:     **end while**
15: **end procedure**

Pairing: Committing to two time series is obviously more costly in terms of processing than committing to one. Thus, the next step of the process is to represent the two time series I, Z with a single one of the same length. This happens by using the Cantor pairing [23], [24], which can represent a pair of integers in the range $[0, 2^N]$ with a single integer in the range $[0, 2^{2N}]$.

The Cantor pairing is based on Equation 5.

$$Pair(x, y) = \frac{x^2 + x + 2x * y + 3y + y^2}{2} \quad (5)$$

The pairing step in our algorithm produces a new time series F where $F[i] = Pair(I[i], Z[i])$

Commitment generation: As soon as the F time series is constructed, the next step is to calculate the coefficients c_F of the polynomial P_F passing through all data points $(0, F_0), ..., (m-1, F_{m-1})$ using the Newton interpolation technique.

$$c_F = [c_0, c_1..., c_{m-1}] = N_F(F_0, F_1, ..., F_{m-1}) \quad (6)$$

The next step is the creation of the two corresponding KZG commitments for the polynomial produced, by using the public parameters pp, as already discussed in Section III-B.

$$comm_F = gen\_comm(c_F), \quad (7)$$

This commitment c_F is stored in an immutable storage (e.g. a smart contract sc). The prover p also stores the pair of polynomial coefficients and commitment $c_F, comm_F$ in storage s that is accessible only by him.

B. Proof generation

Proof request: The role of the verifier involves requesting a subset S of values from the initial time series X. This request is made by presenting a set of corresponding indices $i \in [a, b]$ where $a, b \leq n$ and $a < b$.

Proof generation: The prover must transform the range of requested indices in the initial time series X to the

corresponding indices (a', b') in the shorter time series I, Z which are identical to the indices in F. The new indices a', b' correspond to the points in I that are just before and just after the requested subset.

$$a' = max(i), I[i] < X[a] \quad (8)$$

$$b' = min(i), I[i] > X[b] \quad (9)$$

For time series $Y = [3, 3, 3, 3, 5, 5, 5, 5, 5, 5, 7, 7, 7]$, $I = [1, 5, 11, 13]$ and $Z = [3, 5, 7, 7]$, if the requested range was $a, b = 6, 8$ then the transformed indices would be $a', b' = 2, 3$.

After calculating the updated indices, the prover creates a proof for the subset of time series F and for the range $[a', b']$. He retrieves from storage s the polynomial coefficients c_F and the commitment $comm_F$ and evaluates the polynomial P_F in the indices $[a', b']$ and generates the vectors $values_F = v_1, v_2, ..., v_\lambda$.

Consequently, the prover generates the KZG batch proof for the points between the indices (a, b), based on the coefficients c of polynomial P as follows:

$$proof_F = proof\_gen(c_F, (a', b')), values_F) \quad (10)$$

C. Proof validation

This process allows the verifier to efficiently verify the integrity of specific subsets of the initial time series by connecting them to the commitments produced from the shorter time series generated by RLE.

Proof Validation: The prover sends the generated proofs $proof_F$ and the subset of values $values_F$ from the F time series to the verifier. The verifier then validates the proof using the commitment $comm_F$ produced for the time series F and verifies that the requested subset has remained intact. The verifier can then validate the proofs submitted using the KZG proof validation operation.

$$res_F = proof\_val(comm_F, proof_F, (a', b')) \quad (11)$$

The result res_F may be true or false.

Upon successful validation of the proofs, the verifier can be confident that the subset of the original time series, as represented by the shorter time series, remains unchanged and maintains its integrity.

Depairing: After the verifier verifies the integrity of the F time series, the next step is to depair the elements, according to the Cantor scheme, and retrieve the I, Z time series. The depair function for the paired value z returns x, y according to the following Equations.

$$w = \lfloor \frac{\sqrt{8 * z + 1} - 1}{2} \rfloor \quad (12)$$

$$t = \frac{w^2 + 2}{2} \quad (13)$$

979-8-3503-8368-3/23 $31.00 © 2023 IEEE

$$x = z - t \qquad (14)$$

$$y = w - y \qquad (15)$$

The verifier uses the F time series to produce the I, Z time series.

$$I[i], Z[i] = depair(F[i]), \forall i \in [a', b'] \qquad (16)$$

Decompressing:

The last step is the decompresion process which returns the corresponding subset of the initial smoothed time series Y.

Algorithm 2 Operation of RLE decompression

1: **procedure** DE-RLE(I, Z)
2:     Vectors I,Z contain compressed indices and values
3:     Empty vectors Y initialized
4:     $i \leftarrow 1$
5:     **while** $i < len(I)$ **do**
6:         $j \leftarrow i$
7:         **while** $j < I[i+1]$ **do**
8:             $Z[j] \leftarrow Y[i]$
9:         **end while**
10:    **end while**
11: **end procedure**

The end result is a sub-time series that may be slightly different (longer) than the initial Y constructed during the compression due to the compression process.

V. EXPERIMENTS

In this section, we present the experiments and the results that emerged, which involved the use of the proposed compression algorithm on real-world time series data. The experiments were designed to investigate and validate the hypotheses formulated in the previous sections. We conducted a comprehensive set of experiments in the supply chain domain to gain insight into the behavior within the proposed solution.

A. Experimental Setup

We analyze real-world time series data with multiple measurements from a variety of sensors recorded from monitoring IoT devices in the supply chain context. In particular, there were eight types of sensors, each recording different types of values. The anonymized data was supplied by ITrack, a telematics service provider in Greece. The main intuition behind using real-world data was to have a realistic setup with regards to the measurement deviation between measurements. For all sensors, we used 287 measurements with a sampling rate of 1 sample per 5 minutes. For each set of measurements, we calculate the entropy of the values. We chose three sets of measurements to present on the basis of their entropy to reduce the amount of information presented in the current work. We chose the data series with the lowest and highest entropy, as well as the data series with an intermediate level of entropy.

ble_temp4	ble_humi_4	ble_humi_3
2.6926731541629807	1.122648215461937	0.33492306848458225

TABLE I
ENTROPIES OF THE PRESENTED SENSORS

The experiments were conducted on a machine with the following specifications:

- CPU: Intel® Core™ i7-8700K @ 3.7GHz
- RAM: 16GB @ 3200 MHz

B. Results

For each data set, we calculated the interpolation time in order to generate the commitment without using the compression technique. We then altered the size of the window w when performing the RLE compression algorithm. Also, for each window size, we tested different values of the error threshold $loss_{thr}$. To be more specific, for each data series, we calculated the interpolation time of the compressed data for all combinations between window widths $w \in [5, 10]$ and loss threshold values of $loss_{thr} \in [0.4, 0.2]$. The results of the experiments are depicted below.

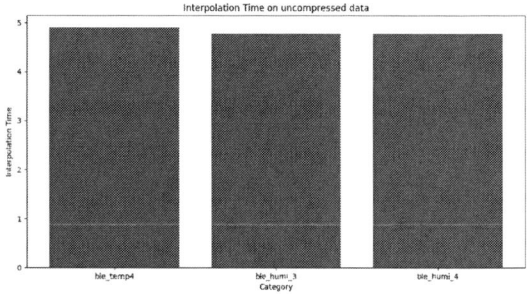

Fig. 1. Interpolation time on uncompressed data

In Figure 1 the interpolation time of the uncompressed data is presented for each sensor. For the data from the sensor *ble_temp4* we measured a slightly higher interpolation time, as this was the series with the highest entropy value. Overall, the interpolation time for all three sensors was roughly the same. This is because the sizes of the data series were the same for all three sensors (287 measurements).

Figure 2 presents the results of the tests performed on the data series generated by the sensor *ble_temp4*(highest entropy ≈ 2.693). The first graph presents the interpolation time for each window size while $loss_{thr} = 0.4$. The interpolation time for the RLE compressed data peaked at 4.75 seconds with the size of the window $w = 6$. In this case, two commitments are generated (one commitment for the compressed indices and the second for the values corresponding to the compressed indices). RLE compression followed by Cantor pairing (RLE-P) peaked at 2.3 seconds with window size $w = 5$. The fourth graph represents the average error in each window for each threshold. The highest average error occurs when $loss_{thr} = 0.4$ is as expected.

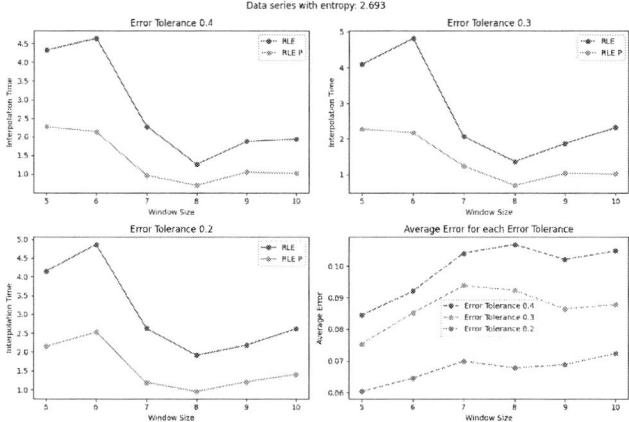

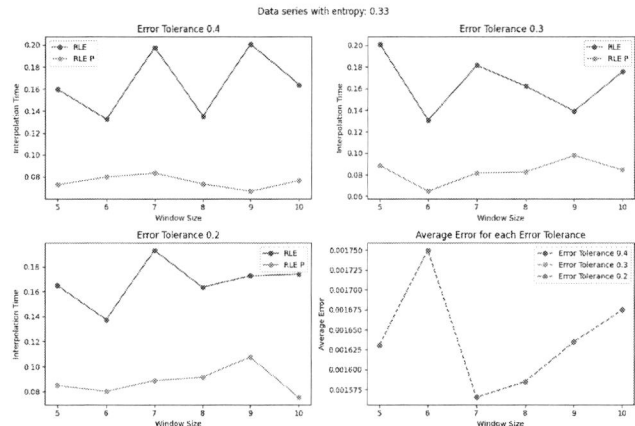

Fig. 2. Interpolation time with compressed data on *ble_temp4* sensor

Fig. 4. Interpolation time with compressed data on *ble_humi_3* sensor

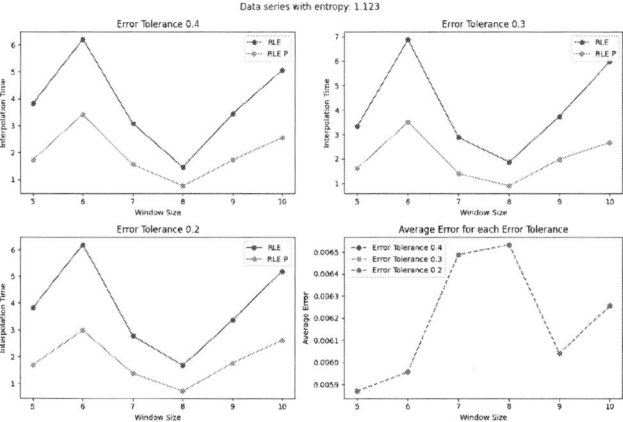

Fig. 3. Interpolation time with compressed data on *ble_humi_4* sensor

In Figure 3, the results of the tests on the data set generated by the sensor *ble_humi_4* are presented. The entropy of this particular data set is ≈ 1.123. The highest interpolation time appeared on the RLE-only compressed data: 7 seconds with the size of the window $w = 6$ and the $loss_{thr} = 0.3$ shown in the second plot. The RLE with pairing peaked for the same window size at ≈ 3.5 seconds. An interesting observation is in the last plot. The average error remains the same regardless of the thresholds we chose in each size of the window. This is because the entropy is not high enough to produce errors that exceed the minimum threshold (0.2) we set.

Figure 4, represents the results of the data set that has the lowest entropy ≈ 0.33. The proposed method managed to compress the data set into 7 indices. This can be observed in any of the plots. Regardless of the thresholds, the interpolation time for both the RLE compressed data and the RLE with pairing did not exceed 0.25 seconds. The noticeable observation mentioned before appeared on this data set, too. The highest average error appeared on $w = 6$ with value ≈ 0.0018.

VI. CONCLUSIONS

We have proposed an approach that enables the compression of time series, so that it is feasible to produce flexible commitments, allowing partial proving, faster, and with less resource requirements. The main steps towards that goal is the smoothing, the compression, and the pairing of the initial time series before generating the KZG commitment. On the basis of the resources of the device on which the commitment is generated, the smoothing window size and the error tolerance can be adapted accordingly. Either time on energy usage restrictions may impose different choices.

As part of our future work, we will test the approach in a variety of different scenarios with respect to devices producing the commitment and the nature of the data series being compressed. We will use a number of different edge devices and different real-world data series from different domains.

REFERENCES

[1] A. Katsika, K. Papageorgiou, A. Fakis, V. Plagianakos, and G. Spathoulas, "An efficient and lightweight commitment scheme for iot data streams," in *2023 19th International Conference on Distributed Computing in Smart Systems and the Internet of Things (DCOSS-IoT)*, 2023, pp. 461–468.

[2] F. Eichinger, P. Efros, S. Karnouskos, and K. Böhm, "A time-series compression technique and its application to the smart grid," *The VLDB Journal*, vol. 24, no. 2, pp. 193–218, Apr. 2015. [Online]. Available: https://doi.org/10.1007/s00778-014-0368-8

[3] L. Wen, K. Zhou, S. Yang, and L. Li, "Compression of smart meter big data: A survey," *Renewable and Sustainable Energy Reviews*, vol. 91, pp. 59–69, 2018. [Online]. Available: https://www.sciencedirect.com/science/article/pii/S1364032118301849

[4] A. Gómez-Brandón, J. R. Paramá, K. Villalobos, A. Illarramendi, and N. R. Brisaboa, "Lossless compression of industrial time series with direct access," *Computers in Industry*, vol. 132, p. 103503, 2021. [Online]. Available: https://www.sciencedirect.com/science/article/pii/S016636152100110X

[5] D. W. Blalock, S. Madden, and J. V. Guttag, "Sprintz: Time series compression for the internet of things," *CoRR*, vol. abs/1808.02515, 2018. [Online]. Available: http://arxiv.org/abs/1808.02515

[6] J. Azar, A. Makhoul, R. Couturier, and J. Demerjian, "Robust iot time series classification with data compression and deep learning," *Neurocomputing*, vol. 398, pp. 222–234, 2020.

[7] S.-J. Lee, G.-Y. Cho, F. Ikeno, and T.-R. Lee, "Baqalc: Blockchain applied lossless efficient transmission of dna sequencing data for next generation medical informatics," *Applied Sciences*, vol. 8, no. 9, 2018. [Online]. Available: https://www.mdpi.com/2076-3417/8/9/1471

[8] T. Pelkonen, S. Franklin, J. Teller, P. Cavallaro, Q. Huang, J. Meza, and K. Veeraraghavan, "Gorilla: A fast, scalable, in-memory time series database," *Proceedings of the VLDB Endowment*, vol. 8, no. 12, pp. 1816–1827, 2015.

[9] G. Chiarot and C. Silvestri, "Time series compression: a survey," *CoRR*, vol. abs/2101.08784, 2021. [Online]. Available: https://arxiv.org/abs/2101.08784

[10] A.-l. Mohammed and I. Emary, "Comparative study between various algorithms of data compression techniques," *Lecture Notes in Engineering and Computer Science*, vol. 2167, 10 2007.

[11] A. Birajdar, H. Agarwal, M. Bolia, and V. Gupte, "Image compression using run length encoding and its optimisation," in *2019 Global Conference for Advancement in Technology (GCAT)*, 2019, pp. 1–6.

[12] J. Ziv and A. Lempel, "A universal algorithm for sequential data compression," *IEEE Transactions on Information Theory*, vol. 23, no. 3, pp. 337–343, 1977.

[13] H. S. Mogahed and A. G. Yakunin, "Development of a lossless data compression algorithm for multichannel environmental monitoring systems," in *2018 XIV International Scientific-Technical Conference on Actual Problems of Electronics Instrument Engineering (APEIE)*, 2018, pp. 483–486.

[14] S. Santoso, E. Powers, and W. Grady, "Power quality disturbance data compression using wavelet transform methods," *IEEE Transactions on Power Delivery*, vol. 12, no. 3, pp. 1250–1257, 1997.

[15] J. Khan, S. Bhuiyan, G. Murphy, and J. Williams, "Pmu data analysis in smart grid using wpd," 04 2014, pp. 1–5.

[16] J. Lin, E. Keogh, S. Lonardi, and B. Chiu, "A symbolic representation of time series, with implications for streaming algorithms," in *Proceedings of the 8th ACM SIGMOD Workshop on Research Issues in Data Mining and Knowledge Discovery*, ser. DMKD '03. New York, NY, USA: Association for Computing Machinery, 2003, p. 2–11. [Online]. Available: https://doi.org/10.1145/882082.882086

[17] A. Notaristefano, G. Chicco, and F. Piglione, "Data size reduction with symbolic aggregate approximation for electrical load pattern grouping," *Generation, Transmission & Distribution, IET*, vol. 7, pp. 108–117, 02 2013.

[18] T. Blu, P. Thévenaz, and M. Unser, "Linear interpolation revitalized," *IEEE transactions on image processing : a publication of the IEEE Signal Processing Society*, vol. 13, pp. 710–9, 06 2004.

[19] Y. Wang, Q. Chen, C. Kang, Q. Xia, and M. Luo, "Sparse and redundant representation-based smart meter data compression and pattern extraction," *IEEE Transactions on Power Systems*, vol. 32, pp. 2142 – 2151, 05 2017.

[20] C. Yu, P. Mirowski, and T. K. Ho, "A sparse coding approach to household electricity demand forecasting in smart grids," *IEEE Trans. Smart Grid*, vol. 8, no. 2, pp. 738–748, 2017. [Online]. Available: https://doi.org/10.1109/TSG.2015.2513900

[21] A. Kate, G. Zaverucha, and I. Goldberg, "Constant-size commitments to polynomials and their applications," 12 2010, pp. 177–194.

[22] V. Nikolaenko, S. Ragsdale, J. Bonneau, and D. Boneh, "Powers-of-tau to the people: Decentralizing setup ceremonies," *Cryptology ePrint Archive*, 2022.

[23] A. L. Rosenberg, "Efficient pairing functions—and why you should care," *International journal of foundations of computer science*, vol. 14, no. 01, pp. 3–17, 2003.

[24] F. Rossini, "Cantor pairing in a reversible programming language," in *Companion Proceedings of the 3rd International Conference on the Art, Science, and Engineering of Programming*, 2019, pp. 1–2.

979-8-3503-8368-3/23 $31.00 © 2023 IEEE

Embedded Platforms for Trusted Edge Computing towards Quality Assurance along the Supply Chain

Vasileios Tenentes[*], Athanasios Xynos[*], Christos Zonios[*], Asimina Koutra[*], Christina Dilopoulou[*],
Konstantinos Tsampiras[*], Yiorgos Tsiatouhas[*], Daniele Rossi[†]

[*]Department of Computer Science & Engineering, University of Ioannina, Greece
[†]Department of Information Engineering, University of Pisa, Italy
Emails: [*]{tenentes, a.xynos, c.zonios, a.koutra, ch.dilopoulou, cs04508, tsiatouhas}@uoi.gr, [†]daniele.rossi1@unipi.it

Abstract—Internet-of-Things (IoT) applications for the traceability of sensitive products are crucial for the quality assurance along the supply chain. Embedded systems involved in such applications must offer protection against adversaries with administrative privileges, which may exploit their elevated access to violate regulatory compliance. In this paper, we examine Platform Security Architecture (PSA) certified platforms for trusted computing that offer Root-of-Trust (RoT) mechanisms to protect assets against adversaries with administrative privileges. We propose two trusted edge computing IoT gateways that secure the traceability of products during the supply chain distribution based on Cortex-M and Cortex-A ARM processors. The gateways are compared in terms of their provided security services and serviceability as well as their power and computational efficiency. The Cortex-M based gateway exhibits smaller attack surface and higher power efficiency than the Cortex-A; however its software stack is limited. Furthermore, we present recent research outcomes that reduce the attack surface in trusted edge computing platforms, and we point out that FPGA-based platforms based on an open microprocessor, such as the RISC-V, could be an alternative direction for secure gateways that combine the necessary computational power with a reduced attack surface.

Index Terms—supply chain traceability, quality assurance, trusted computing, embedded systems, edge computing

I. INTRODUCTION

The Internet of Things (IoT) and edge computing have significantly advanced traceability systems within the supply chain [1]. These systems enable product and process identification and enhance security, assure quality, and improve overall efficiency [2]. In this direction, the European Commission introduced relevant legislation that mandates traceability procedures in food [3] and pharmaceuticals [4] sectors. These regulations extend to the cold supply chain in which maintaining specific temperature conditions of products along the supply chain is crucial to their quality.

Traceability systems are designed to track and record the flow of goods and information across various levels. They consist of various subsystems including embedded platforms that provide the necessary tools to develop, manage, and ensure the reliable and secure operation of a traceability system [5]. These platforms are responsible for for data collection, processing, and decision-making along the supply chain, as

This research has been co-financed by the European Regional Development Fund of the European Union and Greek national funds through the Operational Program Competitiveness, Entrepreneurship and Innovation, under the call RESEARCH – CREATE – INNOVATE (project code:T2EDK–02836).

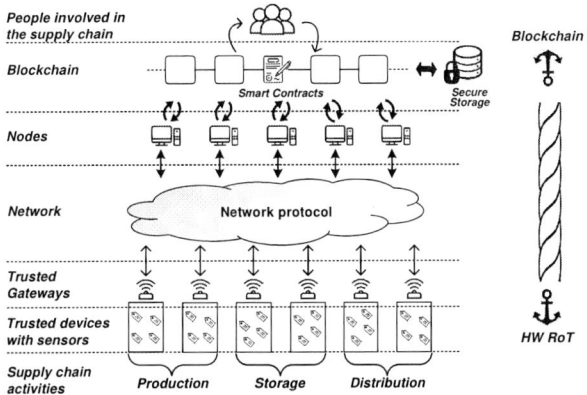

Fig. 1. Blockchain-based with hardware RoT traceability system.

shown in Fig. 1. At the topmost level, people are engaged in the supply chain through nodes, such as computers and servers, which store collected information through the network infrastructure. Blockchain-based systems at these higher layers have emerged for preventing tampering with stored data. For instance, in the domain of the Internet of Medical Things (IoMT) [6], they can safeguard against information disclosure and tampering, ensuring the integrity and confidentiality of sensitive medical data. Similarly, in the Agri-food supply chain domain, these systems ensure transparency and immutability of the stored transaction records [7]. However, at the lower layers embedded platforms reside, which are usually gateways that collect data from devices equipped with sensors and transmit them to the higher levels. These gateways and devices are vulnerable to adversaries with administrative privileges. Such adversaries may exploit their elevated access to compromise data on gateways and sensors, and potentially harm process integrity and violate regulatory compliance. To address this security challenge, a promising approach is to leverage Trusted Execution Environments (TEEs) [8]. TEEs rely on hardware Root-of-Trust (RoT) circuitry, such as immutable storage and Physical Unclonable Functions (PUFs), that verifies the integrity and authenticity of the TEE; and trusted firmware, which can be utilized for the development of security services to safeguard sensitive data and processes from threats and attacks, including those posed by adversaries with administrative privileges.

The Platform Security Architecture (PSA) certified [9] is a security certificate for trusted computing platforms that offers guidelines to vendors for designing RoT mechanisms for IoT platforms in order to support trusted computing. Furthermore, it provides reference implementation of security services to software developers.

In this paper, we examine Platform Security Architecture (PSA) certified platforms for trusted computing that offer Root-of-Trust (RoT) mechanisms to protect assets against adversaries with administrative privileges. We propose two trusted edge computing IoT gateways that secure the traceability of products during the supply chain distribution based on Cortex-M and Cortex-A ARM processors. The gateways are compared in terms of their provided security services and serviceability as well as their power and computational efficiency. The Cortex-M based gateway exhibits smaller attack surface and higher power efficiency than the Cortex-A; however its software stack is limited. Furthermore, we present recent research outcomes that reduce the attack surface in trusted edge computing platforms, and we point out that FPGA-based platforms based on an open microprocessor, such as the RISC-V, could be an alternative direction for trusted edge computing gateways with custom embedded security services and a reduced attack surface.

The remainder of the paper is organized as follows: In Section II, we present hardware platforms vendors and trusted firmware that support trusted computing, along with an exploration of the reference security services that are provided for these platforms. In Section III, we compare the two gateways and we present the implemented gateways. In Section IV, we present recent research works that enhance the security of the examined systems. In Section V we conclude the paper.

II. Background

A. The Platform Security Architecture Certificate

Platforms security certificates, such as the PSA Certified [9], the GlobalPlatform [10], and the Common Criteria [11], exist for various types of platforms. We focus on the PSA Certified certificate, which targets platforms for IoT systems [12].

Introduced by Arm Holdings in 2017 [13], PSA Certified emerges as a comprehensive security certification scheme tailored for IoT hardware, software, and devices. Offering a wide range of specifications, including threat models, security analyses, hardware and firmware architecture, and an open-source firmware reference implementation [14], PSA Certified stands as a fundamental security component for both software and device manufacturers [15], as shown in Table I.

Three assessment levels of PSA Certified with progressively increasing security assurances [13] exist:

Level 1: it ensures IoT compliance to baseline security requirements for chip platforms. This includes non-universal default passwords, tools for vulnerabilities report, and security services, such as secure storage, firmware protection, secure firmware update and software integrity.

TABLE I
CONSORTIUM OF CHIP MANUFACTURERS AND SYSTEM SOFTWARE
PROVIDERS THAT HAVE ADOPTED PSA [16]

Company	Level	Sector	Company	Level	Sector
Cypress Semic.	2	Chip Mfr.	Renesas Elec.	2	Chip Mfr.
Infineon	2	Chip Mfr.	Silicon Labs	3	Chip Mfr.
Microchip Tech.	1	Chip Mfr.	Goodix	1	Chip Mfr.
Nordic Semic.	1	Chip Mfr.	STMicro	3	Chip Mfr.
Nuvoton	1	Chip Mfr.	Unisoc	1	Chip Mfr.
NXP Semic.	2	Chip Mfr.	Winbond	2	Chip Mfr.
NXM Labs	1	SW platform	Veridify	1	SW platform
Express Logic	1	SW platform	RT-Thread	1	SW platform
FreeRTOS	1	SW platform	Sequitur Labs	1	SW platform
OneOS	1	SW platform	Zephyr OS	1	SW platform

Level 2: it offers protection against various risks like remote attacks, rogue device impersonation, and abuse of firmware, update and secure storage. It supports additional security services, such as secure initialization, cryptographic algorithms, software isolation and remote attestation.

Level 3: it offers protection against hardware and data manipulation, undetected memory content alteration, side-channel attacks and physical probing on the chip's surface.

B. Trusted firmware and security services

The Trusted Firmware [17] provides a reference implementation of secure software for processors that implement Arm's A-Cortex and M-Cortex architectures, and it is the foundation for applications running on TEEs. Some of the most well-known reference implementations of trusted firmware are:

For Cortex-M: Trusted Firmware-M (TF-M) [18] consists of modules for secure booting, cryptographic services, secure storage, and isolation between secure and non-secure worlds

For Cortex-A: Trusted Firmware-A (TF-A) consists of modules for secure booting, runtime services for TEEs, and activity monitoring [19] and is compliant to Armv7-A and Armv8-A architectures. Open Portable Trusted Execution Environment (OP-TEE) [20] is a TEE secure kernel that allows the development and integration of secure services and applications. Note that trusted Firmware Reference Monitor Module (TF-RMM) [21] offers additional compartmentalization, however it is available only at Armv9 architectures, which are not suitable yet for IoT applications due to their cost. Reference security services that are available in trusted firmware include:

Cryptographic hardware acceleration (S_1): hardware accelerators that execute fast cryptographic operations.

True random number generators (S_2): provide a foundation for secure key generation and cryptographic processes, ensuring unpredictability and robustness.

Cloning protection and authentication based on Physical Unclonable Functions (PUFs) (S_3): provides an extra layer of security by relying on the inherent variability of PUF responses to authenticate the system and guard against unauthorized access and cloning attempts.

979-8-3503-8368-3/23 $31.00 © 2023 IEEE

Secure and trusted storage (S_4): safeguards data storage, protecting sensitive information and encryption keys.

Secure booting (S_5): guarantees the system integrity during booting, preventing tampering or unauthorized modifications.

Remote attestation (S_6): allows systems to prove their hardware and firmware integrity to remote parties, confirming configuration, security measures, and standards compliance.

Protection against unauthorized access to test ports (S_7): services that verify system integrity by monitoring for authorized access to critical control/testing ports, such as the JTAG.

Resiliency to remote attacks (S_8): establishes a defense mechanism against attacks originating from remote sources.

Resiliency to physical attacks (S_9): ensures robust protection against physical tampering or intrusions aimed at compromising the physical components of the system.

III. PROPOSED TRUSTED EMBEDDED SYSTEMS FOR THE TRACEABILITY DURING SUPPLY CHAIN DISTRIBUTION

The embedded systems involved for the traceability of the supply chain during distribution is depicted in Fig. 2. The devices depicted are sensors for monitoring essential product parameters located to both containers and the truck, and a gateway for collecting the data from the sensors and transmit them towards higher system layers.

A. IoT sensors and Network in the cold supply chain

In our systems, we focus on the case of cold supply chain, which represents the series of actions and equipment applied to maintain a product within a specified low-temperature range from harvest and production to consumption [22]. Table II presents the devices utilized for sensors in this context. Active beacons are responsible for gathering information for temperature, humidity and barometric pressure and typically employ wireless protocols such as BLE, NFC, and GSM. We examined the Ela BLUE PUCK RHT [23], Blukki [24] and Pysense active beacons [25]. We also used a GPS Pmod sensor [26]. Notably, all examined sensors maintain low power consumption, minimal cost, and compact sizes, while delivering high levels of autonomy.

For networking between the gateways and the higher layers of the system, Message Queuing Telemetry Transport (MQTT) [27] is used, shown in Fig. 2. MQTT is an established publish/subscribe messaging transport system, facilitating data exchange across various industries, including automotive, construction, telecommunications, oil and gas.

B. Platforms for Trusted Computing at the Gateway

We have examined the platforms shown in Table III based on the Cortex-M and Cortex-A processors. For Cortex-A, we examine the V2M-Juno r2, which is a development platform for evaluating secure embedded applications on a Cortex-A processor with Armv8 ISA. This system is similar to mobile phones, and it allows the development and evaluation of security countermeasures. We also consider the Raspberry Pi, which is a popular solution for embedded systems. However, it requires an external co-processor unit, such as the

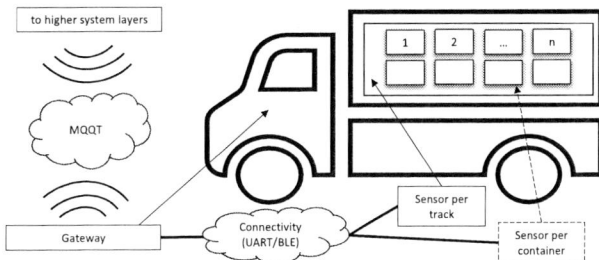

Fig. 2. Embedded systems for the traceability during distribution.

TABLE II
ACTIVE BEACONS AND GPS SENSOR

Type	Platform	Connectivity	Pwr. Cons.	Autonomy	Size	Cost
BLE beacon	Ela BUE PUCK RHT	NFC, BLE	Low	High	1	Low
	Blukii	BLE	Low	High	2	Low
GPS module	Pmod	UART/pmod	Low	High	1	Low
Esp32-based BLE beacon	Pycom/Pysense	BLE, GSM, UART	Low	High	2	Low

TABLE III
PLATFORMS EXAMINED FOR TRUSTED EDGE COMPUTING

CPU Type	Platform	Conn.	Pwr.	Autonomy	Size	Cost	LVL
Cortex-A ARM	V2M-Juno r2	GPIO, UART	High	Low	5	High	-
	Raspberry Pi	BLE, WiFi	High	Low	3	Medium	-
Cortex-M ARM	**STM32L 562E-DK**	BLE	Medium	Medium	2	Low	3
	LPC55S 69-EVK	GPIO, UART	Medium	Medium	3	Low	2
	Hani-IoT	BLE	Medium	Medium	2	Low	1
Cortex-A ARM + FPGA	SK-KR26 0-G	GPIO, UART	High	Low	4	High	2
Infineon TPM	**SLB96 70VQ2.0**	SPI	Rpi addon for RoT capabilities				2

Infineon Trusted Platform Module (TPM) [28], to provide hardware RoT capabilities. For Cortex-M, we examine the STM32L562E, the LPC55S69, and the Hani-IoT platforms, which are based on microcontrollers and incorporate the new secure hardware extensions of Arm's TrustZone-M. Furthermore, we examine Cortex-A platforms with FPGAs, such as the SK-KR260-G, which supports Cortex-A with TEE with an FPGA that is suitable for developing customized hardware solutions. The board has an embedded Infineon TPM peripheral for RoT capabilities. We note that Cortex-M based platforms execute only real-time or bare-metal software, and are known for their moderate power consumption and low cost. However, software development for extending the capabilities of such platforms can be more expensive and requires specialized knowledge in embedded systems. In contrast, a Cortex-A based platform is capable of running operating systems, such as Linux and Android, and can be easier for the software development and maintenance of complex security services.

979-8-3503-8368-3/23 $31.00 © 2023 IEEE

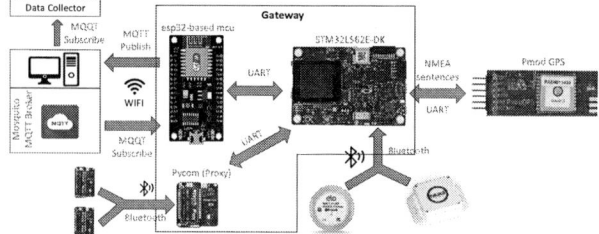

Fig. 3. Cortex-M based trusted gateway for the traceability during distribution.

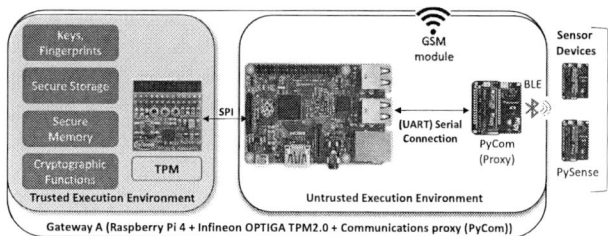

Fig. 4. Cortex-A based trusted gateway for the traceability during distribution.

However, these devices along with FPGA-based platforms fall within the medium to high cost range, and consume more power resulting in reduced system autonomy.

C. Proposed Trusted Gateways

We implement two gateways for our traceability system: one on a Cortex-M and one a Cortex-A based platform, shown in Figs. 3 and 4, respectively. For the Cortex-M, we used the STM32L562E-DK platform (Level 3 PSA-certified), and for the Cortex-A, we used the Raspberry Pi connected to the Infineon TPM via SPI (Level 2 PSA-certified) for RoT.

Cortex-M based system: We implement the system using the STM32CUBE IDE software for programming and setting up the device. A secure software project includes two directories: Secure (S), which contains minimal secure boot and security state switching logic, and Non Secure (NS), which contains the application logic. We utilize the BlueNRG-MS library in the NS source code in order to connect and interact with the BLE sensors. In order to implement secure cryptographic functions, we utilize the Mbed-TLS library in the S source code, and create entry points for specific functions that can be called from NS code. Specifically, we implement three different security operations using the RoT functionalities from the NS world. RSA key pairs are used for singing and verification. Integrity of collected data is assured using signing. Verification is performed on the higher layers. The gateway upon booting generates an RSA key pair, and exports the public key to be sent to the broker for future verification. Then, it reads the sensor data periodically, and signs them with its private key. Data are transmitted using on transit AES encryption to the higher system layers through a proxy esp32 based device over WiFi (or LTE). For this implementation, the MQTT communication infrastructure is supported by the NodeMCU V3, an ESP8266-based WiFi-enabled microcontroller, which handles data transmission. Additionally, Node-RED, installed on a computer, is used to establish virtual publishers and subscribers for MQTT communication for demonstration purposes. The MQTT broker, Mosquitto [29], is employed, and a Python code with the required libraries handles the MQTT subscriber role and message decryption.

Cortex-A based system: We implement the gateway functionality on a Raspberry Pi® 4 Model B platform (RPi). We integrate the Infineon OPTIGA™ TPM 2.0 SLI 9670 evaluation board for the RPi as a TPM. The TPM is responsible for secure key generation, storage and encryption. It ensures that each device has unique read-only authentication and attestation keys. The system runs Raspberry Pi OS, using Debian Bullseye 11, with secure booting implemented using U-Boot [30]. We use the TPM Software Stack (TSS) [31] and the TPM tools [32] provided on GitHub. The U-Boot bootloader utilizes the TPM for hashing the image that is about to boot and verifying that the image has not changed. The entire gateway image is hashed and signed using the hardware RoT keys that are only accessible within the TPM. This also assures hardware authenticity verification. Linux build tools, packages and software libraries are used, such as the TSS library. Attestation of this gateway is implemented by following the Infineon Remote Attestation application note [33]. It receives a challenge by the verifier, which is signed together with the software and hardware versions as well as hardware keys only accessible within the TPM. The signed attestation report is then transmitted back to the verifier. Once verified, the gateway is given the Trusted status.

IV. ATTACK SURFACE EVALUATION AND OPTIMIZATION

A. Evaluation and limitations of the reference systems

In Table IV, the devices used in the developed embedded systems are depicted. Each column corresponds to a security service from the services presented in Paragraph II-B. An "A" indicates the availability of a security service for a specific platform, an "NA" the non-availability and a "UC" the availability under conditions. The serviceability of systems labeled as "UC", their serviceability may also be less optimal as they mandate extra custom hardware and software.

The Cortex-M based gateway (STM32L562E-DK) encompasses every available security service. In contrast, the Cortex-A based gateway (Raspberry Pi) lacks protection against unauthorized access to test ports, as well as resiliency to remote and physical attacks. The main reason is that the TPM is an SPI co-processor and its software stack is limited. Beacon data need to pass through the Rpi, which remains unprotected. This highlights the need for integrated unauthorized access mechanisms to test ports. Moreover, the necessity to address vulnerabilities, like potential information leakage through side channels due to the absence of certification for resilience to remote and physical attacks, emerges as a critical concern. Despite these vulnerabilities, the Cortex-A gateway supports cloning protection and authentication based on PUFs, secure and trusted storage, secure booting, and remote attestation.

979-8-3503-8368-3/23 $31.00 © 2023 IEEE   194

TABLE IV
SECURITY SERVICES AVAILABLE IN PLATFORMS AND DEVICES OF THE
TRACEABILITY SYSTEM DURING DISTRIBUTION

Device	S_1	S_2	S_3	S_4	S_5	S_6	S_7	S_8	S_9
Cortex-A based Raspberry Pi	A	A	UC	UC	UC	UC	NA	NA	NA
Cortex-M based STM32L562E-DK	A	A	A	A	A	A	A	A	A
Devices with Sensors	UC	UC	UC	NA	NA	NA	NA	NA	NA

Devices with sensors are limited to providing cryptographic hardware acceleration, true random number generators, and cloning protection and authentication based on PUFs only under specific conditions. To this end, low cost, strong and cost-effective PUF designs that can be integrated in microcontrollers are required.

B. New security services to reduce the attack surface

The Artemis system architecture [34]–[36] is presented in Fig. 5. Artemis is a blockchain-based traceability system based on embedded trusted edge computing gateways with hardware RoT capabilities. Artemis employs a hardware-assisted blockchain that runs at the higher layers for authenticating the gateways, and another one that runs on the trusted gateways that offers two new services, described in the following.

Decentralized continuous attestation of gateways (S_{10}): This service uses the distributed ledger between the gateways and a smart contract among the trusted gateways. Through the contract, the blockchain network verifies periodically the gateways without the need for a centralized attestation verifier.

Continuous authentication of sensors based on PUFs (S_{11}): allows gateways to continuously authenticate connected devices and to detect connectivity changes in real time. A software implementation of a weak SRAM PUF [37] is currently deployed to existing active beacons and a novel low cost strong in-memory computing SRAM PUF design has been proposed for future low cost devices [38].

In Table V, the security services and the assets involved in the two implemented systems are depicted. Each row shows a security service from those outlined in Paragraph II-B, together with the implemented security services S_{10} and S_{11}. A "P" indicates that an asset is protected by a specific security service, "NP" signifies non-protection, and "UC" denotes protection under specific conditions. The designation "D" indicates protection by new hardware designs that have been proposed, while "I" indicates protection by software implementation on existing designs. In the Cortex-M based system, the gateway hardware and software assets are protected by most security services, however the device sensors are protected only under conditions. The reason is that there are not low cost strong authentication mechanisms for low cost devices that are used for sensors. For this reason, a low cost strong in-memory computing SRAM PUF [38], [39] that is suitable for low cost devices has been designed, and methods to strengthen the reliability of in-memory computing SRAMs have also been proposed [40], [41]. In the Cortex-A based system, the device

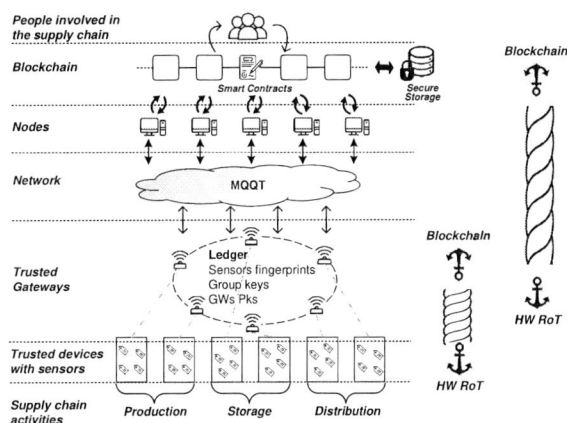

Fig. 5. Artemis system architecture.

TABLE V
SECURITY SERVICES AND PROTECTED ASSETS OF THE GATEWAYS

Security Services	Cortex-M based STM32L562E-DK (LVL 3 PSA)			Cortex-A based Raspberry Pi + PTM (LVL 2 PSA)			
	Protected Assets			Protected Assets			
	Device Sensors	Gateway HW	Gateway SW	Device Sensors	Gateway HW	Gateway SW	Ledger (Keys and FPs)
S_1 (D)	UC+D	P	P	UC+D	P	P	P
S_2	UC+D	P	P	UC+D	P	P	P
S_3 (I+D)	UC+I+D	P	P	UC+I+D	P	P	P
S_4	NP	P	P	NP	P	P	P
S_5	NP	P	P	NP	P	P	P
S_6	NP	P	P	NP	P	P	P
S_7 (D)	NP+D	P	P	NP+D	UC+D	UC+D	UC+D
S_8 (RES)	NP	P	P	NP	RES	RES	UC
S_9 (RES)	NP	P	P	NP	RES	RES	UC
S_{10} (I)	Not implemented			NP	P+I	P+I	P+I
S_{11} (I)				P+I	P+I	P+I	P+I

sensors are further protected by the new S_{11} service, which allows continuous authentication based on implemented on software of a weak SRAM PUF [37]. Moreover, to strengthen all devices defenses against unauthorized test port access and to further enhance cryptographic hardware acceleration, an energy-efficient, high-throughput SHA256 core has been proposed [42], [43]. Furthermore, Resiliency Evaluation Software (RES) for Cortex-A based cryptosystems against power side-channel attacks has been proposed [44], [45].

V. CONCLUSION

In this paper, we examined Platform Security Architecture (PSA) certified platforms for trusted computing that offer Root-of-Trust (RoT) mechanisms to protect assets against adversaries with administrative privileges. We proposed two trusted edge computing IoT gateways that secure the traceability of products during the supply chain distribution based on

979-8-3503-8368-3/23 $31.00 © 2023 IEEE

Cortex-M and Cortex-A ARM processors. The gateways are compared in terms of their provided security services and serviceability as well as their power and computational efficiency. The Cortex-M based gateway exhibits smaller attack surface and higher power efficiency than the Cortex-A; however its software stack is limited. Furthermore, we presented recent research outcomes that reduce the attack surface in trusted edge computing platforms, and we pointed out that FPGA-based platforms based on an open microprocessor, such as the RISC-V, could be an alternative direction for reducing the attack surface of secure gateways towards custom security services on custom hardware. However, it should be noted that FPGA-based platforms consume more power compared to microcontrollers, therefore they might only be suitable for integration environments with no strict power constraints.

REFERENCES

[1] M. Jaberidoost, S. Nikfar, A. Abdollahiasl, and R. Dinarvand, "Pharmaceutical supply chain risks: a systematic review," *DARU Journal of Pharmaceutical Sciences*, vol. 21, no. 1, p. 69, Dec. 2013.

[2] T. Pizzuti, G. Mirabelli, G. Grasso, and G. Paldino, "MESCO (MEat Supply Chain Ontology): An ontology for supporting traceability in the meat supply chain," *Food Control*, vol. 72, pp. 123–133, Feb. 2017.

[3] S. Sarpong, "Traceability and supply chain complexity: Confronting the issues and concerns," *European Business Review*, vol. 26, May 2014.

[4] E. Commission, "COMMISSION DELEGATED REGULATION (EU) 2016/ 161," *Official Journal of the European Union*, 2016.

[5] "Specification for security management systems for the supply chain, Standard ISO 28000:2007," 2007.

[6] T. Yaqoob, H. Abbas, and M. Atiquzzaman, "Security Vulnerabilities, Attacks, Countermeasures, and Regulations of Networked Medical Devices—A Review," *IEEE Commun. Surv. Tutorials*, vol. 21, no. 4, pp. 3723–3768, 2019.

[7] M. P. Caro, M. S. Ali, M. Vecchio, and R. Giaffreda, "Blockchain-based traceability in Agri-Food supply chain management," in *2018 IoT Vertical and Topical Summit on Agriculture*, May 2018, pp. 1–4.

[8] M. Sabt, M. Achemlal, and A. Bouabdallah, "Trusted Execution Environment: What It is, and What It is Not," in *2015 IEEE Trustcom/BigDataSE/ISPA*, vol. 1, Dec. 2015, pp. 57–64.

[9] "PSA Certified: IoT Security Framework and Certification." [Online]. Available: https://www.psacertified.org/

[10] "Global Platform." [Online]. Available: https://globalplatform.org/

[11] "About The Common Criteria : New CC Portal." [Online]. Available: https://www.commoncriteriaportal.org/ccra/index.cfm

[12] "Arm unveils security certification testing for IoT devices," Feb. 2019. [Online]. Available: https://venturebeat.com/business/arm-unveils-security-certification-testing-for-iot-devices/

[13] "Arm partners with testing labs to provide IOT security certification." [Online]. Available: https://www.zdnet.com/article/arm-partners-with-testing-labs-to-provide-iot-security-certification/

[14] O. Media, "Momentum Builds for PSA Certified." [Online]. Available: https://embeddedcomputing.com/application/industrial/industrial-iot/momentum-builds-for-psa-certified

[15] "Google and others back Internet of Things security push," Jul. 2019. [Online]. Available: https://www.engadget.com/2017-10-23-google-arm-internet-of-things-security.html

[16] "PSA Certified," Jul. 2023. [Online]. Available: https://en.wikipedia.org/w/index.php?title=PSA_Certified&oldid=1163426874

[17] "Trusted Firmware - Open Source Secure Software." [Online]. Available: https://www.trustedfirmware.org/

[18] "Trusted Firmware-M v1.8.1 documentation." [Online]. Available: https://tf-m-user-guide.trustedfirmware.org/introduction/readme.html

[19] "Trusted Firmware-A documentation." [Online]. Available: https://trustedfirmware-a.readthedocs.io/en/v2.8/index.html

[20] "About OP-TEE — OP-TEE documentation documentation." [Online]. Available: https://optee.readthedocs.io/en/latest/general/about.html

[21] "Realm Management Monitor documentation." [Online]. Available: https://tf-rmm.readthedocs.io/en/latest/

[22] "Cold chain," Oct. 2023. [Online]. Available: https://en.wikipedia.org/w/index.php?title=Cold_chain&oldid=1182235564

[23] "Temperature & humidity Bluetooth sensor - Blue PUCK RHT." [Online]. Available: https://elainnovation.com/en/catalogue/blue-puck-rht-en/

[24] "Sensor Beacon • Hardware • blukii," Feb. 2019. [Online]. Available: https://www.blukii.com/en/hardware/sensor-beacon/

[25] "Pysense." [Online]. Available: https://docs.pycom.io/datasheets/expansionboards/pysense/

[26] "Pmod HYGRO: Digital Humidity and Temperature Sensor - Digilent." [Online]. Available: https://digilent.com/shop/pmod-hygro-digital-humidity-and-temperature-sensor/

[27] "MQTT - The Standard for IoT Messaging." [Online]. Available: https://mqtt.org/

[28] "OPTIGA™ TPM - Trusted Platform Module - Infineon Technologies." [Online]. Available: https://www.infineon.com/cms/en/product/security-smart-card-solutions/optiga-embedded-security-solutions/optiga-tpm/

[29] "Eclipse Mosquitto," Jan. 2018. [Online]. Available: https://mosquitto.org/

[30] "u-boot/u-boot," Nov. 2023, original-date: 2014-11-12T13:29:02Z. [Online]. Available: https://github.com/u-boot/u-boot

[31] "Overview," Nov. 2023, original-date: 2015-06-30T16:21:57Z. [Online]. Available: https://github.com/tpm2-software/tpm2-tss

[32] "tpm2-tools," Nov. 2023, original-date: 2015-08-21T01:07:28Z. [Online]. Available: https://github.com/tpm2-software/tpm2-tools

[33] Infineon, "Setup and operation guide for optiga tpm backed remote attestation on a linux system," https://github.com/Infineon/remote-attestation-optiga-tpm, 2020.

[34] "Artemis project." [Online]. Available: https://artemis.cs.unipi.gr/en/

[35] V. Malamas, T. K. Dasaklis, T. G. Voutsinas, and P. Kotzanikolaou, "Blockchain service layer for erp data interoperability among multiple supply chain stakeholders," in *2023 9th International Conference on Control, Decision and Information Technologies (CoDIT)*. IEEE, 2023, pp. 145–150.

[36] D. Koutras, V. Malamas, P. Kotzanikolaou, and T. Dasaklis, "A risk assessment methodology for supply chain tracking services," in *2023 International Conference On Cyber Management And Engineering (CyMaEn)*. IEEE, 2023, pp. 555–559.

[37] T. Ashur *et al.*, "A Privacy-Preserving Device Tracking System Using a Low-Power Wide-Area Network," in *Cryptology and Network Security*, S. Capkun and S. S. M. Chow, Eds. Cham: Springer International Publishing, 2018, vol. 11261, pp. 347–369.

[38] A. Xynos, V. Tenentes, and Y. Tsiatouhas, "SiCBit-PUF: Strong in-Cache Bitflip PUF Computation for Trusted SoCs," in *2023 IEEE European Test Symposium (ETS)*. Venezia, Italy: IEEE, May 2023, pp. 1–6.

[39] A. Xynos and V. Tenentes, "MetaSPICE: Metaprogramming SPICE Framework for the Design Space Exploration of PUF Circuits," in *2023 12th International Conference on Modern Circuits and Systems Technologies (MOCAST)*. Athens, Greece: IEEE, Jun. 2023, pp. 1–4.

[40] C. Dilopoulou and Y. Tsiatouhas, "BTI Aging Influence in SRAM-based In-Memory Computing Schemes and its Mitigation," in *2023 30th International Conference on Electronics, Circuits and Systems (ICECS)*, Instabul, Turkey, 2023, pp. 1–5.

[41] ——, "BTI Aging Influence and Mitigation in Neural Networks Oriented In-Memory Computing SRAMs," in *2023 12th International Conference on Modern Circuits and Systems Technologies (MOCAST)*, Athens, Greece, 2023, pp. 1–4.

[42] A. Koutra and V. Tenentes, "High Throughput and Energy Efficient SHA-2 ASIC Design for Continuous Integrity Checking Applications," in *2023 IEEE European Test Symposium (ETS)*. Venezia, Italy: IEEE, May 2023, pp. 1–6.

[43] ——, "Multi-Vt based Energy Efficiency Optimization for Double SHA ASIC Designs towards a Sustainable Bitcoin Network," in *2023 30th International Conference on Electronics, Circuits and Systems (ICECS)*, Instabul, Turkey, Dec. 2023, pp. 1–5.

[44] C. Zonios and V. Tenentes, "REVOLVER: A Zero-Step Execution Emulation Framework for Mitigating Power Side-Channel Attacks on ARM64," in *2022 IEEE 28th International Symposium on On-Line Testing and Robust System Design (IOLTS)*, Sep. 2022, pp. 1–6.

[45] K. Nomikos *et al.*, "Evaluation of Hiding-based Countermeasures against Deep Learning Side Channel Attacks with Pre-trained Networks," in *2022 IEEE International Symposium on Defect and Fault Tolerance in VLSI and Nanotechnology Systems (DFT)*, Jul. 2022, pp. 1–6, iSSN: 2765-933X.

Uninterrupted Trust: Continuous Authentication in Blockchain-Enhanced Supply Chains

Vangelis Malamas
Department of Informatics
University of Piraeus
Piraeus, Greece
0000-0001-9238-6796

Dimitris Koutras
Department of Informatics
University of Piraeus
Piraeus, Greece
0000-0002-9154-8340

Panayiotis Kotzanikolaou
Department of Informatics
University of Piraeus
Piraeus, Greece
0000-0002-8771-9020

Abstract—In today's complex global supply chains, ensuring the integrity, authenticity, and security of products as they traverse through multiple stages is of paramount significance. The growing integration of IoT devices for product monitoring throughout the supply chain amplifies the need for enhanced transparency and data security. Currently, supply chain participants operate with relative autonomy, each maintaining proprietary systems for logging product data. Achieving interoperability among these stakeholders often presents numerous challenges, with the trustworthiness of shared data resting largely on the integrity of the data custodian overseeing the sharing and preservation of a portion of the data. Cryptographic methods have been frequently adopted to cater to the granular access requirements of shareable data. However, conventional cryptographic methodologies occasionally fall short in delivering real-time authentication and trust verification. To address this critical gap, blockchain technology has emerged as a promising solution, offering immutable and transparent record-keeping capabilities. While blockchain's integration addresses several modern supply chain system demands, the need to combine it with continuous authentication processes for the integrated IoT devices persists. This paper presents a novel architecture for ensuring trust and security throughout the supply chain process by leveraging both blockchain and Physical Unclonable Functions (PUF) integrated into the IoT devices. Specifically, the proposed architecture deploys a private blockchain —where each stakeholder maintains a node of the network-and PUF-equipped IoT devices for continuous authentication and monitoring.

Index Terms—Blockchain, Continuous authentication, Supply chain, IoT, PUF

I. INTRODUCTION

In the evolving landscape of digital transformation, the intertwining of blockchain technology with supply chain management has emerged as a beacon of trust and transparency. The fusion offers security and traceability, ensuring that every transaction is recorded in an immutable ledger. However, with the increasing complexity and dynamic nature of modern supply chains, arises a paramount need for the development of new continuous authentication mechanisms. Traditional one-time authentication methods can no longer keep pace with the real-time demands and rapidly changing environments these supply chains operate within.

Recent developments in blockchain technology and its integration into supply chain (SC) have gained significant traction [1]–[3]. Specifically, the transparency and traceability provided by blockchain, due to its ability to establish an unchangeable transaction record, enhance the tracking of product within the SC. By leveraging cryptographic methods, blockchain guarantees that the data remains unmodified, ensuring heightened security for critical SC details. By cutting out middlemen and promoting instantaneous data exchanges, blockchain minimizes the resources needed for information transfer in the SC while addressing the need for trust in this competitive environment. Furthermore, the integration of Smart Contracts, which act as a predifined set of rules agreed among the participants can streamline many processes, leading to efficiency gains and cost savings.

In addition, the need to link authentication mechanisms to the hardware and specifically to the IoT devices used in supply chain systems has led to the development of Physical Unclonable Functions (PUF). A PUF leverages the distinctive attributes of a device, originating from variations in its manufacturing process, to create fingerprints and keys. PUFs are employed to create protocols for authentication that are supported by hardware. In such protocols, an entity such as a gateway, possessing a fingerprint that matches the device's physical traits, can challenge the device to authenticate itself. This is done by using those traits to produce a legitimate response.

There are two primary types of Physical Unclonable Functions (PUFs). The first type, known as Strong PUFs, are more expensive to implement and can be utilized for both authenticating devices and generating cryptographic keys. However, Strong PUFs might not be feasible for Internet of Things (IoT) devices across all areas of application. In situations where Strong PUFs are impractical, Weak PUFs can be employed as an alternative. Weak PUFs are less costly to deploy and consume less energy, but they offer a comparatively lower level of security. Nevertheless, by integrating Weak PUFs with other random elements, like shared secret keys, they can also facilitate robust device authentication backed by hardware security assurances. This means that any attempts to clone or spoof would necessitate physically tampering with the device [4]–[6].

Although authentication supported by hardware can safeguard against remote cloning and spoofing attacks, several unresolved issues remain, particularly in environments with multiple authorities and domains such as SC. IoT devices

979-8-3503-8368-3/23 $31.00 © 2023 IEEE

typically authenticate with gateways intermittently, usually at login, creating an opportunity for attackers with physical access to manipulate a device. An attacker might execute Sybil attacks, wherein they replicate a compromised device and use the duplicates to simultaneously connect to several gateways. Detecting such attacks in real-time can be challenging, especially in distributed settings where gateways might be part of different network domains or under separate authorities.

A. Motivation and contribution

In these highly complex supply chain ecosystems, maintaining the integrity, authenticity, and security of products is more than critical [7]. The main reason is the increasing integration of IoT devices for product monitoring has heightened the demand for improved transparency and data security.

Currently, supply chain stakeholders often operate with relative autonomy, maintaining separate systems for recording product data. Achieving interoperability among these stakeholders poses challenges, and the trustworthiness of shared data relies heavily on the integrity of the data custodian responsible for managing and preserving a portion of the data. While cryptographic methods have been employed to address data access requirements and security concerns, they sometimes fall short in providing real-time authentication and trust verification. This limitation becomes especially apparent as supply chains grow in complexity.

IoT devices facilitate real-time asset tracking, monitoring, and alerting in supply chain management, thereby streamlining processes and minimizing disruptions. Sensors attached to products or containers might interact with various gateways overseen by distinct authorities, such as vehicles, warehouses, or points of sale, to relay tracking and monitoring information. This makes them susceptible to a variety of cyberattacks. The significance of maintaining the integrity and confidentiality of the data exchanged between sensors and gateways can be crucial, depending on the nature of the supply chain, to ensure the quality and safety of assets. Therefore, it is necessary to continuously authenticate IoT devices and ensure secure communication between gateways and these devices, despite frequent connectivity changes.

Current mechanisms for continuous authentication are designed for application settings where both the devices and the gateways are under the same authority. This setup simplifies aspects such as trust management, real-time inventory tracking, and centralized detection of Sybil attacks. However, in the more complex environment of SC, which involves multiple authorities—where devices might connect to gateways managed by different authorities at various times—there is a need for continuous authentication architecture that ensures: (a) distributed trust management, allowing interoperability among the stakeholders for devices and data managed by different authorities and (b) synchronization and real-time inventory tracking to prevent attacks across multiple network domains.

To tackle these difficulties, we propose an innovative continuous authentication architecture that combines blockchain technology and Physical Unclonable Functions (PUF) integrated in IoT devices. Within this architectural framework, every stakeholder involved in the supply chain is responsible for maintaining a node on a private blockchain network. This network serves as a repository for all the essential data required for the distributed continuous authentication mechanism. The integration of PUF-equipped IoT devices with the blockchain network enhances the supply chain data's integrity and trustworthiness by enabling continuous authentication and monitoring capabilities. We propose a novel PUF-assisted continuous authentication architecture which enables gateways to authenticate IoT devices and identify changes in connectivity status in real time. The protocol is designed to be independent of the specific PUF used, which can range in security robustness (either weak or strong PUF).

B. Paper structure

The rest of the paper is structured as follows. Section II provides an overview of the literature on existing hardware-assisted and continuous authentication mechanisms applicable to ecosystems sharing characteristics with the supply chain. Section III describes our proposed blockchain-based continuous authentication architecture for the supply chain. In Section IV, we delve into the security and privacy aspects of the proposed architecture, while Section V summarizes the key findings and explores avenues for future research.

II. LITERATURE REVIEW

Although the current literature is limited, the recent years have witnessed a growing interest in integrating blockchain technology into the supply chain ecosystem. Numerous scholarly articles have highlighted the potential of blockchain technology to efficiently manage information flow among supply chain participants, particularly in the context of globally interconnected networks.

Various hardware-assisted IoT authentication protocols have been suggested in existing literature, such as those referenced in [8]–[10]). However, these protocols are not designed for continuous authentication, and as a result, they are unable to provide real-time inventory management and protection against Sybil attacks in environments with multiple authorities. An intriguing method is put forth by [11], where chaotic transmission is employed over PUF. Another noteworthy approach is presented by [12], which utilizes Masked Authentication Messaging (MAM) in Restricted mode to securely store the keys of Internet of Medical Things (IoMT) devices, derived from PUF, in the IOTA Tangle. This technique ensures the integrity of the devices, and its efficacy is confirmed using PUFchain 3.0, showcasing swift and reliable authentication.

Several studies have highlighted the importance of continuous authentication for IoT devices, as these devices are often the the weakest link in a security system and are commonly exploited by cyber attackers. A protocol for periodic continuous IoT authentication is introduced by [13], where an authentication token is generated during the static authentication phase and subsequently used in the continuous authentication phase. However, this protocol is not equipped to counter Sybil attacks. [14] put forth a continuous authentication protocol leveraging Channel State Information, although this protocol focuses solely on ID-based attacks and comes with a significant computational overhead. Another study by [15] presents a context-aware continuous authentication model, which relies on the consistent ambient context information perceived by devices in the same location for continuous authentication. This model assumes the reliability of sensors used for context inference. Moreover, each of the mentioned solutions assumes a high level of trust in third-party servers involved in the authentication process. [16] outline a three-layer authentication model (comprising End devices layer, Edge layer, and Cloud layer) for device-to-device authentication. Additionally, [17] propose a lightweight continuous authentication protocol for medical Wireless Body Area Networks (WBAN), utilizing physiological signals to reduce energy consumption while ensuring authenticity.

Other studies aim to minimize trust assumptions by disseminating authentication data among devices through blockchain technology. [18] propose a blockchain-based continuous authentication protocol within a Zero Trust framework that decentralizes trust. However, its vulnerability to network-layer remote spoofing attacks remains, as it doesn't leverage trusted hardware components for robust integrity protection. On the other hand, hardware-assisted solutions limit the potential for exploitation to adversaries with physical access to the device. [19] introduce a protocol centered on PUF for authentication and data integrity that utilizes blockchain to store challenge-response pairs (CRPs), enabling continuous authentication of field sensors in Supervisory Control and Data Acquisition (SCADA) systems. However, this protocol is confined to SCADA systems under a single authority, making it unsuitable for multi-authority environments. [20] propose a pairwise continuous authentication protocol that refreshes the CRPs after each round. [21] present the PUFDCA protocol, which uses a static authentication mechanism followed by continuous authentication based on device location. [22] introduce a blockchain-based authentication system for Industrial Internet of Things (IIoT) that uses PUF for device identification. However, the use of a centralized certificate authority for issuing device certificates hinders decentralization and requires substantial trust. [23] suggest a fog nodes layer to overcome the resource constraints of IoT devices in handling computation-intensive tasks for a group of IoT devices. This protocol establishes an encrypted channel through initial mutual authentication between IoT and fog nodes and integrates blockchain with a Proof of Authority consensus algorithm for continuous authentication.

The authors in [24] introduce a hybrid consortium blockchain leveraging Physical PUF for IIoT authentication. This system allows various devices to generate unique PUFs that constitute the consortium blockchain and accommodates multiple endorsers, potentially including several authorities. However, it falls short in supporting dynamic connectivity changes and continuous authentication. The same study pro-

TABLE I
AUTHENTICATION PROTOCOL COMPARISON

Papers	Hardaware assisted	Blockchain based	Continuous authentication
[13]	✗	✗	✓
[14]	✗	✗	✓
[28]	✗	✗	✓
[15]	✗	✗	✓
[16]	✗	✗	✓
[17]	✗	✗	✓
[24]	✓	✓	✗
[25]	✓	✓	✗
[26]	✓	✓	✗
[4]	✓	✓	✗
[27]	✓	✓	✗
[18]	✗	✓	✓
[19]	✗	✓	✓
[20]	✗	✓	✓
[21]	✗	✓	✓
[22]	✗	✓	✓
[23]	✗	✓	✓
Our protocol	✓	✓	✓

poses an authentication protocol for sensor nodes to authenticate to medical providers using device IDs, timestamps, and PUF-generated challenge-response pairs. Despite its merits, the protocol lacks defenses against Sybil attacks and is designed for single-authority environments, making it unsuitable for multi-authority scenarios. [25] present PUFchain 2.0, a hardware-assisted blockchain that merges PUF technology with blockchain, offering a scalable solution for IoMT devices to authenticate through embedded PUF modules. [26] propose an alternative approach that doesn't rely on PUF. They develop a blockchain-based authentication system incorporating AES, SHA-256, and HMAC on a microcontroller board, resulting in a significant boost in processing speed. [4] introduce PUFchain, a pioneering blockchain that integrates hardware security primitives, specifically PUFs, and features a unique consensus algorithm termed "Proof of PUF-Enabled Authentication" (PoP). Lastly, [27] present a blockchain-based authentication framework for IoT that provides decentralized mutual authentication and robust privacy protection through trusted hardware components.

As shown in Table I, the majority of existing authentication protocols are incapable of supporting the requirements of a complex, multi-authority ecosystem such as supply chain. Most of them are designed for centralized environments in which all devices are administered by a single authority ([13], [14], [18], [19], [21]–[24]). Moreover, some of the existing protocols that can support multi-authority environments with dynamic connectivity changes, such as [15], [20], [29], rely on strong trust assumptions, which can lead to system compromise because they maintain a single point of failure, most commonly a cloud server.

979-8-3503-8368-3/23 $31.00 © 2023 IEEE

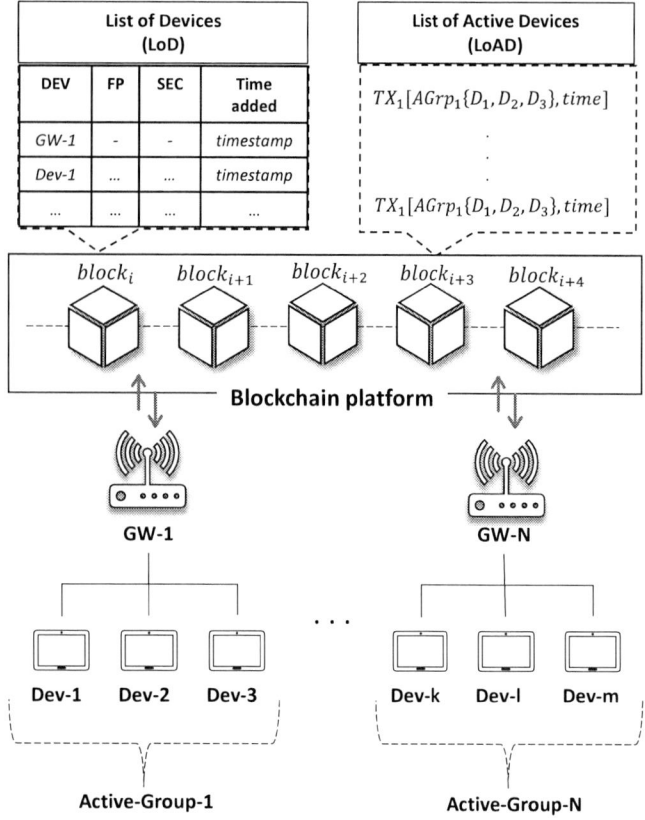

List of Devices (LoD)			
DEV	FP	SEC	Time added
GW-1	-	-	timestamp
Dev-1	...	...	timestamp
...	...	...	...

List of Active Devices (LoAD)
$TX_1[AGrp_1\{D_1, D_2, D_3\}, time]$
$\vdots$
$TX_1[AGrp_1\{D_1, D_2, D_3\}, time]$

Fig. 1. High level architecture

III. A BLOCKCHAIN-BASED CONTINUOUS AUTHENTICATION ARCHITECTURE FOR SUPPLY CHAIN

In this section, we describe the components of the proposed architecture. Initially, we provide a high-level overview of the architecture. Subsequently, we introduce the system actors and elucidate the services the system supports. Finally, we expound on the smart contracts underpinning the aforementioned services.

A. High-level description of the proposed architecture

The goal of the proposed architecture is to ensure the security and trustworthiness of devices and their interactions within a multi-authority network like the supply chain network. From this point of view, the proposed architecture aims to address the following key issues: (a) the continuous authentication of devices that may frequently connect to different gateways; and (b) the real-time inventorying of devices across all gateways, thereby mitigating the risk of compromised nodes and preventing Sybil attacks from cloned devices. An overview of the proposed architecture is depicted in Fig.1.

The architecture operates by initiating a private blockchain network, formed collectively by all trusted gateways. This private blockchain serves as a secure medium for gateways to share and maintain the *List of Devices (LoD)*, a list of recognized gateways and devices. The *LoD* includes details such as unique identifiers and fingerprints of all enrolled IoT

devices. In case where the PUF is insufficient for robust authentication, an additional secret may be exchanged during the device enrollment phase to strengthen security.

Subsequently, another list, the *List of Active Devices (LoAD)*, is also maintained on the blockchain. The *LoAD* allows gateways to consistently record and monitor the devices connected to them across various time intervals. Each gateway periodically broadcasts a new challenge to all devices in range. Devices must successfully respond using their PUF and, if applicable, other secret information shared by the gateway. In doing so, each gateway continuously (re)authenticates its connected devices ensuring that only legitimate devices are authenticated and can participate in the network, thereby safeguarding the supply chain from potential cyber threats.

In addition, each gateway publishes a new blockchain transaction to update the *LoAD* with the list of devices connected at the current time. By continuously checking the *LoAD* before adding a new device to their updated list, gateways have access to real-time device inventories for all other gateways and are therefore protected against Sybil attacks.

B. Initialization phase

During the initialization phase it is presumed that every participating authority, which includes owners of gateways and IoT devices, consents to execute the blockchain protocol and holds an authority public key certificate. The gateways of trusted authorities band together to form a private blockchain network, with each gateway functioning as a node and connecting to others in a peer-to-peer setup. This distributed ledger enables the dynamic sharing and updating of the previously defined Lists of Devices (LoD) and Active Devices (LoAD). Furthermore, the public key certificates of all authorities are incorporated into the blockchain, enabling authorities to mutually verify their validity.

Gateways are registered by trusted administrators who possess access to the authority's private key. Initially, both the LoD and LoAD are initialized as empty lists and added to the genesis block (block-0). A device may be incorporated into the LoD by a trusted gateway, with the objective of establishing a trusted relationship between the gateway and the device. This is achieved by securely exchanging secret information that can later be utilized for the device's authentication.

The steps of the initialization phase unfolds as follows: First, the gateway generates a random value, which is then transmitted to the device to initialize the state of the PUF circuitry, ensuring that the output is both unpredictable and unique. Note that in the case of a weak PUF, the gateway may select an optional secret, which is then securely saved on the device. The aim of the optional secret is to inject sufficient entropy in order to enable a vast number of challenge-response pairs. If not necessary, this step can be omitted or alternatively used as a secondary authentication factor. Subsequently, a fingerprint is derived from the device, serving as a response generator to each presented input. This fingerprint, along with the device ID, is stored in the blockchain to facilitate the generation of the necessary response during the authentication phase.

C. Continuous authentication phase

During this phase, a legitimate gateway with access to the *LoD* periodically broadcast challenges to all active devices within its range, prompting them to authenticate themselves based on their PUF. The responses received by the gateway are subsequently forwarded for verification to the *Continuous Authentication Smart Contract (CASC)* stored on the blockchain. Initially, the smart contract examines the last group published by the specific gateway to identify any changes, such as devices joining or leaving. In the case that the group remains the same no further action is taken, otherwise the smart contract proceeds by retrieving the fingerprint previously stored in the *LoD*, based on the device ID. It then generates the expected response based on the presented challenge and compares this with the response received from the device to ascertain its authenticity. If the comparison yields a match, another function within the smart contract consults the *LoAD* to check if the device is already connected to a different gateway, thereby safeguarding against Sybil attacks from tampered devices . Upon successful verification of all connected devices, the smart contract forms a new group and updates the *LoAD* with a new transaction. A flow of the procedure is depicted in Fig. 2

IV. Security and Privacy

The continuous authentication architecture proposed in this work is meticulously designed to address several security and privacy concerns inherent to the Internet of Things (IoT) ecosystems, particularly those pertinent to supply chain systems. The architecture leverages a private blockchain network formed by trusted gateways, ensuring that the data related to device authentication (registered in the LoD) and active connections (registered in the LoAD) are stored in a secure, tamper-evident, and immutable manner. The use of PUFs in devices adds a layer of security by generating unique responses to challenges, thereby facilitating continuous and robust device authentication. Additionally, the architecture introduces an optional secret exchanged during device enrollment when the PUF is deemed insufficiently strong. This secret is securely stored on the device side and is designed to add sufficient entropy to the exchanged information, ensuring exponentially many challenge-response pairs and enhancing authentication strength. By continuously and periodically challenging devices for authentication, the architecture prevents Sybil attacks and ensures that any tampered or cloned device is swiftly identified and isolated.

The architecture ensures that device identifiers and fingerprints stored on the blockchain are used solely for the purpose of authentication, thereby maintaining user privacy. The private blockchain network ensures that only trusted gateways have access to this data, safeguarding against unauthorized access and potential privacy breaches. By employing smart contracts, the verification process is automated, ensuring that

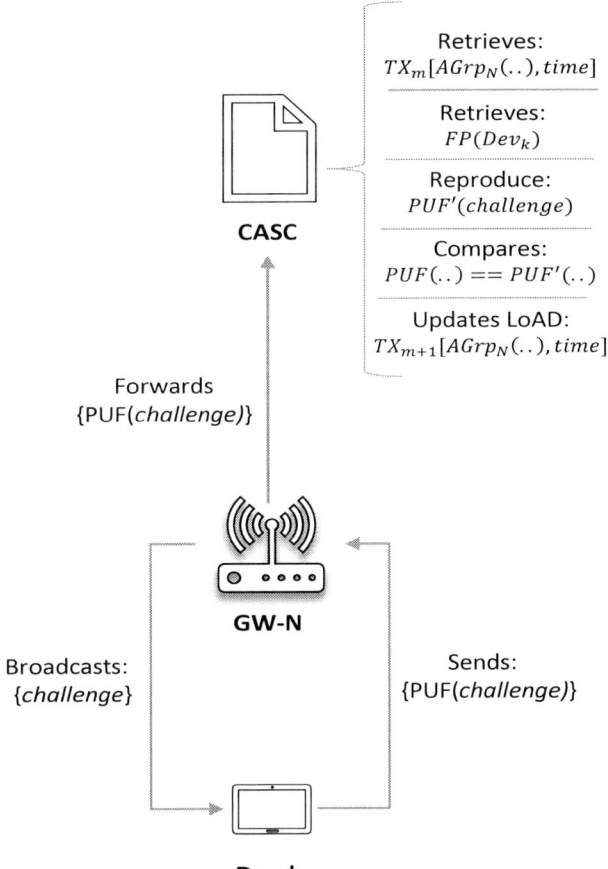

Fig. 2. Device authentication procedure

the data is processed without manual intervention and reducing the risk of privacy violations. Moreover, the smart contract checks the *LoAD* to prevent a device from being connected to multiple gateways simultaneously, ensuring the integrity of device connections while preserving privacy.

V. Conclusions

In conclusion, the proposed continuous authentication architecture provides an innovative solution to enhance security and privacy in supply chain systems. Addressing the critical challenges of authentication and real-time inventorying in environments susceptible to Sybil attacks and clone devices, the architecture ensures the continuous authentication of devices even when they connect to different gateways at different times.

The architecture leverages a private blockchain network constituted by trusted gateways, which facilitates the secure and immutable storage of device data in the *LoD* and *LoAD*. By using PUFs, the architecture ensures that devices generate unique and unpredictable responses to authentication challenges. In addition where PUFs might be inadequate, an optional secret is employed to bolster the authentication process, enhancing the entropy and facilitating a multitude of challenge-response pairs.

Finally, the architecture employs a smart contract, specifically the *CASC*, to automate the verification process, further ensuring security while minimizing the impact of potential human errors. These smart contracts efficiently manage device authentication by comparing expected responses with actual responses and safeguarding against Sybil attacks by checking device connections across gateways.

Prioritizing both security and privacy, the architecture ensures that device data is protected against unauthorized access and tampering while also maintaining user privacy. By adopting this innovative approach, the proposed architecture stands as a resilient solution in addressing the multifaceted challenges associated with secure and continuous authentication in IoT-based supply chain systems.

ACKNOWLEDGMENT

This research has been co-financed by the European Union and Greek national funds through the Operational Program Competitiveness, Entrepreneurship and Innovation, under the call RESEARCH – CREATE – INNOVATE (project code:T2EDK-02836) - ARTEMIS

REFERENCES

[1] S. E. Chang and Y. Chen, "When blockchain meets supply chain: A systematic literature review on current development and potential applications," *IEEE Access*, vol. 8, pp. 62 478–62 494, 2020.

[2] M. Pournader, Y. Shi, S. Seuring, and S. L. Koh, "Blockchain applications in supply chains, transport and logistics: a systematic review of the literature," *International Journal of Production Research*, vol. 58, no. 7, pp. 2063–2081, 2020.

[3] V. Malamas, G. Palaiologos, P. Kotzanikolaou, M. Burmester, and D. Glynos, "Janus: Hierarchical multi-blockchain-based access control (hmbac) for multi-authority and multi-domain environments," *Applied Sciences*, vol. 13, no. 1, p. 566, 2022.

[4] S. P. Mohanty, V. P. Yanambaka, E. Kougianos, and D. Puthal, "Pufchain: A hardware-assisted blockchain for sustainable simultaneous device and data security in the internet of everything (ioe)," *IEEE Consumer Electronics Magazine*, vol. 9, no. 2, pp. 8–16, 2020.

[5] F. Amsaad and S. Köse, "A secure lightweight hardware-assisted charging coordination authentication framework for trusted smart grid energy storage units," *SN Computer Science*, vol. 2, no. 6, pp. 1–15, 2021.

[6] P. Mall, R. Amin, A. K. Das, M. T. Leung, and K.-K. R. Choo, "Puf-based authentication and key agreement protocols for iot, wsns and smart grids: A comprehensive survey," *IEEE Internet of Things Journal*, 2022.

[7] D. Koutras, V. Malamas, P. Kotzanikolaou, and T. Dasaklis, "A risk assessment methodology for supply chain tracking services," in *2023 International Conference On Cyber Management And Engineering (CyMaEn)*. IEEE, 2023, pp. 555–559.

[8] W. Liang, S. Xie, J. Long, K.-C. Li, D. Zhang, and K. Li, "A double puf-based rfid identity authentication protocol in service-centric internet of things environments," *Information Sciences*, vol. 503, pp. 129–147, 2019.

[9] M. Ebrahimabadi, M. Younis, and N. Karimi, "A puf-based modeling-attack resilient authentication protocol for iot devices," *IEEE Internet of Things Journal*, 2021.

[10] K. Lounis and M. Zulkernine, "T2t-map: A puf-based thing-to-thing mutual authentication protocol for iot," *IEEE Access*, vol. 9, pp. 137 384–137 405, 2021.

[11] H. Zhao and L. Njilla, "Hardware assisted chaos based iot authentication," in *2019 IEEE 16th International Conference on Networking, Sensing and Control (ICNSC)*, 2019, pp. 169–174.

[12] V. K. V. V. Bathalapalli, S. P. Mohanty, E. Kougianos, B. K. Baniya, and B. Rout, "Pufchain 3.0: Hardware-assisted distributed ledger for robust authentication in the internet of medical things," in *Internet of Things. IoT through a Multi-disciplinary Perspective*, L. M. Camarinha-Matos, L. Ribeiro, and L. Strous, Eds. Cham: Springer International Publishing, 2022, pp. 23–40.

[13] Y.-H. Chuang, N.-W. Lo, C.-Y. Yang, and S.-W. Tang, "A lightweight continuous authentication protocol for the internet of things," *Sensors*, vol. 18, no. 4, p. 1104, 2018.

[14] B. Yu, C. Yang, and J. Ma, "Continuous authentication for the internet of things using channel state information," in *2019 IEEE Global Communications Conference (GLOBECOM)*. IEEE, 2019, pp. 1–6.

[15] N. Yu, J. Ma, X. Jin, J. Wang, and K. Chen, "Context-aware continuous authentication and dynamic device pairing for enterprise iot," in *Internet of Things–ICIOT 2019: 4th International Conference, Held as Part of the Services Conference Federation, SCF 2019, San Diego, CA, USA, June 25–30, 2019, Proceedings 4*. Springer, 2019, pp. 114–122.

[16] A. Badhib, S. Alshehri, and A. Cherif, "A robust device-to-device continuous authentication protocol for the internet of things," *IEEE Access*, vol. 9, pp. 124 768–124 792, 2021.

[17] T. Wan, L. Wang, W. Liao, and S. Yue, "A lightweight continuous authentication scheme for medical wireless body area networks," *Peer-to-Peer Networking and Applications*, vol. 14, no. 6, pp. 3473–3487, Nov. 2021.

[18] L. Meng, D. Huang, J. An, X. Zhou, and F. Lin, "A continuous authentication protocol without trust authority for zero trust architecture," *China Communications*, vol. 19, no. 8, pp. 198–213, 2022.

[19] A. O. Gomez Rivera, D. K. Tosh, and U. Ghosh, "Resilient sensor authentication in scada by integrating physical unclonable function and blockchain," *Cluster Computing*, pp. 1–15, 2022.

[20] K. Goutsos and A. Bystrov, "Lightweight puf-based continuous authentication protocol," in *2019 International Conference on Computing, Electronics & Communications Engineering (iCCECE)*. IEEE, 2019, pp. 229–234.

[21] S. Alshomrani, S. Li *et al.*, "Pufdca: A zero-trust-based iot device continuous authentication protocol," *Wireless Communications and Mobile Computing*, vol. 2022, 2022.

[22] D. Li, R. Chen, D. Liu, Y. Song, Y. Ren, Z. Guan, Y. Sun, and J. Liu, "Blockchain-based authentication for iiot devices with puf," *Journal of Systems Architecture*, vol. 130, p. 102638, 2022.

[23] F. H. Al-Naji and R. Zagrouba, "Cab-iot: Continuous authentication architecture based on blockchain for internet of things," *Journal of King Saud University-Computer and Information Sciences*, vol. 34, no. 6, pp. 2497–2514, 2022.

[24] W. Wang, Q. Chen, Z. Yin, G. Srivastava, T. R. Gadekallu, F. Alsolami, and C. Su, "Blockchain and puf-based lightweight authentication protocol for wireless medical sensor networks," *IEEE Internet of Things Journal*, vol. 9, no. 11, pp. 8883–8891, 2021.

[25] V. K. V. V. Bathalapalli, S. P. Mohanty, E. Kougianos, B. K. Baniya, and B. Rout, "PUFchain 2.0: Hardware-Assisted robust blockchain for sustainable simultaneous device and data security in smart healthcare," *SN Computer Science*, vol. 3, no. 5, p. 344, Jun. 2022.

[26] J. M. V. Santos, J. E. V. Pascua, and N. M. C. Tiglao, "Hardware-accelerated blockchain-based authentication for the internet of things," in *Cognitive Radio Oriented Wireless Networks and Wireless Internet*, H. Jin, C. Liu, A.-S. K. Pathan, Z. M. Fadlullah, and S. Choudhury, Eds. Cham: Springer International Publishing, 2022, pp. 283–295.

[27] C. Huang and K. Yan, "A blockchain based fast authentication framework for iot networks with trusted hardware," in *2020 IEEE 22nd International Conference on High Performance Computing and Communications; IEEE 18th International Conference on Smart City; IEEE 6th International Conference on Data Science and Systems (HPCC/SmartCity/DSS)*, 2020, pp. 1050–1056.

[28] S. W. Shah, N. F. Syed, A. Shaghaghi, A. Anwar, Z. Baig, and R. Doss, "Lcda: lightweight continuous device-to-device authentication for a zero trust architecture (zta)," *Computers & Security*, vol. 108, p. 102351, 2021.

[29] K. Qian, Y. Liu, X. He, M. Du, S. Zhang, and K. Wang, "Hpcchain: A consortium blockchain system based on cpu-fpga hybrid puf for industrial internet of things," *IEEE Transactions on Industrial Informatics*, 2023.

Edge Artificial Intelligence in Large-Scale IoT Systems, Applications, and Big Data Infrastructures

Aristeidis Karras*, Anastasios Giannaros*, Christos Karras*,
Konstantinos C. Giotopoulos†, Dimitrios Tsolis‡, Spyros Sioutas*
*Department of Computer Engineering and Informatics, University of Patras, Patras, Greece
{akarras, giannaros, c.karras, sioutas}@ceid.upatras.gr
†Department of Management Science and Technology, University of Patras, Patras, Greece
kgiotop@upatras.gr
‡Department of History and Archaeology, University of Patras, Patras, Greece
dtsolis@upatras.gr

Abstract—As the Internet of Things (IoT) landscape grows, with estimates exceeding 75 billion devices by 2025, effective data management and processing become primary challenges. Traditional cloud-centric models may struggle under this large data volume. This research presents Edge AI as an innovative solution, integrating artificial intelligence directly at data sources like sensors and cameras. This ensures real-time analytics and decision-making, promoting responsive and tailored actions. Our literature review details Edge AI's distinct characteristics and applications. The interaction of Edge AI with large-scale IoT domains is critically examined, emphasizing their combined potential. Within big data infrastructures, a comparative study contrasts Edge AI and cloud-based AI, investigating processing speeds, optimization techniques, and essential metrics. The inherent limitations of Edge AI and current challenges are also discussed. In summary, Edge AI offers notable improvements in operational efficiency, data privacy, and bandwidth use. As IoT continues its rapid expansion, the strategic deployment of Edge AI becomes crucial, leading to a future where data is not just collected but smartly utilized.

Index Terms—Edge Computing, Edge Artificial Intelligence, Edge AI, Cloud-based AI, IoT, IoT Systems, Big Data, Scalability

I. INTRODUCTION

Edge AI's integration into Internet of Things (IoT) systems symbolizes a significant advancement in digital technology, emphasizing decentralized and intelligent computing. By enabling real-time data processing at the source, Edge AI addresses the latency and bandwidth limitations inherent in cloud-based models. This approach has widespread applications across various fields, such as enhancing traffic management in smart cities, revolutionizing patient monitoring in healthcare, optimizing energy consumption in smart homes, facilitating predictive maintenance in industries, and improving agricultural yield through precision farming.

The introduction of specialized hardware and custom software solutions, as noted in [1], is critical in ensuring the successful incorporation of Edge AI into large-scale IoT systems. These innovations are designed to meet the computational requirements of complex AI models while maintaining energy efficiency and optimized performance. Developments in neural network optimization and the emergence of AAIoT, discussed

in [2], are indicative of the progress made in overcoming challenges associated with machine learning computation and minimizing response time in multi-layered IoT systems.

In the evolving landscape of large-scale IoT systems, addressing cybersecurity challenges is of utmost importance. The interconnected nature of these systems makes them vulnerable targets, exposing them to many cyber threats and necessitating the deployment of rigorous security measures. The implementation of Edge AI emerges as a potent solution in this context. As explained in [3], using Edge AI for cybersecurity allows for the real-time detection of anomalies and unusual patterns, thereby allowing for immediate responsive actions to mitigate potential risks. This capability is fundamental in safeguarding the integrity of data and ensuring the confidentiality of information transmitted within the network. Moreover, the integration of Edge AI enhances the resilience of IoT systems, strengthening them against sophisticated cyber-attacks and unauthorized intrusions, which is crucial for maintaining trust and reliability in interconnected digital ecosystems.

Additionally, the notion of Edge Intelligence extends the capabilities of decentralized computing [4]. It encompasses the integration of AI, data analytics, and localized processing, contributing to the development of adaptive and responsive networks [5]. Edge Intelligence plays a vital role in reducing dependence on centralized data processing, fostering the creation of more efficient and resilient systems.

The relationship between Edge AI and Large-Scale IoT Systems, Applications, and Big Data Infrastructures is transformative, fostering an environment conducive to innovation and enhanced operational efficacy. The ability of Edge AI to process data close to where it is generated is not only crucial for reducing latency and conserving bandwidth but also key in unlocking new possibilities in real-time decision-making and analytics. This immediacy in data processing is essential for applications such as autonomous vehicles [8], industrial automation, and emergency healthcare services. For instance, a study presented in [7] exemplifies the innovative integration of emotion detection in smart homes, highlighting the subtle relationship between IoT systems and residents and the potential for adaptive environments.

979-8-3503-8368-3/23 $31.00 © 2023 IEEE

Furthermore, the incorporation of Edge AI into large-scale IoT systems signifies a step toward achieving greater autonomy in various sectors. It enables the realization of Smart Cities, where urban infrastructures become more efficient and adaptive to the needs of the populace, and Industrial IoT, where machinery becomes self-aware, leading to optimized resources and maintenance. In the agriculture sector and applications [6], the precision provided by Edge AI's real-time data processing enables more accurate monitoring of environmental variables, leading to better crop yield and resource utilization.

The remainder of this study is structured as follows: Section II provides a literature review, discussing Edge AI's characteristics II-A and applications II-B. Section III examines the integration of large-scale IoT domains with Edge AI use-cases. Section IV explores Edge AI within Big Data infrastructures, with subsections covering a comparison with Cloud-based AI IV-A, efficiency in processing IV-B, influential factors IV-C, optimization methods IV-D, relevant metrics IV-E. Subsequent to these, IV-F pinpoints the limitations of Edge AI, followed by the challenges and open issues in IV-G. The paper concludes in Section V by highlighting insights for future research.

II. Literature Review

A. Characteristics of Edge AI

The rapid evolution of Edge AI is underscored by a set of key characteristics that differentiate it from traditional computing paradigms. These attributes not only address the challenges inherent to expansive IoT systems but also herald capabilities in data-device interplay. Key features include:

- **Decentralization:** A foundational aspect of Edge AI is its emphasis on decentralized processing. Deviating from prevalent cloud-centric methodologies, it promotes localized computation proximal to the data source. Such a shift facilitates rapid data processing, minimizes latency, and alleviates the computational demands on central servers, thereby enhancing overall system resilience.
- **Real-time Analytics:** The need for quick data analysis, clear in sectors from autonomous transportation to industrial automation, is met by Edge AI. By avoiding the need for remote server-based analysis, it equips devices with the ability for on-the-spot data processing. This speed is not only beneficial for user experience but becomes crucial in situations requiring swift decision-making.
- **Scalability:** The decentralized architecture of Edge AI inherently caters to the scalability challenges posed by an ever-growing IoT ecosystem. By distributing computational tasks, it ensures system robustness and latency reduction as the nexus of interconnected devices expands.
- **Privacy and Security:** Edge AI offers robust data protection measures. By reducing extensive data transit over networks, it inherently limits exposure to potential security breaches. Advanced methodologies such as federated learning further fortify this stance by enabling algorithmic training devoid of direct data access.
- **Bandwidth Conservation:** Bandwidth efficiency is accentuated in the Edge AI framework. Localized process-ing ensures only pivotal data or analyses are relayed, thereby preventing network congestion and ensuring resource optimization.
- **Energy Efficiency:** Particularly pertinent for battery-operated IoT apparatus, energy efficiency is a hallmark of Edge AI. Leveraging localized computation and drawing upon energy-optimized AI algorithms, it substantially reduces power consumption, augmenting the operational lifespan of mobile or remotely deployed devices.

B. Edge AI Applications

The integration of Edge AI into large-scale Internet of Things (IoT) systems marks a technological shift with far-reaching implications. The versatility of Edge AI is underscored by its application across a spectrum of functionalities, addressing challenges in large-scale IoT deployments, ranging from data management, enhancement of automation and transmission, to the resolution of scalability and security concerns, and advancement in path planning methodologies.

1) **Distributed MI: Enhanced Data Management and Automation.** Edge AI, by advancing towards the IoT end-device, serves to address the critical demands of latency and network bandwidth in large-scale IoT deployments. The enrichment of the intelligence layer facilitates services to the overlying application layer, consequently offering end-users enhanced control and accessibility of intelligence services. This shift incentivizes the development of versatile solutions, extending their reach to a variety of devices, as explored in [9].

2) **Master-Aided Edge Computing Systems: Optimal Computation and Communication.** A distinct approach involving a master-aided distributed computing system encompasses multiple edge computing nodes and a master node. This system optimizes output function computation at edge nodes, utilizing a coded scheme to minimize communication latency by harnessing the computation and communication capabilities of all nodes. This methodology, detailed in [10], achieves optimal results under varying computing and storage abilities at the master.

3) **Scalable Edge Computing: Addressing Congestion and Scalability.** Edge computing introduces computation offloading to empower resource-constrained devices. However, large-scale offloading induces congestion and presents scalability challenges. The deployment of a cross-entropy-based scalable edge computing framework, integrating IoT devices, edge servers, and the cloud, offers a remedy. Utilizing an IoT clustering technique and a latency-critical computation offloading algorithm, this framework ensures parallel utilization of edge resources and exhibits superior results compared to optimization techniques, as presented in [11].

4) **Profit-Optimized Computation Offloading: Maximizing Edge Node Profit.** Formulating a joint optimization problem focused on task offloading, partitioning, and user associations to edge nodes seeks to maximize

979-8-3503-8368-3/23 $31.00 © 2023 IEEE

the profit of edge nodes. The application of a novel deep-learning algorithm, GSPSA, demonstrates superior performance in addressing this high-dimensional mixed-integer nonlinear program, achieving higher profits while meeting user task latency requirements, as substantiated by real-life data-based experimental results in [12].

5) **End-to-End Collection and Analysis of IoT Security Data: Strengthening Security Infrastructure.** The development of a scalable and configurable data collection infrastructure for data-driven IoT security is paramount. This infrastructure integrates state-of-the-art technologies for large-scale data collection, streaming, and storage, offering a foundation for the application of effective security analytics algorithms. The system identifies threats, vulnerabilities, and related attack patterns in complex IoT deployments, contributing to the enhancement of security measures, as detailed in [13].

6) **Path Planning System by Grid Caching: Enhancing Response Latency.** The introduction of a Cloud-Edge collaboration-based Path Planning algorithm by Grid Caching (PPGC) processes path requests at edge servers in unmanned surface vehicle areas, optimizing path sharing cache and using a grid-based path matching algorithm. The proposed system reduces the average response latency compared to cloud-based path planning systems, as evidenced by experimental results in [14].

C. State-of-the-Art Techniques in Edge Computing

Edge computing is a paradigm that focuses on using processing resources available at the network's edge. This methodology offers new opportunities for learning using Edge Computing principles. SOTA techniques in Edge Computing:

- **TinyML on Microcontroller Edge Devices.** Y. Zhang et al. conducted research on keyword spotting on microcontrollers [15]. Their findings indicate that machine learning tasks can be performed on microcontrollers, even those with limited resources. This has implications for creating edge devices that are cost-effective, use less power, and prioritize privacy.
- **Compute-in-Memory Analog Edge Accelerators.** C. Zhou et al. proposed compute-in-memory accelerators that use analog techniques [16]. These accelerators perform computations within the memory, which can reduce energy use and latency in edge devices.
- **Hybrid Edge-Node Schemes.** Zonios et al. studied hybrid systems that combine the computational power of edge devices with the storage abilities of cloud nodes [17]. Their research suggests such systems can be energy efficient and maintain privacy in smart home settings.
- **Hybrid Cloud-Edge Schemes.** A. Kag, et al. explored the concept of efficient edge inference through selective querying [18]. Their findings show that selective queries can reduce communication needs and energy usage, important considerations for the development of affordable and privacy-focused edge devices.

- **Federated Edge Intelligence with Edge Caching Mechanisms.** The study conducted by Karras et al. leverages Federated Learning (FL) for enhanced user privacy in edge intelligence and caching [4]. Two distinct algorithms within a peer-to-peer framework address challenges like imbalanced data. A notable contribution is the client-balancing Dirichlet sampling algorithm for improved model training. Edge caching, symbolized by the caching factor α, minimizes communication overhead and latency. To refine the machine learning processes, Bayesian Optimization is applied, streamlining the search for optimal hyperparameters in edge scenarios.

III. INTEGRATION OF LARGE-SCALE IOT DOMAINS AND EDGE AI USE-CASES

The integration of Edge AI with Large-Scale IoT has revolutionized decision-making and user experiences across domains. This synergy not only introduces transformative applications and real-time optimizations but also sets the foundation for adaptive, intelligent ecosystems, unlocking the full potential of IoT in various sectors.

Table I outlines this integration, highlighting the advancements in different IoT domains, the corresponding applications of Edge AI, and the expected improvements and future developments resulting from their integration. This detailed overview underscores the interdependent relationship between IoT and Edge AI, emphasizing their joint role in developing intelligent and adaptable systems for the future.

IV. EDGE AI IN BIG DATA INFRASTRUCTURES

In the context of Big Data infrastructures, Edge AI signifies the implementation of artificial intelligence algorithms and models on peripheral devices, such as IoT devices, edge servers, and gateways. This strategy facilitates local processing and analysis of data, thereby minimizing the distance to the data source. The advantages of this approach are manifold, including a reduction in latency, enhanced privacy and security, and efficient utilization of network bandwidth.

- **Real-time Big Data and Data Analytics at the Edge:** The evolution of network architectures is essential to provide real-time Big Data and data analytics capabilities at the edge of the network. This advancement is crucial to prevent network saturation and delays. Moreover, it is imperative to consider energy efficiency for sustainable and effective deployment across various domains of human activity [19].
- **Data Locality in Edge Environments:** Conventional Big Data processing solutions often fall short in managing data locality efficiently and struggle with processing small and frequent events, which are characteristic of Edge environments. Solutions that prioritize data locality are instrumental in mitigating the performance implications resulting from data transfers between disparate data centers [20], [21].
- **Integration of AI in IoT Devices for Edge Analytics:** The incorporation of AI into IoT devices, exemplified

TABLE I: Combination of IoT Domains and Edge AI Applications

Domains	IoT Innovations	Edge AI Applications	Enhancements & Future Prospects
Smart Cities	Sustainable and Engaged Urban Living	Adaptive Traffic Management and Congestion Prediction	Urban Sustainability and Citizen Engagement through Adaptive Public Services
Industrial IoT (IIoT)	Dynamic Manufacturing and Supply Chain Optimization	Comprehensive Predictive Maintenance and Anomaly Detection	Enhanced Worker Safety and Resource Optimization through Real-Time Adjustments
Agricultural IoT	Intelligent Irrigation and Drone-Based Crop Health Assessment	Disease Detection using Multi-Spectral Imaging and Preemptive Intervention	Sustainable Farming through Autonomous and Data-Driven Agricultural Practices
Healthcare IoT	Telemedicine and Real-Time Global Collaboration	Personalized Patient Monitoring and Predictive Health Insights	Advancements in Remote Healthcare and Collaborative Medical Research
Connected Vehicles	Integration with Urban IoT Systems and Weather Adjustments	Collision Avoidance in Dynamic Environments and Multi-Modal Data Analysis	Vehicular Communication with Urban Infrastructure for Traffic Mitigation
Smart Grids	Renewable Energy Integration and Real-Time Consumer Feedback	Energy Consumption Prediction and Distribution Strategy Refinement	Fostering Energy-Saving Habits and Integration of Diverse Renewable Energy Sources
Smart Homes	Anticipatory Resident Needs and Adaptive Systems	Intrusion Detection and Security Recommendations based on Behavioral Pattern Analysis	Residents' Behavioral Prediction for Personalized and Secure Living Environments
Retail IoT	Customer Behavior Analysis and Augmented Reality Shopping	Inventory Management using Socio-Economic Indicators and Real-Time Store Layout Adjustment	Enhanced Shopping Experience through Personalization and Real-Time Analytics

by smart meters in Advanced Metering Infrastructures (AMIs), is pivotal for enabling micro-analytics at the edge. This integration facilitates high-performance analytics and Edge computing capabilities, thereby allowing instant data verification at the source and transmitting pertinent real-time data to utilities via the Internet [22].

- **Efficient Management of IT Infrastructures for Edge Analytics:** Innovative tools such as Pangea are instrumental in automatically generating appropriate execution environments for deploying analytic pipelines across Edge, fog, cloud, or on-premise environments. This capability is vital for minimizing latency and optimizing the utilization of hardware and software resources [23].

- **Computing Platforms for Big Data Analytics in Specific Domains:** In specialized domains such as Electric Vehicle (EV) infrastructures, a variety of computing platforms, including distributed cloud computing and edge/fog computing, are integral to supporting Big Data analytics activities. These platforms, each with their unique features, present opportunities, trends, and challenges for successful integration, thereby addressing the specific requirements of diverse domains [24].

Ultimately, the integration of Edge AI into Big Data infrastructures addresses the challenges of real-time extensive data processing. This is crucial for energy efficiency, data locality, and domain-specific requirements, paving the way for technological advancements.

A. Comparison of Edge AI and Cloud-based AI in Big Data Processing

In this section, we focus on at the differences between Edge and Cloud methods, especially how they work with Big Data. We want to see how each method performs in

considering factors like real-time analysis, scalability, and energy efficiency. We'll also discuss the challenges in managing these systems. It's crucial to comprehend the similarities and differences between Edge AI and Cloud-based AI in today's evolving landscape. Through extensive research studies, this analysis aims to provide a comprehensive understanding of these technologies. By doing so, it will assist stakeholders in making well-informed decisions and aligning their strategies with the requirements of Big Data processing environments.

The table II below illustrates the strengths and weaknesses of both approaches, providing insights into their suitability for different scenarios and requirements. Readers will find detailed points on real-time processing, scalability, privacy, security, energy efficiency, and infrastructure management, along with associated references.

B. Balancing Efficiency in Big Data Processing: Edge AI and Cloud-Based AI

In addressing the demands of Big Data processing speed, Edge AI emerges as a significant frontrunner, enabling real-time data processing and analysis directly at the network's edge, which facilitates swift decision-making processes. This immediacy is pivotal, especially in applications where time sensitivity is paramount, and any delay in data processing can lead to impactful consequences [25]. Conversely, cloud-based AI can potentially encounter higher latency, attributed to the necessity of transferring data to and from the cloud for processing, which can pose challenges in applications requiring instantaneous responses [31].

Acknowledging these characteristics, researchers have proposed edge-cloud collaborative processing as a promising avenue for enhancing the efficiency of Big Data processing in IoT systems. This innovative approach envisions dividing the processing workload of a Neural Network model between

TABLE II: Comparison of Edge AI and Cloud-based AI in Big Data Processing

	Edge AI	Cloud-based AI
Pros	• Real-time data processing and decision-making [25]. • Optimized bandwidth; reduced data transfer cost and delay [26]. • Enhanced privacy and security through localized data processing [27]. • Scalable and flexible; adaptable to varying data volumes and demands [28]. • Energy-efficient algorithms on constrained devices [29].	• High processing power and storage for large-scale tasks [30]. • Centralized management and resource allocation. • Enhanced parallel processing and scalability through distributed computing [30].
Cons	• Limited processing power and storage compared to cloud [30]. • Challenges in managing distributed edge devices.	• Higher latency due to data transfer to/from cloud [26]. • Increased network bandwidth usage and cost [26]. • Potential privacy and security risks with sensitive data [27]. • Higher energy consumption compared to Edge AI [29].

the edge and the cloud, aiming to address challenges related to network congestion, privacy concerns, and computational load [29]. However, this method is not devoid of challenges, particularly regarding the substantial volume of the edge part's output that necessitates transmission to the cloud. To mitigate this, scholars have introduced data transmission reduction strategies, which leverage the similarities in stationary objects to significantly diminish the volume of data transmitted in edge-cloud collaborative frameworks [29].

Moreover, within the scope of applications such as traffic tracking and detection, Edge AI-based sensors have demonstrated remarkable capabilities. Not only do they collect voluminous traffic data, but they also alleviate the communication network's bandwidth and reduce the workload on the cloud or server-side [32]. This is accomplished by performing processing, storage, and extraction of valuable data at the network's edge prior to transferring it to a centralized server, epitomizing the concept of Artificial Intelligence on The Edge [32].

In manufacturing, integrating Edge AI and Cloud-based AI is essential. A key study [33] outlines an innovative architecture for collaborative edge and cloud processing. This architecture uses hierarchical gateways at the edge to support latency-sensitive applications and ensure real-time responses. The study introduces "AI-Mfg-Ops" (AI enabled Manufacturing Operations), emphasizing the role of software in manufacturing control and decision-making, which helps in the quick operation and upgrading of cloud manufacturing systems.

Another notable study [34] discusses the importance of private on-premise edge clouds, especially when connected with private 5G networks, for secure, real-time communication in production plants. The study suggests that off-premise edge clouds, integrated with AI and real-time video streaming systems, are economically viable, particularly for small and medium-sized enterprises (SMEs), enabling remote quality assurance in distributed production sites. This highlights the potential benefits of edge clouds in industrial settings.

This analysis examines the strengths and limitations of Edge AI and Cloud-based AI in Big Data processing. It emphasizes the role of current technological advancements and academic research in optimizing both paradigms. The goal is to promote an effective ecosystem where both Edge AI and Cloud-based

AI meet the dynamic needs of the digital age.

C. Factors Influencing Processing Speed in Edge AI

The processing speed of Edge AI in Big Data processing is not uniform and can be affected by various influential factors. Each factor holds significance in determining how efficiently an Edge AI system can process data.

1) **Data Volume:** Larger volumes of data necessitate prolonged processing times, particularly when the Edge device is constrained by computational resources [36].
2) **Data Complexity:** The nature of the data also plays a crucial role, with complex data types like images or videos requiring more computational power compared to text data [36].
3) **Edge Device Capabilities:** The computational capabilities of the Edge device, including processor speed and memory, directly impact processing speed [36].
4) **Network Latency:** The reliance on network communication for data processing in Edge AI systems means that any delay in data transmission can hinder the overall processing speed [36].
5) **Edge AI Model Complexity:** The complexity of the AI model being employed can also dictate the processing speed, with more intricate models demanding additional computational resources [35].
6) **Data Pre-processing:** The efficacy of data pre-processing techniques can influence the overall processing speed, where optimized methods can significantly reduce computation and memory requirements [35].

TABLE III: Factors Influencing Processing Speed in Edge AI

Factor	*Description*	*Ref.*
Data Volume	Impact of the amount of data on processing time	[36]
Data Complexity	Variance in processing speed due to data nature	[36]
Edge Device Capabilities	Influence of device computational power	[36]
Network Latency	Effect of data transmission delay	[36]
Edge AI Model Complexity	Role of AI model intricacy in processing	[35]
Data Pre-processing	Impact of pre-processing efficiency	[35]

D. Optimizing Big Data Processing with Edge Intelligence

The integration of Edge computing and AI stands as a cornerstone in enhancing Big Data processing, providing a foundation for real-time insights, robust security, and optimal resource allocation, thereby marking a transformative advancement in the capabilities of organizations.

1) **Efficiency Through Proximity**
   The strategic placement of Edge computing reduces latency, enabling the swift application of AI for immediate insights and informed decisions. This is vital for applications where time is of the essence, such as healthcare monitoring and autonomous navigation.

2) **Strengthened Security and Privacy**
   By processing data locally, Edge AI significantly reduces the exposure of sensitive information to external networks, enhancing the overall integrity and security of data within organizational boundaries.

3) **Optimized Bandwidth and Reduced Costs**
   Edge computing and AI collaboratively filter data at the source, ensuring the transfer of only essential information to the cloud, thus mitigating network congestion and lowering associated operational costs.

4) **Adaptable and Scalable Architecture**
   Edge AI facilitates dynamic adaptation to varying demands through efficient resource allocation, offering organizations the flexibility to modify their infrastructure in response to the ever-evolving Big Data environment.

5) **Sustainability in Operations**
   Incorporating energy-conscious strategies, Edge AI effectively balances performance and energy usage, ensuring the durability of devices and promoting environmentally responsible operations.

6) **Forward-Looking Analytics**
   The utilization of AI for predictive analytics allows Edge computing to anticipate and adjust to future resource needs, thereby ensuring the resilience and adaptability of Big Data infrastructures.

7) **Customized Solutions Through Context Awareness**
   The inherent adaptability of Edge AI enables the development of tailored solutions across a variety of applications, ensuring optimal performance through informed decision-making in data processing.

The evolving synergy between Edge computing and AI is instrumental in reshaping Big Data infrastructures, positioning organizations to explore new realms of insight and functionality, and exemplifying the ongoing shift in digital advancement.

E. Edge AI Metrics for Big Data Infrastructures

When evaluating Edge AI's effectiveness in Big Data environments, it's pivotal to introduce metrics tailored to the unique challenges and opportunities of this convergence. Below are advanced metrics designed to offer a more nuanced understanding of the performance and capabilities of Edge AI deployments in Big Data infrastructures:

1) Edge AI Model Update Frequency: This metric evaluates how frequently the Edge AI model is updated based on new data, ensuring that the AI remains current and effective.

$$U_F = \frac{N_{\text{updates}}}{\Delta t} \tag{1}$$

where:

- U_F is the model update frequency.
- N_{updates} is the number of model updates.
- Δt is the time period.

2) Local vs. Cloud Decision Efficiency: Given the dynamic nature of data processing in Edge AI, this metric evaluates the system's ability to decide whether processing should be local or offloaded to the cloud:

$$L_C = \frac{N_{\text{efficient decisions}}}{N_{\text{total decisions}}} \tag{2}$$

where:

- L_C is the local vs. cloud decision efficiency.
- $N_{\text{efficient decisions}}$ refers to instances where the most resource-efficient decision was made.
- $N_{\text{total decisions}}$ is the total number of local vs. cloud processing decisions.

3) Dynamic Resource Allocation Efficiency: This metric evaluates the Edge AI system's capability to dynamically allocate resources based on the current data load and processing requirements:

$$R_A = \frac{N_{\text{optimal allocations}}}{N_{\text{total allocations}}} \tag{3}$$

where:

- R_A is the resource allocation efficiency.
- $N_{\text{optimal allocations}}$ is the number of instances optimal resource allocation was achieved.
- $N_{\text{total allocations}}$ is the total number of resource allocation decisions made.

4) Edge AI Drift Detection Rate: Model drift occurs when the model's predictions gradually become less accurate. This metric evaluates the system's capability to detect and address model drift:

$$D_R = \frac{N_{\text{detected drifts}}}{N_{\text{actual drifts}}} \tag{4}$$

where:

- D_R is the drift detection rate.
- $N_{\text{detected drifts}}$ is the number of detected model drifts.
- $N_{\text{actual drifts}}$ is the actual number of model drifts.

F. Limitations of Edge Artificial Intelligence

While Edge Artificial Intelligence (Edge AI) holds much promise in various fields, it is crucial to examine its inherent limitations carefully. A detailed study of these challenges, outlined in Table IV, provides a more balanced view, helping guide thoughtful advancements in this growing domain.

979-8-3503-8368-3/23 $31.00 © 2023 IEEE

TABLE IV: Limitations of Edge Artificial Intelligence

Limitation	Description
Hardware Constraints	Edge devices, due to their form factor and purpose, often have restricted computational power, memory, and storage, which can sometimes curtail the complexity of AI models that can be run on these devices without compromising on performance.
Model Optimization	Optimization techniques like model pruning and quantization, developed to fit complex AI models into edge devices, often reduce accuracy, making them unsuitable for certain critical applications.
Data Limitations	Edge AI operates on localized data, which might not offer holistic data diversity centralized systems access, leading sometimes to suboptimal or biased model training.
Security Concerns	Edge devices can present a broader attack surface, making ensuring consistent security measures across all edge devices a logistical challenge.
Lifecycle Management	Keeping the AI modules on edge devices consistently updated is challenging, especially with devices spread across diverse and remote locations.
Interoperability	Ensuring that Edge AI solutions are universally compatible and interoperable, given the diversity of edge devices, becomes a significant challenge.
Latency Variability	The latency in Edge AI can still vary significantly based on the specific edge device's capabilities, network conditions, and other environmental factors.
Limited Redundancy	Edge devices might lack the redundant systems found in centralized cloud infrastructure, leading to data loss or system failures if an edge device encounters an issue.
Cost Implications	The initial setup, maintenance, and updates for Edge AI devices can be capital-intensive, especially for large-scale deployments.
Scalability Concerns	The physical deployment, maintenance, and individual management of numerous devices can pose challenges as the scale grows.

G. Challenges & Open Issues

The introduction of Edge AI into the IoT landscape marks a significant advancement, but also brings forth numerous challenges and unresolved issues. Addressing these requires collaboration across disciplines and a blend of computational and domain knowledge. The continuous development of Edge AI underscores the need to navigate these challenges to assess its viability in large-scale IoT systems. A detailed overview of these challenges and issues can be found in Table V, which clarifies each challenge and offers insight into the various dimensions of integration complexity.

TABLE V: Challenges and Open Issues in Edge AI

Challenges & Open Issues
Standardization and Protocol Development: The diversity of edge devices necessitates standardized protocols for device management, security, and model deployment, the absence of which complicates integration.
Distributed Intelligence Coordination: Achieving cohesive decision-making across multiple edge devices, especially in real-time, poses significant challenges.
Resource Allocation: The effective allocation of computational resources among a variety of devices with different capabilities is imperative.
Data Veracity and Lineage: Ensuring the authenticity of data and tracing its origin are critical in a decentralized setup like Edge AI.
Evolutionary AI Models: There is a necessity for AI models on the edge that can autonomously adapt to environmental changes.
Decentralized Security Frameworks: Addressing vulnerabilities at the device level and risks associated with model deployment necessitates comprehensive security measures.
Edge-Centric Data Markets: The prospect of data monetization at the edge demands the development of regulated and transparent data markets.
Domain-Specific Challenges: Each domain presents unique challenges for Edge AI integration, particularly in terms of regulatory compliance and integration with existing systems.
Digital Twin Integration: Real-time synchronization between digital twins and edge devices in Edge AI presents considerable challenges.
Multimodal Data Fusion: The need for algorithms capable of effectively combining diverse types of data at the edge is crucial for extracting valuable insights.

Addressing these challenges and open issues is paramount,

requiring a synergy of computational expertise and domain-specific insights. The resolution of these issues will be a determining factor in the efficacy of Edge AI within large-scale IoT systems.

V. CONCLUSION

This study has explored the multifaceted integration of Edge Artificial Intelligence (Edge AI) in Large-Scale IoT systems, illustrating its transformative impact across various domains. The amalgamation of Edge AI and IoT has been shown to significantly enhance Big Data infrastructure, optimizing efficiency, reducing latency, and fostering real-time, informed decision-making. This study highlights how the integration of various IoT domains and Edge AI applications is notably transforming industries, including smart cities, healthcare, agriculture, retail, manufacturing, and energy.

VI. FUTURE DIRECTIONS IN EDGE AI FOR IOT

As we stand on the threshold of a new era in technological evolution, the advent of 6G, Ultra Nano Things, and LPWAN technologies is poised to significantly reshape the digital landscape. 6G is set to redefine communication with unparalleled reliability and efficiency, thus fulfilling the ever-increasing demands for seamless connectivity and broader bandwidth. Ultra Nano Things is unveiling a multitude of innovative applications, particularly in sectors like healthcare and environmental monitoring, by utilizing the microscopic scale to bring forth novel functionalities. Meanwhile, the advancement of LPWAN technologies is ensuring enhanced connectivity in remote regions, thereby promoting inclusivity and equitable access to the digital world.

These emerging technologies not only signify substantial progress in connectivity and innovation but also mark the dawn of a future where the collaboration of diverse technologies amplifies the capabilities of digital ecosystems. The union of

979-8-3503-8368-3/23 $31.00 © 2023 IEEE

6G, Ultra Nano Things, and LPWAN is laying the path toward a future that is more interconnected, intelligent, and sustainable, thus opening new avenues for research, exploration, and technological breakthroughs.

REFERENCES

[1] Sipola, T., Alatalo, J., Kokkonen, T. and Rantonen, M., 2022, April. Artificial intelligence in the IoT era: A review of edge AI hardware and software. In 2022 31st Conference of Open Innovations Association (FRUCT) (pp. 320-331). IEEE.

[2] Zhou, J., Wang, Y., Ota, K. and Dong, M., 2019. AAIoT: Accelerating artificial intelligence in IoT systems. IEEE Wireless Communications Letters, 8(3), pp.825-828.

[3] Kuzlu, M., Fair, C. and Guler, O., 2021. Role of artificial intelligence in the Internet of Things (IoT) cybersecurity. Discover Internet of things, 1, pp.1-14.

[4] Karras, A., Karras, C., Giotopoulos, K.C., Tsolis, D., Oikonomou, K. and Sioutas, S., 2023. Federated Edge Intelligence and Edge Caching Mechanisms. Information, 14(7), p.414.

[5] Karras, A., Karras, C., Giotopoulos, K.C., Tsolis, D., Oikonomou, K. and Sioutas, S., 2022, September. Peer to peer federated learning: Towards decentralized machine learning on edge devices. In 2022 7th South-East Europe Design Automation, Computer Engineering, Computer Networks and Social Media Conference (SEEDA-CECNSM) (pp. 1-9). IEEE.

[6] Karras, A., Karras, C., Drakopoulos, G., Tsolis, D., Mylonas, P. and Sioutas, S., 2022, June. SAF: a peer to peer IoT LoRa system for smart supply chain in agriculture. In IFIP International Conference on Artificial Intelligence Applications and Innovations (pp. 41-50). Cham: Springer International Publishing.

[7] Karras, C., Karras, A., Drakopoulos, G., Tsolis, D., Mylonas, P. and Sioutas, S., 2022, September. A LoRa-Based Emotion Estimation Scheme for Smart Home Automated Actions Using ELMs. In 2022 IEEE International Smart Cities Conference (ISC2) (pp. 1-7). IEEE.

[8] Giannaros, A., Karras, A., Theodorakopoulos, L., Karras, C., Kranias, P., Schizas, N., Kalogeratos, G. and Tsolis, D., 2023. Autonomous Vehicles: Sophisticated Attacks, Safety Issues, Challenges, Open Topics, Blockchain, and Future Directions. Journal of Cybersecurity and Privacy, 3(3), pp.493-543.

[9] Ramos, E., Morabito, R. and Kainulainen, J.P., 2019. Distributing intelligence to the edge and beyond [research frontier]. IEEE Computational Intelligence Magazine, 14(4), pp.65-92.

[10] Chen, H., Long, J., Ma, S., Tang, M. and Wu, Y., 2023. On the Optimality of Data Exchange for Master-Aided Edge Computing Systems. IEEE Transactions on Communications, 71(3), pp.1364-1376.

[11] Babar, M., Khan, M.S., Habib, U., Shah, B., Ali, F. and Song, D., 2021. Scalable edge computing for IoT and multimedia applications using machine learning. Human-centric Computing and Information Sciences, 11.

[12] Yuan, H., Hu, Q., Bi, J., Lü, J., Zhang, J. and Zhou, M., 2023. Profit-optimized Computation Offloading with Autoencoder-assisted Evolution in Large-scale Mobile Edge Computing. IEEE Internet of Things Journal.

[13] Roukounaki, A., Efremidis, S., Soldatos, J., Neises, J., Walloschke, T. and Kefalakis, N., 2019, June. Scalable and configurable end-to-end collection and analysis of IoT security data: Towards end-to-end security in IoT systems. In 2019 Global IoT Summit (GIoTS) (pp. 1-6). IEEE.

[14] Yan, L., Chen, H., Tu, Y., Zhou, X. and Drew, S., 2022, October. PPGC: A Path Planning System by Grid Caching based on Cloud-Edge Collaboration for Unmanned Surface Vehicle in IoT Systems. In 2022 IEEE 19th International Conference on Mobile Ad Hoc and Smart Systems (MASS) (pp. 74-80). IEEE.

[15] Zhang, Y., Suda, N., Lai, L. and Chandra, V., 2017. Hello edge: Keyword spotting on microcontrollers. arXiv preprint arXiv:1711.07128.

[16] Zhou, C., Redondo, F.G., Büchel, J., Boybat, I., Comas, X.T., Nandakumar, S.R., Das, S., Sebastian, A., Gallo, M.L. and Whatmough, P.N., 2021. AnalogNets: ML-HW co-design of noise-robust TinyML models and always-on analog compute-in-memory accelerator. arXiv preprint arXiv:2111.06503.

[17] Zonios, C. and Tenentes, V., 2021, July. Energy Efficient Speech Command Recognition for Private Smart Home IoT Applications. In 2021 10th International Conference on Modern Circuits and Systems Technologies (MOCAST) (pp. 1-4). IEEE.

[18] Kag, A., Fedorov, I., Gangrade, A., Whatmough, P. and Saligrama, V., 2022, September. Efficient Edge Inference by Selective Query. In The Eleventh International Conference on Learning Representations.

[19] Cárdenas, R., Arroba, P. and Risco Martin, J.L., 2022. Bringing AI to the edge: A formal M&S specification to deploy effective IoT architectures. Journal of Simulation, 16(5), pp.494-511.

[20] Corodescu, A.A., Nikolov, N., Khan, A.Q., Soylu, A., Matskin, M., Payberah, A.H. and Roman, D., 2021, November. Locality-aware workflow orchestration for big data. In Proceedings of the 13th International Conference on Management of Digital EcoSystems (pp. 62-70).

[21] Corodescu, A.A., Nikolov, N., Khan, A.Q., Soylu, A., Matskin, M., Payberah, A.H. and Roman, D., 2021. Big data workflows: Locality-aware orchestration using software containers. Sensors, 21(24), p.8212.

[22] Ogu, R.E., Ikerionwu, C.I. and Ayogu, I.I., 2021, February. Leveraging artificial intelligence of things for anomaly detection in advanced metering infrastructures. In 2020 IEEE 2nd International Conference on Cyberspac (CYBER NIGERIA) (pp. 16-20). IEEE.

[23] Miñón, R., Diaz-de-Arcaya, J., Torre-Bastida, A.I. and Hartlieb, P., 2022. Pangea: an MLOps tool for automatically generating infrastructure and deploying analytic pipelines in edge, fog and cloud layers. Sensors, 22(12), p.4425.

[24] Hussain, M.M., Beg, M.S., Alam, M.S., Krishnamurthy, M. and Ali, Q.M., 2018, August. Computing platforms for big data analytics in electric vehicle infrastructures. In 2018 4th International Conference on Big Data Computing and Communications (BIGCOM) (pp. 138-143). IEEE.

[25] Zhang, X., Qi, L. and Yuan, Y., 2022. Convergency of AI and Cloud/Edge Computing for Big Data Applications. Mobile Networks and Applications, 27(6), pp.2292-2294.

[26] Rexha, H. and Lafond, S., 2021, May. Data collection and utilization framework for edge AI applications. In 2021 IEEE/ACM 1st Workshop on AI Engineering-Software Engineering for AI (WAIN) (pp. 105-108). IEEE.

[27] Rexha, H. and Lafond, S., 2021. Data Collection and Acceleration Infrastructure for FPGA-based Edge AI Applications. arXiv preprint arXiv:2103.06518.

[28] Gao, C., 2023. Efficiency of artificial intelligence automatic control system and data processing unit based on edge computing technology. International Journal of Emerging Electric Power Systems, 24(4), pp.519-528.

[29] Elouali, A., Mora, H. and Gimeno, F.J.M., 2022, December. Similarity-based data transmission reduction solution for edge-cloud collaborative AI. In Proceedings of the 2022 5th Artificial Intelligence and Cloud Computing Conference (pp. 42-48).

[30] Wang, X., Ren, L., Yuan, R., Yang, L.T. and Deen, M.J., 2022. Qtt-dlstm: A cloud-edge-aided distributed lstm for cyber-physical-social big data. IEEE Transactions on Neural Networks and Learning Systems.

[31] Bringmann, O., Ecker, W., Feldner, I., Frischknecht, A., Gerum, C., Hämäläinen, T., Hanif, M.A., Klaiber, M.J., Mueller-Gritschneder, D., Bernardo, P.P. and Prebeck, S., 2021, September. Automated HW/SW co-design for edge AI: state, challenges and steps ahead. In Proceedings of the 2021 International Conference on Hardware/Software Codesign and System Synthesis (pp. 11-20).

[32] Minh, H.T., Mai, L. and Minh, T.V., 2021, November. Performance evaluation of deep learning models on embedded platform for edge ai-based real time traffic tracking and detecting applications. In 2021 15th International Conference on Advanced Computing and Applications (ACOMP) (pp. 128-135). IEEE.

[33] Yang, C., Lan, S., Wang, L., Shen, W. and Huang, G.G., 2020. Big data driven edge-cloud collaboration architecture for cloud manufacturing: a software defined perspective. IEEE access, 8, pp.45938-45950.

[34] Safari, P., Shariati, B., Przewozny, D., Chojecki, P., Fischer, J.K., Freund, R., Vick, A. and Chemnitz, M., 2022, July. Edge Cloud Based Visual Inspection for Automatic Quality Assurance in Production. In 2022 13th International Symposium on Communication Systems, Networks and Digital Signal Processing (CSNDSP) (pp. 473-476). IEEE.

[35] Liao, C.Y., Li, C.Y. and Fang, W.C., 2021, September. AI-based Emotion Recognition System with Tensor Decomposition Optimized Pre-processing. In 2021 IEEE International Conference on Consumer Electronics-Taiwan (ICCE-TW) (pp. 1-2). IEEE.

[36] Rahman, M.A., Tari, Z., Zhu, D., Piccialli, F. and Wang, X., 2020. IEEE Access Special Section Editorial: Cloud-Fog-Edge Computing in Cyber–Physical–Social Systems (CPSS). IEEE Access, 8, pp.222859-222864.

Synthesis-Embedded Verification

Michael Dossis
Dept. of Computer Science
University of Western Macedonia
Kastoria, Greece
mdossis@uowm.gr

Abstract—The proliferation of computing systems and subsystems combined with the need for high performance, product security and low energy have forced the development of tools that generate hardware systems and hardware components automatically from higher-level of algorithmic descriptions. The technology background behind this phenomenon is called High-Level Synthesis (HLS) and it has been an active research topic for at least the last 3 decades. HLS hasn't been established as a de facto design approach in industry because of problems with the quality of the generated hardware descriptions and the lack of complete automation and compatibility with all the programming constructs that sometimes require manual rewriting of the whole module in order for the HLS tool to be able to process it. One greater problem is that often the available HLS tools generate bugs in the generated hardware which are difficult to trace and correct, as the generated architecture usually contains hundreds of FSM states. Moreover, verification tasks and flows which are disjoint from the synthesis flows often fails to find all bugs and it is difficult to learn and run easy as in the case of formal verification. Here, we discuss a unique verification strategy that it is integrated in the synthesis flow and it doesn't require any verification skills at all, but only how to compile and run C programs, which are driven by run-time options activated by keystrokes. Moreover, because the verified model is the same as this of the synthesis process, it reassures that the verified module is exactly the same as the generated hardware, so the same regression tests can run as in the high-level specification C/Ada program model. The generated cycle accurate simulators are C programs that model the execution of the formal synthesized FSMs. Because of the synthesis transformation formal model with logic inference that we use, the whole synthesis/verification flow of our approach is formal.

Keywords—formal verification, high-level synthesis, E-CAD, cycle-accurate simulation

Data availability

The data that support the findings of this study are not openly available due to reasons of patent protecting data, and are available from the corresponding author upon reasonable request in a controlled access repository where relevant.

I. Introduction (Heading 1)

The emergence of embedded computing systems and components, the promotion of IoT-based solutions and even the appetite for energy of large computing systems have refortified the current trend for automatic high-level design and development flows that generate custom hardware which is order of magnitude times faster, smaller, and low energy – hungry in comparison with typical microprocessor-based solutions. Sometimes the output of what we call High-level Synthesis (HLS) is implemented in the form of hardware accelerator boards, which are plugged in the computing system's bus, increasing in this way dramatically its performance. Some other times a small to medium-size FPGA undertakes the run of critical tasks of such an embedded system. In all cases the complexity of such components and

subsystems is increasing, and thus the requirement for robust and well verifiable HLS tools and design flows.

Therefore, there is a need for automated and formal verification flows that are modeled in the same way as the internal synthesis hardware model and that can deliver in an automatic and formal way, verification of hundreds of states in a straightforward and easy to get accustomed to way. The main contribution of this paper is exactly this. In this way, the presented and discussed verification process that is presented here, is automatically modelled and automatically generated from the internal formal FSM model that is used in the our tools which synthesize into coprocessor or accelerator components. Moreover, since the HLS framework is completely based on formal transformations the whole synthesis to simulation flow becomes formal. Our contribution also continues to that the generated verification models are in a way cycle-accurate simulators (CAS), coded in ANSI-C and which match the FSM-controlled coprocessors, one-to-one module at a time. Therefore, there is no training and learning period of complex mathematical formalisms, but just how to compile and execute C programs. Moreover, since the synthesis starting point is in high-level program code, flexible and fast to execute high-level testbenches can compare automatically the synthesized results with those of the high-level specification programs.

The generated cycle-accurate C programs correspond one to one with the design's hardware modules, which in turn correspond to the input ADA high-level routines. Therefore, the same test vectors can be used at the two verification levels, and it is easy to compare, manually or automatically the two level results. Also, faults simulation can be provided by altering the C code of the generated accelerator and inserting faults. The source of the generated cycle-accurate simulator is the internal FSM model of the synthesized coprocessor, therefore, the verification flow is formal.

Next section analyses existing work, mainly on formal verification. The section after this introduces the technology of the CCC hardware synthesizer. Next section analyses the formal background of the paper's verification strategy. The section after this discusses experimental results and finally the last section draws useful conclusions.

II. Existing Work and Background

Based on And-Inverter Graphs (AIGs), which allowed to work with constraints in both Synthesis and Verification, the ABC tool exercised AIGs, rewriting, sequential SAT sweeping, retiming, interpolation, etc., as new means to deal with large logic cones and an efficient method for solving Boolean problems in both the above tasks [1]. However, the origin of this methodology has been rather focusing on Boolean networks and much less on large, hierarchical and complex conditioned FSM models.

A formulation based on based on a dependency graph (DG), and conditions expressed with linear temporal logic

979-8-3503-8368-3/23 $31.00 © 2023 IEEE

(LTL)/past LTL (PLTL) properties, is extended to provide a systematic and automatic method for sequential clock-gating synthesis and verification [2]. However, this method suffers from large redundancy in the Sequential equivalence checking (SEC) task, which has to be removed in order for the method to be practical.

In [3], the well-known technique from High-level Synthesis, speculation is used in applying sequential equivalence checking, although this is tuned to optimize a particular type of designs, and it is not directly applicable to the general set of complex conditional and hierarchical models. The problem of state explosion of the model checking algorithms, with the increase of system components, has forced the authors of [4] to introduce bounded model checking techniques which combine model checking with satisfiability solving.

The authors in [5] argue that proof construction is unnecessary in the case of finite-state concurrent systems, and can be replaced by a model-theoretic approach which will mechanically determine if the system meets a specification expressed in propositional temporal logic. The authors claim that their model checker, is similar to the global flow analysis algorithms used in compiler optimization, and has complexity linear in both the size of the structure and the size of the specification. A different approach concerns the combination of a SAT-solver with induction for safety property checking of hardware in a real design flow [6].

A combination of equivalence checking problem theorems with clock gating, retiming, and redundancy removal, leads to SEC problems, which are problematic for a general, state-of-the-art SEC engine. However, a new approach completed verification of all examples and was about 30 times faster on the 3 cases where the conventional SEC engine was able to prove the problem within one hour [7]. An algebraic approach to functional verification of gate-level, integer arithmetic circuits is reported in [8]. This method is based on extracting a unique bit-level polynomial function computed by the circuit directly from its gate-level implementation and the experiments were done with the ABC system.

In order to efficiently verify gate level arithmetic circuits, the authors in [9] utilized a computer algebra based approach so that the circuit and the specification are modeled in polynomial system and the verification problem is formulated as polynomial reduction techniques using Groebner basis of circuit polynomial. To deal with the costly Groebner basis computation as well as intensive polynomial reduction, they introduced a canonical decision diagram named Horner Expansion Diagram (HED), derived a suitable term order to represent and manipulate polynomials and found repetitive components based on automata. To evaluate the effectiveness of their verification technique, they applied it to very large arithmetic circuits including multipliers. The method proved efficient and scalable for any size of complex multiplier circuits, nevertheless it seems that the method doesn't apply in complex, hierarchical FSM-controlled sequential circuits.

The authors in [10] propose a design flow for modular equivalence checking, high level synthesis, and optimization; consider hidden monomials to factorize those polynomials which do not have any common monomials; improve the previous optimization heuristics to eliminate multi-operand common sub-expressions as much as possible; implement all these algorithms on top of the modular Horner expansion

diagram package. Again here the method is not suitable for complex sequential engines.

A mutation-based debugging method is discussed in [11] which is suitable for verifying datapath-dominated multimedia and embedded applications. In order to deal with the problem of verifying sequential circuits, which is not feasible with contemporary techniques, the authors in [12] introduce an algebraic geometry which: i) implicitly tests the sequential arithmetic circuit over multiple (k) clock-cycles; and ii) represents the function computed by the state-registers of the circuit, canonically, as a multi-variate word-level polynomial over Sequential Galois field (F_2^k). This method was validated with verification of sequential multiplier circuits. However, it is doubtful that it is applicable to any arbitrary complex FSM-driven sequential design.

[13] presents a model checking algorithm with a strategy similar to a human. The algorithm is suitable for safety properties and it is particularly industrious. The algorithm generates lemmas with a structure similar to logic clauses which are inductive to previous lemmas and some assumptions in a stepwise manner. A subset of the generated lemmas that converge include a one-step strengthening of the induction of a given property. Lemmas generated by Humans are more generic in their form than the above generated clauses. In this manner, the machine-human analogy breaks down at some point of the process; nevertheless, in order to conclude to a proof the incremental generation of the above lemmas is similar. State-based designs can be tested as well, but not in fully automatic mode.

The authors in [14] aim in calculating the probability of bug detection, with MatLab models, taking into account the temporal behavior of a hardware design. However, their model is decoupled from any formal model of synthesis, therefore, their method is susceptible to produce diverting results depending on the simulation method and model.

Fournier et al. [15] calculate the probability of bug detection with different test strategies, such as directed, random and round-robin, but their approach doesn't consider the temporal behavior of a design under simulation, it only applicable to combinational sets. Therefore, it cannot be used for verifying complex FSM-based architectures.

In [16], a probabilistic analysis framework is proposed for bug detection, which considers the temporal behavior of system design. The procedure is complex as the analysis has time complexity quadratic to the number of coverage elements and linear to the number of simulation cycles. The authors in [17] presented an approach of performing soft-error failure rate analysis of arbitrary sequential circuit designs. This approach relates the failure rate of FFs with the behaviors of applications run on it. The technique is developed using sequential equivalence checking techniques, with minimal requirements on designers and is an automatic and complete evaluation technique. The circuit reliability analysis is based on fault simulation, by comparing parallel simulations of the fault netlist and the "golden" netlist, therefore it requires a high number of simulations and fault simulations for different sets of stimuli.

A RTL FSMD transformation procedure for equivalence checking of scheduling based on deep sequences or paths was proposed in [18]. The authors use FSMD transformations to facilitate equivalence checking with instances of combinatorial operations. However, this method doesn't allow

979-8-3503-8368-3/23 $31.00 © 2023 IEEE   212

processing of deep Synthesis transformations such as loop pipelining and loop unrolling. Nevertheless, it is the only known method that uses heuristic transformations to validate FSMD sequences (e.g. paths). Also, in the paper in [18] there is no indication of the time complexity of this method.

III. THE CCC HARDWARE SYNTHESIS FRAMEWORK

The two main phases of the CCC synthesizer [19] are the frontend and the backend compiler. The frontend compiler is capable of translating programs in a number of programming language programs (including ADA) into the Intermediate Tables Format (ITF). ITF is a formal syntax which resembles and implements the form of Prolog facts. Therefore, each generated ITF is a form of database which models the translated program algorithms into this formal syntax. The frontend compiler is implemented with the aid of compiler-compiler techniques with a formal language subset syntax for the parser generator, which makes it formal.

The backend compiler is based on logic predicates and therefore is implemented in formal Horn clauses. The whole translation, scheduling and optimizations process is formal since it is based on first-order logic. The formal internal model of each generated hardware model is that of the FSMD. There are finite N states which belong to the set S of states, the set T of transactions which map current states CS to nest states NS on the total set of states S.

Each state $s \in S$ and each transition $t \in T$ where

$$T \subseteq S \times S \qquad (1)$$

and CS,NS are orders

(ordered sets or paths) of states S   (2)

Also initial state and final state

$$is, fs \in S, CS, NS \qquad (3)$$

While the FSMD is executed a number of input signals and memories are read by a finite number of operations OP

$$op \in OP \qquad (4)$$

When the system reaches the final state $fs \in S,NS$, then the number of output registers and memories have reached their final result values.   (5)

The whole formal model of the optimized and generated FSMD is described by the above five conventions. Ii is important to note that the generation of the hardware HDL code, as well as the CAS C code are extracted from the above formal model of the generated and optimized FSMD. The elementary information from which the FSMD is extracted is the notion or entity of state entity. A state entity is based on the above five rules (1) to (5) and it contains:

- The current state, *cs*.

- The next state when there is no conditions applied, *ns*.

- The condition when there is any: *tc*.

- The list of operations that execute in parallel in this state, *ops*.

- The true/false conditional execution of operations, when there is a condition, *tops, fops*.

- The true/false conditional next states, when there is a condition, *tns, fns*.

A schedule is complete when it contains all such state entities, including the initial state and final state, *is, fs* entities. This formal model is followed from the beginning of the hardware optimization throughout all the synthesis flow, thus both the synthesis and embedded verification flow are formal.

The same formal internal model of the optimized FSMD is used to extract the ANSI-C, cycle-accurate simulators as it is demonstrated in Fig. 1, bellow.

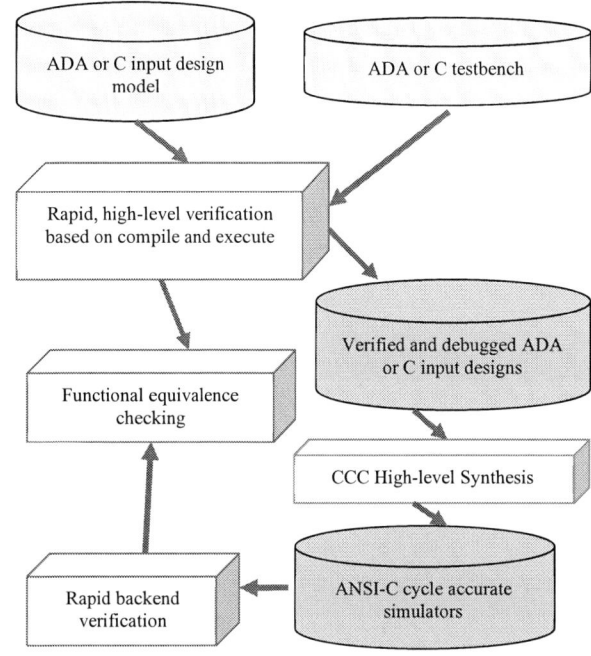

Fig. 1.   Synthesis-embedded verification flow in the CCC framework

Each automatically generated CAS is coded in ANSI-C and is compilable by an as well automatically generated compile script. The generated executable provides to the user buttons for resetting the FSM, writing values to inputs, advancing to the next state, reading the values of all registers and outputs, printing the values of all arrays in external text files, running the whole path of the states up to the final state, and finally exiting the CAS.

A. The synthesis formal rules and constructive predicates

The high-level synthesis and optimization engine of the backend compiler is constructed with Logic Programming predicates. The engineered scheduling algorithm is called PARCS. These follow the format of the Horn clauses of the Formal Logic relations like the ones described in the following paragraphs.

The backend compiler reads and incorporates the ITF statements into its internal knowledge-base of logic predicates and relations. Therefore, the ITF "guides" this knowledge-base of rules and inference engine, in such a way that the latter infers the appropriate "conclusions" during transformations of the source programs into scheduled, functionally-equivalent FSMs. To implement these transformations, the back-end compiler consists of definite clauses [2] of the following form:

$$A_0 \leftarrow A_1 \wedge A_2 \ldots \wedge A_n \text{ (where } n \geq 0) \qquad \text{(form 1)}$$
where $\leftarrow$ and $\wedge$ are the logical implication and conjunction symbols (A $\leftarrow$ B means that if B applies then A applies)

respectively, and $A_0, \ldots, A_n$ are atomic formulas (Prolog clauses) of the form:

   predicate_symbol(Var_1, Var_2, ..., Var_N)   (form 2)

where the positional parameters Var_1,...,Var_N are either variable names (as in the "inference rules"- Prolog clauses), or constants (as in the ITF statements' facts) [20].

B. The synthesis scheduler algorithm

The following text-box abstracts the embedded scheduler in the backend, HLS framework.

```
1. start with the initial schedule (inc. the
   special external port configuration)
2. Current PARCS state <- 1
3. Get the 1st state (is) and make it the current
   state (cs <- is)
4. Get the next state (ns)
5. See if the operations of the next state (ns
   ops) do not have dependencies of any kind
   with the current state operations (cs ops)
6. If there are no dependencies, then absorb the
   next state into the current PARCS state; (ns
   ops ∈ cs ops). If there are dependencies then
   absorb the current state into the PARCS state
   (PARCS cs <- cs), store the PARCS state,
   PARCS state <- PARCS state + 1; make next
   state the current state (ns <- cs)
7. If next state = conditional then call the
   conditional (true/false branch) processing
   predicates, else continue
8. If there are more states to process (cs |=
   fs) then goto step 4, otherwise (cs = fs)
   finalize the current PARCS state (PARCS cs
   ops <- all cs ops) and terminate
```

Fig. 2.   Pseudo-Code or the synthesis-optimisation algorithm

Of course, in reality, the code of the backend compiler, including the HDL writers and CAS writers are much more sizeable and complex than the code in Fig. 2 (about 150000 lines of Prolog code lines).

The flow of the scheduler follows the flow of the input algorithms as they are encapsulated in the ITF facts. Because all complex control flow structures are dissolved in simple and (if necessary enclosed) if-then-else blocks, with any necessary control hierarchy, there are no limitations in the folded control structures that PARCS can optimise. This is not true for most of the existing industrial and academic synthesizers (e.g. known problem for while-like control constructs that the latter cannot optimize or report).

The above are major contributions of the work reported here and in conjunction with the benefits of a formal synthesis and verification philosophy makes this paper's approach unique.

IV. Synthesis and Verification Experiments

Experiments with five synthesis/verification benchmarks are discussed in this section, in order to prove the concept of this research. These are an insertion sorting algorithm, a procedure to return true if an array is sorted, a binary gap algorithm to return the length of all zeros surrounded by ones, the calculation of the smallest index S^N that it's so far indexed array elements are all within the range 0-S^{N-1}, and a more complex line drawing procedure from the area of computer graphics. All of the five benchmarks are originally coded in procedural ADA and then synthesized into hardware with the CCC High-level Synthesis Framework.

```
for I in 0..Size loop
  Key := ARRAYvar(i);
  j := i - 1;
  TMP2 := ARRAYvar(j);
  while (j >= 0) and (Key < TMP2) loop
    TMP1 := ARRAYvar(j);
    ARRAYvar(j + 1) := TMP1;
    j := j - 1;
    TMP2 := ARRAYvar(j);
  end loop;
  ARRAYvar(j + 1) := Key;
  ...
```

Fig. 3.   Code or the insertion sorting procedure core

```
-- now instantiate the device-under-test
InsertionFunc (Array_out,Array_in, Size);

  -- now print the results
for I in 0..9 loop
  TEMPP := Array_out(I);
  put(my_output_file, TEMPP);
  put(my_output_file, " ");
  new_line(my_output_file);

  -- now compare with reference data
  get(my_reference_file, my_int);
  if (my_int /= TEMPP) then
    test_passed := false;
    put(" result "); put(I);
    put(" mismatch "); new_line;
    put(" actual = "); put(TEMPP);
    put(" reference = "); put(my_int);
    new_line;
  end if;

end loop;
...
```

Fig. 4.   High level testbench of the insertion sorting algorithm

A. Insertion Sorting Algorithm

The excerpt core of the procedure is shown in listing Fig. 3. From the code listing in Fig. 3, it is easy to spot the while loop that is the loop core of the sorting mechanism. Please note here that no other known HLS tool accepts and optimizes into hardware while loops, as it was explained in the formal synthesis flow in section III. The result of the sorted array is stored in array variable ARRAYvar.

The high level testbench in ADA which automatically compares the high level ADA model simulation with that of the generated hardware model is listed in Fig. 4. It consists of a loop that goes through the generated array data and compares it with reference data from the cycle-accurate simulation. If it finds differences between the two it issues an alarm for every difference found.

After the generation of the VHDL for the main procedure, the automatic generation of the cycle-accurate simulator in ANSI-C follows. It is extracted from the same formal internal FSM model with the synthesis of the VHDL model. The cycle-accurate simulator (CAS) offers all options for fast simulation of the FSMD. It provides buttons for initializing the FSMD, provide values for inputs, advancing to the next

979-8-3503-8368-3/23 $31.00 © 2023 IEEE

state, running all FSMD schedule until the last state, and printing the values of all registers and flash the contents of all output arrays to external files. The execution of the CAS of the insertion sorting algorithm is shown in Fig. 3.

Fig. 5 is a snapshot of showing the registers' values, when the whole FSMD has reached the final state. Before this, by pressing p(rint arrays) CAS stored the contents of the array variable ARRAYvar into the arrayvar.txt file, and in this way allowing direct comparison with the high level ADA model execution. The whole execution path from the ADA compilation down to CAS execution took less than 1 minute.

Fig. 5. Execution of the CAS of the insertion sorting FSMD model

B. Determining when an array is sorted

The core code of the main procedure is shown in Fig. 6.

```
i := 1;
SortedCond := TRUE;
while (i < Size) and SortedCond loop
    tmp1 := Array_in(i - 1);
    tmp2 := Array_in(i);
    SortedCond := tmp1 <= tmp2;
    i := i+ 1;
end loop;    -- while
...
```

Fig. 6. The code to determine if Array_in is sorted

The testbench code for the sorted procedure is trivial, it just executes the above code and check at the end the value of the Boolean SortedCond (see code of Fig. 6). Fig. 7 shows the CAS execution snapshot and the end of the FSMD.

Fig. 7. CAS end of execution of the sorted procedure

In Fig. 7, it is easy to see from the value of the Boolean sortedcond, or sortedfunc that the input array was sorted.

```
-- Skip leading 0s from left to right.
TMP1 := Array_in(bit);
bit := 0;
while (TMP1 = 0) and (bit <= Size_1) loop
  NonZeroBit := Bit;
  bit := bit + 1;
  TMP1 := Array_in(bit);
end loop;

--  Could not find any leading 1.
if NonZeroBit = -1 then
  LargestGap := 0;
end if;

--  locate largest gap (sequence of 0s).
while (bit <= Size_1) loop
  TMP2 := Array_in(bit);
  bit := bit + 1;
if (TMP2 = 0) then
  gap := gap + 1;
else
  if (LargestGap < gap) then
    LargestGap := Gap;
    gap := 0;
  end if;
end if;
end loop;
return LargestGap;
```

Fig. 8. The largest 0s gap of an integer representation

C. Detecting the largest zeros gap in binary representation

The next application detects the maximal sequence of consecutive zeros that is surrounded by ones at both ends in the binary representation of an integer N. Initially integer N is given in a form of an array which contains the binary digits of its binary representation. The code is given in Fig. 8.

The code of the high level test is trivial and therefore it is not shown here, it just calls the main function of the binary gap and it returns the gap. However the same input array of bits is used in the high level and the CAS testbench. The execution of the generated CAS is shown in Fig. 9.

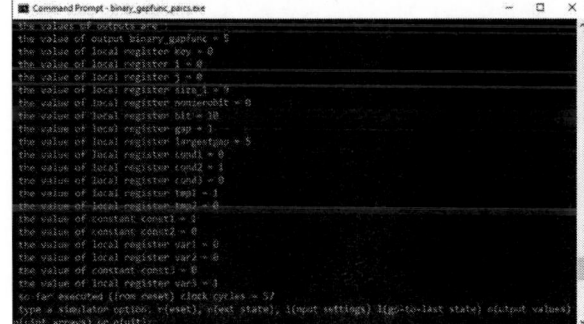

Fig. 9. CAS execution close to last state of the binary gap, with option o(utput values) to show the values of registers (please see binary_gapfunc).

The same array with a max gap of 5 was used in both the high-level and the CAS simulations.

979-8-3503-8368-3/23 $31.00 © 2023 IEEE

D. Smallest Index that contains/covers all indexed values of the array

This benchmark calculates the smallest possible index N, so that all array elements are within the range 0..N-1. The core of the ADA code is shown in Fig. 10.

```
    smallest := 0;

    N_1 := N-1;
    for I in 0..N_1 loop
     Occur(i) := 0;
    end loop;
-- Preserve position of last
-- modification.
     for I in 0..N_1 loop
            TMP1 := Array_in(i);
            TMP2 := Occur(TMP1);
            IF TMP2 = 0 THEN
              TMP2 := TMP2 + 1;
              SMALLEST := i;
            END IF;
        end loop;
...
```

Fig. 10. Smallest containing array index

The smallest "covering" index is held by variable smallest, and the algorithm's core is mainly a for loop which doesn't feature any issues for the High-Level Synthesis flow. As in the previous benchmarks, here also the CCC Synthesis flow was followed. The synthesized model produced also the CAS in ANSI-C.

The high level testbench just calls the procedure of the algorithm, so it is trivial and it is not shown here. The execution of the generated CAS is shown in Fig. 9.

Fig. 11. The CAS execution of the smallest covering index and results

E. A Computer Graphics Line-Drawning algorithm

This benchmark contains a repetitive procedure which pixel by pixel calculates the points of a straight line and which can be drawn on the screen of a computer graphics application. The ADA code core is depicted in Fig. 12.

The array LINEPOINTS() contains the coefficients of the horizontal and vertical position of each point of the calculated line. The user just provides the X,Y dimensions of the start and finish points of the line, and the above procedure draws the straight line that connects them.

The high level testbench of the linedraw procedure reads the coefficients of the straight line points and compares them with

```
    DX := FINISHP.X - STARTP.X;
    DY := FINISHP.Y - STARTP.Y;
    LINEPOINTS(FIRST) := STARTP;   -- first point
    XC := STARTP.X;
    YC := STARTP.Y;
    if DX >= 0 then
      if DX >= DY then
        for J in 1..DX-1 loop
          if ERROR < 0 then
            XC := XC + 1;
            I := I + 1;
            LINEPOINTS(I) := (XC, YC);
            ERROR := ERROR + DY;
          else
            XC := XC + 1;
            YC := YC + 1;
            I := I + 1;
            LINEPOINTS(I) := (XC, YC);
            ERROR := ERROR + DY - DX;
          end if;
        end loop;
      else
        for J in 1..DY-1 loop
          if ERROR < 0 then
            XC := XC + 1;
            YC := YC + 1;
            I := I + 1;
            LINEPOINTS(I) := (XC, YC);
            ERROR := ERROR + DY - DX;
          else
            YC := YC + 1;
,,,
```

Fig. 12. Part of the core of the line-drawing procedure ADA code

the line produced by CAS or Modelsim on VHDL. In all cases, there were no mismatches between the two views of the straight line. The CAS execution is viewed in Fig. 13.

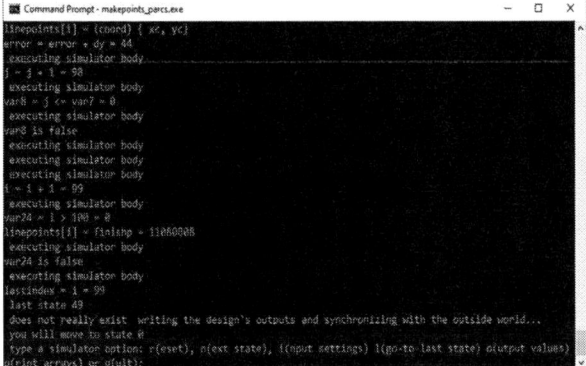

Fig. 13. The linedraw CAS execution

CONCLUSIONS AND FUTURE WORK

This paper discussed a novel formal flow of Synthesis-embedded verification, where the formal verification flow is integrated with the HLS one. Verification by the user is done via running the result of the flow, which includes the automatic and rapid generation of cycle-accurate simulators (CAS). As far as the author is concerned there are no other known formal techniques or tools of this kind, that they embed the verification data automatically generated by the synthesis core, as it happens in CCC. The Synthesis transformational optimizations are based on logic Horn's predicates and the frontend compiler is developed with compiler-compiler techniques. Thus, the whole transformation from software programs to hardware modules is formal. The synthesizer generates a provably-correct FSMD (synthesizable) HDL of

979-8-3503-8368-3/23 $31.00 © 2023 IEEE

program accelerators or coprocessors. The same formal model is used in the extraction and writer of the CAS.

Many examples were verified with this flow, nevertheless for reasons of space and time five smaller ones were presented in this paper as proof of concept for the formality of the generated CAS. In all cases the CAS proved the correct functionality of the generated coprocessor. CAS is driven with a number of next-state options and data observing techniques and it only requires the GNU-C tools for compiling it. It contains all the options for a successful simulator of this kind. Compilation and execution are done in a matter of seconds.

The main contributions of this work include the formal verification, embedded in synthesis flow, and the formal model of the compiled FSMD. The same formal model is consistent in its form and process from the beginning of the synthesis flow until the generation of the optimized FSM-controlled coprocessors. Moreover, it is a fact that this flow is rapid, and it doesn't require the knowledge of complex mathematical model checking or equivalence checking techniques. Another important contribution is the degree of automation which is achieved (even the CAS compilation is done via automatically generated batch command line file) using this paper's methodology. Also, apart from basic ANSI-C programming no other hardware or simulation – related skills are required by the user of the tools.

Potential applications of the CCC tool and the CAS testbenches include all hardware design areas. Some are the development of embedded systems, FPGAs, ASICs, SOCs, aviation and space electronics, safety-critical systems, real time systems, nano-electronics, all battery-run systems, hardware accelerators, massively parallel systems, video and audio processing hardware and all types of hardware found and any type of smart systems.

Future work includes generalization of the input programs with the acceptance of other forms (MatLab, C++, Python, etc) and more automation in the point of cross-checking the high-level simulation results with that of the low level, functional CAS results.

ACKNOWLEDGEMENT

This paper was partially supported by the University of Western Macedonia.

REFERENCES

[1] R. Brayton and A. Mishchenko, "ABC: An academic industrial-strength verification tool" in Computer Aided Verification CAV 2010, Berlin, Germany:Springer, 2010.

[2] U. Nilsson, and J. Maluszynski, Logic, Programming and Prolog, John Wiley & Sons Ltd., 2nd Edition, 1995.

[3] Yu-Yun Dai and Robert K. Brayton, "Verification and Synthesis of Clock-Gated Circuits", IEEE Transactions On Computer-Aided Design of Integrated Circuits and Systems, Vol. 38, No. 2, February 2019.

[4] R. K. Brayton, N. Een, and A. Mishchenko, "Using speculation for sequential equivalence checking," in Proc. Int. Workshop Logic Synth., 2012, pp. 139–145.

[5] E. Clarke, A. Biere, R. Raimi, and Y. Zhu. "Bounded model checking using satisfiability solving", Proc. Formal Methods in System Design (FMSD), vol. 19(1), Kluwer 200

[6] E.M. Clarke, E.A. Emerson, and A.P. Sistla, "Automatic verification of finite-state concurrent systems using temporal logic specifcations," ACM Transactions on Programming Languages and Systems, Vol. 8, No. 2, pp. 244–263, 1986.

[7] Mary Sheeran, Satnam Singh, and Gunnar St°almarck. Checking safety properties using induction and a SAT-solver. In Warren A.

Hunt, Jr. and Steven D. Johnson, editors, Formal Methods in Computer-Aided Design, Third International Conference, FMCAD 2000, Austin, TX, USA, November 1-3, 2000, Proceedings, volume 1954 of Lecture Notes in Computer Science, pages 74–91, Heidelberg Berlin, 2000. Springer-Verlag

[8] Y.-Y. Dai, K.-Y. Khoo, and R. Brayton, "Sequential equivalence checking of clock-gated circuits," in Proc. ACM Design Autom. Conf., San Francisco, CA, USA, 2015, pp. 1–6.

[9] Cunxi Yu, Walter Brown, Duo Liu, André Rossi, and Maciej Ciesielski, "Formal Verification of Arithmetic Circuits by Function Extraction", IEEE Transactions On Computer-Aided Design Of Integrated Circuits And Systems, vol. 35, no. 12, December 2016.

[10] F. Farahmandi, B. Alizadeh, and Z. Navabi, "Effective combination of algebraic techniques and decision diagrams to formally verify large arithmetic circuits," in Proc. IEEE Comput. Soc. Annu. Symp. VLSI (ISVLSI), Tampa, FL, USA, 2014, pp. 338–343.

[11] B. Alizadeh, and M. Fujita, "Modular Datapath Optimization and Verification based on Modular-HED", IEEE Transactions on Computer-aided design (TCAD), vol. 29, no. 9, pp. 1422-1435, September 2010.

[12] B. Alizadeh, "A Formal Approach to Debug Polynomial Datapath Designs ", in ASPDAC, 2012, pp. 683-688.

[13] X. Sun, P. Kalla, T. Pruss, and F. Enescu, "Formal verification of sequentia Galois field arithmetic circuits using algebraic geometry," in Proc. Design Autom. Test Europe Conf. Exhibit. (DATE), Grenoble, France, 2015, pp. 1623–1628.

[14] A. Bradley, "Sat-based Model Checking Without Unrolling", in Verification, Model Checking and Abstract Interpretation (VMCAI), pp. 70–87, 2011.

[15] Mingming Zhang, Shuqin Geng , Wensi Wang, Xiaohong Peng, Menghao Chu, Shengyuan Zhou, Zhonghou Zhang, Hang Lu, Pengkun Li, and Ronghao Zhu, "Probabilistic Analysis for Sequential Circuits Verification Using Markov Chains", IEEE Transactions on Computer-aided design (TCAD), vol. 68, no. 1, pp. 481-485, January 2021.

[16] L. Fournier, A. Ziv, E. Kutsy, and O. Strichman, "A probabilistic analysis of coverage methods," ACM Trans. Design Autom. Electron. Syst., vol. 16, no. 4, pp. 1–20, Oct. 2011.

[17] M. Zhou, W. N. N. Hung, X. Song, M. Gu, and J. Sun, "Temporal coverage analysis for dynamic verification," IEEE Trans. Circuits Syst. II, Exp. Briefs, vol. 65, no. 1, pp. 66–70, Jan. 2018.

[18] Tun Li, Qinhan Yu, Hai Wan, and Sikun Li, "Application Specified Soft-Error Failure Rate Analysis Using Sequential Equivalence Checking Techniques", Tsinghua Science And Technology, ISSN l11007-0214 10/14, pp. 103–116, DOI: 10.26599/TST.2018.9010136, vol. 25, no. 1, February 2020.

[19] Michael Dossis, "Automated Extraction of Hardware Accelerators via an Intelligent Knowledge-based System", International Journal of Intelligent Information Processing (IJIIP), vol. 1, no. 2, pp. 14-31, December 2010.

[20] B. Lin, and S. Vercauteren, "Synthesis of Concurrent System Interface Modules with Automatic Protocol Conversion Generation", Proc. of ACM ICCAD 1994, pp. 101-108, San Jose, CA, Nov 1994.

[21] Raul Acosta Hernandez, Marius Strum,and Wang Jiang Chau, "Transformations on the FSMD of the RTL Code with Combinational Logic Statements for Equivalence Checking of HLS", in 16th IEEE Latin-American Test Symposium (LATS), pp. 1-6, doi: 10.1109/LATW.2015.7102518 25-27, March 2015.

[22] Kuehlmann, A., Paruthi, V., Krohm, F., Ganai, M.K., "Robust Boolean reasoning for equivalence checking and functional property verification". IEEE Trans. CAD, vol. 21, no 12, 1377–1394 (2002).

Study and Development of a High-Speed Fused Filament Fabrication 3D Printer

Ioannis Christodoulou
School of Mechanical Engineering
National Technical University of Athens
Athens, Greece
ichristodoulou@mail.ntua.gr

Vasiliki Alexopoulou
School of Mechanical Engineering
National Technical University of Athens
Athens, Greece
valexopoulou@mail.ntua.gr

Angelos P. Markopoulos
School of Mechanical Engineering
National Technical University of Athens
Athens, Greece
amark@mail.ntua.gr

Abstract—**Fused Filament Fabrication (FFF) is a very well-known Additive Manufacturing Method for the development of highly customized, lightweight and with complex geometry products. Nevertheless, yet the achieved printing speeds of this method are rather slow, making it unaffordable for mass production. In the current paper, a customized FFF printer is presented which is based on an advanced electromechanical system, which achieves to increase the printed speed over 500% compared to conventional FFF printers, without sacrificing the quality and the mechanical strength of the fabricated product.**

Keywords— Fused Deposition Modeling (FDM), High-Speed Manufacturing, Printing Speed, Advanced Electromechanical Systems, ABS.

I. INTRODUCTION

Fused Filament Fabrication (FFF) is an Additive Manufacturing method, which is widely used for the fabrication of components out of various thermoplastic materials, such as acrylonitrile butadiene styrene (ABS), nylon, polyetheretherketone (PEEK), polylactic acid (PLA), polypropylene (PP), polyvinyl alcohol (PVA), etc [1].

During FFF, the fabricated object is built layer-by-layer. Specifically, melted thermoplastic filament is forced through a nozzle over the building bed and it is deposited on the previously solidified layer, with which it gets bonded [2]. This technique enables the fabrication of highly customized objects with very complex geometries, which is the most important advantage of FFF [3]. Other benefits of FFF are fabrication of lightweight products, low cost and small size equipment, no need for tools and reduced material waste [4].

All the above benefits have made FFF a very promising method for the development of products needed in aerospace, electronics, biomedical, construction and automotive industries [5]. Nevertheless, the reasons why this method is not yet widely used in the industry are mainly the low mechanical strength of its products [6] and the high printing time, which make it unaffordable for mass production [7]. Thus, when researchers try to increase the printing speed of an FFF machine, they should be very careful not to sacrifice any further the quality and the mechanical strength of the printed object.

Several studies have explored the potential for achieving high-speed printing in FFF. By optimizing various parameters and adopting innovative approaches, researchers have reported significant improvements in printing speeds while maintaining acceptable part quality. For instance, Czyżewski et al. [8] demonstrated that using a larger diameter nozzle of 0.8 [mm] and 1.2 [mm] instead of the standard 0.4 [mm] nozzle in FFF printing can double the printing speed without compromising part quality. This nozzle modification allows for faster extrusion rates, leading to less printing time. In the pursuit of faster printing, Chauvette et al. [9] investigate the use of multi-nozzle systems in FFF. By employing four nozzles simultaneously, they demonstrated a printing speed of up to 250 [mm/s], with a flow rate of approximately 319.4 [mm³/s], showcasing the potential of parallel extrusion for high-speed FFF. Examining the influence of layer height on build time, Yang et al. [10] suggests that a three-fold increase in layer height causes a four-fold decrease in printing time. Furthermore, Kopets et al. [11] investigates the relation between printhead positioning rate and detail quality of FFF parts. The printing speeds were ranging from 15 to 175 [mm/s]. These speeds indicate the achievable rates when optimizing process parameters for specific layer thicknesses. In another study, Allen et al. [12] examines the impact of different kinematics and printing strategies on FFF printing speed. By using a parallel robot instead of the traditional cartesian kinematics robot, they achieved printing speeds of up to 300 [mm/s] while maintaining acceptable part quality.

The current paper presents the high-speed FFF printer developed by the researchers of the Laboratory of Manufacturing Technology of the School of Mechanical Engineering of National Technical University of Athens (NTUA). The novelty of this FFF printer is its advanced level electromechanical system that allows excellent control to the nozzle movement and the filament deposition, which allows it to reach speeds up to 300 [mm/s] combined with minimum losses of quality and mechanical strength of the fabricated product. To prove this, the current FFF printer has been compared to other conventional 3D Printers, according to various ISO-based benchmark tests. The results of these comparisons are also given in the current paper.

II. EXPERIMENTAL METHODS

The specifications of the in-house built FFF-based technology 3D Printer are shown in the Table 1 below:

Table 1. Machine Specifications

Table 1.	Machine Specifications
Kinematics (type)	Core-XY
Max Flow-rate [mm³/s]	17
Max Printing Speed [mm/s]	260-320
Max Travel Speed [mm/s]	600-1200
Max Acceleration [mm²/s]	25000
Max Deceleration [mm²/s]	25000
Printing volume [LxWxH mm³]	320mmx320mmx350mm
Voltage [V]	48

979-8-3503-8368-3/23 $31.00 © 2023 IEEE

The design of the printer chassis originated from the Flying Bear Reborn 2 3D Printer chassis, with several modifications implemented. The essential design considerations for the specific 3D printer design encompass the selection of the kinematics type, the evaluation of the board's output electric power capabilities, and the determination of the maximum flowrate of the specified nozzle. The significance of kinematics lies in its impact on both the power demands of the motors responsible for the movement of the printing head, as well as the dynamic characteristics of the system. In this proof-of-concept, the chosen kinematics is Core-XY kinematics because to its involvement of both motors in the printhead movement, except for a 45-degree angle on the XY plane. This suggests that in the majority of movements, the necessary force for achieving the desired acceleration is distributed among the motors. This phenomenon can also be observed through the equations of motion of the system provided below:

$$X(t) = X_m(t) + Y_m(t) \qquad (1)$$

$$Y(t) = X_m(t) - Y_m(t) \qquad (2)$$

where X(t) represents the position of the print head in the X-axis as a function of time in [mm], X_m(t) represents the position of the X-axis motor (or motors) as a function of time in [mm], Y(t) represents the position of the print head in the Y-axis as a function of time, Y_m(t) represents the position of the Y-axis motor (or motors) as a function of time, both also in [mm]. Given the motion equations above, one can derive the force equations for each axis as a ratio to the acceleration of each axis:

$$A_x(t) = \frac{d^2 X(t)}{dt^2} \qquad (3)$$

$$A_y(t) = \frac{d^2 Y(t)}{dt^2} \qquad (4)$$

Thus, the force equations are,

$$F_x(t) = M_{printhead} * A_x(t) \qquad (5)$$

$$F_y(t) = M_{printhead} * A_y(t) \qquad (6)$$

Where, A_x(t) and F_x(t) represents the acceleration in [mm/s²] and the force in N, implied on the printhead in the X-axis, and A_y(t) and F_y(t) in Y-axis accordingly. $M_{printhead}$ represents the mass of the printhead in kg. In Figure 1 such system can be visualized.

Proceeding to the electrical/electronic design, the speeds and acceleration, as can be seen from Table 1 are in average 290 [mm/s] and 25000 [mm/s²] accordingly. Thus, a powerful enough electric system is needed to have the necessary output power for the motors to run in such speeds and accelerations. The board used is a BigTreeTech Octopus Pro board that supports up to 60 [V] input voltage giving the ability to the researchers to increase the current in each motor. In this proof-of-concept a 48 [V] power supply was used with 5 [A]

supply current capability. Based on this, the electric grid schematic can be seen in Figure 2 below.

Fig. 1. Core-XY kinematics equations of motion.

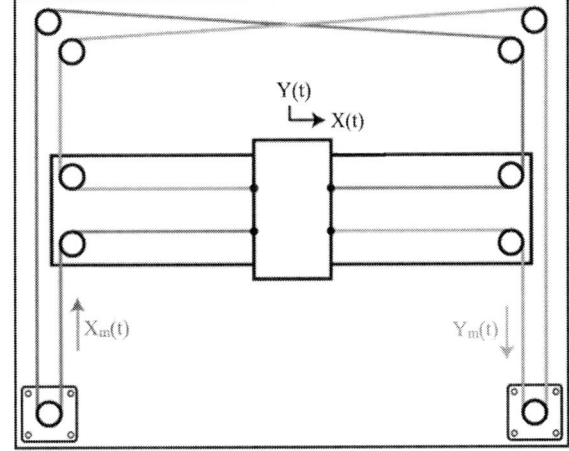

Fig. 2. Electrical grid schematic showing the board used.

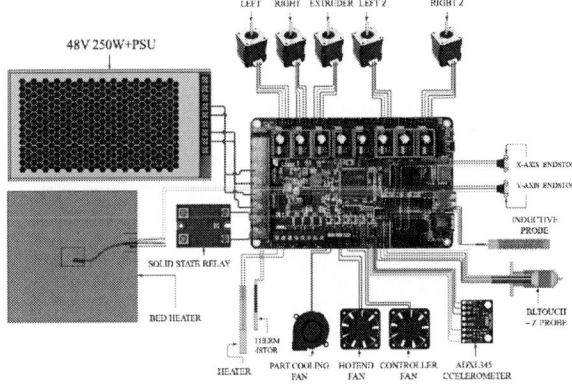

The software and firmware used in this project are based on Klipper Project . *Klipper* is an open-source firmware for 3D printers that offers an alternative to conventional firmware options such as *Marlin* or *Repetier*. What distinguishes *Klipper* is its distinctive methodology in printer control. Instead of doing intricate computations and motion planning on the microcontroller of the printer, Klipper redistributes the majority of these responsibilities to a higher-capacity host computer, such as a *Raspberry Pi* or a personal computer. The utilization of a distributed architecture enables Klipper to attain elevated levels of precision and enhanced printing rates. This is made possible by offloading sophisticated calculations and generating smoother motion commands to the host computer. *Klipper* is known for its extensive compatibility with many 3D printers and its high level of customizability, rendering it a favored option among proficient users aiming to enhance performance and achieve enhanced adaptability in their 3D printing configurations. In addition, *Klipper* presents a range of sophisticated functionalities that distinguish it from conventional 3D printer firmware. One notable characteristic is the presence of Input Shaper, a tool designed to mitigate vibrations and minimize the occurrence of resonance-induced artifacts in the context of 3D printing. Through the examination of the mechanical attributes of the printer, Input Shaper employs filters to the motion directives, so yielding prints that are more refined and of superior quality. *Klipper* offers a wide range of motor driver changes, enabling users to finely alter motor currents, microstepping, and other parameters in order to improve the performance of the printer. The degree of control provided is of great use in attaining accurate motion and addressing concerns such as overheating or skipped steps. Moreover, the modular architecture of Klipper renders it exceptionally versatile, enabling seamless integration with various 3D printer setups and supplementary components, like multiple extruders or supplementary axes.

In the present work *Input Shaper* feature was implemented on the machine using the Portable Input Shaper by FYSETC giving maximum frequencies of X and Y-axis at 28400 [mm/s²] and 26600 [mm/s²] accordingly using *mzv* algorithm as suggested by the *Input Shaper* feature. Furthermore, motor driver adjustments were crucial to achieve such accelerations as not only the motors' current was increased but also the active braking of the electric motors was also turned off as it would consume a lot of power without any significant advantage. The X-axis motor driver section of the *Klipper* configuration file can be seen in the Figure 3 below,

Fig. 3. X-axis motor driver section of the Klipper configuration file.

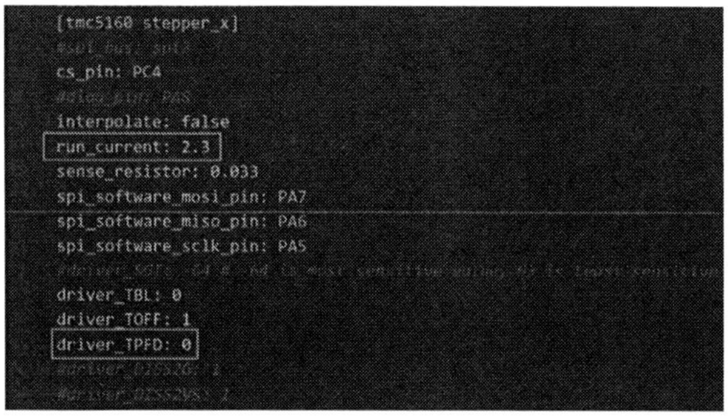

By default, *Klipper* configures the TMC drivers to operate in "spreadCycle" mode. If the driver is compatible with the "stealthChop" feature, it can be activated by incorporating the parameter *stealthchop_threshold: 999999* into the TMC configuration section. In a broad context, it can be observed that the spreadCycle mode offers superior torque and positional accuracy compared to the *stealthChop* mode. Nevertheless, it is worth noting that the utilization of stealthChop mode has the potential to result in a notable reduction in audible noise levels in certain printers. Tests conducted to compare different modes have revealed that when utilizing *stealthChop* mode, there is an observed increase in "positional lag" of around 75% of a full-step during constant velocity movements. For instance, on a printer with a rotation distance of 40 [mm] and 200 steps per rotation, the divergence in position during constant speed movements rose by approximately 0.150 [mm]. Nevertheless, the potential delay in acquiring the desired position may not result in a notable print imperfection, and some individuals may favor the more subdued operation of *stealthChop* mode.

To achieve optimal positional accuracy, it is recommended to utilize the *spreadCycle* mode and deactivate interpolation by setting the interpolate parameter to *false* in the TMC driver configuration, as shown in the Figure 3. When the configuration is adjusted in this manner, it is possible to enhance the microstep setting in order to minimize the audible noise that occurs during the movement of the stepper. Furthermore, current of the motors was set to 2.3 [A] by changine the *run_current* parameter in the configuration file. The increase in the current has as an effect the increase of the temperature of the motors, thus extra cooling was necessary which was achieved by installing heatsinks on the motors. Last but not least, the driver_TPFD parameter was set to zero to turn off the active braking of the motors between phase change of the rotors rotation. These settings enabled the researchers to push higher the accelerations of the printhead.

Same for Y-axis, *interpolate* was set to false, *run_current* to 2.3 A and *driver_TPFD* to zero.

Regarding the flow rate, the nozzle should be able to supply enough material for the print to continue smoothly on such high speeds. That is given by the maximum flow rate of a nozzle which in this case is 17 [mm³/s], giving the researchers the ability to achieve speeds of 260-320 [mm/s] with layer heights of 0.24 to 0.18 [mm] accordingly. The flow rate equation for FFF-technology 3d Printers can be seen in the equations 7 below,

$$Q = \frac{V}{t} \tag{7}$$

where, Q is the flow rate in cubic millimeters per second [mm³/s], V is the volume of material extruded in cubic millimeters [mm³], t is the time taken to extrude that volume in seconds [s].

The volume (V) is calculated by multiplying the cross-sectional area of the extruded filament (usually determined by the nozzle size) by the length of filament extruded (determined by the printing speed):

$$V = A * L \tag{8}$$

where, A is the cross-sectional area of the extruded filament in square millimeters [mm²] and L is the length of filament extruded in millimeters [mm].

The printer that will be subjected to comparison with the high-speed FFF machine is the Flashforge Creator 3 3D Printer. This particular printer is classified as a professional-grade 3D Printer, with a suggested printing speed of 60 [mm/s] and a recommended acceleration of 500 [mm/s²]. Figure 4 depicts the CAD of the High Speed FFF machine at the top and the Flashforge Creator 3 3D Printer at the bottom.

To allow for a straightforward comparison between the two different 3D printers, two artifacts were manufactured using each machine, following the requirements of the ISO ASTM 52902-2021 standard. Based on the ISO ASTM 52902-2021 standard, the specimens selected for the current study (refer to Figure 5) contain a linear artifact and a circular accuracy artifact of medium resolution.

Fig. 4. Top. CAD of High Speed FFF 3D Printer. Bottom. Flashforge Creator 3

Fig. 5. Top. 1. Linear artifact 2. Circular accuracy artifact Bottom. Artifacts' printing orientation on the build plate of the 3D printer. Origin's X-axis is depicted with red line, Y-axis with green and Z-axis with blue color line.

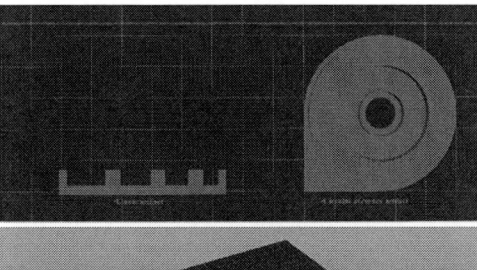

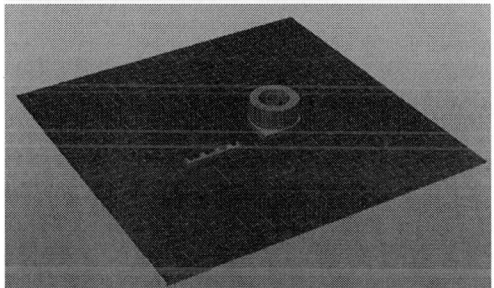

The print settings for each machine can be seen in the Table 2 below.

Lastly, it is worth noting that each specimen was duplicated, resulting in a total of eight printed specimens, with four produced by each printer. The measurements were conducted using a digital handheld caliper that adhered to the ISO ASTM 52902-2021 standard, which specifies a maximum permissible error (MPE) of 0.02 [mm]. To mitigate potential human errors, each specimen was measured in three distinct regions.

Table 2. Printing settings for each printer.

Table 2.	High Speed 3D Printer	Flashforge Creator 3
Printing Speed (mm/s)	260	60
Travel Speed (mm/s)	760	80
Acceleration/Deceleration (mm/s^2)	25000	500
Layer height (mm)	0.18	0.18
Nozzle Temperature (° Celcius)	240	230
Bed Tmperature(° Celcius)	100	100
Slicer Software	Prusa Slicer	Flashprint
Infill Density	15%	15%
Infill pattern	Triangles	Triangles

III. RESULTS AND DISCUSSION

The measured parameters for the linear artifacts include the thickness of the protrusions and the spacing between them. As for the circular accuracy artifacts, the measured parameters pertain to the various diameters. The selection of these dimensions was based on their direct correlation with the printing speed, namely inside the XY plane of the printer.

The mean values for measurements of each linear artifact can be seen in the Table 3 below.

Table 3. Mean values for measurements of linear artifacts on both printers.

Table 3.	Nominal Dimension	High Speed Mean Values	Low Speed Mean Values
Protrusion Spacing	5	4.86	4.97
	7.5	7.42	7.48
	10	9.91	9.89
	12.5	12.42	12.45
Protrusion thickness	2.5	2.68	2.63
	5	5.12	5.08
	5	5.13	5.04
	5	5.12	5.07
	2.5	2.64	2.60

Same for circular accuracy artifacts, mean values of measurements can be seen in the Table 4 below.

To enhance understanding of the results, Figure 6 presents the differences between the measured values and the nominal dimensions.

A positive value indicates that the measured measurements exceeded the nominal dimensions, whereas a negative value signifies the contrary. As depicted in Figure 6, the disparities observed in the measurements obtained from specimens produced by the Flashforge Creator 3 exhibit a greater proximity to the X axis, indicating a higher level of precision compared to the High-Speed printer.

Table 4. Mean values for measurements of circular accuracy artifacts on both printers.

Table 4.	Nominal Dimension	High Speed Mean Values	Low Speed Mean Values
	50	49.69	49.81
	47	46.86	46.74
Diameters	30	29.35	29.58
	16	15.07	14.66
	14	13.84	13.73

Fig. 6. Linear artifacts' measurement differences from nominal dimensions.

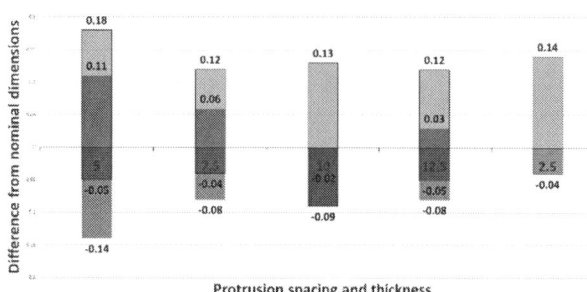

Though according to the study, such deviation from nominal dimensions should not be considered a sufficient justification for not utilizing the High-Speed 3D printer. Further on Figure 7 the same difference between circular accuracy artifacts can be seen.

Fig. 7. Linear artifacts' measurement differences from nominal dimensions.

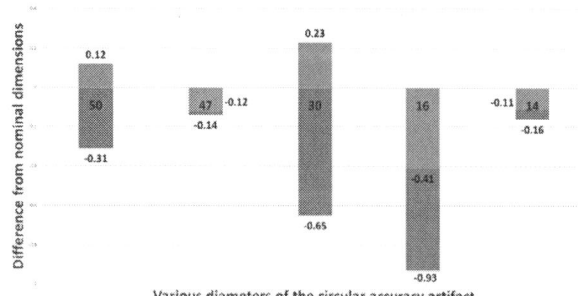

Same as with the linear artifacts, Creator 3 printer seems to have better accuracy than the High Speed one. The last step involves conducting a visual comparison of the specimens produced by both printers, as depicted in Figure 8. For better visualization circular accuracy artifacts were chosen as their dimensions are significant enough for the defects to be easily shown on camera.

Fig. 8. Top. High Speed Circular Accuracy Artifact. Bottom. Low Speed Circular Accuracy Artifact. Good characteristics of each specimen are depicted with green arrows and defects with red arrows.

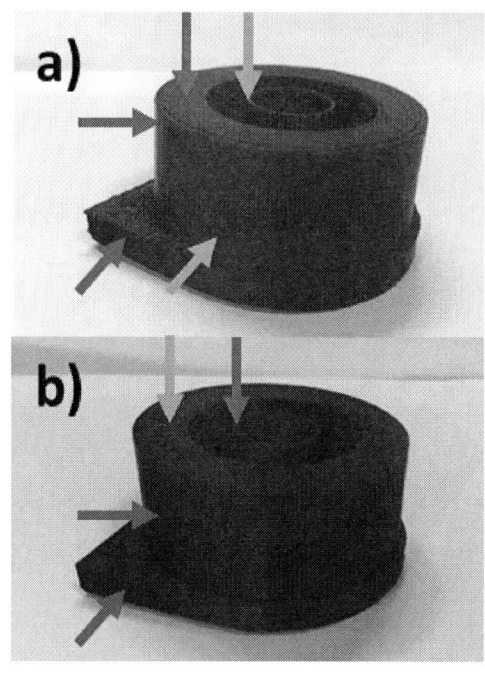

In terms of time-based comparison, it was observed that the High-Speed 3D Printer required a total of 2 minutes and 13 seconds to complete the printing process for the linear artifact, whereas it took 23 minutes to print the circular accuracy artifact. On the other hand, the Flashforge Creator 3 took 12 minutes to print the linear artifact and 118 minutes to print the circular accuracy artifact. The High-Speed 3D printer was able to complete the print for the linear artifact in just 16.6% of the time it took the Creator 3 printer. Similarly, for the circular artifact, the High-Speed 3D printer finished the print in 19.5% of the time it took the Creator 3 printer, as can be seen in Table 5.

Table 5. Time to finish the print for both printers as well as time reduction for the High Speed 3D Printer.

Table 5.	High Speed Printer	Creator 3	%reduction in time
Linear artifact	2 minutes and 13 seconds	12 minutes	83.4%
Circular accuracy artifact	23 minutes	118 minutes	80.5%

Lastly, from the current work advantages and disadvantages of the High Speed FFF 3D Printer can be derived and presented in the Table 6 below.

IV. CONCLUSION

In conclusion, the development and initial implementation of this high-speed 3D printer hold significant promise within the realm of additive manufacturing. This endeavor has involved meticulous design, innovative engineering, and a preliminary phase of testing, paving the way for future exploration. Notable among the anticipated advantages is the potential reduction in print time, a factor that may profoundly impact the efficiency and productivity of 3D printing processes across various industries. While the tool is yet to be fully tested, this research lays the foundation for future investigations and refinements. As we peer into the horizon, it becomes evident that this high-speed 3D printer has the potential to make substantial contributions to the fields of design, production, and customization. With further scrutiny and widespread adoption, it may usher in a new era of possibilities for professionals and manufacturers, reshaping conventional paradigms and advancing the frontiers of manufacturing technology. Figure 9 below displays the radar graphs of both printers, facilitating a straightforward comparison between the two.

Table 6. Advantages, Disadvantages and Mitigation of them.

Advantages	Drawbacks	Mitigation
Significant time reduction (80%)	Not good top surfaces	Calibration and Print settings changes (using ironing technique)
Smoother edges and polygons	Wrapping	Better adhesion in the build surface (use of raft)
Stiffer single wall prints		

Fig. 9. Radar graph for High-Speed 3D Printer and Low Speed 3D Printer (Flashforge Creator 3) emphasizing in some important aspects of printer selection.

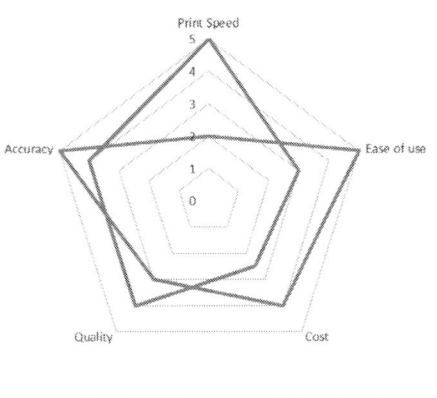

REFERENCES

[1] O. A. Mohamed, S. H. Masood, and J. L. Bhowmik, "Optimization of fused deposition modeling process parameters: a review of current research and future prospects," *Adv. Manuf.*, vol. 3, no. 1, pp. 42–53, 2015, doi: 10.1007/s40436-014-0097-7.

[2] P. C. Sai and S. N. Yeole, "Fused Deposition Modeling," no. December, 2001, doi: 10.1201/9780203910795.ch8.

[3] P. K. Penumakala, J. Santo, and A. Thomas, "A critical review on the fused deposition modeling of thermoplastic polymer composites," *Compos. Part B Eng.*, vol. 201, no. August, p. 108336, 2020, doi: 10.1016/j.compositesb.2020.108336.

[4] K. Rajan, M. Samykano, K. Kadirgama, W. S. W. Harun, and M. M. Rahman, *Fused deposition modeling: process, materials, parameters, properties, and applications*, vol. 120, no. 3–4. Springer London, 2022. doi: 10.1007/s00170-022-08860-7.

[5] J. Yang and Y. Liu, "Energy, time and material consumption modelling for fused deposition modelling process," *Procedia CIRP*, vol. 90, no. March, pp. 510–515, 2020, doi: 10.1016/j.procir.2020.02.130.

[6] A. Bellini and S. Güçeri, "Mechanical characterization of parts fabricated using fused deposition modeling," *Rapid Prototyp. J.*, vol. 9, no. 4, pp. 252–264, 2003, doi: 10.1108/13552540310489631.

[7] A. Boschetto and L. Bottini, "Accuracy prediction in fused deposition modeling," *Int. J. Adv. Manuf. Technol.*, vol. 73, no. 5–8, pp. 913–928, 2014, doi: 10.1007/s00170-014-5886-4.

[8] P. Czyżewski, D. Marciniak, B. Nowinka, M. Borowiak, and M. Bieliński, "Influence of Extruder's Nozzle Diameter on the Improvement of Functional Properties of 3D-Printed PLA Products," *Polymers (Basel).*, vol. 14, no. 2, 2022, doi: 10.3390/polym14020356.

[9] J. F. Chauvette, D. Brzeski, I. L. Hia, R. D. Farahani, N. Piccirelli, and D. Therriault, "High-speed multinozzle additive manufacturing and extrusion modeling of large-scale microscaffold networks," *Addit. Manuf.*, vol. 47, no. June, p. 102294, 2021, doi: 10.1016/j.addma.2021.102294.

[10] L. Yang, S. Li, Y. Li, M. Yang, and Q. Yuan, "Experimental Investigations for Optimizing the Extrusion Parameters on FDM PLA Printed Parts," *J. Mater. Eng. Perform.*, vol. 28, no. 1, pp. 169–182, 2019, doi: 10.1007/s11665-018-3784-x.

[11] E. E. Kopets, D. A. Protasova, V. S. Andreev, I. I. Loginov, K. A. Kurtova, and A. D. Skuratov, "Relation between 3D Printer Printhead Positioning Rate and Detail Quality," *Proc. 2022 Conf. Russ. Young Res. Electr. Electron. Eng. ElConRus 2022*, pp. 700–703, 2022, doi: 10.1109/ElConRus54750.2022.9755569.

[12] R. J. A. Allen and R. S. Trask, "An experimental demonstration of effective Curved Layer Fused Filament Fabrication utilising a parallel deposition robot," *Addit. Manuf.*, vol. 8, pp. 78–87, 2015, doi: 10.1016/j.addma.2015.09.001.

Assessing Swapping Policies as a Detailed Placement Approach

Ioannis Arvanitakis ◎
Dept. of Informatics and Telecommunications
University of Thessaly
Lamia, Greece
ioarvani@uth.gr

George K. Kranas ◎
Computer Science & Biomedical Informatics Dept.
University of Thessaly
Lamia, Greece
gekranas@uth.gr

Michael Dossis ◎
Computer Science Dept.
University of Western Macedonia
Kastoria, Greece
mdossis@uowm.gr

Athanasios Kakarountas ◎
Computer Science & Biomedical Informatics Dept.
University of Thessaly
Lamia, Greece
kakarountas@uth.gr

Antonios N. Dadaliaris ◎
Dept. of Informatics & Telecommunications
University of Thessaly
Lamia, Greece
dadaliaris@uth.gr

Abstract—**Detailed placement plays a crucial role in the physical design of integrated circuits, impacting factors such as area, power consumption, and performance. Swap-based placement algorithms have gained significant attention in recent years due to their effectiveness in improving the quality of placement solutions. This paper presents a comprehensive study of swap-based detailed placement algorithms for standard cell design, highlighting their principles, advantages, challenges, and recent advancements. In this paper we present results of swap-based algorithms, using different approaches for standard cells selection and how these affect the final solution.**

Index Terms—**placement, detailed placement, standard cell design, swap based policies.**

I. Introduction

The field of integrated circuit (IC) design has been continually evolving to meet the increasing demands for higher performance, lower power consumption, and smaller form factors. Placement is the step of physical design that could affect the overall quality of the design for the following routine of the back-end flow and meet the design's closure. There are three stages that compile the placement process, (i) global placement, (ii) legalization and (iii) detailed placement. Global placement provides a rough estimation of how the cells should be dispersed in a specified area in order to meet some constraints, i.e. wirelength, timing, power. Legalization solves local problems and corrects the overlaps caused by global placement. Detailed placement further improves the legalized solution.

Detailed placement is a critical phase in the physical design process, determining the exact locations of standard cells on an IC layout, greatly influencing the chip's final performance and manufacturability. Swap-based algorithms are used mostly on legalization and detailed placement, in order to refine the circuit on a localized scale (in a greedy manner) and have emerged as a promising approach to optimize the placement of standard cells, aiming to minimize various objectives such as wirelength, power consumption, and timing violations.

This paper provides an exploration of different swapping methods considering detailed placement. In our approach we will study swaps, of same sized standard cells, identifying parameters that could result to improved performance from global to a local perspective. We begin by outlining the fundamental concepts and motivations behind swap-based placement. Subsequently, we delve into various algorithmic techniques employed in swap-based placement, followed by a discussion of challenges and recent advancements.

Swap-based placement algorithms optimize the placement of standard cells by iteratively swapping their positions while adhering to specific constraints. The fundamental principles of swap-based placement can be summarized as follows:

- Objective Function: A cost function, typically based on wirelength, timing, and other design objectives, guides the optimization process. The objective function quantifies the quality of a placement solution, encouraging the algorithm to improve it iteratively.
- Local Moves: Swap-based algorithms rely on local moves, such as cell-cell swaps or cell-macro swaps. These moves aim to optimize the placement without altering the overall structure of the design.
- Swap-based algorithms iteratively apply local moves to improve the placement quality. The algorithm stops when convergence criteria are met or a predefined number of iterations are reached.

Various techniques and algorithms have been developed to implement swap-based detailed placement. Some of the notable methods include:

- Greedy Algorithms: Greedy algorithms perform cell swap-based on heuristics. Examples include Kernighan-Lin (KL) [1] and Fiduccia-Mattheyses (FM) [2] algorithms. Furthermore, some already implemented greedy techniques can be adjusted via swap-based schemes to improve their overall functionality [3].

979-8-3503-8368-3/23 $31.00 © 2023 IEEE

- Simulated Annealing: Simulated annealing employs a probabilistic approach to explore the solution space. Temperature-controlled annealing schedules allow the algorithm to escape local optima [4] [5].
- Genetic Algorithms: Genetic algorithms use evolutionary principles to search for optimal placements. They incorporate operators like mutation, crossover, and selection to evolve solutions [6], [7].
- Machine Learning-Based Approaches: Recent advancements involve machine learning techniques to predict beneficial swaps. Neural networks and reinforcement learning have shown promise in improving placement quality [8] [9].

Swap-based detailed placement algorithms face several challenges, including:

- Scalability: Efficient handling of large-scale designs is a significant challenge. Parallel processing and hierarchical techniques are often employed to mitigate this issue.
- Multi-Objective Optimization: Balancing conflicting objectives like wirelength and timing is complex. Multi-objective optimization algorithms are required to explore trade-offs effectively.
- Legalization: Ensuring that swapped placements adhere to design rules and constraints is essential. Legalization techniques are integrated into the swap-based algorithms.
- Runtime: Reducing placement runtime while maintaining solution quality is critical for practical use. Algorithmic optimizations and parallelization help address this challenge.

Recent advancements in swap-based placement algorithms for standard cells include:

- Hybrid Algorithms: Combining different techniques such as simulated annealing and genetic algorithms to leverage their strengths [10].
- Reinforcement Learning: Utilizing reinforcement learning to guide the search process and learn effective swaps [11], [12].
- Parallelization: Leveraging modern hardware, including GPUs and distributed computing, to accelerate placement algorithms [13], [14].
- Global-Local Approaches: Integrating global placement with swap-based techniques for improved convergence and quality.

II. SWAPPING POLICIES

To investigate the impact of swaps on cells of equal size, our study employs a categorization approach for algorithms, ranging from global to more localized strategies. This categorization encompasses three primary dimensions:

- Scope of Cell Swapping: Algorithms are categorized based on whether they perform swaps on all cells, cells within the same net, or cells within the same row.
- Order of Examination: The sequence in which cells are considered for swaps varies between strategies. Cells may be examined in ascending order of connectivity (from

lowest to highest) or descending order of connectivity (from highest to lowest).
- Sufficient Swap Condition: We evaluate different implementation scenarios, considering whether the algorithm continues swapping until all potential swaps have been explored and the best wirelength is achieved, or if it halts upon encountering the first available swap that improves wirelength.

The metric we use for the cost function is only the total wirelength, as the swaps are performed between equal sized cells and no overlap may occur.

In summary, our study explores various permutations of these dimensions, resulting in the following distinct implementation strategies:

- Swap Best Available (SBAF), starting with those with the Fewest connections.
- Swap Best Available (SBAM), starting with those with the Most connections.
- Swap Best Available same Net (SBNF), starting with those with the Fewest connections.
- Swap Best Available same Net (SBNM), starting with those with the Most connections.
- Swap Best Available same Row (SBRF), starting with those with the Fewest connections.
- Swap Best Available same Row (SBRM), starting with those with the Most connections.
- Swap First Available (SFAF), starting with those with the Fewest connections.
- Swap First Available (SFAM), starting with those with the Most connections.
- Swap First Available same Net (SFNF), starting with those with the Fewest connections.
- Swap First Available same Net (SFNM), starting with those with the Most connections.
- Swap First Available same Row (SFRF), starting with those with the Fewest connections.
- Swap First Available same Row (SFRM), starting with those with the Most connections.

Through these comprehensive approaches, our study aims to provide insights into the effectiveness of cell swaps in optimizing wirelength across various conditions and strategies.

A. Swap Best Available

In this approach, cells are sorted based on their connectivity, both in ascending and descending order. For each cell in the sorted list, the algorithm searches for the next cell of the same size and swaps them. If the resulting wirelength is improved, the current cell becomes a candidate. The algorithm then proceeds to swap back and evaluate the next cell. After examining all candidate cells, valid swaps are executed, or the algorithm moves on to the next cell in the sorted list if no valid candidates are found.

B. Swap Best Available same Net

Similar to the previous method, cells are sorted by connectivity. However, in this case, the algorithm specifically looks for cells of the same size within the same net to swap. It follows the same procedure of evaluating wirelength improvements and selecting candidate cells, subsequently swapping them if valid candidates are found.

C. Swap Best Available same Row

Again, cells are sorted by connectivity, but the focus here is on finding cells of the same size within the same row for swapping. The wirelength improvement check, candidate selection, and swapping procedure are consistent with the previous methods.

D. Swap First Available

This approach involves sorting cells by connectivity and considering both ascending and descending orders. For each cell in the sorted list, the algorithm seeks the next cell of the same size and immediately swaps them if it improves wirelength. It proceeds to the next cell without considering multiple candidates.

E. Swap First Available same Net

Similar to the previous method, this approach sorts cells by connectivity. It then searches for the next cell of the same size within the same net and performs an immediate swap if it enhances wirelength. The algorithm continues to the next cell without evaluating multiple candidates.

F. Swap First Available same Row

In this variant, cells are sorted based on connectivity. The algorithm focuses on locating cells of the same size within the same row for direct swapping if it improves wirelength. It proceeds to the next cell without considering multiple candidates.

These algorithm categories encompass a range of strategies for cell swapping, each with its own criteria and procedures for enhancing wirelength optimization. Examples of pseudocodes are provided in Algorithm 1 and Algorithm 2

Algorithm 1 Swap Best Available Fewest Connections

$CellList$ in ascending connections order
while $cell$ in $CellList$ **do**
    while $candidateCell$ in $CellList$ **do**
        if $candidateCell.Size = cell.Size$ **then**
            $swapCells(candidateCell, cell)$
            if $newWirelength \geq oldWirelength$ **then**
                $bestCandidate \leftarrow candidateCell$
            end if
            //swap back cells
            $swapCells(candidateCell, cell)$
        end if
    end while
    if $bestCandidate$ exists **then**
        $swapCells(candidateCell, cell)$
        $oldWirelength \leftarrow newWirelength$
    end if
end while

Algorithm 2 Swap First Available Same Row Most Connections

$CellList$ in descending connections order
while $cell$ in $CellList$ **do**
    $row \leftarrow cell.Row$
    while $candidateCell$ in row **do**
        if $candidateCell.Size = cell.Size$ **then**
            $swapCells(candidateCell, cell)$
            if $newWirelength \geq oldWirelength$ **then**
                $break$
            end if
            //swap back cells
            $swapCells(candidateCell, cell)$
        end if
    end while
end while

III. EXPERIMENTAL RESULTS

All algorithms are implemented in C++ and the experiments run on a linux-based virtual machine upon an AMD Ryzen 5500U (4GHz), Our algorithms were applied upon designs of the ISPD98 benchmark suite [15] that were legalized using the mPL5 [16] technique. We employed swap-based detailed placement strategies to enhance the design further. The key metrics captured for each design included:

- A. Wirelength Improvement: We measured the reduction in wirelength achieved by our algorithms.
- B. Number of Swaps: We counted the total number of swap operations performed during the optimization.
- C. Wirelength Improvement per Swaps: This metric represented the ratio of wirelength improvement to the number of swap operations.

A. Wirelength Improvement

Wirelength is a common optimization objective for any placement related heuristics. It measures the length of the wire that connects cells within a design, forming its nets. There are various metrics used to calculate wirelength, with the most frequent that of half-perimeter wirelength (HPWL). HPWL measures the half perimeter of the wire needed to connect the cells of its nets forming a rectangle. Summing up the HPWL of each net derives the wirelength of a design.

We present our findings regarding wirelength improvement in two tables: Table I summarizes cases with the best available swaps, while Table II presents the first available swaps. In both tables, cells are sorted in descending order based on their connections. The tables depict the Wirelength difference between the benchmarks' initial and final HPWL after the application of our detailed placement scheme. Depending on a design's scale these units represent any standard length measurement units of the metric system (i.e. nanometers, micrometers etc.).

Our analysis, as depicted in Figure 1, revealed that the global strategies "best available" and "first available" consistently delivered the best final wirelength results. However, it's worth noting that these strategies also exhibited the highest runtime requirements, as shown in Figure 2. This is due to the fact that these algorithms have as input the whole design, thus the solution space exploration is bigger than the others. Additionally, Figure 3 illustrates that sorting cells by the number of connections, whether most or fewer, had a minimal impact on the algorithm results.

TABLE I: Wirelength Improvement Best Available Most Connections (length units)

design	SBAM	SBNM	SBRM
ibm01	4550	1589	2010
ibm02	24068	8406	11939
ibm03	24480	5852	10259
ibm04	38573	4759	12790
ibm05	24044	4514	10728
ibm06	22027	4253	10743
ibm07	46690	9979	19473
ibm08	191135	87419	91614
ibm09	72036	18470	29042
ibm10	158633	41982	52226

TABLE II: Wirelength Improvement First Available Most Connections (length units)

design	SFAM	SFNM	SFRM
ibm01	4542	1581	2055
ibm02	24546	8419	11798
ibm03	24448	5808	10231
ibm04	38731	4706	12988
ibm05	23887	4438	10677
ibm06	21972	4277	10660
ibm07	47049	9943	19462
ibm08	186514	87398	94944
ibm09	72472	18515	29050
ibm10	159291	42449	52194

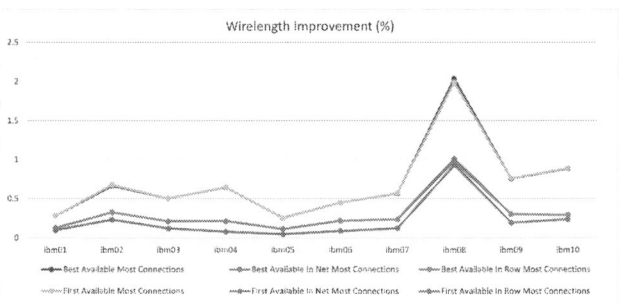

Fig. 1: Percentage improvement of all strategies sorting cells in descending connections order.

B. Swaps

Tables III and IV provide insights into the number of swap operations performed during the experiments. Notably, global strategies such as SBA/SFA(M/L) resulted in a higher number of swaps. Conversely, when we constrained the algorithm to search within more restricted contexts, such as within the same nets, fewer swaps were performed. Interestingly, the order in

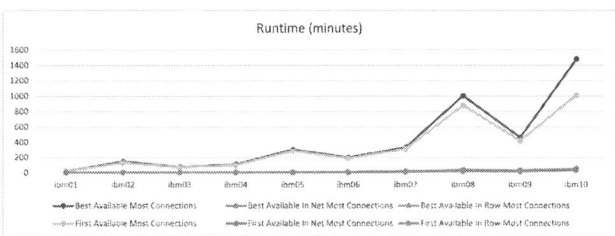

Fig. 2: Execution runtime of every approach.

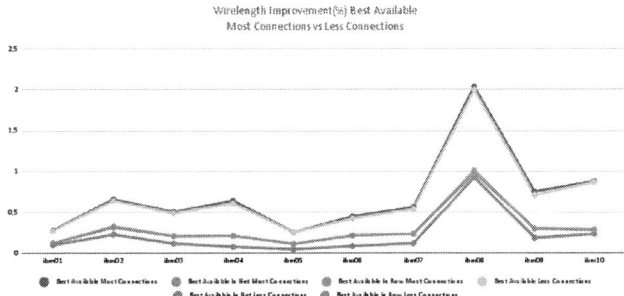

(a) sorted by most connections

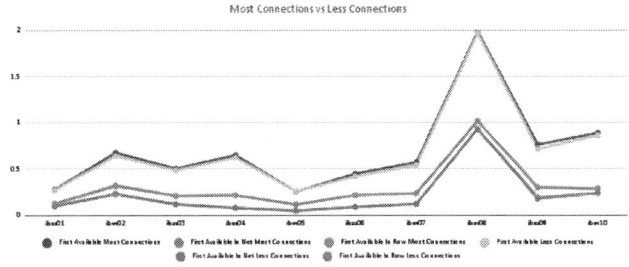

(b) sorted by fewer connections

Fig. 3: Wirelength improvement most connections vs less connections comparison for all strategies.

which cells were sorted based on their connections, whether ascending or descending, had a negligible effect on the number of swaps, as the values were closely aligned.

TABLE III: Number of Swaps Performed Most Connections

design	SBAM	SBNM	SBRM	SFAM	SFNM	SFRL
ibm01	794	290	414	796	292	415
ibm02	3486	998	1871	3528	989	1890
ibm03	3326	813	1754	3340	812	1747
ibm04	4521	756	1896	4526	751	1867
ibm05	3714	773	1724	3740	784	1702
ibm06	4001	948	2236	3978	946	2259
ibm07	6743	1512	3250	6760	1506	3250
ibm08	11711	2836	5178	11735	2864	5151
ibm09	9279	2281	4458	9228	2278	4501
ibm10	14706	3863	5726	14645	3842	5657

979-8-3503-8368-3/23 $31.00 © 2023 IEEE

TABLE IV: Number of Swaps Performed Fewer Connections

design	SBAL	SBNL	SBRL	SFAL	SFNL	SFRL
ibm01	807	295	411	796	297	415
ibm02	3486	1003	1861	3528	996	1877
ibm03	3326	778	1769	3340	789	1759
ibm04	4521	746	1876	4526	749	1870
ibm05	3714	771	1712	3740	777	1704
ibm06	4001	956	2214	3978	951	2245
ibm07	6743	1481	3215	6760	1477	3230
ibm08	11711	2765	5142	11735	2779	5132
ibm09	9279	2292	4363	9228	2285	4359
ibm10	14706	3838	5686	14645	3813	5633

C. Wirelength Delta/Swaps

Figures 4a and 4b present the wirelength improvement over the number of swaps executed upon the benchmarks. They shed light on the effectiveness of swaps in reducing wirelength when constrained to a local environment, such as within the same net or within the same row. These figures demonstrate that when swaps are limited to a local context, are proportionally effective in minimizing wirelength compared to those performing upon the whole design. Considering the time needed for global adjustments, local schemes deliver an analogous result quality, while being timely efficient due to the smaller area of effect (less swaps).

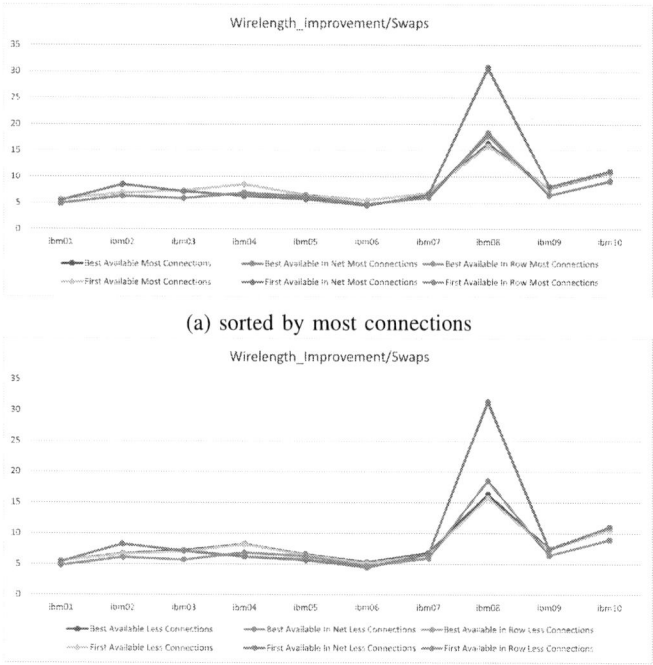

(a) sorted by most connections

(b) sorted by fewer connections

Fig. 4: Wirelength improvement/Swaps for all strategies.

IV. CONCLUSIONS

In this paper we presented a study for swap-based detailed placement. Initially we categorized various algorithms that perform this technique and continued by developing our own strategies. The results have provided valuable insights into the performance of our placement algorithms. Global strategies like "best available" and "first available" consistently achieved the best final wirelength results, albeit with increased runtime. Restricting swaps to local environments, such as within the same net or row, proved to be an effective approach for wirelength reduction. Additionally, the order in which cells were sorted by connection count had limited influence on the number of swaps. These findings contribute to a deeper understanding of our algorithms' effectiveness and efficiency in optimizing designs.

For future work, we would like to test more strategies, such as advanced heuristics and optimization techniques that could lead to further improvements. This might involve incorporating machine learning models, reinforcement learning, or genetic algorithms into the placement process to adapt to various design scenarios. Moreover leveraging parallel computing resources to accelerate placement algorithms is an important direction for future work. Implementing parallelization techniques can help address the growing complexity of modern integrated circuits. Last but not least developing algorithms that can dynamically adapt to the specific characteristics of a design, such as netlist structure or cell types, could lead to more tailored and efficient placement strategies.

REFERENCES

[1] B. W. Kernighan and S. Lin. An efficient heuristic procedure for partitioning graphs. *The Bell System Technical Journal*, 49(2):291–307, 1970.

[2] Charles M Fiduccia and Robert M Mattheyses. A linear-time heuristic for improving network partitions. In *Papers on Twenty-five years of electronic design automation*, pages 241–247. 1988.

[3] Antonios Dadaliaris, George Kranas, Panagiotis Oikonomou, George Floros, and Michael Dossis. Exploiting net connectivity in legalization and detailed placement scenarios. *Information*, 13(5):212, 2022.

[4] Maogang wang, Xiaojian Yang, and M. Sarrafzadeh. Dragon2000: standard-cell placement tool for large industry circuits. In *IEEE/ACM International Conference on Computer Aided Design. ICCAD - 2000. IEEE/ACM Digest of Technical Papers (Cat. No.00CH37140)*, pages 260–263, 2000.

[5] C. Sechen and A. Sangiovanni-Vincentelli. Timberwolf3.2: A new standard cell placement and global routing package. In *23rd ACM/IEEE Design Automation Conference*, pages 432–439, 1986.

[6] K. Shahookar and P. Mazumder. A genetic approach to standard cell placement using meta-genetic parameter optimization. *IEEE Transactions on Computer-Aided Design of Integrated Circuits and Systems*, 9(5):500–511, 1990.

[7] George K Kranas, Taxiarchis G Kouskouras, Vasileios Dimitriadis, Michael Dossis, Panagiotis Oikonomou, and Antonios N Dadaliaris. A novel genetic algorithm for i/o pad planning retaining former cell positions. In *2020 5th South-East Europe Design Automation, Computer Engineering, Computer Networks and Social Media Conference (SEEDA-CECNSM)*, pages 1–5. IEEE, 2020.

[8] Yibo Lin, Shounak Dhar, Wuxi Li, Haoxing Ren, Brucek Khailany, and David Z. Pan. Dreampiace: Deep learning toolkit-enabled gpu acceleration for modern vlsi placement. In *2019 56th ACM/IEEE Design Automation Conference (DAC)*, pages 1–6, 2019.

[9] Kevin Immanuel Gubbi, Sayed Aresh Beheshti-Shirazi, Tyler Sheaves, Soheil Salehi, Sai Manoj PD, Setareh Rafatirad, Avesta Sasan, and Houman Homayoun. Survey of machine learning for electronic design automation. In *Proceedings of the Great Lakes Symposium on VLSI 2022*, GLSVLSI '22, page 513–518, New York, NY, USA, 2022. Association for Computing Machinery.

[10] Ahmed Elkarashily, Mohamed Ahmed, Omar Salah, and Abdullah Ezzat. Vlsi placement using modified parallel simulated annealing, 06 2020.

[11] Anthony Agnesina, Kyungwook Chang, and Sung Kyu Lim. Vlsi placement parameter optimization using deep reinforcement learning. In *Proceedings of the 39th international conference on computer-aided design*, pages 1–9, 2020.

[12] Azalia Mirhoseini, Anna Goldie, Mustafa Yazgan, Joe Jiang, Ebrahim Songhori, Shen Wang, Young-Joon Lee, Eric Johnson, Omkar Pathak, Sungmin Bae, et al. Chip placement with deep reinforcement learning. *arXiv preprint arXiv:2004.10746*, 2020.

[13] Ahmad Al-Kawam and Haidar M. Harmanani. A parallel gpu implementation of the timber wolf placement algorithm. In *2015 12th International Conference on Information Technology - New Generations*, pages 792–795, 2015.

[14] Yibo Lin, Wuxi Li, Jiaqi Gu, Haoxing Ren, Brucek Khailany, and David Z. Pan. Abcdplace: Accelerated batch-based concurrent detailed placement on multithreaded cpus and gpus. *IEEE Transactions on Computer-Aided Design of Integrated Circuits and Systems*, 39(12):5083–5096, 2020.

[15] Charles J. Alpert. The ispd98 circuit benchmark suite. In *Proceedings of the 1998 International Symposium on Physical Design*, ISPD '98, page 80–85, New York, NY, USA, 1998. Association for Computing Machinery.

[16] Tony Chan, Jason Cong, and Kenton Sze. Multilevel generalized force-directed method for circuit placement. In *Proceedings of the 2005 international symposium on physical design*, pages 185–192, 2005.

979-8-3503-8368-3/23 $31.00 © 2023 IEEE

ARM64 Architecture: A Review in Virtualization Technology and Cloud Computing Maturity, in the context of Environmental Sustainability

Georgios Lambropoulos
Department of Informatics
University of Piraeus
Piraeus, Greece
g.lambropoulos@hotmail.com

Sarandis Mitropoulos
Regional Development Department
Ionian University
Lefkada, Greece
smitropoulos@ionio.gr

Christos Douligeris
Department of Informatics
University of Piraeus
Piraeus, Greece
cdoulig@unipi.gr

Abstract—**ARM64 emerges as a new transformative force that intersects with technology, sustainability and security, aiming to drastically reshape the current digital landscape. This paper provides a review of the ARM64 architecture, exploring its impact on various aspects of modern computing. Specifically, this review covers ARM64's architectural principles, its role and maturity in virtualization technology, its integration in the cloud computing landscape, its alignment with environmental and social considerations and the adoption challenges and considerations it presents.**

Index Terms—**ARM64, Virtualization, Cloud Computing, ESG**

I. Introduction

In the ever-evolving landscape of modern computing, virtualization has been established as the main underlying technology for cloud computing, server consolidation, and efficient resource utilization. In this dynamic landscape, the ARM64 (AArch64) architecture has emerged as a new transformative force, that due to its energy efficiency and scalability properties is able to reshape traditional datacenters, cloud computing infrastructures, edge computing deployments and the majority of connected devices. While the vast majority of datacenters and computing infrastructures are historically designed around the x86 architecture, characterized by its Complex Instruction Set Computing (CISC) principles and focus on high performance, the ARM64 architecture brings a major shift in processor design. Based on the Reduced Instruction Set Computing (RISC) principles, ARM64 prioritizes reduced power consumption, pointing towards a new era of energy-efficient computing.

The goal of this paper is to investigate the ARM64 architecture from multiple angles. Specifically, it explores the various aspects of ARM64, from its architectural foundations, its virtualization capabilities and its influence on cloud computing. It also explores how the ARM64 architecture aligns with environmental and social considerations addressing global sustainability challenges. Furthermore, it addresses the

main concerns regarding ARM64 architecture adoption and presents practical mitigation proposals.

The paper is organized as follows: Section 2 presents an overview the of ARM processors architecture along with indicative examples of System-on-Chip (SoCs) implementations and their practical applications. Section 3 provides a review on the maturity of the Virtualization technology on ARM64 architecture along with the latest advances in the fields of hypervisors, containerization orchestration platforms and emulation tools. Section 4 presents the current state regarding the adoption of the ARM64 architecture by major cloud providers along with the services currently provided and the concerns they present. Section 5 presents the challenges that derive from the necessity for environmental, social, and corporate governance (ESG) compliance and the contribution of ARM64 architecture towards a more energy-efficient and sustainable computing era. Section 6 presents the challenges and concerns of the ARM64 adoption in the context of compatibility, security, training and investment. Finally, Section 7 summarizes the findings of this review and suggests potential future work.

II. ARM64 Architecture and SoCs

The ARM64 architecture provides an alternative compared to traditional computing models commonly based on the x86 architecture. Based on the RISC principles, ARM64 processors are designed to prioritize simplicity in instruction sets and operate with reduced power consumption [1].

Specifically, one of ARM processors fundamental characteristics is the load-store architecture. This architecture defines that data processing instructions operate almost exclusively on registers, leading to minimized memory access overhead and overall performance enhancement. Additionally, ARM64 processors also feature a large number of general-purpose registers that significantly reduce the need for intense and frequent memory access, which further improves speed and efficiency [2]. Scalability is another key feature of ARM64 architecture. The customization of parameters such as pipeline

979-8-3503-8368-3/23 $31.00 © 2023 IEEE

depth and cache sizes, facilitate a variety of different computing needs. This adaptability is crucial in a world where computing requirements vary widely [2].

ARM architecture can be traced back to Acorn Computers in the 1980s, with the ARM1 processor marking its inception. Subsequent iterations, including ARMv8, introduced 64-bit capabilities alongside the traditional 32-bit ARM architecture (AArch32). Today, ARM64's applications cover a wide spectrum, from mobile devices and embedded systems to servers, supercomputers, and consumer electronics. System on Chip (SoC) designs are a crucial aspect of ARM64 architecture. A SoC integrates CPU cores, memory, GPU, I/O interfaces, and other specialized accelerators into a compact and power-efficient package. This integration offers several advantages such as power efficiency, compact form factor, performance and scalability [3]. Indicative examples of ARM64 SoCs showcase the diversity and versatility of this architecture across a wide variety of computing devices and applications, each bringing its unique strengths and capabilities. A few indicative examples of such designs are:

The Qualcomm Snapdragon Series of ARM64 SoCs has gained a significant share in the mobile and tablet market, offering a balance of performance and energy efficiency. These SoCs frequently integrate ARM Cortex-A CPUs along with Adreno GPUs, delivering powerful processing capabilities while maintaining an increased battery life [4]. This combination of processing power and efficiency has made Snapdragon SoCs one of the most popular choices for Android-based mobile devices. They are also known for their integrated Long-Term Evolution (LTE) modems, which enables integrated connectivity capabilities in smartphones and tablets. Additionally, Microsoft's collaboration with Qualcomm has led to ARM64-based processors powering devices like the Microsoft Surface Pro X. The integration of ARM64 architecture into Microsoft's Surface series, offers a seamless experience combined with Windows 10 and 11 on ARM, leveraging the efficiency of the ARM64 while maintaining compatibility with a wide range of applications designed for the x86 architecture [5].

Apple's involvement into the ARM64 architecture, particularly with its M1 and M2 SoCs, represents a very significant technological milestone. The M1 and M2 Apple Silicon chips are now employed as the core of Apple's MacBook and iPad series products, pointing towards a significant shift targeting to enhanced performance and energy efficiency [6]. Specifically, the M1 and M2 series, has gained widespread recognition for their exceptional processing power and application performance while maintaining high energy efficiency levels [7].

NVIDIA Tegra SoCs have been established in the gaming and automotive sectors. These SoCs combine ARM CPU cores with NVIDIA Graphics Processing Units (GPUs), featuring powerful computational and graphical results. In gaming, Tegra SoCs are able to deliver a powerful gaming experience, featuring advanced graphics capabilities. Beyond gaming, Tegra SoCs have also a significance presence in the automotive industry, powering infotainment systems, advanced driver assistance and even autonomous driving features [8].

The Ampere Altra SoC family is gaining exponential recognition in the ARM64 server market. Ampere Altra SoCs are especially designed for datacenter and cloud applications, featuring high performance and scalability. Ampere Altra processors feature up to 80 ARMv8 cores, enabling them to handle a variety of enterprise workloads and to adequately support technologies such as virtualization [9]. Their energy efficient design and scalability are positioning Ampere Altra SoCs as a key consideration in the design of new infrastructures and datacenters.

The Raspberry Pi series is considered as an indicative example of ARM-based SoCs. Specifically, due to its compact and cost-effective design, is a perfect example of the adaptability that ARM64 architecture may provide. Raspberry Pi boards are equipped with ARM-based processors and offer a platform ideal for the creation of a wide range of projects. Their applications span across the educational sector, home automation and embedded systems [10]. By providing an affordable and flexible platform, Raspberry Pi is an ideal example that showcases the flexibility and accessibility of ARM64 technology, enabling innovation and establishing a new ecosystem for developers and creators.

III. VIRTUALIZATION ON ARM64

The implementation of virtualization technology offers significant advantages and opportunities in enhancing efficiency and reducing operational and maintenance expenditures in the organizations that adopt it. Specifically, enterprises that have virtualized their datacenters have found that it enables a more effective utilization of their resources, leading to faster and higher-quality service delivery and improved cost-effectiveness of IT operations [11]. Over time, the virtualization technology has adequately matured to feature dependable tools for consolidation of workloads, flexible administration and disaster recovery, and eventually become a standard in the field of enterprise datacenters. Today, virtualization permits the aggregation of multiple virtualized server instances into a centralized manageable infrastructure. Such an infrastructure serves as a unified environment for consolidating all available hardware resources within a datacenter into a single resource pool, from which they may be assigned to various virtual machines. This approach facilitates a more efficient distribution of total hardware resources, achieving high utilization rates and decoupling installed operating systems from specific hardware configurations [12]. However, it is essential to acknowledge that while virtualization evolved to eventually become an industry standard in x86-based datacenters, addressing and studying the advances in ARM64 technology is becoming increasingly crucial, especially in the context of integrating ARM technology into pre-existing infrastructures. Recent advances in the ARM64 virtualization technology play a major role in achieving cross-architecture compatibility and support for a variety of workloads. Different virtualization approaches apply to specific use cases, enhancing the architecture's adaptability and flexibility.

In Type-1 hypervisors, ongoing development is directed towards adapting pre-existing platforms traditionally associated with x86 virtualization to the ARM64 architecture. One such example is the VMware's ESXi on ARM Fling. This endeavor, while still in experimental phase, aims to extend enterprise-grade virtualization to ARM64 platforms, entering in the new era of cross-architecture infrastructures [13]. Another example is Microsoft Hyper-V. Hyper-V, which is currently supported on the Microsoft Windows 10 and 11 operating systems for ARM64, which provides the same features as its x86 counterpart [14]. In the same field, yet another example is Xen, a well-established virtualization platform that has made considerable steps towards the ARM64 architecture. Xen features efficient virtualization for ARM64-based servers, particularly in high-performance computing (HPC) implementations [15].

Containerization is a method involving a packaging process for applications and their dependencies in lightweight and isolated virtualized environments, commonly known as containers. These containers may be easily migrated, simplifying the process of software deployment across various computing environments, ranging from testing servers to large production datacenters. Containerization technologies, including widely-used platforms like Docker and Kubernetes, are extensively adopted in recent software deployment practices [16]. In terms of containerization, the ARM64 architecture has been proven to be considerably efficient. Specifically, Docker features native support for ARM, allowing the execution of containerized applications on ARM64-based systems. Additionally, Kubernetes, which is regarded as a leading container orchestration platform, also provides support for ARM64, simplifying the deployment of cloud-native applications [17]. Moreover, the extensive adoption level of these technologies on ARM64 reflects their adequate level of maturity, underlining their significance in a variety of projects and industry sectors.

In the field of emulation, QEMU (Quick Emulator), a well-established emulator, features various architectures, including ARM64. Specifically, QEMU provides a variety of virtualization tasks, such as enabling the execution of x86 virtual machines on ARM64 hosts and vice versa for development and testing purposes [18]. QEMU is regarded as a valuable tool for cross-platform virtualization and emulation. Even though QEMU is not classified as a hypervisor, it is a well-established and mature tool currently supporting ARM64 emulation and virtualization.

While the above described examples may provide a clear overview on the current advances in the field of ARM64 virtualization, it is essential to consider the maturity of these technologies. The landscape of ARM64 virtualization is rapidly evolving, and the maturity of each presented approach varies significantly. As development, testing, and industry adoption increases, ARM64 virtualization will continue to mature, until it becomes an essential part of the computing industry landscape.

IV. ARM64 IN CLOUD COMPUTING

ARM64 architecture directly aligns with the fundamental principles of cloud computing, characterized by scalability, efficiency and cost-effectiveness. As cloud providers seek innovative ways to optimize their resource utilization and energy efficiency, ARM64-based servers emerge as a new appealing solution. One significant development towards this direction, is the availability of ARM64-based instances by major service providers. These instances are suitable for facilitating a wide range of services, from web hosting to advanced data analytics by leveraging the energy efficiency of ARM64 architecture, so to deliver cost effective solutions [19]. Specifically, Amazon Web Services (AWS) provides ARM64-based instances known as "Graviton processors". These instances are designed to offer significant cost savings and energy efficiency, making them an attractive choice for a variety of cloud services [20]. In the same direction, Microsoft Azure offers ARM64 instances in its cloud infrastructure as a part of the Azure HBv2 virtual machines, demonstrating the ARM64's ability to address diverse cloud computing requirements [21]. Additionally, Google Cloud Platform (GCP) provides ARM64 virtual machines, known as "N2D". These VMs feature a balance of performance and cost-effectiveness, making them suitable for a variety of cloud workloads, including web applications, microservices, and data analysis tasks [22].

The ARM64 diverse ecosystem features SoCs from various manufacturers that promote competition and innovation in the field of cloud computing. This diversity enables cloud providers to fine-tune their infrastructures to meet specific customer needs, providing flexibility and efficiency. Moreover, the adaptability of the ARM64 SoCs makes them ideal for edge computing, enabling real-time processing and reduced latency for crucial for applications like IoT, data analytics and content management. However, it is essential to acknowledge that the integration of ARM64 to cloud environments introduces the same considerations as discussed earlier. Specifically, compatibility with x86-based infrastructures, virtualization adoption challenges and the need for ARM64-specific software adaptation are the main points that cloud providers must carefully consider so to benefit from the full potential that the ARM64 architecture can bring in cloud computing.

As major cloud providers continue to expand their ARM64 offerings and optimize their infrastructure for diverse workloads, ARM64's role in cloud computing will exponentially grow. The combination of energy efficiency, scalability, and flexibility places the ARM64 architecture as a valuable asset in the cloud computing landscape. Nevertheless, the adoption of ARM64 introduces new complications in resource management, as cloud providers need to balance ARM64 and x86 resources efficiently. Towards this direction, the adaptability of ARM64 SoCs for edge computing and IoT devices introduce new infrastructural challenges [23]. While the ARM64 processors are strategically aligned with edge computing requirements, it is important to acknowledge that the nature of edge computing requires various adaptations in

979-8-3503-8368-3/23 $31.00 © 2023 IEEE

the datacenter design. Specifically, edge computing relies on distributed deployments, often at remote locations, to enable data processing closer to its source. Due to this fact, datacenters should be further optimized to support decentralized infrastructures and ensure low-latency data processing across numerous smaller nodes.

V. ARM64 AND ENVIRONMENTAL, SOCIAL, AND GOVERNANCE (ESG) CONSIDERATIONS

Over the last decade, the adoption of the ARM64 architecture has gained an additional significance not only from a technological point of view but also from the specter of Environmental, Social, and Governance (ESG) considerations. ESG includes a set of criteria designed to measure the impact of a company or technology in the environment, the society and corporate governance. Today, the importance of ESG factors has grown exponentially, mostly driven by the urging challenges posed by climate change and sustainability [24]. ESG factors include a set of metrics and standards designed to evaluate the performance of a company or technology in areas beyond traditional financial metrics . They are categorized in three main dimensions [25]:

1. Environmental (E): These criteria are used to evaluate a company's impact on the environment. They are mainly focused on issues such as carbon monoxide emissions, energy efficiency, resource management and sustainability practices. Because of the urgency posed by global climate crisis, environmental considerations are currently placed in the center of attention for ESG evaluations.

2. Social (S): These criteria are used to evaluate the social factors impacting a company's relationship with customers, staff, external suppliers and the communities in which it resides and operates. Specifically, these criteria are focusing on areas such as inclusion, human rights, diversity and community engagement.

3. Governance (G): Governance criteria are concerned with internal policies, leadership, and decision-making structures of the companies. This dimension specifically focuses on issues like board independence, shareholder rights, ethical business practices and executive compensations.

A. The role of ARM64 in the Context of ESG

As mentioned before, the escalating climate change crisis has placed the importance of ESG in the center of attention. Due to the global temperature rising and the extreme weather events that become more frequent, there is an urgent imperative to reduce the overall carbon footprint of technology and industry [26]. The ARM64 architecture can play a crucial role in addressing ESG concerns, particularly in the environmental dimension due to its unique architectural characteristics. Specifically, the ARM64 processors design prioritizes power efficiency, making them suitable for battery-powered devices, datacenters, and edge computing implementations. Their unique energy efficiency aligns with ESG's environmental goals by reducing energy consumption and as a result carbon footprint. Additionally, the ARM64 SoC's scalability enables the creation of energy-efficient, compact systems. These systems may significantly contribute to sustainability by reducing the environmental impact of hardware manufacturing, by reducing electronic wastes and by promoting new resource-efficient designs. In the context of cloud computing and datacenters, ARM64's energy efficiency leads to reduced operational costs and lower overall environmental impact, providing a more sustainable alternative to the datacenters' notorious energy consumption. Finally, ARM64's suitability for edge computing implementations enable real-time data processing with reduced latency, enhancing the efficiency of IoT applications, resulting on minimizing the energy consumption of centralized datacenter infrastructures [27].

VI. CHALLENGES AND CONCERNS OF ARM64 ADOPTION

The increased adoption rate of the ARM64 architecture in the computing industry brings new challenges and concerns. ARM64 brings a number of serious advantages, it also presents several issues that organizations must carefully consider. One of the most important concerns is software compatibility. Specifically, the ARM64 architecture leads to a departure from the traditionally used and well-established x86 architecture. As a result, operating systems, applications and software created for x86 systems are incompatible with ARM64-based platforms. This disparity in software ecosystems brings many challenges for organizations that may rely on legacy applications or proprietary software without ARM64 support. To address this issue, organizations need to considerably invest in software re-engineering or emulation, which is proven to be expensive, resource intensive and time consuming. In terms of application compatibility, even though operating systems for ARM64, such as Microsoft Windows, may include the ability to run x86 software in emulation mode [28], their performance still needs to be thoroughly investigated.

Additionally, the integration of the ARM64 to existing datacenters may raise serious service compatibility issues with the existing infrastructure. The majority of datacenters are built based on x86 systems and their provided services may not be able to facilitate ARM64 systems demands. Due to this fact, a careful planning and investment in the aspects of both hardware and software should be undertaken, to ensure a smooth integration process. The coexistence of x86 and ARM64-based systems in the same datacenter also introduces management complexities, as administrators should maintain two different architectures by potentially using different management tools and processes. While the ARM64 architecture is known for its energy efficiency, its adoption introduces new security challenges. As ARM64 becomes increasingly relevant, it also attracts increased attention from malicious actors, working on exploits and vulnerabilities specific to it. Moreover, ARM64-based systems may not benefit from the same level of provided security solutions as their x86 counterparts. Organizations should seriously research and invest in security countermeasures, early threat detection systems, and attack response strategies specifically regarding ARM64,

979-8-3503-8368-3/23 $31.00 © 2023 IEEE   233

so to assure the effective and efficient protection of their infrastructures [29].

Another serious challenge is the need for ARM64 specific training and expertise. IT administrators and developers that are currently familiar with the x86 architecture may require training and skill development to be able to effectively manage and optimize ARM64 systems. This upskilling requires additional costs and time, particularly for organizations that have already substantially invested in x86-based training and technology. Furthermore, the ARM64 adoption requires changes in procurement and vendor management relationships. Specifically, organizations need to investigate new partnerships with ARM64 hardware and software providers that could potentially disrupt already established vendor relationships. Vendor support plans for ARM64 may also vary, affecting the availability of critical updates and supporting services. Finally, while the ARM64 systems are known for their energy efficiency and cost-effectiveness, the required initial investment for migrating to ARM64 architecture, including hardware, software and training spending, is substantial and cannot be overlooked. Organizations need to carefully weigh these expenditures against the long-term benefits, such as the energy savings and the improved performance, as to determine the overall Return on Investment (ROI) [30].

VII. Conclusions and Future work

In this paper, the impact of the ARM architecture on modern computing has been explored. ARM64's RISC-based design was presented as an appealing alternative to traditional x86 architecture, due to its inherited prioritization on simplicity and energy efficiency. Adaptability is achieved through a variety of different SoCs, each of them designed to assist in different and specific computing areas. Indicative examples include Qualcomm's Snapdragon Series for mobile devices and tablets, Apple's M1 and M2 chips for Apple devices, NVIDIA's Tegra SoCs for gaming and automotive applications, the Ampere Altra SoC series for datacenters and the widely appreciated Single Board Computer (SBC) Raspberry Pi.

Virtualization technology on ARM64 is evolving and maturing in a rapid pace, already featuring in experimental level, hypervisors such as VMware's ESXi, Microsoft Hyper-V and Xen, enhancing adaptability and compatibility across different types of workloads, while containerization technologies, such as Docker and Kubernetes are already well-established and constantly adopted. In the field of Cloud Computing, major cloud providers such as AWS, Azure, and Google Cloud have already begun offering ARM64-based services, capitalizing the energy efficient nature of ARM64 architecture. Nevertheless, this shifting towards ARM64 in the cloud computing industry, introduces serious considerations such as software compatibility and infrastructure adaptation issues.

Moreover, the alignment with Environmental, Social, and Governance (ESG) criteria is a crucial development for ARM64, due to the ever-growing importance of sustainability and energy efficiency. Its impact on reducing energy consumption in datacenters and edge computing implementations,

places ARM64 as a highly sustainable choice. However, ARM64 adoption brings up new challenges and concerns. Considerations such as software compatibility, management complexities in heterogeneous infrastructures, security threats, new skills development and vendor relationships need to be further addressed and discussed. Concluding, ARM64 architecture emerges as a transformative force within the computing industry, offering an energy efficient alternative for a wide range of applications. Nevertheless, it is important to note that its complete adoption potential depends on the continuous refinement and standardization on both hardware and software levels, so to align with the evolving requirements of the current technological landscape. Based on the above study, a potential next step for future research could be the assessment of virtualization technology solutions offered for ARM64-based infrastructures, so to assess their ability to seamlessly operate in actual datacenter environments. A research like that, could produce valuable comparative data and analysis between ARM64 and x86 implementations. Additionally, another field for future research could be the employment of ARM-based SBCs in edge computing scenarios, leveraging their small-form factor and low power consumption requirements for minimum carbon footprint implementations.

References

[1] S. Dandamudi, *Guide to RISC Processors: For Programmers and Engineers.* Springer Science Business Media, 2005.

[2] J. Goodacre and A. Sloss, "Parallelism and the ARM instruction set architecture," *Computer*, vol. 38, no. 7, pp. 42–50, 2005.

[3] S. Furber, *ARM System-on-Chip Architecture.* Pearson Education, 2000.

[4] "Qualcomm Snapdragon 660 and 662 Mobile Platforms," Qualcomm. (2023), [Online]. Available: https://www.qualcomm.com/content/dam/qcomm-martech/dm-assets/documents/sdm-sda660-product_brief_87-pu779-1.pdf.

[5] "Qualcomm Application Processors," Qualcomm. (2023), [Online]. Available: https://www.qualcomm.com/products/technology/processors/application-processors.

[6] "Apple Announces M2 SoC (Apple Silicon) Updated for 2022," Apple Inc. (2022), [Online]. Available: https://www.anandtech.com/show/17431/apple-announces-m2-soc-apple-silicon-updated-for-2022.

[7] "Apple announces Mac transition to Apple Silicon," Apple Inc. (2022), [Online]. Available: https://www.apple.com/newsroom/2020/06/apple-announces-mac-transition-to-apple-silicon/.

[8] "Tegra X1 Whitepaper," NVDIA Corporation. (2022), [Online]. Available: http://international.download.nvidia.com/pdf/tegra/Tegra-X1-whitepaper-v1.0.pdf.

[9] "Ampere Altra Family Product Brief," Ampere Semiconductor Company. (2023), [Online]. Available: https://amperecomputing.com/briefs/ampere-altra-family-product-brief.

[10] Jolles, Jolle W., "Broad-scale applications of the Raspberry Pi: A review and guide for biologists," *Methods in Ecology and Evolution*, vol. 12, no. 9, pp. 1562–1579, 2021. DOI: https://doi.org/10.1111/2041-210X.13652.

[11] Lambropoulos, Georgios and Douligeris, Christos and Mitropoulos, Sarandis, "A Review on Cloud Computing services, concerns, and security risk awareness in the context of Digital Transformation," Sep. 2021. DOI: 10.1109/SEEDA-CECNSM53056.2021.9566267.

[12] Lambropoulos, Georgios and Mitropoulos, Sarandis and Douligeris, Christos, "Improving Business Performance by Employing Virtualization Technology: A Case Study in the Financial Sector," *Computers*, vol. 10, no. 4, 2021, ISSN: 2073-431X.

[13] "ESXi-Arm Edition," VMware Flings. (2023), [Online]. Available: https://flings.vmware.com/esxi-arm-edition.

[14] "Hyper-V technology overview," Microsoft Corporation. (2022), [Online]. Available: https://learn.microsoft.com/en-us/windows-server/virtualization/hyper-v/hyper-v-technology-overview.

[15] "Xen ARM with Virtualization Extensions," Xen Project. (2022), [Online]. Available: https://wiki.xenproject.org/wiki/Xen_ARM_with_Virtualization_Extensions.

[16] Younge, Andrew J and Pedretti, Kevin, "HPC at Sandia: Exploring the Virtualization and Containerization of ARM64 Processors for Future HPC Workloads.," Sandia National Lab.(SNL-NM), Albuquerque, NM (United States), Tech. Rep., 2018.

[17] R. Sörensen, *An evaluation of edge deployment models for Kubernetes*, 2023.

[18] J.-S. Ma, H.-Y. Kim, and W. Choi, "Kvm-qemu virtualization with arm64bit server system," in *Cloud Computing: 6th International Conference, CloudComp 2015, Daejeon, South Korea, October 28-29, 2015, Revised Selected Papers 6*, Springer, 2016, pp. 334–343.

[19] Q. Jiang, Y. C. Lee, and A. Y. Zomaya, "The power of ARM64 in public clouds," in *2020 20th IEEE/ACM International Symposium on Cluster, Cloud and Internet Computing (CCGRID)*, IEEE, 2020, pp. 459–468.

[20] "Amazon Elastic Container Service on Arm," Amazon Web Services. (2022), [Online]. Available: https://docs.aws.amazon.com/AmazonECS/latest/developerguide/ecs-arm64.html.

[21] "Azure Virtual Machines with Ampere Altra Arm-based Processors," Microsoft Azure. (2022), [Online]. Available: https://azure.microsoft.com/en-us/updates/public-preview-arm64based-azure-vms-can-deliver-up-to-50-better-priceperformance/.

[22] "ARM on Compute Engine," Google Cloud. (2021), [Online]. Available: https://cloud.google.com/compute/docs/instances/arm-on-compute.

[23] P. Bodmann, G. Papadimitriou, D. Gizopoulos, and P. Rech, "The Impact of SoC Integration and OS Deployment on the Reliability of ARM Processors," in *2021 IEEE International Symposium on Performance Analysis of Systems and Software (ISPASS)*, IEEE, Mar. 2021, pp. 223–225.

[24] M. Camilleri, "Environmental, Social and Governance Disclosures in Europe," *Sustainability Accounting, Management and Policy Journal*, vol. 6, no. 2, pp. 224–242, 2015.

[25] "ESG Explained: What Is ESG?" Deloitte. (2021), [Online]. Available: https://www2.deloitte.com/ce/en/pages/global-business-services/articles/esg-explained-1-what-is-esg.html.

[26] "ESG EU Regulatory Change and Its Implications," Harvard Law School Forum on Corporate Governance. (2023), [Online]. Available: https://corpgov.law.harvard.edu/2023/02/18/esg-eu-regulatory-change-and-its-implications/.

[27] K. Seidenfad, M. Greiner, J. Biermann, and U. Lechner, "Emissions on the Edge," in *Innovations for Community Services: 23rd International Conference, I4CS 2023*, Springer Nature, Aug. 2023, p. 123.

[28] "Apps on ARM: x86 emulation," Microsoft Docs. (2022), [Online]. Available: https://learn.microsoft.com/en-us/windows/arm/apps-on-arm-x86-emulation.

[29] H. Zeyu, X. Geming, W. Zhaohang, and Y. Sen, "Survey on edge computing security," in *2020 International Conference on Big Data, Artificial Intelligence and Internet of Things Engineering (ICBAIE)*, IEEE, 2020, pp. 96–105.

[30] A. Botchkarev and P. Andru, "A Return on Investment as a Metric for Evaluating Information Systems: Taxonomy and Application," *Interdisciplinary Journal of Information, Knowledge, and Management*, vol. 6, pp. 245–269, 2011.

Efficient Categorization of Pneumonia Diagnosis Using Low-Power Embedded Devices

Theodora Sanida[1], Maria Vasiliki Sanida[2], Argyrios Sideris[1], Michael Dossis[1] and Minas Dasygenis[1]

[1]*Department of Electrical & Computer Engineering, University of Western Macedonia,* 50131, Kozani, Greece

[2]*Department of Digital Systems, University of Piraeus* 18534, Piraeus, Greece

thsanida@uowm.gr, sanidasilia@gmail.com, asideris@uowm.gr, mdossis@uowm.gr, mdasyg@ieee.org

Abstract—The lungs are integral to facilitating the respiratory processes crucial for human survival. However, lung infections can pose severe threats to human health. In medical diagnostics, numerous tools and techniques exist for assessing and diagnosing lung-related issues. Among these, Chest X-ray images have emerged as a widely favoured choice due to their distinct advantages, such as accessibility and the ability to provide valuable insights into pulmonary conditions. This work is dedicated to developing a robust multiclass category system for Chest X-ray images. A notable feature of this work is its emphasis on the efficient execution of the categorization work on a low-power heterogeneous embedded device, demonstrating the potential for practical applications in resource-constrained environments. One of this study's key highlights lies in exploring different optimization algorithms and their impact on category accuracy. The study conducts comprehensive experiments utilizing four prominent optimizers: Adam, Adamax, RMSprop, and SGD. Through these experiments, we observed that the proposed modified design attained the highest rate of accuracy, 97.16% when the Adam optimizer was employed. This outcome underscores the significance of optimizer selection in developing accurate and reliable diagnostic models for Chest X-ray image categorization, ultimately contributing to advancements in medical imaging and healthcare.

Index Terms—Low-power embedded device; Categorization; Pneumonia diagnosis; Optimization algorithms; Medical imaging.

I. INTRODUCTION

Chest X-rays are pivotal in identifying a broad spectrum of lung-related diseases. They are a fundamental and frequently employed tool in the medical field, owing to a combination of factors that make them highly advantageous. One of the primary strengths of chest X-rays is their affordability, making them accessible across a wide range of healthcare environments, from large hospitals to smaller clinics. Another significant advantage of chest X-rays is their widespread availability. In many medical structures, X-ray machines are a standard component of the diagnostic toolkit, enabling healthcare providers to promptly obtain crucial insights into a patient's lung health. This accessibility is particularly vital in emergencies where immediate medical decisions are required [1]. In contrast, more progressive imaging techniques like computed tomography (CT) scans, and magnetic resonance imaging (MRI) offer higher-resolution images with greater anatomical detail. However, they come with certain drawbacks that make

them less suitable for certain scenarios. CT scans, for instance, expose patients to higher ionising radiation levels than X-rays. Moreover, CT and MRI examinations are more expensive than chest X-rays and typically require longer examination times, which can be a limiting factor in time-sensitive situations [2].

Within hospitals, chest radiography has assumed a vital role in diagnosing infectious pneumonia. This diagnostic technique has proven invaluable in the early identification and monitoring of lung infections and is pivotal in patient care. However, it is important to acknowledge that in the early phases of pneumonia, the radiographic attributes captured on chest X-rays may lack the clarity and distinctiveness to simplify their interpretation. Computer Aided Diagnosis (CAD) methods have emerged as a powerful solution in response to these challenges. These systems harness the capabilities of artificial intelligence (AI) algorithms to support healthcare providers and help ensure that potential cases of pneumonia are identified accurately, even in their early stages [3], [4]. CAD systems do not seek to replace the expertise of radiologists and clinicians but complement it as assistants, augmenting the capabilities of healthcare professionals. So, CAD systems contribute to enhanced overall efficiency in healthcare delivery, particularly in the realm of pneumonia diagnosis, where time can be a critical factor in patient care [5], [6].

This work presents two significant contributions that advance the field of medical image categorization using low-power embedded devices:

- Firstly, we employ cutting-edge technology by training DenseNet121 CNN with transfer knowledge procedures. We introduce a novel layer head on top of the model, which is instrumental in extracting discriminative features from the input images, allowing for more precise categorization. Additionally, we conduct comprehensive experiments utilizing four prominent optimizers (Adam, Adamax, RMSprop, and SGD).

- Secondly, our method categorizes chest X-rays into 3 distinct types (normal, COVID-19, and pneumonia) in low-power embedded devices, demonstrating the potential for applications in resource-constrained conditions.

This work's remaining parts are structured as follows: We give comparable investigation studies in Part II, and Part III investigates the materials and methods operated in our strategy.

979-8-3503-8368-3/23 $31.00 © 2023 IEEE

Part IV explains the experiment's outcomes, and Part V summarises our study outcomes and future investigation.

II. RELATED WORK

In the previous two years, many researchers worldwide have created and published many analyses to detect and diagnose infectious pneumonia. Many of these investigators have employed various artificial intelligence methods to examine and analyze X-ray images to determine infections. The power of deep learning strategies yields better outcomes than standard machine learning procedures and has constructed them as the most famous technique for categorizing images. In this part, we focus on the studies that use new methods to diagnose infectious pneumonia based on deep learning procedures [7].

CoroNet is presented in [8], based on the Xception model, and was specifically designed to categorize chest X-ray images. The collection employed for training and evaluation consisted of 327 images representing viral pneumonia, 330 images depicting bacterial pneumonia, 284 images showcasing COVID-19 issues, and 310 images of normal chest X-rays. CoroNet underwent training in 80 epochs—the outcomes for all four categories reached an accuracy rate of 89.60%.

Hafeez et al. [9] developed a specialized CNN prognosis system tailored to analyze chest X-ray images. This system was meticulously crafted and corresponded with 2 well-established pre-trained models, AlexNet and VGG16. The primary objective was to assess their performance in categorizing X-ray images into 4 types: normal, bacterial bacteria, virus bacteria, and COVID-19. The custom CNN prediction system exhibited an accuracy rate of 89.855%. In contrast, VGG16 achieved an accuracy of 89.015%, while AlexNet demonstrated a slightly higher accuracy of 89.155%.

In contrast to the previous authors, we present a modified version of the deep learning design that can be operated by an embedded real-time device that operates on minimal power. In addition, we trained our model on a dataset that included more than 33k different images. We are investigating a variety of optimization strategies and the effect that they have on category accuracy. The research involves extensive tests using four different optimizers (Adam, Adamax, RMSprop, and SGD). In our study, one of our primary objectives is to design a model that is capable of achieving outstanding performance in the classification of the three types of lung illnesses that are considered to be the most important (pneumonia, COVID-19, and normal), all while maintaining a modest size number of parameters.

III. MATERIALS AND METHODS

A. Collection of dataset

The COVID-QU-Ex collection [10] was used as the work's experimental dataset, which consists of 3 types: non-COVID infection, normal, and COVID-19. Patients with non-COVID infection cases represent 11,263 samples, healthy conditions represent 10,701 samples, and COVID-19 conditions represent 11,956 samples. Figure 1 depicts a sample of two different disorders that may damage the lungs and normal instance.

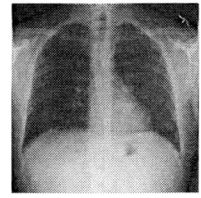 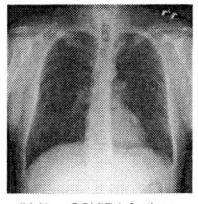 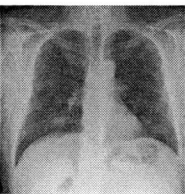

(a) Normal   (b) Non-COVID infections (Viral or Bacterial Pneumonia)   (c) COVID-19

Fig. 1. The COVID-QU-Ex collection [10] sample.

We randomly separated the collection into training/validation (80%) and testing (20%). The number of each type operated for training/validation/testing is summarised in Table I. In the collection, each image's resolution is 256 pixels per flank, and we changed the size to 224×224 pixels.

TABLE I
THE NUMBER OF EACH CATEGORY FOR TRAINING/VALIDATION/TESTING.

Category	Number of Images	Training Images	Validation Images	Test Images
Normal (Healthy)	10,701	6849	1712	2140
COVID-19	11,956	7658	1903	2395
Non-COVID infections (Viral or Bacterial Pneumonia)	11,263	7208	1802	2253
Total	33,920	21,715	5417	6788

B. The proposed approach using DenseNet121

The primary objective of the transfer learning method is to leverage the knowledge attained from one specific task or domain and apply it to another task or domain that exhibits similarity. This method stands out for its time and resource utilization efficiency, primarily because the pre-trained weights already encapsulate valuable insights and features. Doing so significantly reduces the training duration required, making it a highly effective strategy for various machine and deep learning tasks. In our case, we formulated a deep learning design by modifying the architecture of the well-established DenseNet121 [11] model.

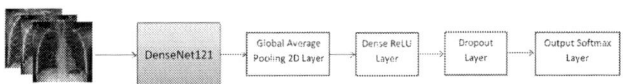

Fig. 2. The overall mechanism that has been proposed for DenseNet121 model.

First, we added a global average pooling layer, which plays a vital role in reducing the spatial dimensions of the feature maps, allowing for a more compact representation of the information. Following that, we incorporated a dense layer, improving the design's capability to capture complex patterns and connections within the data. A dropout layer 0.3 is used during training to mitigate overfitting and improve generalization. Finally, we operated a softmax activation function for the outcome layer, enabling the design to produce probability distributions over the target categories. Figure 2 illustrates our model's overall structure and components.

C. Performance measurements

Four measurements (1) to (4) were utilised to assess the performance of the deep transfer learning design. Measurement of accuracy displays the proportion of accurate predictions to all examined instances. The ratio of properly identified positive instances to all positive samples analysed is calculated as recall. Precision is computed as the percentage of accurately predicted positive representatives divided by the whole number of optimistic forecasts. Finally, the F1-Score determines the harmonic mean of recall and accuracy.

$$\text{Accuracy} = \frac{AP + AN}{AP + WN + WP + AN} \star 100\% \quad (1)$$

$$\text{Recall} = \frac{AP}{AP + WN} \star 100\% \quad (2)$$

$$\text{F1-Score} = \frac{2 \star (\text{ Precision } \star \text{ Recall })}{(\text{ Precision } + \text{ Recall })} \star 100\% \quad (3)$$

$$\text{Precision} = \frac{AP}{AP + WP} \star 100\% \quad (4)$$

Equations (1) to (4)'s symbols AP, AN, WP, and WN stand for the number of actual positives, actual negatives, wrong positives, and wrong negatives, respectively.

IV. EXPERIMENTAL OUTCOMES

The NVIDIA Jetson AGX Xavier [12] was operated to train the proposed design using Keras Tensorflow backend in Python. Noteworthy, the testing data were kept entirely separate from the training process. The model training process involved utilizing four distinct optimization algorithms, Adam, Adamax, RMSprop, and SGD, each executed for a total of 30 epochs. A batch size of 16 was employed to efficiently process and update the model's weights during this training process.

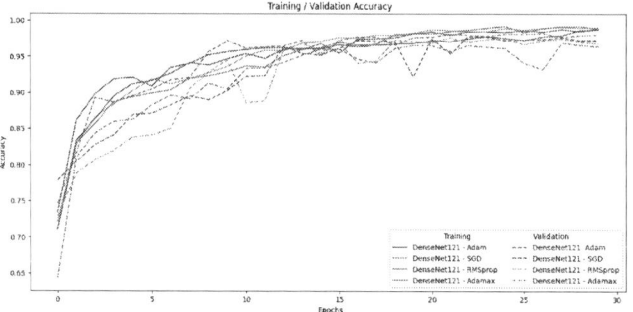

Fig. 3. Training and validation accuracy of four distinct optimization algorithms for 30 epochs.

Figure 3 illustrates four distinct optimization algorithms' training and validation accuracy, providing a comprehensive view of how these algorithms perform across the training process. So, by comparing the accuracy curves, we can discern

which algorithm exhibits the most stable and efficient learning behaviour. Figure 4 depicts our experimentation's training and validation loss throughout 30 epochs. The loss curves offer critical insights into the model's ability to minimize error and generalize to unrecognized data.

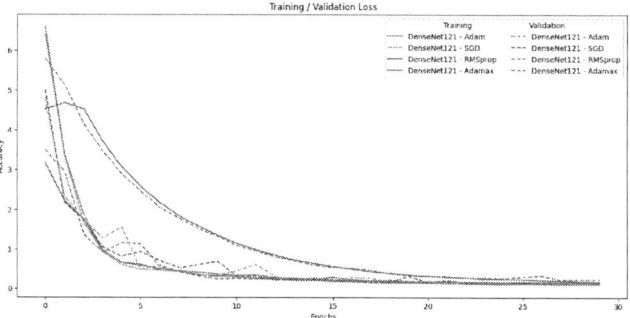

Fig. 4. Training and validation loss of four distinct optimization algorithms for 30 epochs.

Table II shows the evaluation parameters for the four optimizers (Adam, Adamax, RMSprop, and SGD). Among the four optimization algorithms examined in this study, Adam stands out as a clear frontrunner regarding its performance metrics. Notably, it achieves the highest F1-score, boasting an impressive score of 97.13%. In the realm of precision, both Adam and SGD shine with excellent precision scores of 97.16% and 96.44%, respectively. Moreover, when considering the average accuracy metric, Adam again emerges as the leader, achieving an accuracy score of 97.16%. So, Adam's consistent excellence across all evaluated metrics positions it as an exceptionally compelling choice for optimizing deep learning models.

TABLE II
PERFORMANCE METRICS OF FOUR OPTIMIZERS (ADAM, ADAMAX, RMSPROP, AND SGD)

Optimizer	Average F1-score (%)	Average precision (%)	Average recall (%)	Average accuracy (%)
Adam	97.13	97.16	97.12	97.16
Adamax	95.50	95.50	95.57	95.52
RMSprop	96.02	96.03	96.10	96.04
SGD	96.45	96.44	96.50	96.46

Table III shows the classification reports for the DenseNet121 model for the four optimizers (Adam, Adamax, RMSprop, and SGD). Comparing the outcomes of classification reports for DenseNet121 with different optimizers, we can discern noteworthy variations in the model's performance across these optimization techniques. Firstly, when utilizing the Adam optimizer, the model consistently demonstrated high precision, recall, and F1-scores for all three categories, including COVID-19, Non-COVID, and Normal. This optimizer yielded an impressive overall F1-score of 0.9713, indicative of a well-rounded categorization performance. In contrast, the Adamax optimizer used slightly lower precision and recall for COVID-19 cases but maintained strong overall F1-scores.

979-8-3503-8368-3/23 $31.00 © 2023 IEEE

The F1-scores for Non-COVID and Normal cases remained robust, with an average F1-score of 0.9550. Moving to the RMSprop optimizer, the model exhibited remarkably high precision for COVID-19 cases and strong F1-scores across the board. Although it achieved a high average F1-score of 0.9602, it is important to note that RMSprop demonstrated a slight decrease in precision for Non-COVID cases compared to other optimizers. Lastly, the SGD optimizer yielded consistent and robust precision, recall, and F1-scores for all three categories, with a strong overall F1-score of 0.9645. While it showed slightly lower precision for COVID-19 cases than Adam and RMSprop, it maintained excellent recall for COVID-19 and Non-COVID cases.

TABLE III
CLASSIFICATION REPORTS FOR DENSENET121 MODELS WITH DIFFERENT OPTIMIZERS

Optimizer	Category	Precision	Recall	F1-Score
Adam	COVID-19	0.9865	0.9770	0.9817
	Non-COVID	0.9521	0.9791	0.9654
	Normal	0.9762	0.9575	0.9667
Adamax	COVID-19	0.9860	0.9436	0.9644
	Non-COVID	0.9524	0.9512	0.9518
	Normal	0.9265	0.9724	0.9489
RMSprop	COVID-19	0.9943	0.9441	0.9685
	Non-COVID	0.9584	0.9605	0.9594
	Normal	0.9282	0.9785	0.9527
SGD	COVID-19	0.9917	0.9507	0.9708
	Non-COVID	0.9552	0.9738	0.9644
	Normal	0.9462	0.9706	0.9582

The credibility and efficacy of the proposed design are rigorously evaluated by subjecting it to a comprehensive comparison with the latest and most cutting-edge plans in the field. The outcomes of this thorough evaluation, as showcased in Table IV, unequivocally affirm the remarkable performance of the proposed design. This superior performance not only enhances the reliability of the proposed system but also bolsters its suitability for tasks demanding the highest level of accuracy.

TABLE IV
COMPARISON OF OUTCOMES WITH PREVIOUS STUDIES

Work	Collection	Training epochs	Accuracy (%)
[8]	330 bacterial pneumonia, 310 normal, 284 COVID-19, and 327 viral pneumonia	80	89.60%
[9]	79 virus pneumonia, 79 bacterial pneumonia, 78 COVID-19, and 28 viral pneumonia	60	89.855%
This work	11.263 pneumonia, 11.956 COVID-19, and 10.701 normal	30	97.16%

V. CONCLUSIONS AND FUTURE WORK

Lung diseases can be severe and require timely diagnosis to ensure effective treatment. In the context of this study, the research focused on leveraging transfer learning strategies to categorize radiography images, particularly for lung-related conditions. The DenseNet121 was employed as a model, and we compared the categorization efficiency achieved using various optimization algorithms, shedding light on the impact of optimization choices on model performance. The findings from this research revealed that DenseNet121, when paired with the Adam optimizer, demonstrated a level of accuracy, achieving an impressive 97.16%. In further research, we want to verify our model by integrating more types of medical data. This expansion will encompass various modalities such as clinical data, patient histories, genetic information, or even data from more advanced imaging techniques like CT scans or MRI. So, by incorporating a broader range of medical data, we can create more comprehensive and holistic models that offer even greater accuracy and insights into lung disease diagnosis.

ACKNOWLEDGMENTS

This work was partially supported by the University of Western Macedonia.

REFERENCES

[1] S. Sharma and K. Guleria, "A systematic literature review on deep learning approaches for pneumonia detection using chest x-ray images," *Multimedia Tools and Applications*, pp. 1–51, 2023.

[2] T. Sanida, A. Sideris, A. Chatzisavvas, M. Dossis, and M. Dasygenis, "Radiography images with transfer learning on embedded system," in *2022 7th South-East Europe Design Automation, Computer Engineering, Computer Networks and Social Media Conference (SEEDA-CECNSM)*. IEEE, 2022, pp. 1–4.

[3] W. Liu, Z. Ni, Q. Chen, and L. Ni, "Attention-guided partial domain adaptation for automated pneumonia diagnosis from chest x-ray images," *IEEE Journal of Biomedical and Health Informatics*, 2023.

[4] T. Sanida, I.-M. Tabakis, M. V. Sanida, A. Sideris, and M. Dasygenis, "A robust hybrid deep convolutional neural network for covid-19 disease identification from chest x-ray images," *Information*, vol. 14, no. 6, p. 310, 2023.

[5] D. P. Mannepalli and V. Namdeo, "A cad system design based on hybrid-multiscale convolutional mantaray network for pneumonia diagnosis," *Multimedia Tools and Applications*, vol. 81, no. 9, pp. 12 857–12 881, 2022.

[6] T. Sanida, A. Sideris, D. Tsiktsiris, and M. Dasygenis, "Lightweight neural network for covid-19 detection from chest x-ray images implemented on an embedded system," *Technologies*, vol. 10, no. 2, p. 37, 2022.

[7] H. Bhatt and M. Shah, "A convolutional neural network ensemble model for pneumonia detection using chest x-ray images," *Healthcare Analytics*, vol. 3, p. 100176, 2023.

[8] A. I. Khan, J. L. Shah, and M. M. Bhat, "Coronet: A deep neural network for detection and diagnosis of covid-19 from chest x-ray images," *Computer methods and programs in biomedicine*, vol. 196, p. 105581, 2020.

[9] U. Hafeez, M. Umer, A. Hameed, H. Mustafa, A. Sohaib, M. Nappi, and H. A. Madni, "A cnn based coronavirus disease prediction system for chest x-rays," *Journal of Ambient Intelligence and Humanized Computing*, pp. 1–15, 2022.

[10] A. M. Tahir, M. E. Chowdhury, A. Khandakar, T. Rahman, Y. Qiblawey, U. Khurshid, S. Kiranyaz, N. Ibtehaz, M. S. Rahman, S. Al-Maadeed *et al.*, "Covid-19 infection localization and severity grading from chest x-ray images," *Computers in biology and medicine*, vol. 139, p. 105002, 2021.

[11] A. Raza, M. S. Alshehri, S. Almakdi, A. A. Siddique, M. Alsulami, and M. Alhaisoni, "Enhancing brain tumor classification with transfer learning: Leveraging densenet121 for accurate and efficient detection," *International Journal of Imaging Systems and Technology*, 2023.

[12] S. Shi, Q. Jiang, X. Jin, W. Wang, K. Liu, H. Chen, P. Liu, W. Zhou, and S. Yao, "A comparative analysis of near-infrared image colorization methods for low-power nvidia jetson embedded systems," *Frontiers in Neurorobotics*, vol. 17, p. 1143032, 2023.

The role of social media before and during a holiday travel: generational and gender differences

Ifigeneia Mylona
Dept. of Management Science and Technology, IHU
Kavala, Greece
ORCID: 0000-0003-4880-8132

Dimitrios Amanatidis
Dept. of Informatics
University of Western Macedonia
Kastoria, Greece
ORCID: 0000-0001-6667-9237

Michael Dossis
Dept. of Informatics
University of Western Macedonia
Kastoria, Greece
ORCID: 0000-0002-1863-3119

Irene (Eirini) Kamenidou
Dept. of Management Science and Technology, IHU
Kavala, Greece
ORCID: 0000-0002-8213-5843

Spyridon Mamalis
Dept. of Management Science and Technology, IHU
Kavala, Greece
ORCID: 0000-0003-3035-6385

Abstract— This article discusses the use of social media by different generations in destination marketing. The study analyses the use of social media by travelers at all stages of a trip. It considers Kimolos, a small Aegean Sea Island as a case study. This research aimed to examine the degree of social media influence by travelers before and during their trip to the island. Other elements such as the most popular social media, the reason for their use by travelers and their role as a source of information were also analyzed.

Keywords—social media, RStudio, data analysis, marketing, generations X Y Z, Baby Boomers, Kimolos

I. INTRODUCTION

Social media are changing the ways that information is transmitted across societies and tourists all over the world. Different cohorts of tourists may use social media in different ways. Cohorts are groups of individuals who are born during a specific time interval of 20 years and travel through life together [1]. Within the above theoretical framework, this study examines four Greek generational cohorts, the Baby Boomers (BB) born between 1946 – 1964, Generation X, consisting of people born between 1965 – 1980 Generation Y (or millennials according to many scholars) who were born between 1981 – 1996 and Generation Z, people that were born during 1997 – 2012 [2]. This paper presents an initial study in order to examine the use of social media before and during holidays for tourists in Kimolos Island. More specifically the aim of the study is to examine the frequency of using different social media platforms by different gender and/or generational cohort, while planning their holiday travel and during their holiday.

II. LITERATURE REVIEW

A. Social media, tourism and generations

At the beginning of the 21st century, most enterprises and organisations used the Internet for one-way communication, which is comparable to traditional promotion tools, such as advertising [3,4]. Social media platforms seem to be popular with consumers and businesses as they provide new opportunities for interactivity and connectivity for both [5]. According to Richter and Koch [6], social media is defined as an online application, platform, or medium that can facilitate interactions, collaborations, and content sharing. As Thevenot [7] claims, social media popularity increases, users acquire more power while, at the same time, the authority of marketers and institutions recesses. Social networks enable users to post and propose ideas about new services [8]. Social media's technological nature encourages active participation of citizens in the public sphere, especially through likes, shares and comments [9]. Users can create content and affect other users by making comments and posts. The user's activities on social media are broadly classified as either contribution, or consumption tasks and usually maximum users are consumption oriented than contributors to the social media [10].

Social media play a significant role both on the demand and on the supply side of tourism allowing destinations to interact directly with visitors via various internet platforms and monitor and react on visitors´ opinions and evaluations of services [11]. Leung, Law, Van Hoof, and Buhalis [12], highlight the strategic importance of social media for the successful promotion of tourism in terms of competitiveness. Another work from the same year [13] studies Scandinavian tourists' perceptions in Mallorca, differentiating between Web 1.0 and Web 2.0 information sources. The authors found high motivation for sharing tourism experiences and knowledge on sites with a high richness of social cues. A year later, in their subsequent study, Munar and Jacobsen [14] found out that Scandinavian social network sites such as Facebook were the most popular among the Scandinavian Mallorca tourists, who liked to share their visual experiences on social media. Pinto [15] claims that there is a close relationship between social networks and tourist destinations and visitors based on low cost and efficiency. Stavrianea and Kavoura [16] in their work about the use of social media from tourists in Greece found that, Facebook was the second most important medium that people use when collecting material about places they want to visit. In addition, Facebook was found to be the third source that respondents strongly agreed that they pay attention to advertisements, with the official websites/blogs to be the first source and travel guidebooks and travel magazines to follow, leaving traditional media far behind.

Social media influenced by social networking is pressing suppliers and buyers who value more and more the opinions, reviews and referrals of fellow travelers [17]. Social media will play an increasingly important role in marketing activities in the field of tourism. In fact, they are already used to build the brand of the region, company and attractions, as well as build relationships with tourists before, during and after a tourist stay [14].

979-8-3503-8368-3/23 $31.00 © 2023 IEEE

In terms of generations there is a differentiation between cohorts in the use of technology. Most of the Baby Boomers were not much influenced by advertising [19]. According to AARP [50], 85% of BB travelers use the Internet for travel planning. According to the same research [20] 80% of BB likes to share their memories by using digital methods, such as sending pictures via text messages or Facebook posts. The Baby Boomers are characterized as "digital immigrants" [21]. Generation X use online social media to facilitate their need to exchange and share information and as the they feel nostalgia, is the generation that uses interactions at the level of exchange in online social media to communicate with others [22] Generation X is the generation that is using Facebook at a higher rate than the other generations, generation Y users use social media to interact with others, they are "digital natives" and use the Internet when planning and booking their tourist trips [23, 24,18]. Online reviews had been an influential source of information for Generation Y. They use to Share experiences and communicate in social networks sites, write and read online reviews and blog posts. [25]. Generation Z members navigate the digital world, using social media for more varied purposes [18,19] and represent a major challenge to destination marketers [26].

B. Kimolos island

Kimolos, is a small island located in the Central Aegean Sea. It is part of the South Aegean Volcanic Arc [27]. Kimolos is close to the northwest of the larger Milos Island, separated by a one-km-wide channel. The only village is Chorio, located on a hill on the west of the island. There are also smaller settlements such as Psathi (port), Goupa, Kara, Prasa, Aliki, and Bonatsa near the sea and offer themselves for a swim. Despite its small size, Kimolos has a great number of churches [28]. Over the last decade, there has been a rise in the island's touristic development, and the number of visitors has increased significantly. Small hotels have been built and more restaurants have opened [28].

III. MATERIALS AND METHODS

The research took place in July and August 2023. An aided questionnaire was utilized in different places of the island and the participants were asked to answer the questionnaire, confirming that they have not previously answered it in a different place. The questionnaire design was based on similar studies [29, 30]. Participants were Greek citizens that have visited Kimolos island. The questionnaire in this study comprises 37 questions, including demographic data. The questionnaire items fell into one of three thematic classes, with the first concerning the users' general point about the satisfaction of tourists in Kimolos' island, the second being related with Kimolos and social media and the final one contained demographic data. Some of the questions require the respondents to indicate their level of agreement on a 5-point Likert scale for several different social media platforms, e.g., Facebook, Instagram etc. Thus, in constructing the respective dataset, these questions essentially expand to more questions, in order to record all responses for the different platforms. The complete dataset consists of 83 column-variables and 130 responses. However, in this work we limit focus on the 22 variables (detailed later) that correspond to the Research Questions and respective Hypotheses:

- RQ1: How does the frequency of using different social media platforms vary with respect to members of

different gender and/or generational cohort, while planning their holiday travel?

Hypothesis 1.1. There are significant variations in how frequently, users of different age and/or gender, rely on social media before making a trip.

- RQ2: How does the frequency of using social media for a specific reason vary with respect to members of different gender and/or generational cohort, while being on their touristic destination?

Hypothesis 2.1. There are significant variations in the frequency and reasons why, users of different age and/or gender, use social media while having their holidays.

IV. STATISTICAL ANALYSIS

RQ1 refers to the time before the trip takes place and there are two related questions. The first question examines the frequency that four (4) different social media platforms (TripAdvisor, Trivago, Booking, AirBnB) are used for making pre-travel reservations, e.g., mainly travel and accommodation related reservations, as this is the main concern of holiday travelers. The second question is related to the frequency that ten (10) different social media platforms (Facebook, Instagram, YouTube, TikTok, Twitter, TripAdvisor, Google Maps, Trivago, Booking, AirBnB) are utilized for seeking general pre-travel information on the touristic destination. RQ2 refers to the time while actually being on holiday. There is one related question that asks the participants on how frequently they have used social media in general, without discriminating the specific platform, but indicating the particular reason out of six (6) possible (communicating with relatives and friends, information seeking, uploading photos and videos, making comments, location sharing, activity reservation). Thus, our dataset variables sum to 22 (=4+10+6+2), with last two for generation and gender. The Spearman correlation matrix is given in Fig. 1. It can be seen that most variables seem to have a positive pairwise association, except for gender which has a negative association.

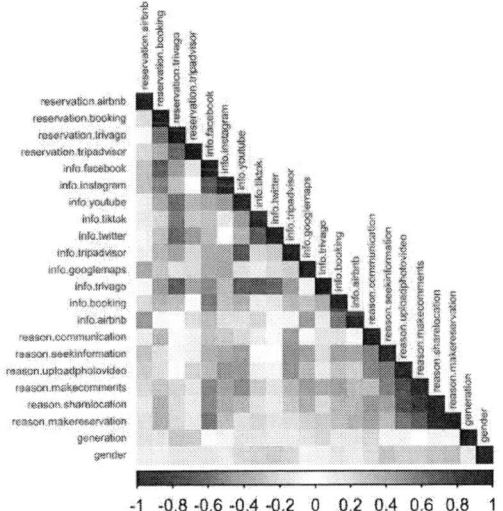

Fig. 1. Spearman correlation matrix

In order to determine whether there are significant differences between the different age and gender subgroups with respect to the different social media platform used before travel and different reason while on holiday we employed the following statistical analyses in R:

- One-way ANOVA to compare the means of different social media platform frequencies (before travel) and the means of different reasons for their usage (during travel) with respect to generation.

- Student t-test (actually ANOVA has been also confirmed to produce identical results) to compare the means of different social media platform frequencies (before travel) and the means of different reasons for their usage (during travel) with respect to gender.

- Two-way ANOVA to compare mean differences as before but with respect to both age and gender, as the two independent variables.

The respondents' sample consists of 58 male (coded as 1) and 72 female (coded as 2). According to their cohort, there are 32 Baby Boomers (code 1), 36 members of Generation X (code 2), 32 Millennials (Y – code 3) and 30 members of Generation Z (code 4). ANOVA requires some assumptions to be met in order to proceed with a reliable interpretation of results. Here, inter-group and intra-group independency are obvious and normality is guaranteed as the number of observations for each group is $n \geq 30$ [31]. Normality can also be inspected visually by e.g., a QQ-plot (an example is shown later) or a normality test (Shapiro-Wilk or Kolmogorov-Smirnov). In case that the normality test fails, even perhaps after a data transform (logarithmic or Box-Cox), the Kruskal-Wallis test can be applied instead of ANOVA. Homoscedasticity (equal group variances) can also be inspected visually, or with Levene or Bartlett tests. As an example, Fig. 2 displays the diagnostic plots for *information.youtube* by *generation* and *gender* model. The rule of thumb is that the mean of residuals should be horizontal and centered on zero, except for scale-location where it should be centered on one and the QQ-plot should have an approximate slope of one. In case that equality of variances is not guaranteed, another version of ANOVA is suggested, Welch ANOVA, which nevertheless requires normality. The non-parametric Kruskal-Wallis test does not require normality nor homoscedasticity, compares subgroup medians rather than means and can efficiently handle outliers as well.

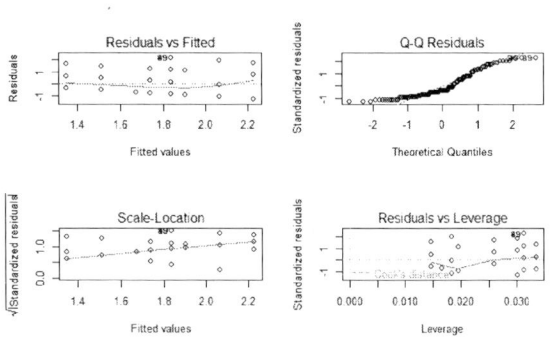

Fig. 2. Normality and homoscedasticity diagnostic plots

The modeling of our 20 dependent variables by generation and/or gender was subsequently carried out and the respective ANOVA tests produced the following results summarized on Table 1. Only the eleven significant, at the .05 level (*) or less, cases are reported. Moreover, the table displays the model with the best fit, according to Akaike Information Criterion

(AIC). At this point, before proceeding with post-hoc tests we decided to consider only the models with one independent variable, generation or gender. The reason is that two-way models would have a decreased subgroup size and this might violate the normality assumption.

Table 1. Significant modelling cases

Variable	generation	gender	gnr + gnd
reservation.trivago	*** (AIC)		***/
reservation.tripadvisor	**	.	**/ (AIC)
information.youtube	*	*	*/* (AIC)
information.tiktok		** (AIC)	/**
information.tripadvisor		* (AIC)	/*
reason.communicate	.	**	*/* (AIC)
reason.seakinfo		*** (AIC)	/***
reason.uploadphoto		** (AIC)	/**
reason.comment		** (AIC)	/**
reason.shareloc		** (AIC)	/**
reason.reserve		*** (AIC)	/***

With a first glance at the table, we can infer that generation moderates the frequency of use of Trivago (F=13.03, p=0.000439) and TripAdvisor (F=6.873, p=0.00981) for making pre-travel reservations. There are no significant differences in the frequency of reservations with AirBnB and Booking. The use of YouTube for pre-travel information is also varied by generation (F=5.272, p=0.0233), but by gender as well (F=6.164, p=0.0143). In fact, the two-way ANOVA model seems to have a slightly better fit (followed by the gender-based model) as seen for example on the following Fig. 3 and Fig. 4. However, as stated previously, we opted to consider only models with one independent variable, i.e., either generation or gender alone.

```
            Df Sum Sq Mean Sq F value Pr(>F)
generation   1    5.13   5.127   5.446 0.0212 *
gender       1    4.93   4.932   5.239 0.0237 *
Residuals  127  119.55   0.941
---
Signif. codes:  0 '***' 0.001 '**' 0.01 '*' 0.05 '.' 0.1 ' ' 1
```

Fig. 3. YouTube use by generation and gender

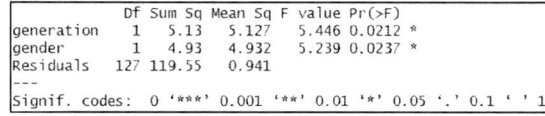

Fig. 4. AIC model best fit

Going back to Table 1 results, we can also infer that gender moderates the use of TikTok (F=9.711, p=0.00226) and TripAdvisor (F=5.709, p=0.0183) for seeking pre-travel information. Pre-travel use (for reservations or information) of the other social media platforms is insignificant by either generation or gender. Finally, it is interesting that gender alone is the dominant predictor for all six reasons that social media are used during holiday. Specifically, significant variations with respect to gender were found in the reasons that users exploit social media in general while on holiday: communicating with relatives and friends (F=6.965, p=0.00934), information seeking (F=13.37, p=0.000372), uploading photos and videos (F=8.473, p=0.00425), making comments (F=10.82, p=0.0013), location sharing (F=9.791,

979-8-3503-8368-3/23 $31.00 © 2023 IEEE

p=0.00217) and activities reservations (F=12.9, p=0.000467). A very useful R package, *report*, facilitates a rapid and user-friendly way of interpreting and reporting results. The package output also includes information on effect size and their interpretation according to [32]. As an example, for the case of *reason.makereservation* by *gender*, the package gives (Fig. 5):

```
The ANOVA (formula: reason.makereservation ~ gender) suggests that:

  - The main effect of gender is statistically significant and medium
(F(1, 128) = 12.90, p < .001;
Eta2 = 0.09, 95% CI [0.03, 1.00])

Effect sizes were labelled following Field's (2013) recommendations.
```

Figure 5. Output from the "report" package

Subsequently, we performed Tukey's Honestly Significant Difference (Tukey HSD) post-hoc test, in order to discriminate between the four generational cohorts and the two genders' means. Example results for *reservation.trivago* are shown on Fig. 6, where the differences between subgroups 3-1 (Generation Y's mean significantly larger than Baby Boomers' mean) and 4-1 (Generation Z' mean significantly larger than Baby Boomers' mean) are evident:

```
   Tukey multiple comparisons of means
     95% family-wise confidence level

Fit: aov(formula = reservation.trivago ~ as.factor(generation), data = dat)

$`as.factor(generation)`
        diff        lwr      upr     p adj
2-1  0.65625 -0.19424627 1.506746 0.1901478
3-1  0.90625  0.03109656 1.781403 0.0393507
4-1  1.20625  0.31663023 2.095870 0.0032170
3-2  0.25000 -0.60049627 1.100496 0.8698896
4-2  0.55000 -0.31537489 1.415375 0.3520464
4-3  0.30000 -0.58961977 1.189620 0.8162437
```

Fig. 6. Post-hoc test for Trivago reservations

A summary of the remaining post-hoc tests is given on Table 2 for the significant cases. The first three rows correspond to prediction of the respective variable by generation, while the last eight rows correspond to prediction of the respective variable by generation. The diff column displays the difference in means between the subgroups, as ordered. The p adj column reports on the p-values, adjusted for multiple comparisons. We can infer that generation Z members' frequency on *information.youtube* averages higher than generation Y's (0.7 points) and generation X's (0.8 points). In the case of *reservation.tripadvisor*, Zers score an average 0.9 points higher than BBers. In the case of *reservation.trivago*, Zers score 1.2 points higher than BBers, while Zers score 0.9 points higher than BBers. Finally, it seems that male participants score on average higher than female participants in the cases of *information.tiktok* (0.3 points), *information.tripadvisor* (0.5 points) and on all cases of social media usage during holidays, with differences lying in the interval [0.7, 1.0].

Table 2. Tukey HSD post-hoc tests

Variable	Subgroups	p adj	Diff
reservation.trivago	3-1,4-1 (gnr)	0.039,0.003	0.9,1.2
reservation.tripadvisor	4-1 (gnr)	0.037	0.9
information.youtube	4-2,4-3 (gnr)	0.006,0.024	0.8,0.7
information.tiktok	1-2 (gnd)	0.002	0.3
information.tripadvisor	1-2 (gnd)	0.018	0.5
reason.communicate	1-2 (gnd)	0.009	0.8
reason.seakinfo	1-2 (gnd)	0.0004	1
reason.uploadphoto	1-2 (gnd)	0.04	0.9
reason.comment	1-2 (gnd)	0.001	0.8
reason.shareloc	1-2 (gnd)	0.002	0.7
reason.reserve	1-2 (gnd)	0.0005	0.9

V. CONCLUSION

A questionnaire was designed and distributed to Greek holiday travelers, with the objective of exploring the possible differences according to generational cohort and gender. The case study considers Kimolos, a small island of the Aegean Sea. The dataset constructed consists of twenty plus two variables that aim to measure the pre-travel frequency in use of different social media platforms, for making reservations (four platforms) and seeking information (ten platforms), as well as measure the frequency and the different reason (out of six) that drives participants, during their holiday, to engage in any social media platform.

The statistical analysis involved the use of ANOVA in R. Several significant differences are reported; generation moderates the frequency of use of Trivago (F=13.03, p=0.000439) and TripAdvisor (F=6.873, p=0.00981) for making pre-travel reservations. There are no significant generational differences in the frequency of reservations with AirBnB and Booking. The use of YouTube for pre-travel information is also varied by generation (F=5.272, p=0.0233), but by gender as well (F=6.164, p=0.0143). Gender moderates also the use of TikTok (F=9.711, p=0.00226) and TripAdvisor (F=5.709, p=0.0183) for seeking pre-travel information. Pre-travel use (for reservations or information) of the other social media platforms is insignificant by either generation or gender.

Gender alone is the dominant predictor for all six reasons that social media are used during holiday. Specifically, significant variations with respect to gender were found in the reasons that users exploit social media in general while on holiday: communicating with relatives and friends (F=6.965, p=0.00934), information seeking (F=13.37, p=0.000372), uploading photos and videos (F=8.473, p=0.00425), making comments (F=10.82, p=0.0013), location sharing (F=9.791, p=0.00217) and activities reservations (F=12.9, p=0.000467). Other studies [33] though found that women much more often than men declare a more favourable attitude towards information posted on social media and check SM opinions about places they want to visit.

Post-hoc tests (Tukey HSD) were also carried out to discriminate between the generational cohorts and infer the relation between their average scores and for gender as well. Generation Z members' frequency on *information.youtube* averages higher than generation Y's (0.7 points) and generation X's (0.8 points). In the case of *reservation.tripadvisor*, Zers score an average 0.9 points higher than BBers. In the case of *reservation.trivago*, Zers score 1.2 points higher than BBers, while Zers score 0.9 points higher than BBers. Finally, it seems that male participants score on average higher than female participants in the cases of *information.tiktok* (0.3 points), *information.tripadvisor* (0.5 points) and on all cases of social media usage during holidays, with differences lying in the interval [0.7, 1.0].

Concluding Generation Z is more positive in seeking information in social media during it holidays and before as generations X, Y and Baby Boomers seems to be more skeptical and critical of information posted on social networks. What comes out from all the above is that tourists understood the importance of social media during their travelling and their seeking for information about the touristic destination, especially male and generation Z participants.

As Di Pietro claims [34], tourists use social media applications according to different modalities and behave towards them during the decisional process, as well as in the post‐travel (post‐consumption) phase. Users nowadays generate content online very quickly while at the same time they receive, process and share a great amount of information either from organizations or their peers.

VI. LIMITATIONS

This research has some unavoidable limitations. First of all, as the language of the questionnaire was Greek it was only handed out to Greek people. Second, the size of the sample that was collected was relatively small. A larger sample would validate the results of this paper. Lastly, the questionnaire was handed out only in one place, Kimolos island.

REFERENCES

[1] I. (E.) Kamenidou, S. Mamalis, A. Stavrianea, and I. Mylona, "Differences in Generational Cohort Satisfaction from a Public Hospital Medical Personnel: Insights from Generation Cohorts X, Y, and Z", In: Tsounis N, Vlachvei A (eds) *Advances in Quantitative Economic Research.* ICOAE Springer Proceedings in Business and Economics, Springer, Cham, pp 409-423, 2021. doi: 10.1007/978-3-030-98179-2_28

[2] "The 7 Generations: What do we know about them?, 2023", available at: https://journeymatters.ai/7-generations/

[3] S. Agresta, and B.B. *Bonin, Perspectives on Social Media Marketing" Course Technology,* Cengage Learning PTR: Boston, MA, USA, 2010.

[4] P.R. Berthon, L.F. Pitt, K. Plangger, and D. Shapiro, "Marketing meets Web 2.0, social media, and creative consumers: Implications for international marketing strategy", *Bus. Horiz.* vol.55, pp.261–271, 2012.

[5] D. Amanatidis, I. Mylona, S. Mamalis, and I. (E.) Kamenidou, "Social media for cultural communication: A critical investigation of museums' Instagram practices", *Journal of Tourism, Heritage & Services Marketing,* vol. 6, no. 2, pp. 38-44, 2020. Available at: http://dx.doi.org/10.5281/zenodo.3836638

[6] A. Richter, and M. Koch "Social software-status", Quo Und Zukunft. Fakultat Fur Informatik, Universitat der Bundeswehr Munchen, 2007

[7] G. Thevenot, "Blogging as a social media", *Tourism and Hospitality Research,* vol.7, no.3/4, pp.282-289, 2007

[8] M. Sigala, "Social networks and customer involvement in new service development (NSD): the case of www.mystarbucksidea.com", *International Journal of Contemporary Hospitality Management,* vol.24, no. 7, pp.966-990, 2012.

[9] S. Karekla, G. Gioltzidou, K. Kenterelidou, F. Galatsopoulou, M. Touri, I. Kostarella, and A. Skamnakis, Communication for Development and Social Change in Greece: An Emerging Field. *The Step of Social Sciences,* vol.20, no.75, pp.3-30, 2022.

[10] G. PrakashYadav, and J. Rai "The Generation Z and their social media usage: A review and a research outline", *Global journal of enterprise information system,* vol.9, no.2, pp.110-116, 2017.

[11] A. Kiráľová and A. Pavlíčeka, "Development of social media strategies in tourism destination" *Procedia-Social and Behavioral Sciences,* vol.175, pp.358-366, 2015.

[12] D. Leung, H.A Lee, R. Law and D. Buhalis, "Adopting Web 2.0 technologies on chain and independent hotel websites: A case study of hotels in Hong Kong" In: R. Law, M. Fuchs, F. Ricci (Eds). *Information and communication technologies in tourism,* New York:Springer-Wien, 2011.

[13] A.M., Munar and J.K.S. Jacobsen, "Trust and Involvement in Tourism Social Media and Web-Based Travel Information Sources",

Scandinavian Journal of Hospitality and Tourism, vol.13, no.1, pp.1-19, 2013

[14] A.M., Munar, and J.K.S. Jacobsen, "Motivations for sharing tourism experiences through social media", *Tourism Management,* vol.43, pp.46-54, 2014.

[15] M.Pinto, "influência das redes sociais na perceção e escolha de um destino turístico na Geração Y", Instituto Universitário de Lisboa, Dissertação de Mestrado, 2016.

[16] A.Stavrianea, and A. Kavoura, "Social media's and online user-generated content's role in services advertising", *In AIP conference proceedings,* vol.1644, no.1 pp. 318-324, 2015.

[17] J.Miguéns, R.Baggio, and C. Costa, "Social media and tourism destinations: TripAdvisor case study", *Advances in tourism research,* vol.26, no.28, pp.1-6, 2008.

[18] B. Hysa, A. Karasek, I. Zdonek, Social media usage by different generations as a tool for sustainable tourism marketing in society 5.0 idea, *Sustainability,* vol.13, no.3, 1018, 2021.

[19] P. Naidooa, P. Ramseook-Munhurrunb and N.V. Seebaluckc, "Janvierd Procedia, S. Investigating the Motivation of Baby Boomers for Adventure Tourism", *Soc. Behav. Sci.,* vol.175, pp.244–251, 2015.

[20] AARP Reaserch. Travel Trends, January. Available online: https://www.aarp.org/content/dam/aarp/research/surveys_statistics/life-leisure/2019/2020-travel-trends.doi.10.26419-2Fres.00359.001.pdf, 2020

[21] M. Prensky, "Digital natives, digital immigrants", *On the Horizon,* vol.9 no.5,pp.1–6, 2001

[22] A.Euajarusphan, "Online Social Media Usage Behavior, Attitude, Satisfaction, and Online Social Media Literacy of Generation X, Generation Y, and Generation Z", *PSAKU International Journal of Interdisciplinary Research,* vol.10, no.2, pp.44-58, 2021.

[23] V. Giarla, Generational social media: How social media influences the online and in-person relationships of gen X, Gen Y And Gen Z (Doctoral dissertation), USA, Salem State University, 2019

[24] N. Nuzulita and AP Subriadi, "The role of risk - benefit and privacy analysis to understand different uses of social media by Generations X, Y, and Z in Indonesia", *The Electronic Journal of Information Systems in Developing Countries,* vol.86, no,3, pp.121-22, 2020. Available at: doi: 10.1002/isd2.12122

[25] C.V. Priporas, N. Stylos, and A. K. Fotiadis. 2017. "Generation Z Consumers' Expectations of Interactions in Smart Retailing: A Future Agenda", *Computers in Human Behavior,* vol. 77, pp.374–81, 2017 Available at doi:10.1016/j.chb.2017.01.058.

[26] R. Pauliene and K. Sedneva "The influence of recommendations in social media on purchase intentions of generations Y and Z", *Organizations and markets in emerging economies,* vol.10 no.2, pp.227-256, 2019.

[27] D. Fragoulis, E. Chaniotakis, and M.G. Stamatakis, "Zeolitic tuffs of Kimolos island, Aegean Sea, Greece, and their industrial potential", *Cement and Concrete Research,* vol.27 pp.889-905, 1997.

[28] I. Mylona, D. Amanatidis, A. Stavrianea, I. (E) Kamenidou, and S. Mamalis, "Promoting Tourists' Destinations in Greece with Social Media: The Case of Kimolos" *International Journal of Economics & Business Administration (IJEBA),* vol. 9, no.1, pp.347-361, 2021

[29] Ly, Bora, "Effect of Social Media in Tourism (Case in Cambodia)", *Journal of Tourism Hospitality,* vol. 9 1 no.424, pp. 1-9, 2020, Available at SSRN: https://ssrn.com/abstract=3603074

[30] E. Stiakakis, and M. Vlachopoulou, "The impact of social media on travelers 2.0", *tourismos,* vol.12, no.3, pp.48-74, 2017.

[31] J.P. Stevens, *Intermediate statistics: A modern approach,* Routledge, 2013.

[32] A. Field, *Discovering statistics using SPSS,* Sage, 2013.

[33] B. Hysa, A. Karasek, A. and I. Zdonek. "Social media usage by different generations as a tool for sustainable tourism marketing in society 5.0 idea", *Sustainability,* vol.13, no.3, pp.1018, 2021.

[34] L. Di Pietro, F. Di Virgilio and E.Pantano, "Social network for the choice of tourist destination: attitude and behavioural intention", *Journal of Hospitality and Tourism Technology,* vol.3, no.1, pp.60-76, 2012.

The role of hashtags in Social Networks: The case of social mobilization in Greece

Georgia Gioltzidou
Dept. of Journalism and Mass Communications
Aristotle University of Thessaloniki
Hellenic Open University
Thessaloniki, Greece
gioltzidou@gmail.com

Michael Dossis
Dept. of Informatics
University of Western Macedonia
Kastoria, Greece
mdossis@uowm.gr

Theodoros Chrysafis
Faculty of Law
Aristotle University of Thessaloniki
Thessaloniki, Greece
theodoros_c@yahoo.com

Ifigeneia Mylona
Dept. of Management Science and Technology
International Hellenic University
Kavala, Greece
imylona@mst.ihu.gr

Fotini Gioltzidou
Dept. of Journalism and Mass Communications
Aristotle University of Thessaloniki
Thessaloniki, Greece
gioltzidoufotini@gmail.com

Dimitrios Amanatidis
Department of Informatics
University of Western Macedonia
Hellenic Open University
Kastoria, Patras, Greece
damanatidis@uowm.gr

Abstract—This article examines the role of hashtags in Social Networks during periods of social mobilization. Although such platforms have been used in the past to organize demonstrations, in recent years, citizens' response to the invitation to participate in demonstrations through hashtags is much greater. The base of the theoretical study is the theory of the network society, developed by the Spanish sociologist Manuel Castells. Our research focuses on the use of hashtags on Twitter during four periods of protests in Greece. In this context, we analyzed more than 4.000 tweets at the content and user level, to identify developing trends and attitudes. The article argues that technology revolutionized social movements in Greece, as Twitter was used to mobilize and organize protest activities and as a means of political communication among users.

Keywords—hashtag, social mobilization, communication, Twitter, social networks

I. Introduction

Modern Western societies have been characterized as information societies [1] or knowledge societies [2], reflecting the central role of science and technology in social life. New technologies are one of the most visible changes in our contemporary life. They have strengthened economies, facilitated transportation and communication, led to major advances in health and education, expanded information and participation, and created new tools for security [3]. But mainly, they changed the way society is organized and the power relations within it.

Twitter, Facebook, and YouTube have created new contexts for activism that do not exist in traditional Media [4]. According to [5], Twitter is a case of a public sphere on the Internet, since it brings together people with similar interests through #hashtags. Essentially, taking advantage of the potential of the network, its users form interest groups of each category. On top of this observation, the authors in [6] add that Twitter (now X) is a global digital news network that broadcasts in real-time [7].

II. Theoretical Framework

A. The network society

Manuel Castells created one of the most ambitious macro-social theories of our time [1], which attempts to explain and interpret power, the economy, and social life in a world transformed by globalization and IT [8, p.2]. Castells in [1] first defines the concept of the network as he argues, that it plays such a central role in characterizing society in the information age.

The idea of an autonomous space where citizens' power will be exercised in a space independent of organized interests and political parties, remains a timeless quest for Castells. The existence of this "autonomous" space, located between cyberspace and urban areas, is expected to allow movements to gather and conquer their goals [9]. He argues that by coordinating the action of individuals through web initiatives, networked social movements could become outnumbered forces that can be structured across multiple settings, converge transnational agendas, and achieve invisible political accountability [9, p.23]. At this point, he introduces the role of Social Networks in organizing the individuals of a society. According to [10], "the ability of networks to introduce new actors and new contents in the process of social organization, with relative independence of the power centers, increased over time with technological change, and, more precisely, with the evolution of communication technologies". In modern times, as he claims, the connection between free communication on Facebook, YouTube, and Twitter and the occupation of urban space created a hybrid public space of freedom that became a main feature of the Tunisian uprising [9]. Castells therefore argues that we live in a hybrid world, a networked world that incorporates both global communication networks and social networks [11].

The fact is that social media is the medium that shows the greatest growth worldwide. There is a dynamic increase in their usage, as they become more and more popular among Internet users. The familiarity of the world in general with the environments and the logic of using these systems is constantly increasing. The more people use this software, the more powerful they become in terms of their capabilities. The most interesting is that social media's power and influence are not only due to the number of people who use them but are multiplied and magnified by the number of connections they create between these people [12, pp.17-18].

B. The #hashtag

The hashtag comes from the union of the words "Hash" (i.e., "symbol") and "Tag" (i.e., "label"). A hashtag is defined by the user by adding the pound sign (#) before a word, which acts as a search word or tag for a message [13]. Whenever a user uses a hashtag to highlight a situation, other users can watch its context and then join the channel and offer their contribution [14]. To avoid overlapping different topics under one tag, users often use multiple hashtags in their posts.

The original idea was to create a "channel label", which would allow user groups to be formed based on their interests and affiliations. Instead of formal groups, however, hashtags would create channels of communication to which users could selectively pay attention [14]. Messina's vision of the "channel tag" has, over time, become embedded as a practice in both the social and communication habits of the Twitter user community. However, it is also in the architecture of the system itself, with the internal intersection of hashtags in the results search and trending topics [14].

Hashtags can be used for a wide range of purposes, particularly focusing on moderating public debate and sharing information on news and political issues [15]. The increasing popularity of the hashtag is demonstrated by the fact that it is now appearing in other forms of communication – from email to print. In this context, the label function in X is judged as one of the features that contribute to targeted communication between citizens. Kwak et al studied the structure of Twitter and concluded that the connections between different users and hashtags are the elements that differentiate it from other Social Media [6]. Having started as ad hoc public discussion groups [15], #hashtags now seem to have transformed into vehicles for smaller audiences, enabling them to participate and interact in the long term [7].

Thanks to hashtags, the reach of posts on Twitter is not necessarily limited to a specific group (such as subscribers, followers, or friends) as it is on other Social Media. Posted messages are public by default and can also be read by visitors (non-users) browsing the site or following the platform feed [16]. Thus, any user can create public posts to start discussions, join already existing discussions, or monitor communication between other users.

C. The role of #hashtag in Social Mobilization

The literature review shows that in times of social and political crisis, there is a rapid use of Social Media. Since 2011, activists have organized and publicized unprecedented protests using Social Media, leading to the so-called Arab Spring. As a result of these mobilizations, governments in Egypt and Tunisia, as well as regimes in Syria, Libya, Yemen, and Bahrain that had been in power for many years, diverged. Since then, in many countries, Social Media has become a tool of "revolution", not only in Arab countries but also in the USA and European countries.

In the research concerning both authoritarian and democratic regimes, we observe many differences but a common basis: the key role of social media in the self-organization of citizens. A common point in the cases of countries with an authoritarian regime is that Twitter provides citizens with access to information that has been denied to them by the country's regime.

In recent years, mobilizations have taken place almost entirely through websites such as social media, which through hashtags allow citizens to quickly create groups to distribute personalized information about events. Moreover, in times of crisis, social media usage seems participatory, as more user accounts start to publish comments, posts, and shares while such events are moving [17]. Researchers in [18], studied the action of Twitter users in almost 200 different cases of political protest in the period 2009-2013. In their study, they examine the concept of an activist and demonstrate that these are essentially users who deviate noticeably from both elite and ordinary users. Persistent activists, as they are called, participate in Twitter discussions under many different #hashtags, regularly and for long periods. To deal with language and national barriers, they rely on their language skills as well as translation tools. Finally, they attempt to describe their demographic and social characteristics: they belong to an older age group than typical Twitter users, have a low income, and their professional background is related to IT industries [18].

Recent research has particularly highlighted the role of the hashtag as a means of coordinating information in public discussions on Twitter. According to [14], the central role of the hashtag in coordinating citizens is particularly evident in contexts of general political discussions during local, state, and national elections, as well as in protests and other activist mobilizations. The number of hashtags in the same message reveals the degree of exposure of the user to public discussions, while also showing his willingness or unwillingness to engage with more users.

III. METHODOLOGY

This study examines the role of hashtags in cases of social mobilization in Greece. Specifically, we attempt to find correlations, if any, between the number of hashtags used in a message and the content and author of the same message.

The sample of the present research concerns periods of social unrest and protests in Greece. During the period of intense economic crisis, massive demonstrations and mass protests in Greece were a frequent phenomenon. The contraction of the Greek economy and high unemployment rates combined with a series of measures such as the reduction of wages and pensions and the increase of taxes, incited unions and citizen movements to declare strikes and organize rallies [19]. The purpose of the demonstrations was to put pressure on the politicians. The choice of each period of the present study depended on the current political situation. The study aimed to select appropriate, popular hashtags, determined by careful observation of Twitter communication. During the days considered by this research, the hashtags selected were in the first position of popular discussion topics (trending topics).

In this research, the following four periods of demonstrations and social mobilizations were selected, due to their intensity and massiveness (Table 1).

TABLE I.   PERIODS OF SOCIAL MOBILIZATION

Period	Dates	Description	Hashtags used
1st	1-22 Nov 2012	1 week before and 2 weeks after the vote on the adoption of the new austerity cycle by the Greek Parliament	7ngr, #syntagma, #greekrevolution, #mnimonio, #12fgr

2nd	17 Mar – 7 Apr 2013	The period of 3 weeks after the announcement of the deposit cut in Cyprus	#kourema, #trapezes, #katatheseis, #cypros, #kypros, #eurogroup, #HandsOffCyprus
3rd	13 Jul – 2 Aug 2015	The interval of 3 weeks between the referendum and the signing of the 3rd Memorandum	#ThisIsACoup
4th	10 – 12 Feb 2015	1 day before and 2 days after the demonstration in support of the Greek government	#mazi

To map the dimensions of who, when, and what is published, the study collected a total of approximately 60,000 tweets, through freely available software (Twitter Archiver, etc.), which had access to Twitter's API (Application Programming Interface). After that, we selected a random sample of 1,000 messages for each period and then we proceeded to random sampling. All tweets were collected in a database that included the user's name, the content of the message, and the number of hashtags.

The next step was to classify Twitter users based on Lotan's study [20, p.1382], which developed a 'classification scheme' based on subject types and arrived at 'after several phases of coding'. Looking at public profiles and user data, we categorized users into seven categories:

1. Accounts belonging to Traditional Media

2. Accounts belonging to alternative Media

3. Accounts belonging to journalists

4. Accounts belonging to bloggers

5. Accounts belonging to politicians

6. Accounts belonging to citizens

7. Other

Regarding the content of the messages, the classification of the texts into categories was based on the method of empirically grounded theory [20], which is interested in "patterns of action and interaction between different types" of users and tweets [21, p.278]. It is a coding process, which "starts from the data and moves from bottom-up" and is called inductive.

After continuous readings of randomly selected published messages per period, we came up with nine main content categories. The categorization was done based on the textual types and mixed types of speech that users use to discuss the topics that interest them. The seven main content categories are:

1. News about the protests

2. News with negative comment/irony

3. News along with a positive comment

4. Critical/negative/sarcastic comments

5. Comments in support of government and/or politicians

6. Message with solidarity content

7. Other

By tracking hashtags, we can assume that we have created a dataset of the most "visible" tweets related to the specific event, "since the purpose of topical hashtags is to contribute to the visibility and discovery of messages" [14, p.19]. This does not mean that we were able to collect all the messages related to the topic, as some users are not familiar with the use of hashtags in the messages they post, while others are not aware of the existence of the central hashtag.

To process the data obtained from the research, we used the Excel pivot table as well as the SPSS statistical program. Specifically, a different data file was created for each period studied. Then, for the needs of the research, certain periods were consolidated to produce separate summary tables. Our purpose was to establish the similarities and differences per crisis period.

IV. RESULTS

A. Message content and #hashtags during times of protest

Firstly, we are interested in the number of hashtags that appear in published messages, as this indicates the degree of willingness or unwillingness of the user to initiate conversations with other Twitter users. In Table 2 we observe that during the periods of protests, Twitter users in Greece use on average about 2.5 hashtags in each message they publish. Their attitude, on the one hand, shows a willingness to engage with more users but also expresses a reluctance to "sacrifice" characters from the message for this purpose.

TABLE II.   NUMBER OF HASHTAGS PER TWEET

	Number of hashtags
Mean	2.51
Minimum	0
Maximum	14

In Table 3 we notice that in all categories, except "Message with solidarity content", there were messages without hashtags. However, the maximum number of hashtags varies widely. It is characteristic, that messages of the content category "News about the protests" have up to 13 hashtags, while the messages "Message with solidarity content" have up to 7 hashtags.

TABLE III.   MIN AND MAX OF HASHTAGS PER CONTENT CATEGORY

Content Category	Min of hashtags	Max of hashtags
Other	0	4
Critical/negative/sarcastic comments	0	11
Comments in support of the government and/or politicians	0	8
News along with a positive comment	0	12
News with negative comments/irony	0	10
Message with solidarity content	1	7
News about the protests	0	13

Fig. 1 helps us to identify the degree of use of hashtags in the case of each content category and to draw important conclusions about whether user behavior changes depending on the content of the messages.

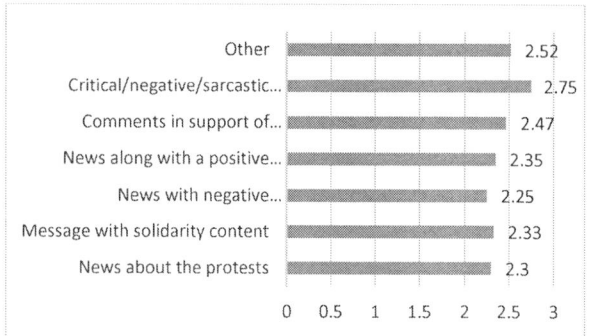

Fig. 1. Number of hashtags per content category

In more detail, we observe that the published messages belonging to the "News about the protests" category have from 0 to 13 hashtags, with an average of 2.3. However, some posts even contain 13 hashtags, a particularly high number. In the case of messages containing news and comments together, we see that the accounts that published messages belonging to the category "News along with a positive comment", have from 0 to 12 hashtags with an average of 2.35. At the same time, the accounts that published during the protest periods messages belonging to the "News with negative comment/irony" category, have an average of 2.25 hashtags.

Regarding the messages classified as commentaries, we notice that the accounts that published messages belonging to the "Critical/negative/sarcastic comments" category have from 0 to 11 hashtags with an average of 2.75, therefore there is a slight increase in the use of hashtags compared to messages containing news. Similarly, in the case of messages belonging to the "Comments in support of government and/or politicians" category, we find an average of 2.47 hashtags - a particularly high percentage. In activist messages, there seems to be a heavy use of hashtags, with an average of 2.33 hashtags, but not more than the tweets of support for the government and politicians.

The strong use of hashtags in messages that contain criticism shows the strong disposition of Twitter users in Greece to comment on political current affairs and the persons who play a leading role in the political scene. On the other hand, at the user level, traditional media appear to believe more in the possibilities offered by hashtags in social media communication and seek their active participation in public online discussions.

B. Users and #hashtags during times of protests

In the second section of this paper, we aim to identify the users who are most likely to start a conversation with other users of the Medium, through the use of hashtags. We are therefore looking at the use of hashtags by each user category, to spread their messages more.

In Fig. 2 we observe an average of 2.51 hashtags per post in the cases of protests, while users used up to 14 hashtags per message.

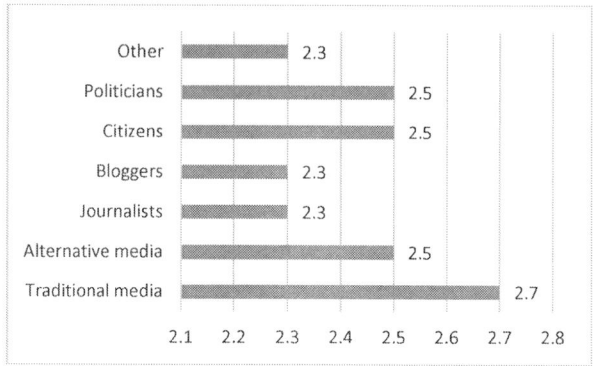

Fig. 2. Number of hashtags per user category

As we observe in Table 4, most user categories use at least one hashtag in the messages they publish. However, there are citizens and politicians, who do not realize the crucial role of hashtags in Twitter communication. As a result, some of them post messages without hashtags, thereby excluding their non-followers from the conversation.

TABLE IV.   MIN AND MAX OF HASHTAGS PER USER CATEGORY

Content Category	Min of hashtags	Max of hashtags
Other	0	4
Politicians	0	5
Citizens	0	10
Bloggers	1	12
Journalists	1	8
Alternative media	1	10
Traditional media	1	13

More specifically, traditional media that published tweets had an average of 2.7 hashtags per message, seeking greater diffusion of their messages. Specifically, they used from 1 to 13 hashtags per message, with an average of 2.7 – a particularly high number for Twitter data in Greece. As for the alternative media that posted tweets, they seem to use fewer hashtags, an average of 2.5 per message. The traditional Media were constant protagonists in the use of hashtags during the periods of demonstrations in Greece.

On the other hand, journalists, along with bloggers, are among the categories with the fewest hashtags per message: they used up to 8 hashtags in their messages, with an average of 2.3. When it comes to politicians, the use of hashtags is also low. Specifically, they posted an average of 2.5 hashtags per message. The third major category of users in this study, citizens, posted messages with an average of 2.5 hashtags per message. Although citizens appear particularly willing to start online discussions with other users, we would expect them to use more hashtags, as their Twitter activity during protests is intense.

C. Conclusions

In the above analysis, we discerned significant evidence for the participation of hashtags in periods of political mobilization. The average number of hashtags used during mass protests and demonstrations is about 2 per message. The frequent use of hashtags in the messages proves that Twitter users in Greece seek meaningful dialogue and for this reason, they publish relevant news and comment on developments. An important finding of the study is that there is an increased usage of hashtags in negative-commentary messages and in messages that support the government – two seemingly conflicting sides of messages.

As far as the users, we find that citizens participate in the activity, while politicians and journalists hesitate about the new communication possibilities opened up to them through Twitter. The above observation leads us to the conclusion that Twitter during the period of mass demonstrations in Greece, turns into a field of intense activity not only for those who wish to support the protesters but also for those who want to oppose them.

Although we do not detect strong differences in other categories which proves that the above choice of users is due to the content of each message, the main conclusion is that the usage of the hashtag as a communication tool seems to be recognized by Twitter users in Greece, whatever category they belong to. Technology revolutionized social movements in Greece, as Twitter was used to mobilize and organize protest activities and as a means of political communication among the users.

In this context, Twitter in Greece perhaps leads to be more participatory and better reflection of citizens' sentiments than mainstream media but remains unable to introduce any political shifts in a deeply troubled political landscape [22, p.149]. The aim is now to study the evolution of the phenomenon over a longer period, with an emphasis on the relationships that develop between users.

REFERENCES

[1] M. Castells, *An introduction to the information age*, City, 1996.

[2] N. Stehr, *Knowledge Societies*, London, England: SAGE Publications, 1994.

[3] UNDP, "Africa Human Development Report 2016 accelerating gender equality and women's empowerment in Africa", (No. 267638), United Nations Development Programme, 2016.

[4] K.M., DeLuca, S. Lawson, and Y. Sun, "Occupy Wall Street on the public screens of social media: The many framings of the birth of a protest movement", *Communication, Culture & Critique*, 5(4), 483-509, 2012.

[5] Y. Benkler, *The Wealth of Networks*, New Haven, CT: Yale University Press, 2006.

[6] H. Kwak, C. Lee, H. Park, and S. Moon, "What is Twitter, a social network or a news media?", In *Proceedings of the 19th International Conference on World Wide Web*, 2010, pp. 591-600.

[7] A. Bruns and T. Highfield, "Is Habermas on Twitter? Social media and the public sphere", In A. Bruns, G. Enli, E. Skogerbø, A.O. Larsson, C. Christensen, *The Routledge Companion to Social Media and Politics*, Routledge, New York, 2016, pp. 56-73.

[8] A. Anttiroiko, *Networks in Manuel Castells' theory of the network society*, 2015.

[9] M. Castells, *Networks of outrage and hope: Social movements in the Internet age*, John Wiley & Sons, 2015.

[10] M. Castells, *The network society: A cross-cultural perspective*, Edward Elgar Publishing, 2004.

[11] M. Castells, *Communication power*, Oxford University Press, UK, 2013.

[12] G. Pleios, "Social Media in Time of Crisis", In Aydemir Okay (Ed.) *Understanding communications in the new media era*, The Journalists and Writers Foundation Press 42, pp. 19-42, 2013.

[13] C. Megele, "Theorizing Twitter chat", *Journal of Perspectives in Applied Academic Practice*, 2(2), 2014.

[14] A. Bruns, J. Burgess, K. Crawford and F. Shaw, *#qldfloods and @QPSMedia: Crisis Communication on Twitter in the 2011 South East Queensland Floods*, Brisbane, 2012.

[15] A. Bruns and J. Burgess, "The use of Twitter hashtags in the formation of ad hoc publics", In *Proceedings of the 6th European Consortium for Political Research (ECPR) general conference*, The European Consortium for Political Research (ECPR), 2011.

[16] A. Bruns and S. Stieglitz, "Metrics for understanding communication on Twitter", In K. Weller, A. Bruns, J. Burgess, M. Mahrt, and C. Puschmann (Eds.), *Twitter and Society*, Peter Lang, New York, 2016, pp. 69-82.

[17] G. Gioltzidou, D. Amanatidis, and I. Mylona, "Natural disaster information dissemination on Twitter: testing against mainstream media coverage", *SafeKozani - 5th International Conference on Civil Protection & New Technology*, 2018, p. 286

[18] M. Bastos, D. Mercea and A. Charpentier, "Tents, tweets, and events: The interplay between ongoing protests and social media", *Journal of Communication*, 65(2), 2015, 320-350.

[19] G. Gioltzidou, "Journalism in the era of Twitter – The case of Greek social mobilization", *EJTA Teachers Conference*, 2018, Greece

[20] G. Lotan, E. Graeff, M. Ananny, D. Gaffney, I. Pearce, and D. Boyd, "The Revolutions Were Tweeted: Information Flows During the 2011 Tunisian and Egyptian Revolutions", *International Journal of Communication*, 2011, pp. 1375-1405.

[21] A. Strauss and J. Corbin, *Grounded theory methodology: An overview*, 1994

[22] G. Gioltzidou, Social media, and political communication in times of crisis (Doctoral dissertation, Aristotle University of Thessaloniki), http://dx.doi.org/10.12681/eadd/51543, 2020, unpublished.

A Multi-class Classification Approach for Anemia Level Prediction with Machine Learning Models

Maria Trigka[*], Elias Dritsas[†], and Phivos Mylonas[*]
[*]Department of Informatics and Computer Engineering,
University of West Attica, Egaleo, Athens, Greece
{mtrigka,mylonasf}@uniwa.gr
[†]Department of Electrical and Computer Engineering
University of Patras, Patras, Greece
dritsase@ceid.upatras.gr

Abstract—**It is a common belief that Artificial Intelligence (AI) and Machine Learning (ML) provide researchers and medical experts with concepts, tools, and techniques to build intelligent systems able to analyze, process and detect hidden patterns, in order to turn data into actionable knowledge for personalized medicine and decision-making (e.g., disease prevention via associated risk prediction, treatment, etc.). The present study seeks an ML solution that will facilitate the recognition of anemia level by investigating the performance of two well-established strategies for multi-class classification tasks; the One-Vs-All (OVA) and One-Vs-One (OVO). Under the specific strategies, two well-known ML models were assumed, namely, Logistic Regression (LR) and Support Vector Machines (SVM). The validation of both strategies was carried out in a publicly available dataset by measuring the weighted average performance in all involved classes that capture the anemia levels; accuracy, precision, recall and Area Under the ROC curve (AUC) were captured and compared for the identification of the dominant model and strategy. After the experimental evaluation, the OVO strategy with the LR model is the main suggestion of the current study achieving an accuracy of 95.05%, precision and recall equal to 0.951 and an AUC of 0.990.**

Index Terms—**Anemia, Multi-class Classification, Prediction, Machine Learning, Data Analysis**

I. INTRODUCTION

Anemia is the condition in which there is a significant deficiency of viable red blood cells that carry oxygen from the lungs to the tissues, as well as carbon dioxide to the lungs for removal through exhalation [1]. There are several types of anemia, some hereditary and some acquired. The most common form of anemia worldwide for both genders and all age groups is iron deficiency anemia [2]. Additionally, other forms of anemia are hemolytic anemia, sickle cell anemia, megaloblastic anemia, pernicious anemia, mediterranean anemia, hemolytic anemia, aplastic anemia, and anemia due to bone marrow disorders [3].

Some causes that contribute to the appearance of anemia are the lack of iron and folic acid, the lack of vitamins, infections or inflammations, various diseases (chronic kidney disease, liver problems or spleen dysfunction), pregnancy, bleeding, bowel disorders (Crohn's disease and celiac disease), heavy menstruation and heredity [4]–[6].

Symptoms of anemia vary depending on its type and severity. The most common, however, are fatigue, irregular pulse, chest pain, paleness, shortness of breath, headache, dizziness or even frozen extremities. More severe types of anemia can cause fainting, severe dizziness and thirst, sweating, cramps and difficulty breathing. Other symptoms may be dark stools, blood in the urine or stools. Finally, burning or numbness in the hands is a symptom of anemia that can be linked to a lack of vitamin B12 [7], [8].

A blood test may be all that is needed if it is obvious what is causing the anemia. This will check the levels of iron, vitamin B12 and folic acid, the level of hemoglobin in the blood, how many blood cells there are, as well as their size and shape. If there is no obvious cause, additional tests such as a myelogram or some imaging method may need to be done [9], [10].

Treatment for anemia depends on the cause, but also on the intensity and type of symptoms. Most of the time, a healthy diet combined with nutritional supplements depending on the element missing from the body is sufficient to balance the issue. In more critical situations, a blood transfusion may be needed to compensate for the lack of red blood cells [11]. In addition, complications can occur if the anemia is severe and if not treated properly it can worsen an existing heart problem such as angina or heart failure [12].

A proper diet, which is rich in nutrients, is extremely important in treating anemia. Foods rich in iron combined with foods with vitamin C are very important for its intake and absorption by the body. Foods rich in iron are liver, red meat and poultry, eggs, green vegetables, nuts, the legumes, dark green leafy vegetables (e.g. spinach), dried fruit, cereals, bread and pasta with added iron [13], [14].

It is important to carry out regular laboratory testing in infants and children when there is a family history, as any alarming symptom should not be left uninvestigated, and there should be regular contact with the attending physician and specialist nutritionist [15].

Nowadays, AI and ML have transformed and revolutionized the way the healthcare sector operates helping institutions, practitioners, specialists and medical experts efficiently store, process and analyse big medical data to identify early signs of diseases, often more accurately than human doctors and at a lower cost. Supervised ML algorithms have been established as important tools in forecasting various chronic condition risk

979-8-3503-8368-3/23 $31.00 © 2023 IEEE

prediction, such as breast cancer tumour type [16], chronic kidney disease [17], diabetes type-2 as classification [18], [19] or time-series tasks for continuous glucose prediction [20], [21], cardiovascular diseases [22]–[24], hypertension [25], [26], SARS-CoV-2 [27], mental confusion [28], lung cancer [29], sleep disorders [30], [31], liver disease [32] metabolic syndrome [33], [34], stroke [35], cholesterol [36], [37] etc.

The main motivation of this study was to present a multi-class classification framework for anemia level prediction that could support medical experts during the screening process of anemia since its occurrence may impact other co-existed conditions. For this purpose, we investigated the performance of two strategies (OVA and OVO) assuming LR and SVM as base models for solving the single binary classification tasks involved in each strategy [38]–[40] targeting a solution that can achieve high sensitivity and separation ability (namely, high recall and AUC). Exploiting a publicly available dataset, the LR classifier under the OVO strategy achieved more consistent outcomes in all classes and in general superior performance than the rest.

The remaining paper is structured as follows. Section II describes related works with the subject under study. Next, in Sections III and IV, a dataset description and analysis of the adopted methodology are described. Besides, in Section V, we discuss the acquired research results. Finally, conclusions and future directions are outlined in Section VI.

II. RELATED WORKS

In this section, we illustrate some relevant works with the study under investigation.

Firstly, [41] focused on the recognition of anemia under general clinical practice conditions. For this purpose, the authors experimented with four different ML models namely, Artificial Neural Networks (ANN), SVM, Naïve Bayes (NB), and Ensemble Decision Trees, and evaluated with various metrics. Also, the models utilized eight different datasets created via particular feature selection techniques. The highest accuracy (85.6%) was achieved using Bagged Decision Trees, followed by Boosted Trees (83.0%) and ANN (79.6%).

Moreover, in [42], the authors aimed to design a model to predict anemia in children under five years of age using complete blood count (CBC) reports. The data were normalized and balanced. Also, feature selection was applied. The RF model showed the biggest accuracy of 98.4%.

In paper [43], five ensemble ML methods namely, Stacking, Bagging, Voting, Adaboost and Bayesian Boosting, were applied in order to detect anemia. After the experiments' evaluation, the Stacking model with the combination of the Decision Tree (DT) and K-Nearest Neighbor as base classifiers and NB as meta classifier achieved a higher accuracy of 92.12%.

In [44], the authors based on CBC data obtained from pathology centres to examine supervised ML techniques and models for the detection of anemia. In comparison to LR, LASSO and Exponential smoothing, the results outlined that the NB model outperforms the other ones in terms of accuracy (92.08%).

Similarly, in [45], supervised ML models, namely, NB, Random Forest (RF) and DT were evaluated and compared to predict anemia based on CBC data. The results showed that the NB model prevailed achieving an accuracy of 96.09%.

A Multi-label Classification technique is applied in the work [46]. The authors attempt to categorize patients with anemia according to its type with the contribution of the machine learning models. The Multilayer Perceptron (MLP) and SVM are nearly always the best classifiers, with MLP being significantly better according to the AUC.

Finally, in [47], multi-class classification algorithms for the diagnosis of anemia were applied. The authors used feature selection with majority voting to identify the key attributes in the input patient data set. Also, the Synthetic Minority Oversampling Technique (SMOTE) was performed on the imbalanced dataset. The experimental results indicated that the MLP model showed the best results.

III. DATASET DESCRIPTION AND ANALYSIS

This section describes the collection method and the features of the publicly available data [48], which was used for the validation of the learning models' performance in the adopted methodology for predicting anemia level.

The CBC test was applied to generate the data set parameters. A Hematology analyzer performed this test following standard operating protocols to determine the prevalence of different types of Anemia treated at the Eureka diagnostic center in Lucknow, India. The diagnostic center performed 4 – 8 CBC investigations a day on average. A collection of data was gathered between September and December 2020, and 1000 CBC investigations were performed. Exclusion criteria: i) adult males and females who are pregnant and older than 15 years old in the study population, ii) infants, young children less than 10 years old and pregnant women. After excluding the above categories of persons from the randomly selected sample of 400 subjects, the final data set consisted of 364 records and 11 features that were used to represent the involved subjects. In Table I, we list these features and provide a statistical description of them.

For characterizing anemia levels (which will be the target classes labels), we based on the haemoglobin (HGB) values [47] and formulated the next rules based on World Health Organization (WHO) criteria [49]:

- Female: HGB$\geq$12 normal, 11$\leq$ HGB $<$ 12 mild, 8 $\leq$ HGB $<$ 11 moderate, HGB$<$8 severe
- Male: HGB$\geq$13 normal, 11$\leq$ HGB $<$ 13 mild, 8 $\leq$ HGB $<$ 11 moderate, HGB$<$8 severe

In addition, in the specific data, anemia prevalence and association with age and gender is depicted in Figure 1. Focusing on the risk factor of age, all anemia levels are observed in all age groups with the severe level consisting of very few instances. As for gender, a small sample of men and women are equally distributed in the severe class. Also, men dominate in the normal and mild classes, while women

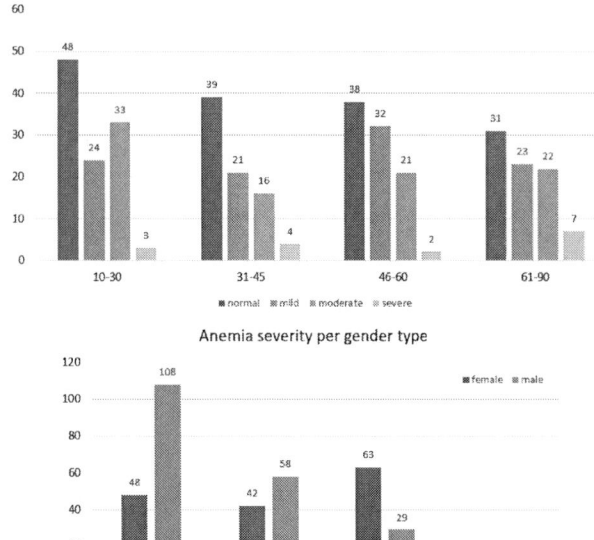

Fig. 1. Anemia prevalence in terms of age and gender type.

stand out (with at least double presence) in the moderate class. The distribution of subjects into four classes is 156 - 42.86% (normal), 100 - 27.47% (mild), 92 - 25.27% (moderate) and 16 - 4.39% (severe). Although the dataset is imbalanced (mainly towards the "sever" class), no technique has been applied to make the class distribution uniform.

IV. METHODOLOGY

The problem of predicting the anemia level will be treated as a multi-class classification and, therefore will be solved by adopting two different strategies: OVA and OVO.

Assuming that M stands for the number of severity levels, in our case M will be equal to 4 (including the normal level). The first strategy (OVA) uses the instances belonging to one of the M classes and solves M binary sub-tasks where, each time, class "yes" consists of the samples of one severity level, while class "no" is formulated by the samples stemming from the rest of classes. On the contrary, the second strategy breaks down and solves $\frac{M(M-1)}{2}$ binary classification sub-tasks. The final decision is based on the outcomes of multiple classifiers. Figure 2 illustrates how the specific strategies are adapted to anemia level prediction.

The binary classification problems will be solved using the learning models LR and SVM. Therefore, in the next paragraphs, we will explain how the specific models achieve the anemia level classification of a subject.

a) Logistic Regression: LR [50] uses a logistic function to model a binary output variable ranging between 0 and 1. Also, logistic regression applies a nonlinear log transformation to the odds ratio, where odds $= \frac{p}{1-p}$ (p is the probability of an event occurring divided by the probability $1 - p$ of an event

not occurring) and then the logit function $logit(p)$ can take any real number in $(-\infty, +\infty)$:

$$
logit(p) = \begin{cases} 0 & \text{if } p = 0.5 \\ < 0 & \text{if } p > 0.5 \\ > 0 & \text{if } p < 0.5 \end{cases} . \quad (1)
$$

If the probability is greater than 0.5, the predictions will be classified as class "no". Otherwise, class "yes" will be assigned.

b) Support Vector Machine: SVM [51] aims to find the best classification function to discern the instances of the two classes in the training data. Once this function is found, a new instance u can be classified by examining the sign of the function $f(u)$; if $f(u) > 0$, the u belongs to the positive class. SVM is a popular kernel-based classification algorithm in pattern recognition. Assuming that u and v are the feature vectors of two instances in the dataset, the most common kernel functions are defined as follows.

- A linear kernel is the simplest function represented by $k(u,v) = u^T v + c$, where the first term is the inner product of features vector u, v and c is an optional constant.
- A polynomial kernel is non-stationary and compatible with problems where all the training data is normalized. It is calculated by

$$
k(u, v) = (\gamma u^T v + c)^d, \quad (2)
$$

where γ, c and d are adjustable parameters that stand for the slope, constant term and polynomial degree, respectively.

- A radial basis kernel function is denoted as

$$
k(u, v) = \exp\left\{ -\gamma \| u - v \|^2 \right\}, \quad (3)
$$

where $\gamma = \frac{1}{2\sigma^2}$ and σ is an adjustable parameter for measuring the performance of the kernel. It measures the similarity between pairs of data points based on their distance in the feature space.

- A sigmoid kernel function is represented by

$$
k(u, v) = tanh(\gamma u^T v + c), \quad (4)
$$

where γ and c are the adjustable parameters. A common value for γ is $\frac{1}{N}$, where N is the number of features in the dataset.

Notation: u^T denotes the transposition of vector u, $\|u\|$ denotes the Euclidean norm of vector u, $u^T v$ term is the inner product of vectors u, v and $tanh(\cdot)$ is the hyperbolic tangent function.

V. RESULTS AND DISCUSSION

In this section, we will show the outcomes of the experiments executed following the strategies OVO and OVA. The performance of these approaches is evaluated assuming the LR and SVM models. In both strategies and considered models, 10-fold cross-validation was applied. In this method, the dataset is separated into 10 different subsets where each

TABLE I
LIST OF FEATURES AND STATISTICAL DESCRIPTION IN THE DATASET.

Feature	Type	Notation	Description	Limitis	Min-Max	Mean± std
Age	Numerical	age	Current age of the patients	11 - 100 (years)	11 - 89 (years)	44.92 ± 18.78
Gender	character	gender	Gender	Male (203) /Female (161)	-	-
Hemoglobin	Numerical	HGB	Level of HGB	11-16 g/dL	4.2-19.6	11.91 ± 2.19
Mean cell volume	numerical	MCV	Level of MCV	80-101 fL	55.7-124.1	87.51 ± 9.33
Mean cell hemoglobin	numerical	MCH	Level of MCH	27-32 pg	14.7-41.4	28.23±3.87
Mean cell hemoglobin concentration	numerical	MCHC	Level of MCHC	31-37 g/dL	23.6-50.2	32.05± 2.80
Red cell distribution width	numerical	RDW	Level of RDW	11 -16	10.6 -29.2	15.12±2.18
Red blood cell count	numerical	RBC	Level of RBC	3.80-4.80 M/uL	1.36-6.90	4.28.6 ±0.82
White blood cell count	numerical	WBC	Level of WBC	3.5-11.5 ths/uL	2 - 42.42	8.86 ± 4.87
Platelet count	numerical	PLT	Level of PLT	150-450 (10^3/uL)	10-660	223.75±99.41
Packed cell volume	numerical	PCV	Level of PCV	36-46	13.1-56.9	36.76±6.83

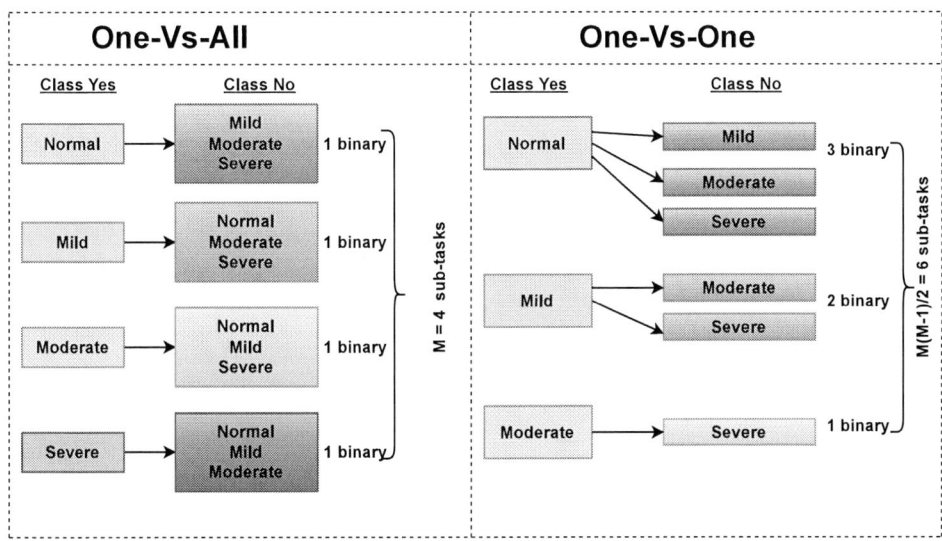

Fig. 2. Multi-class Classification strategies for anemia level prediction.

time one subset constitutes the test set, and the remaining 9 ones are used as training sets in turn. In this way, all the combinations are tested and the performance score is obtained by taking the average of each result.

A. Experiments environment

WEKA software tool [52] was utilized for the execution of the experiments by exploiting the libraries LibSVM and MultiClassClassifier. The former helped us to experiment with different types of kernel functions for the identification of the one that achieves the best performance. The latter provided us with an interface through which we tested the two strategies mentioned above, OVO and OVA. It should be noted that in the OVO case the field "PairWiseCoupling" was set to true.

The system on which the experimental measurements were performed had the following characteristics: 11th generation Intel(R) Core(TM) i7-1165G7 @ 3.2 GHz, 32 GB RAM, Windows 11 Pro, 64-bitOS and x64 processor.

B. Evaluation metrics

The performance evaluation of the investigated models and strategies was based on the accuracy, precision, recall and AUC, demonstrated in Table II under 10-fold cross-validation.

These metrics for the multi-class case are captured as the weighted average of the class-wise score of each metric, namely multiplied by the weight of the class or frequency of the class on the entire dataset divided by the sum of the weights. So, the weighted average metrics [53] are as follows:

$$Precision = \frac{\sum_{i=1}^{K} w_i \frac{TP_i}{TP_i+FP_i}}{\sum_{i=1}^{K} w_i} \quad (5)$$

$$Recall = \frac{\sum_{i=1}^{K} w_i \frac{TP_i}{TP_i+FN_i}}{\sum_{i=1}^{K} w_i} \quad (6)$$

$$AUC = \frac{\sum_{i=1}^{K} w_i AUC_i}{\sum_{i=1}^{K} w_i} \quad (7)$$

$$Accuracy = \frac{\sum_{i=1}^{K} w_i \frac{TN_i+TP_i}{TN_i+TP_i+FN_i+FP_i}}{\sum_{i=1}^{K} w_i}, \quad (8)$$

where TP, TN, FP, FN capture the class-wise true positive/negative and false positive/negative and $w_1 = 156, w_2 = 100, w_3 = 92, w_4 = 16$ stand for the number of instances per class label in the dataset.

TABLE II

MULTI-CLASS CLASSIFICATION STRATEGIES EVALUATION IN TERMS OF PRECISION, RECALL AND AUC.

Class	OVA - LR			OVO - LR			OVA - Polynomial SVM $d = 2$, c=1, $\gamma = 0.1$			OVO - Linear SVM		
	Precision	Recall	AUC	Precision	Recall	AUC	Precision	Recall	AUC	Precision	Recall	AUC
normal	0.974	0.974	0.992	0.962	0.968	0.994	0.762	0.942	0.958	0.943	0.955	0.969
mild	0.788	0.630	0.916	0.939	0.920	0.979	0.589	0.560	0.805	0.896	0.860	0.928
moderate	0.682	0.815	0.934	0.957	0.957	0.994	0.706	0.522	0.877	0.925	0.935	0.982
severe	0.833	0.938	0.992	0.882	0.938	0.992	0.625	0.313	0.785	0.882	0.938	0.995
Weighted Average	0.843	0.838	0.956	0.951	0.951	0.990	0.694	0.703	0.888	0.923	0.923	0.962

C. Results

Table II summarizes the performance of LR and SVM under the two multi-class classification strategies.

Also, as for the SVM model, we investigated which parameters and kernel functions (as the one presented in Section IV) would give us the highest scores. We concluded that the linear SVM was the most efficient presenting high classification performance under the OVO strategy. However, in the case of the OVA strategy, only the polynomial kernel function of $d = 2$ (see Eq.2) gave us an acceptable performance against other parameter settings. However, for other datasets or problems, the optimal value may be different.

Next, we show, in ascending order, the accuracy of the models per strategy. More specifically, the outcomes are the following:

- Accuracy = 70.33% - OVA polynomial SVM ($d = 2, \gamma = 0.1, c = 1$)
- Accuracy = 83.79% - OVA LR
- Accuracy = 92.31% - OVO linear SVM
- Accuracy = 95.05% - OVO LR

As for the accuracy of the LR model, under the OVA strategy, it was 13.46% higher than the SVM's model in the same strategy and 11.26% lower than the LR's model in the OVO strategy.

Focusing on the rest metrics, as for the OVO strategy, the LR noted essentially superior classification scores with average precision and recall equal to 0.951 and an AUC of 0.990. In terms of the OVA strategy, again, LR was the dominant model which indicated average precision and recall equal to 0.843 and 0.838, respectively, and an AUC of 0.956. Comparing the average results of the best-performing model in each strategy, the experiments unveiled the LR as the most appropriate. Moreover, the specific model exhibited constantly high class-wise performance and the highest accuracy of 95.05%.

Also, all models presented high AUC indicating an increased discrimination ability between the classes that capture anemia levels. Regarding the models' sensitivity, the OVA strategy, especially with polynomial SVM, indicated low levels to all class labels except for the normal one, while the severe class presented the worst performance (0.313). It is essential to mention that currently, no resampling technique (e.g., SMOTE) has been applied before the models' training to tackle the issue of class imbalance. However, this issue was efficiently handled by both strategies with the LR model and

OVO with linear SVM. To sum up, OVO with the LR model is the main suggestion of the current analysis for anemia level prediction.

VI. CONCLUSIONS

In conclusion, this study presented a multi-class classification framework for predicting the anemia level. Through experiments and performance evaluation, the OVO strategy resulted as the most efficient both class-wise and on average succeeding accuracy of 95.05%, precision and recall of 0.951 and AUC of 0.951, using the LR to solve the binary classification subtasks. In future work, we aim to deal with feature ranking and selection techniques in order to reduce the models' complexity by selecting the most important features for the anemia level prediction and investigating, to what extent, the models' performance is enhanced or not. Finally, alternative multi-class classification models and/or schemes in combination with class balancing techniques will be assessed and compared to the strategies evaluated here.

ACKNOWLEDGMENT

This research was funded by the European Union and Greece (Partnership Agreement for the Development Framework 2014-2020) under the Regional Operational Programme Ionian Islands 2014-2020, project title: "Indirect costs for project "TRaditional corfU Music PresErvation through digiTal innovation" ", project number: 5030952.

REFERENCES

[1] J. Turner, M. Parsi, and M. Badireddy, "Anemia," in *StatPearls [Internet]*. StatPearls Publishing, 2022.
[2] T. G. DeLoughery, "Iron deficiency anemia," *Medical Clinics*, vol. 101, no. 2, pp. 319–332, 2017.
[3] M. J. Gamit, H. S. Talwelkar *et al.*, "Survey of different types of anemia." *International Journal of Medical Science and Public Health*, vol. 6, no. 3, pp. 493–496, 2017.
[4] P. Bhadra and A. Deb, "A review on nutritional anemia," *Indian Journal of Natural Sciences*, vol. 10, no. 59, pp. 18 466–18 474, 2020.
[5] G. M. Brittenham, G. Moir-Meyer, K. M. Abuga, A. D. Mitra, C. Cerami, R. Green, S.-R. Pasricha, and S. H. Atkinson, "Biology of anemia: a public health perspective," *The Journal of Nutrition*, 2023.
[6] G. Weiss, T. Ganz, and L. T. Goodnough, "Anemia of inflammation," *Blood, The Journal of the American Society of Hematology*, vol. 133, no. 1, pp. 40–50, 2019.
[7] H. Ludwig and K. Strasser, "Symptomatology of anemia," in *Seminars in oncology*, vol. 28. Elsevier, 2001, pp. 7–14.
[8] K. F. Lasch, C. J. Evans, and D. Schatell, "A qualitative analysis of patient-reported symptoms of anemia," *Nephrology Nursing Journal*, vol. 36, no. 6, p. 621, 2009.

979-8-3503-8368-3/23 $31.00 © 2023 IEEE

[9] S. Kundrapu and J. Noguez, "Laboratory assessment of anemia," *Advances in clinical chemistry*, vol. 83, pp. 197–225, 2018.

[10] J. Válka and J. Čermák, "Differential diagnosis of anemia." *Vnitrni lekarstvi*, vol. 64, no. 5, pp. 468–475, 2018.

[11] L. T. Goodnough and A. K. Panigrahi, "Blood transfusion therapy," *Medical Clinics*, vol. 101, no. 2, pp. 431–447, 2017.

[12] I. Anand and P. Gupta, "How i treat anemia in heart failure," *Blood, The Journal of the American Society of Hematology*, vol. 136, no. 7, pp. 790–800, 2020.

[13] S. D. Pawar, S. D. Deore, N. P. Bairagi, V. B. Deshmukh, T. N. Lokhande, and K. R. Surana, "Vitamins as nutraceuticals for anemia," *Vitamins as Nutraceuticals: Recent Advances and Applications*, pp. 253–279, 2023.

[14] A. S. A. Sharourou, M. A. Hassan, M. B. Teclebrhan, H. M. Alsharif, S. A. Alhamad, T. S. Alsinani *et al.*, "Anemia: Its prevalence, causes, and management," *The Egyptian Journal of Hospital Medicine*, vol. 70, no. 10, pp. 1877–1879, 2018.

[15] L. Khan, "Anemia in childhood," *Pediatric annals*, vol. 47, no. 2, pp. e42–e47, 2018.

[16] E. Dritsas, M. Trigka, and P. Mylonas, "Ensemble machine learning models for breast cancer identification," in *IFIP International Conference on Artificial Intelligence Applications and Innovations*. Springer, 2023, pp. 303–311.

[17] E. Dritsas and M. Trigka, "Machine learning techniques for chronic kidney disease risk prediction," *Big Data and Cognitive Computing*, vol. 6, no. 3, p. 98, 2022.

[18] N. Fazakis, O. Kocsis, E. Dritsas, S. Alexiou, N. Fakotakis, and K. Moustakas, "Machine learning tools for long-term type 2 diabetes risk prediction," *IEEE Access*, vol. 9, pp. 103 737–103 757, 2021.

[19] E. Dritsas and M. Trigka, "Data-driven machine-learning methods for diabetes risk prediction," *Sensors*, vol. 22, no. 14, p. 5304, 2022.

[20] E. Dritsas, S. Alexiou, I. Konstantoulas, and K. Moustakas, "Short-term glucose prediction based on oral glucose tolerance test values." in *HEALTHINF*, 2022, pp. 249–255.

[21] S. Alexiou, E. Dritsas, O. Kocsis, K. Moustakas, and N. Fakotakis, "An approach for personalized continuous glucose prediction with regression trees," in *2021 6th South-East Europe Design Automation, Computer Engineering, Computer Networks and Social Media Conference (SEEDA-CECNSM)*. IEEE, 2021, pp. 1–6.

[22] E. Dritsas and M. Trigka, "Efficient data-driven machine learning models for cardiovascular diseases risk prediction," *Sensors*, vol. 23, no. 3, p. 1161, 2023.

[23] E. Dritsas, S. Alexiou, and K. Moustakas, "Cardiovascular disease risk prediction with supervised machine learning techniques." in *ICT4AWE*, 2022, pp. 315–321.

[24] M. Trigka and E. Dritsas, "Long-term coronary artery disease risk prediction with machine learning models," *Sensors*, vol. 23, no. 3, p. 1193, 2023.

[25] E. Dritsas, S. Alexiou, and K. Moustakas, "Efficient data-driven machine learning models for hypertension risk prediction," in *2022 International Conference on INnovations in Intelligent SysTems and Applications (INISTA)*. IEEE, 2022, pp. 1–6.

[26] E. Dritsas, N. Fazakis, O. Kocsis, N. Fakotakis, and K. Moustakas, "Long-term hypertension risk prediction with ml techniques in elsa database," in *Learning and Intelligent Optimization: 15th International Conference, LION 15, Athens, Greece, June 20–25, 2021, Revised Selected Papers 15*. Springer, 2021, pp. 113–120.

[27] E. Dritsas and M. Trigka, "Supervised machine learning models to identify early-stage symptoms of sars-cov-2," *Sensors*, vol. 23, no. 1, p. 40, 2022.

[28] M. Trigka, E. Dritsas, and P. Mylonas, "Mental confusion prediction in e-learning contexts with eeg and machine learning," in *Novel & Intelligent Digital Systems Conferences*. Springer, 2023, pp. 195–200.

[29] E. Dritsas and M. Trigka, "Lung cancer risk prediction with machine learning models," *Big Data and Cognitive Computing*, vol. 6, no. 4, p. 139, 2022.

[30] I. Konstantoulas, E. Dritsas, and K. Moustakas, "Sleep quality evaluation in rich information data," in *2022 13th International Conference on Information, Intelligence, Systems & Applications (IISA)*. IEEE, 2022, pp. 1–4.

[31] I. Konstantoulas, O. Kocsis, E. Dritsas, N. Fakotakis, and K. Moustakas, "Sleep quality monitoring with human assisted corrections." in *IJCCI*, 2021, pp. 435–444.

[32] E. Dritsas and M. Trigka, "Supervised machine learning models for liver disease risk prediction," *Computers*, vol. 12, no. 1, p. 19, 2023.

[33] M. Trigka and E. Dritsas, "Predicting the occurrence of metabolic syndrome using machine learning models," *Computation*, vol. 11, no. 9, p. 170, 2023.

[34] E. Dritsas, S. Alexiou, and K. Moustakas, "Metabolic syndrome risk forecasting on elderly with ml techniques," in *International Conference on Learning and Intelligent Optimization*. Springer, 2022, pp. 460–466.

[35] E. Dritsas and M. Trigka, "Stroke risk prediction with machine learning techniques," *Sensors*, vol. 22, no. 13, p. 4670, 2022.

[36] N. Fazakis, E. Dritsas, O. Kocsis, N. Fakotakis, and K. Moustakas, "Long-term cholesterol risk prediction using machine learning techniques in elsa database." in *IJCCI*, 2021, pp. 445–450.

[37] E. Dritsas and M. Trigka, "Machine learning methods for hypercholesterolemia long-term risk prediction," *Sensors*, vol. 22, no. 14, p. 5365, 2022.

[38] E. Dritsas, M. Trigka, and P. Mylonas, "A multi-class classification approach for weather forecasting with machine learning techniques," in *2022 17th International Workshop on Semantic and Social Media Adaptation & Personalization (SMAP)*. IEEE, 2022, pp. 1–5.

[39] Y. Liu, J.-W. Bi, and Z.-P. Fan, "A method for multi-class sentiment classification based on an improved one-vs-one (ovo) strategy and the support vector machine (svm) algorithm," *Information Sciences*, vol. 394, pp. 38–52, 2017.

[40] E. Dritsas, S. Alexiou, and K. Moustakas, "Copd severity prediction in elderly with ml techniques," in *Proceedings of the 15th International Conference on PErvasive Technologies Related to Assistive Environments*, 2022, pp. 185–189.

[41] T. K. Yıldız, N. Yurtay, and B. Öneç, "Classifying anemia types using artificial learning methods," *Engineering Science and Technology, an International Journal*, vol. 24, no. 1, pp. 50–70, 2021.

[42] A. Dixit, R. Jha, R. Mishra, and S. Vhatkar, "Prediction of anemia disease using machine learning algorithms," in *Intelligent Computing and Networking: Proceedings of IC-ICN 2022*. Springer, 2023, pp. 229–238.

[43] P. T. Dalvi and N. Vernekar, "Anemia detection using ensemble learning techniques and statistical models," in *2016 IEEE International Conference on Recent Trends in Electronics, Information & Communication Technology (RTEICT)*. IEEE, 2016, pp. 1747–1751.

[44] C. Sasikala, M. Ashwin, M. Dharanessh, and M. Dhanabalan, "Curability prediction model for anemia using machine learning," in *2022 8th International Conference on Smart Structures and Systems (ICSSS)*. IEEE, 2022, pp. 1–7.

[45] M. Jaiswal, A. Srivastava, and T. Siddiqui, "Machine learning algorithms for anemia disease prediction: Select proceedings of ic3e 2018," *ResearchGate. DOI*, vol. 10, pp. 978–981, 2019.

[46] C. Bellinger, A. Amid, N. Japkowicz, and H. Victor, "Multi-label classification of anemia patients," in *2015 IEEE 14th International Conference on Machine Learning and Applications (ICMLA)*. IEEE, 2015, pp. 825–830.

[47] R. Vohra, A. Hussain, A. K. Dudyala, J. Pahareeya, and W. Khan, "Multi-class classification algorithms for the diagnosis of anemia in an outpatient clinical setting," *Plos one*, vol. 17, no. 7, p. e0269685, 2022.

[48] "Anemia dataset," https://data.mendeley.com/datasets/dy9mfjchm7/1, (accessed on 30 September 2023).

[49] M. B. Mengesha and G. B. Dadi, "Prevalence of anemia among adults at hawassa university referral hospital, southern ethiopia," *BMC hematology*, vol. 19, pp. 1–7, 2019.

[50] E. Bisong and E. Bisong, "Logistic regression," *Building Machine Learning and Deep Learning Models on Google Cloud Platform: A Comprehensive Guide for Beginners*, pp. 243–250, 2019.

[51] S. Chidambaram and K. Srinivasagan, "Performance evaluation of support vector machine classification approaches in data mining," *Cluster Computing*, vol. 22, pp. 189–196, 2019.

[52] "Weka," https://www.weka.io/, (accessed on 30 September 2023).

[53] T. Kautz, B. M. Eskofier, and C. F. Pasluosta, "Generic performance measure for multiclass-classifiers," *Pattern Recognition*, vol. 68, pp. 111–125, 2017.

Recognition of Greek Orthodox Hymns Using Audio Fingerprint Techniques

Konstantinos Karasavvidis*, Dimitris Kampelopoulos*, Lazaros Moysis†‡,
Achilles D. Boursianis*, Spiridon Nikolaidis*, Panagiotis Sarigiannidis§, Sotirios K. Goudos*

*: ELEDIA@AUTH, School of Physics, Aristotle University of Thessaloniki, 54124,
Thessaloniki, Greece. {kokarasa,dkampelo,snikolaid,bachi,sgoudo}@physics.auth.gr

†: Laboratory of Nonlinear Systems - Circuits & Complexity, School of Physics, Aristotle University of Thessaloniki,
54124, Thessaloniki, Greece. lmousis@physics.auth.gr

‡: Department of Mechanical Engineering, University of Western Macedonia, Kozani, Greece.

§:Department of Electrical and Computer Engineering,
University of Western Macedonia, 501 00 Kozani, Greece. psarigiannidis@uowm.gr

Abstract—Audio fingerprinting was originally developed for music song identification, and over the years has been used for many more cases. With fingerprinting, an equivalent signature of the audio signal is saved in a simple form, so that it can be easily compared with another fingerprint. In this work, an audio fingerprinting algorithm is developed for the recognition of Greek Orthodox hymns. The main difference between music songs and Greek Orthodox hymns is the absence of music instruments, as there is only the chanter's voice. The test dataset consists of 10 hymns, and for each hymn there are 5 different performances, creating a dataset of 50 performances. Several tunable parameters of the fingerprinting technique are tested, like the parameters of the Hamming window, the frequency spectrum, the sample duration, and the sample noise, to find the differences in Greek Orthodox hymns. The recognition results for all cases are positive. The algorithm, in most cases, had a high recognition rate, even with noisy samples.

Index Terms—Audio fingerprinting, Greek Orthodox hymns, audio recognition, signal processing

I. INTRODUCTION

The task of identifying songs is among the most prominent in the field of Music Information Retrieval (MIR). This problem involves identifying the musical origin (song name and artist) of an audio recording of a short length, usually less than 10 seconds (s). There are several identification techniques, and the one considered in this work is audio fingerprinting [1].

Fingerprinting refers to the generation of a compact and minimal signature for an audio track that carries some essential and unique characteristics. This signature can then be used to identify the track, as each signature is distinct. The short size of the signature facilitates fast comparison, making this information useful for fast song identification [2], [3], and other relevant applications, such as plagiarism detection [4], and broadcast monitoring [5], [6].

This research was carried out as part of the project 'Recognition and direct characterization of cultural items for the education and promotion of Byzantine Music using artificial intelligence' (Project code: KMP6-0078938) under the framework of the Action 'Investment Plans of Innovation' of the Operational Program 'Central Macedonia 2014 2020', that is co-funded by the European Regional Development Fund and Greece.

So, addressing this problem efficiently opens up innumerable commercial applications for the music industry, for example regarding music listeners in social places like bars, participants in social or cultural events, music learners who want to improve their performance, religious and cultural tourism, and also the protection of intellectual property, as well as the advertising industry in general. Techniques for audio fingerprinting are constantly improving, with the main goals of developing fingerprints that are robust to audio noise, having fast search techniques for detecting matches in large song datasets, as well as addressing other issues, like desynchronization, and more [7]–[10].

The problem of song identification is usually addressed for published artists, meaning that the given task involves the identification of a studio or live recording of a song from an artist, which has been published in some format, like an album. So, there is always some official source of a song that can be considered the original and can be used as a reference to match a recorded query.

However, this is not the case for all music genres. For example, in religious music and chanting [11]–[16], where the human voice is prominent, the same song can be performed by numerous people, and each professional performer has a certain level of creative freedom to dress their recitation with their own pace, tonality, and emotion. All such performances are considered equally important. Thus, there is no objective way to compare them. People in the audience who are attending each social or religious event can form their own liking for each performer and tend to enjoy the diversity in the recitation performances.

Thus, when considering the problem of song identification for such music genres, the task becomes much more difficult. Here, there is no distinct musical feature for the song (apart from the lyrics) that will remain completely invariant among performances, as the pace, tone, and emotion are prone to variations. Thus, deriving a unique feature that will be prevalent in the fingerprint and common among different performances of the same chant, is not a straightforward task.

979-8-3503-8368-3/23 $31.00 © 2023 IEEE

A. Motivation

Motivated by this problem, this work considers audio fingerprinting for the identification of Greek Orthodox chants. The audio fingerprint algorithm that is created has two functions. First, it can extract multiple audio fingerprints from a hymn and save them in a database. Second, it can extract audio fingerprints from a sample and compare them with the saved fingerprints on the database. In both parts, the same technique and parameters are used for audio fingerprint extraction, so they can be comparable.

The remainder of the work is structured as follows. In Section II, the methodology developed is described. In Section III, the recognition results on the considered dataset are presented. Section IV concludes the work with a discussion on future topics of interest.

II. RECOGNITION OF GREEK ORTHODOX HYMNS USING AUDIO FINGERPRINTS

A. The Dataset

The dataset consists of ten different hymns, and for each of them five performances are available. The performances may be from the same or different chanters, and there is variation in the performances' duration. Furthermore, different tempos may have been used during the performance.

B. Audio Track Pre-Processing

Before processing from the algorithm, the audio files were pre-processed. First of all, the recordings of the hymns were WAV files, which is an uncompressed non-lossy format. Then, all the files were converted to a 44100 Hz sample rate frequency. This frequency is selected because it is the standard for CD-quality audio. Alternatively, any frequency that confirms the Nyquist theorem can be used. Additionally, the silent parts at the beginning and end of every track were manually removed.

C. The Algorithm

The algorithm is divided into two parts. The first part is where the hymns are processed and added to the database, and the second part constitutes the recognition of the unknown sample. The process for adding the hymns to the database goes as follows:

1) The audio file is inserted and the sample rate frequency is checked.
2) For dual-channel audio files, the signal is converted to monophonic by an average of both channels.
3) A Hamming window is applied and the signal is fragmented. The Hamming window size is a parameter that can be changed in the algorithm.
4) Fast Fourier Transform is applied, converting the signal from the time domain to the frequency domain.
5) The spectrogram of the audio file is generated.
6) From every fragment of the signal, an audio fingerprint is generated.
7) Fingerprints are saved in the database.

D. Audio Fingerprint Calculation/Extraction

The algorithm calculates the fingerprint from four frequencies. The spectrum is divided into four frequency ranges, and from every range the algorithm finds the frequency with the maximum amplitude. The four ranges are a parameter that can be changed in the algorithm. The four frequencies are combined and saved to the database as a value, forming the audio fingerprint. A fingerprint is calculated for every fragment created by the Hamming window. The database consists of a table where every row has the fingerprint, the hymn/performance number, and the serial number of the fingerprint. Fig. 1 visually represents how the frequency spectrum is divided into four areas from which fingerprints are calculated.

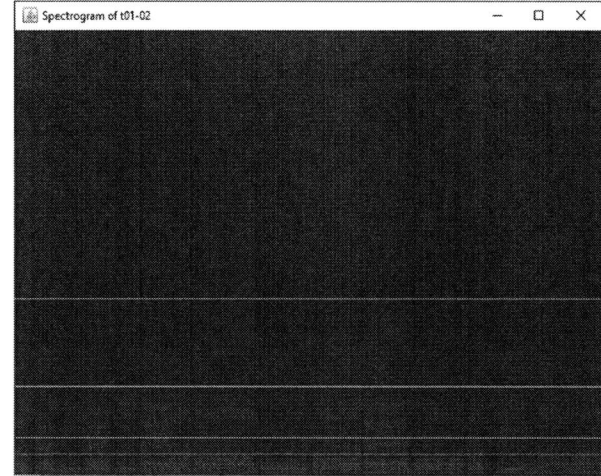

Fig. 1. Spectrogram of the hymn with the 4 frequency ranges of fingerprint calculation.

E. Audio Sample Test Process

For the testing process, the same audio files from the database were used. Every file was inserted directly from the PC into the program. Then, the program cuts the audio file to a specified duration. The duration of the sample is a parameter that can be changed in the algorithm. Specifically, the algorithm reads the duration of the file. Then, a random generator selects the second, which will be the start of the sample file, and cuts the file according to the specified duration. The generator is limited between the start of the file and the end of the file minus the sample duration, so the generated sample always has the specified duration. All the tests are repeated 10 times to eliminate random events that may affect the result.

In the future, another method for inserting the sample audio could be developed using the PC microphone. In this way, the algorithm can be tested with real environment noise.

F. Audio Sample Matching Process

The audio sample matching with the hymn of the dataset is performed by matching the audio fingerprints. The algorithm searches the database to find the same fingerprint values of the sample. The value must be 100% the same. Additionally, in order to be considered a match, the fingerprints, except for those with the same value, must also have the same distance. The distance is calculated by the serial number of the fingerprint. Again, the fingerprints must have the same distance by 100%. The algorithm saves the total number of matches for every hymn of the dataset and at the end returns the name of the hymn with the most matches.

In future work, both the matching of the fingerprint value and the distance of the fingerprint could be used as parameters by setting a minimum percentage limit of the matching.

III. RECOGNITION RESULTS

The performance of the algorithm is tested by changing its available parameters, which are the Hamming window, the frequency ranges for the calculation of fingerprints and the duration of the test sample. Additionally, another test with noisy samples is performed to study the robustness of the algorithm to noise. Finally, considering the processing time required for the algorithm to return the matching results, the processing power is evaluated.

A. Hamming Window

The Hamming window is responsible for the number of fingerprints that are extracted from the audio file. The number of fragments and consequently the fingerprints, result from the equation:

$$n_{fragment} = \frac{R_{sample} \times bytes_{samples} \times channels}{Hamming.window} \left(\frac{fragments}{sec} \right) \quad (1)$$

The duration of every fragment is $1/n_{fragment}$. From the equation it arises that with the increase of the Hamming window, the number of fragments is reduced and the duration per fragment is increased.

Four different Hamming window sizes have been tested, 512, 1024, 2048, and 4096 bytes. For the other parameters, the frequency ranges are between 0Hz - 7kHz, and 8 s samples were used. The samples had no noise. For all cases, the algorithm succeeded in recognizing the correct hymn from the database with an outstanding performance of 100%. From the results, it is observed that the correlation ratio of the sample with the corresponding hymn in the database is dropping with the increase of the Hamming window. This is justified because, as mentioned above, the number of audio fingerprints is consistent with the number of fragments that occur from the size of the window. And with an increasing window, the number of fragments decreases.

B. Frequency Ranges of the Fingerprint

The frequency range test is based on the Bark scale, which consists of 24 bands from 0Hz to 15500Hz with uneven bandwidths. Three different tests occurred for the bands $0-8$, $9-16$, and $17-24$. In each case, two Bark bands are used to correspond in each of the four frequency ranges, that are used for the calculation of fingerprints. So, in the first case, the fingerprint was calculated for frequencies 0Hz - 1080Hz,

in the second case for frequencies 1080Hz - 3150Hz, and in the third case for frequencies 3150Hz - 15500Hz.

For all three cases, the algorithm managed to have 100% correct recognition. But there is a significant difference between the results. For case 1 (bands $1-8$), the correlation between the sample and the database is the highest. The maximum correlation rate is 69% with an average of 46%. There is also a higher correlation with the other tracks in the database, which is undesirable. The margin between the correlations is higher than 35%. On the other hand, the third case ($17-24$ bands) has the lowest correlation. The correlation is very low and there is only 2.5% margin with the unwanted correlation of the other tracks. The second case ($7-16$) is somewhere in the middle. The margin between the correlations is 19%.

The correlation tables can be seen in Fig. 2. In Fig. 2a and Fig. 2b the main diagonal is visually clear, representing the high correlation between the audio sample and the hymn, while in Fig. 2c the main diagonal is not visually clear due to the low correlation between the samples and the hymns. In Fig. 2a, the main diagonal is darker, representing the higher correlation percentage, but there is also a higher correlation with the other hymns.

From the results, we conclude that the lower frequencies contain more information that can be used on the fingerprint than the higher frequencies. Considering that the male voice ranges mainly at frequencies lower than 300 Hz, this coincides with the results.

C. Sample Duration

To test the effect of sample duration on algorithm performance, six different durations were examined, 3, 5, 8, 10 and 12 s. For the other parameters, a Hamming window of 1024 bytes and frequency ranges between 0 Hz $-$ 7 kHz were used. The samples did not have noise.

In all five cases, the algorithm had 100% correct recognition. The average rate of correlation increases with increasing sample duration, which is explained because a longer duration corresponds to more audio fingerprints. The increase rate is slowed for a duration of more than 8 s. The margin is 26.55% for the 3 s case and is going to 28.88% for the 15 s case. For the other cases, the margins are between the two previous values. The margin increases until the 8 s case. For a longer sample duration, the margin has a minor increase. Since the algorithm had correct recognition 100% even for samples of 3 s, in the future, this parameter could be further tested to find the minimum duration for successful recognition.

D. Noisy Samples

For the noisy samples test, ten different tests were performed, with noise levels ranging from 10% to 100% of the maximum signal amplitude, or from 10dB to 0dB Signal-to-Noise Ratio (SNR). White noise was used and introduced to the audio file using a process similar to the process described in Section II-B. Then the noisy audio file followed the same process as described in Section II-E.

979-8-3503-8368-3/23 $31.00 © 2023 IEEE   258

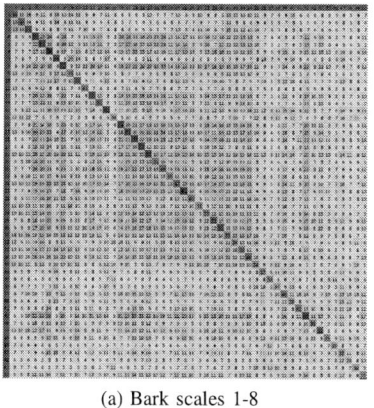

(a) Bark scales 1-8

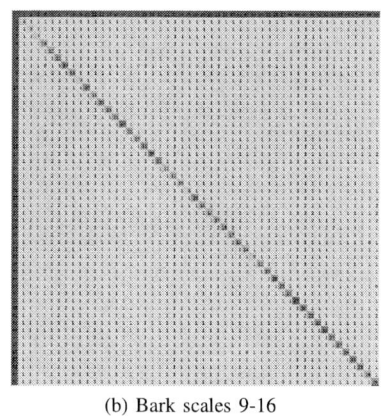

(b) Bark scales 9-16

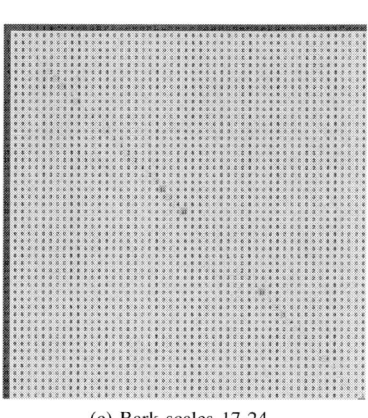

(c) Bark scales 17-24

Fig. 2. Correlation tables of Frequency Ranges of fingerprint calculation.

The results of this test are considerably different for varying noise levels. In Table I, the matching results for all cases are displayed. In the first case with 10% noise or 10dB SNR, the correct recognition drops slightly from 100% to 99.80%. The same holds for 20% noise, the drop is to 98.80%. It is only after a 30% noise that a more significant decrease in recognition is observed. For 50% noise or 3dB SNR, the correct recognition is 72.6%. In this case, the margin between the correlation of the sample and the same hymn in the database and the correlation with the other hymns in the database is 3%. With increased noise, the margin decreases further, making recognition even more difficult. In Fig. 3, the correlation tables for three of the cases, SNR = 10dB (Fig. 3a), SNR = 3dB (Fig. 3b) and SNR = 0dB (Fig. 3c are displayed. In Fig. 3a the main diagonal is clearly visible, representing the high percentage of correlation. In Fig. 3b is visible only a part of the main diagonal, while in Fig. 3c the main diagonal is not visible because the correlation is similar across the dataset.

audio fingerprints of the sample, compare the fingerprints with the ones saved for every hymn in the database, and return the hymn with the most matches. During all tests, the PC only ran the algorithm and all other applications were shut down, so it did not consume additional processing power or RAM memory.

The execution time is similar for all tests, giving an average of 817 milliseconds (ms) in all cases (11500 recognition runs). The range of times is 114 ms with a coefficient of variation $CV = 0.79\%$. This shows that most execution times are close to average.

Table II lists the specifications of the PC that is used for the algorithm and the tests. Specifications are common and can be found even on mobile devices, too. Taking this fact along with the execution time, we can conclude that the algorithm does not require much processing power.

TABLE I
RESULTS OF NOISY SAMPLES

Noise level (%)	SNR (dB)	Correct recognition (%)
10	10.00	99.80
20	6.99	98.80
30	5.23	93.00
40	3.98	84.20
50	3.00	72.60
60	2.22	62.20
70	1.55	51.20
80	0.97	37.60
90	0.46	30.40
100	0.00	23.00

TABLE II
PC SPECIFICATIONS

Operating System (OS)	Windows 10 Home 64 bit 21H2
System Manufacturer	Hewlett-Packard
System Model	HP 15 Notebook PC
BIOS	F.0D
Processor	Intel(R) Core(TM) i5-3230M CPU @ 2.60GHz (4CPUs)
RAM	4096 MB
Storage	500 GB SSD
Pagefile	5862 MB used, 1311 MB available
DirectX Version	12

E. Execution Time / Processing Power

Finally, considering the execution time of the previous tests, the processing power needed for the algorithm is evaluated. The execution time consists of the time it takes for the algorithm to apply the Hamming window and fragment the audio sample, apply the Fourier transformation, extract the

IV. DISCUSSION-CONCLUSIONS

In this work, an audio fingerprinting algorithm is created to recognize Greek Orthodox chants. The algorithm has two functions. The first function is the creation of the audio fingerprint database. And the second function is the recognition of a sample through the database. In both functions, the same process is used for the extraction of the fingerprint. The effect of parameters, Hamming window, fingerprint frequencies, and sample duration was examined on the performance of the al-

979-8-3503-8368-3/23 $31.00 © 2023 IEEE

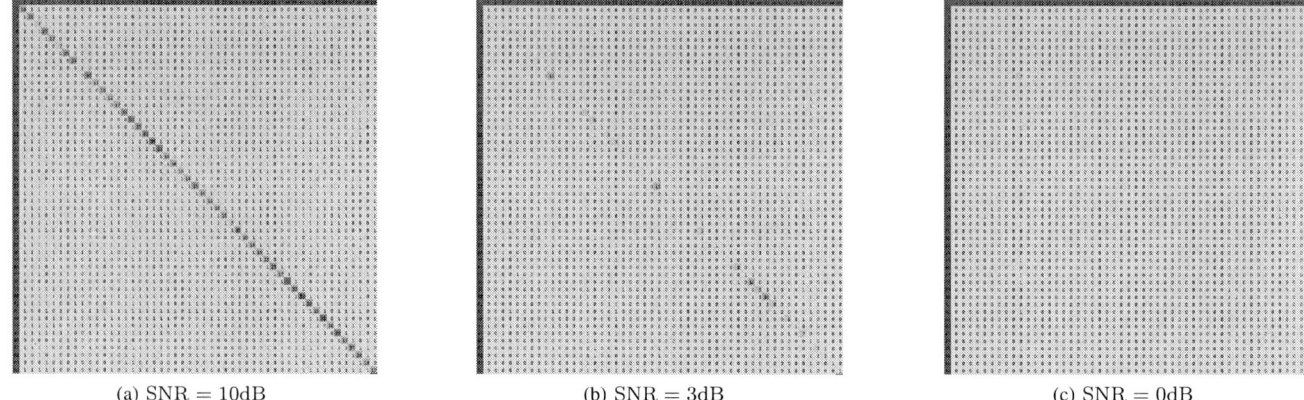

| (a) SNR = 10dB | (b) SNR = 3dB | (c) SNR = 0dB |

Fig. 3. Correlation tables of frequency ranges of the fingerprint calculation

gorithm. Additionally, a noisy sample test has been performed and the execution time of the algorithm has been evaluated.

Regarding the results, the algorithm had excellent results with 100% correct recognition in most tests. In terms of parameters, the Hamming window size did not have a significant effect on the results. On the other hand, the frequencies used for fingerprint extraction can be significant. For the sample duration tests, the algorithm had excellent results even for samples of 3 sec, and this is a parameter that can be tested further. The algorithm was then tested for noise tolerance and the results were positive. Finally, the execution time is less than a second in all the tests, and compared to the specifications of the PC that was used (Table II), we conclude that the algorithm has low processing power requirements.

In general, it has been shown that it is possible to use audio fingerprints for the recognition of Greek Orthodox hymns. The positive result is that the algorithm also worked with noisy samples. But, due to the small database of 10 hymns (or 50 performances) and the fact that most of the test occurred without noisy samples, the algorithm can be parameterized and improved further.

There are numerous topics that can be considered for future studies as expansions of the current work. First, more tunable parameters can be introduced into the algorithm, and their effect can be evaluated.

One consideration is generalizing the method so that it recognizes similarities between all the performances of the same chant. This would make the method more effective, as the characteristics of a fingerprint would remain, to a certain extent, invariant between different performers.

Another consideration would be to test the method on other datasets, where singing voices are prominent. This includes datasets relevant to religious music, chanting, a capella singing, or folk music, in general [17]–[19].

Another experiment of interest is to test the recognition robustness under different types of noises, other than white noise. For example, realistic scenarios would include cell phone recording of a track under urban sounds such as background human mumbling, cheering, walking sounds, wind

sounds, rain, and more. Recognition under such conditions can be tested on-site, or using available datasets of urban sound recordings [20], [21].

Finally, it is of interest to develop a mobile application of the proposed scheme [22], to facilitate its usage by a wide range of people interested in Greek Orthodox music. This can have potential applications in religious tourism, music education, and of course cultural preservation.

REFERENCES

[1] Pedro Cano, Eloi Batlle, Ton Kalker, and Jaap Haitsma, "A review of audio fingerprinting," *Journal of VLSI signal processing systems for signal, image and video technology*, vol. 41, pp. 271–284, 2005.

[2] Avery Wang et al., "An industrial strength audio search algorithm.," in *Ismir*. Washington, DC, 2003, vol. 2003, pp. 7–13.

[3] Wooram Son, Hyun-Tae Cho, Kyoungro Yoon, and Seok-Pil Lee, "Sub-fingerprint masking for a robust audio fingerprinting system in a real-noise environment for portable consumer devices," *IEEE Transactions on Consumer Electronics*, vol. 56, no. 1, pp. 156–160, 2010.

[4] Neetish Borkar, Shubhra Patre, Raunak Singh Khalsa, Rohanshhi Kawale, and Priti Chakurkar, "Music plagiarism detection using audio fingerprinting and segment matching," in *2021 Smart Technologies, Communication and Robotics (STCR)*. IEEE, 2021, pp. 1–4.

[5] Jose Ramon Cerquides, "A real time audio fingerprinting system for advertisement tracking and reporting in fm radio," in *2007 17th International Conference Radioelektronika*. IEEE, 2007, pp. 1–4.

[6] Antonio Camarena-Ibarrola, Edgar Chávez, and Eric Sadit Tellez, "Robust radio broadcast monitoring using a multi-band spectral entropy signature," in *Progress in Pattern Recognition, Image Analysis, Computer Vision, and Applications: 14th Iberoamerican Conference on Pattern Recognition, CIARP 2009, Guadalajara, Jalisco, Mexico, November 15-18, 2009. Proceedings 14*. Springer, 2009, pp. 587–594.

[7] Jaap Haitsma and Ton Kalker, "A highly robust audio fingerprinting system with an efficient search strategy," *Journal of New Music Research*, vol. 32, no. 2, pp. 211–221, 2003.

[8] Sunhyung Lee, Dongsuk Yook, and Sukmoon Chang, "An efficient audio fingerprint search algorithm for music retrieval," *IEEE Transactions on Consumer Electronics*, vol. 59, no. 3, pp. 652–656, 2013.

[9] Neil J Hurley, Félix Balado, Elizabeth P McCarthy, and Guenole CM Silvestre, "Performance of Philips Audio Fingerprinting under Desynchronisation.," in *ISMIR*, 2007, pp. 133–134.

[10] Murat Köseoğlu and Hakan Uyanık, "The effect of different noise levels on the performance of the audio search algorithm," in *2020 International Congress on Human-Computer Interaction, Optimization and Robotic Applications (HORA)*. IEEE, 2020, pp. 1–7.

[11] Paraskevi Kritopoulou, Athanasia Stergiaki, and Konstantinos Kokkinidis, "Optimizing Human Computer Interaction for Byzantine music learning: Comparing HMMs with RDFs," in *2020 9th International*

979-8-3503-8368-3/23 $31.00 © 2023 IEEE

Conference on Modern Circuits and Systems Technologies (MOCAST). IEEE, 2020, pp. 1–4.

[12] Konstantinos Kokkinidis, Theodoros Mastoras, Apostolos Tsagaris, and Panagiotis Fotaris, "An empirical comparison of machine learning techniques for chant classification," in *2018 7th International Conference on Modern Circuits and Systems Technologies (MOCAST).* IEEE, 2018, pp. 1–4.

[13] K Kokkinidis, A Panagi, and A Manitsaris, "Finding the optimum training solution for Byzantine music recognition—A Max/Msp approach," in *2016 5th International Conference on Modern Circuits and Systems Technologies (MOCAST).* IEEE, 2016, pp. 1–4.

[14] Konstantinos-Hercules Kokkinidis and Athanasios Manitsaris, "Intelligent Sensorimotor Learning for Byzantine music," in *4th International Conference on Modern Circuits and Systems Technologies (MOCAST 2015),* pp. 1–5.

[15] Sakib Shahriar and Usman Tariq, "Classifying maqams of qur¢anic recitations using deep learning," *IEEE Access,* vol. 9, pp. 117271–117281, 2021.

[16] Ephrem A Retta, Richard Sutcliffe, Eiad Almekhlafi, Yosef Enku, Eyob Alemu, Tigist Gemechu, Michael A Berwo, Mustafa Mhamed, and Jun Feng, "Kinit Classification in Ethiopian Chants, Azmaris and Modern Music: A New Dataset and CNN Benchmark," *arXiv:2201.08448,* 2022.

[17] Sebastian Rosenzweig, Frank Scherbaum, David Shugliashvili, Vlora Arifi-Müller, and Meinard Müller, "Erkomaishvili Dataset: A curated corpus of traditional Georgian vocal music for computational musicology," *Transactions of the International Society for Music Information Retrieval,* vol. 3, no. 1, 2020.

[18] Tomohiko Nakamura, Shinnosuke Takamichi, Naoko Tanji, Satoru Fukayama, and Hiroshi Saruwatari, "jaCappella Corpus: A Japanese a Cappella Vocal Ensemble Corpus," in *ICASSP 2023-2023 IEEE International Conference on Acoustics, Speech and Signal Processing (ICASSP).* IEEE, 2023, pp. 1–5.

[19] Dimitrios Delviniotis and Georgios Kouroupetroglou, "DAMASKINOS: The prototype corpus of Greek Orthodox ecclesiastical chant voices," in *Proc. Int. Conf. Crossroads\ Greece as an Intercultural Pole of Musical Thought and Creativity,* 2011, pp. 1–14.

[20] Jean-Rémy Gloaguen, Arnaud Can, Mathieu Lagrange, and Jean-Francois Petiot, "Creation of a corpus of realistic urban sound scenes with controlled acoustic properties," in *Proceedings of Meetings on Acoustics.* AIP Publishing, 2017, vol. 30.

[21] Jean-Rémy Gloaguen, Mathieu Lagrange, Arnaud Can, and Jean-François Petiot, "Isolated urban sound database," Apr. 2018, https://doi.org/10.5281/zenodo.1213793.

[22] Nan Zhang, "Mobile music recognition based on deep neural network," *Mobile Information Systems,* vol. 2022, 2022.

Author	Session
Dimitris Gatsios	A1
Georgios Rigas	A1
Georgios Bourazanis	A1
Spyridon Konitsiotis	A1
Ioanna Moustaka	B2
Spyridon Doukakis	B2
Marina Mattheoudakis	B2
Konstantinos Karasavvidis	B5
Dimitris Kampelopoulos	B5
Lazaros Moysis	B5
Achilles Boursianis	B5
Spyridon Nikolaidis	B5
Panagiotis Sarigiannidis	B5
Sotirios Goudos	B5
Michael Dossis	A3
	A5
	B1
	B5
Kosmas Glavas	A1
	A2
Georgios Prapas	A1
Aimilia Ntetska	A1
Pinelopi Adamakidou	A1
Pantelis Angelidis	A1
Markos Tsipouras	A1
Alexandros Zervopoulos	A2
Luís Miguel Campos	A2
Konstantinos Oikonomou	A2
	A4
	B1
Asterios Papamichail	A4
Athanasios Tsipis	A4
Dimitrios Fragkoulis	B1
Fotis Koumboulis	B1
Nikolaos Kouvakas	B1
Antonios Menexis	B1
Nikolaos Baras	A4
Antonios Chatzisavvas	A4
Dimitris Ziouzios	A4
Ioannis Vanidis	A4
Minas Dasygenis	A3
	A4
	B1
	B5
Athanasios Karakostas	A2
Akrivi Vlachou	B2
Aristeidis Karras	A5
	B1
Christos Karras	A5

	B1
Anastasios Giannaros	A5
	B1
Konstantinos Giotopoulos	A5
	B1
Dimitrios Tsolis	A5
	B1
Spyros Sioutas	A5
	B1
Fotios Roumpies	A3
Athanasios Kakarountas	A3
	A5
	B2
	B4
Anastasia Gasidou	B2
Dimitrios Kotsifakos	A4
	B2
	B3
Christos Douligeris	A4
	A5
	B2
	B3
Konstantinos C. Giotopoulos	A5
	B1
Anastasios Papathanasiou	A2
Georgios Germanos	A2
Nicholas Kolokotronis	A2
Euripidis Glavas	A2
Theodora Sanida	A3
	B1
	B5
Maria Vasiliki Sanida	A3
	B1
	B5
Argyrios Sideris	A3
	B1
	B5
Dimitrios Kiriakos	B3
Yannis Psaromiligkos	B3
Gerasimos Magoulas	A3
Spyros Polykalas	A3
Ioannis Christodoulou	A5
Vasiliki Alexopoulou	A5
Angelos Markopoulos	A5
Ifigeneia Mylona	B5
Dimitrios Amanatidis	B5
Irene Kamenidou	B5
Spyridon Mamalis	B5
Georgia Gioltzidou	B5
Theodoros Chrysafis	B5

Fotini Gioltzidou	B5
Odysseas Karadimas	A4
Aikaterini Florou	A4
Spiridoula V. Margariti	A4
Eleftherios Stergiou	A4
Chrysostomos D. Stylios	A4
Ioannis Arvanitakis	A5
George Kranas	A5
Antonios Dadaliaris	A5
Akrivi Krouska	B3
Christos Troussas	B3
Yorghos Voutos	B3
Phivos Mylonas	B3
	B5
Cleo Sgouropoulou	B3
Eleni Seralidou	B2
Theodoros Karvounidis	B2
Konstantinos Kalovrektis	B2
Ioannis Dimos	B2
Apostolos Xenakis	B2
Dimitrios Chatzoulis	A4
Costas Chaikalis	A4
Dimitrios Kosmanos	A4
Kostas Anagnostou	A4
Maria Trigka	B5
Elias Dritsas	B5
Nikiforos Kontopoulos	A4
Konstantinos Korakis	B3
Georgios Drakopoulos	A2
Yagmur Yigit	A3
Leandros Maglaras	A3
Mohamed Amine Ferrag	A3
Naghmeh Moradpoor	A3
Giorgos Lambropoulos	A3
Konstantinos Georgitsaros	B1
Konstantinos Palegkas	A1
Dimitrios Magetos	B3
Panagiotis N. Smyrlis	A1
Odysseas Tsakai	A1
Konstantinos Vogklis	A1
Nikolaos Giannakeas	A1
George Fragulis	A1
Georgios Lambropoulos	A5
Sarandis Mitropoulos	A5
Theodoros Dimitriou	B1
Emmanouil Skondras	B1
Christos Hitiris	B1
Cleopatra Gkola	B1
Ioannis S. Papapanagiotou	B1
Dimitrios J. Vergados	B1

Constantinos Vergopoulos	B1
Stratos Koumantakis	B1
Angelos Michalas	B1
Dimitrios D. Vergados	B1
Daniele Rossi	B4
Nicasio Canino	B4
Stefano Di Matteo	B4
Sergio Saponara	B4
Vasileios Tenentes	A3
	B4
Aggeliki Katsika	B4
Konstantinos Papageorgiou	B4
Alexandros Fakis	B4
Fotis Andritsopoulos	B4
Vassilis Plagianakos	B4
Georgios Spathoulas	B4
Athanasios Xynos	B4
Christos Zonios	B4
Asimina Koutra	B4
Christina Dilopoulou	B4
Yiorgos Tsiatouhas	B4
Vangelis Malamas	B4
Dimitris Koutras	B4
Panayiotis Kotzanikolaou	B4

IEEE
445 Hoes Lane
Piscataway, NJ 08854-4141

ISBN 979-8-3503-8368-3